From the safety and security of his extended Bristol family, the author, Geoffrey Allen, has launched himself into the post-war years; gradually, widening his horizons and seizing every opportunity to indulge his adventurous spirit at home and abroad. An adventurer at heart, with the full realization that even now, he has difficulty in coming to terms with his environment. He embraced the mighty continent of Africa on first acquaintance where he has survived arrest, imprisonment, coups and unrest as well as leading an exciting life. This autobiography's written in a refreshing style and holds much of interest and is a good read.

Painting The Mosque For Christmas?

Geoffrey Allen

AUSTIN MACAULEY PUBLISHERS™

LONDON • CAMBRIDGE • NEW YORK • SHARJAH

A CIP catalogue record for this title is available from the British Library.

ISBN 9781398417748 (Paperback)
ISBN 9781398417755 (ePub e-book)

www.austinmacauley.com

First Published 2023
Austin Macauley Publishers Ltd®
1 Canada Square
Canary Wharf
London
E14 5AA

My grateful thanks to my nephew Tony Allen for his expertise and many contributions.

Table of Contents

Preface

It was a hot, sweaty morning, typical of the West African heat and conditions in which I had worked for some years now, as I drove along the dirt road that ran from our complex of sawmills and moulding shops three miles outside the town of Zwedru, situated in the Grand Jeddah County of North East Liberia, close to the Ivorian border. Secondary species of tall trees reached to the edge of the road and, as it only led to our site, there was little traffic and I made good progress into the town, which had three or four streets lined by one-storey houses and shops, mainly of mud construction, and with the customary corrugated iron roofs rusting and stained from the tropical rains.

We had recently had a storm so the usual dust had been replaced by standing puddles and mud which made driving along the streets quite interesting, especially trying to dodge dogs, chickens, goats and the occasional cow. The inhabitants of Zwedru stayed well clear of the streets in these conditions for fear of being splattered with mud.

I headed towards the Lebanese store, which was my destination, as Christmas was approaching and my errand was to obtain supplies for the few remaining staff on site for the festivities. Most of our married expatriates had gone on leave and the bachelors stayed to maintain a low production period with a skeleton staff. As I pulled up outside the store, I noticed a 2,000-gallon road tanker on the other side of the road. I greeted the driver, who assured me he was going out to our site straight away but I had my doubts.

Our whole logging operation was very large. We logged one million acres of forest in several concessions throughout Grand Jeddah and we maintained a vast fleet of lorries, bush equipment and support vehicles as well as powerful generators for the sawmills and moulding shops. Our daily fuel requirement was 2,000 gallons. To achieve this, a tanker had to leave Monrovia, the capital of Liberia, and make a 200-mile journey over rough dirt roads every day—14,000 gallons a week. The tankers were not all in good condition and by tradition, the drivers had a Seepage Allowance of 180 gallons. This meant, invariably, that a tanker would arrive on our site with 1,820 gallons. On the face of it, this was no problem. My workshop manager would dip the tank, sign for that amount and then we would be credited in Monrovia. But all was not as it seemed!

It only took me a couple of days to realise that it was stretching belief that every tanker that made a delivery lost exactly 180 gallons through seepage from their tanks. Amazing, you might say, that seepage could be so accurate.

I had an excellent personnel manager, who was invaluable to me, and I called on his advice. He immediately informed me that all the tankers stopped just on the outskirts of Zwedru and the driver sold their seepage allowance, for cash, to a local trader who also dealt in fuel but never received a full tanker delivery at his yard.

I knew the trader concerned because, to rub salt into the wound, even though I kept a reserve supply, if the tankers were held up and we ran short, I had to go and buy my fuel back from him at exorbitant prices to keep my operation running. His name was Tinkersee. He was well respected, a big man in the town and a pillar of the local mosque. He had quite a few employees in his businesses, lived in a big two-storey house and was always finely and richly dressed. He seemed to have countless minions and servants all over the town and he always served me with Arabic coffee whenever I visited. He was likeable but very astute and good at business and trading.

I entered the Lebanese store run by two brothers who somehow managed to supply most of my expatriates' needs. They quickly gave me a beer and then commenced to gather the items on my list. It never ceased to amaze me how far and wide the Lebanese community had penetrated into the African continent. They opened stores in the most remote areas and towns and generally provided a good service…and they always provided good coffee and beer, of course.

It was then that I became aware of a slapping sound from next door where the mosque was located. I peered over a stack of mineral crates but my view was blocked so I walked out on to the veranda and around the corner to the mosque. To my utter amazement, there was Tinkersee, dressed in bright blue flowing robes and white shoes, whitewashing the walls of the mosque.

'Tinkersee! What on earth are you doing?' I asked.

'Painting the mosque,' he answered.

'But why are YOU painting the mosque?' I exclaimed.

'For Christmas!' he replied.

Hence the title of this autobiography.

Now I should perhaps explain—for the benefit of my Muslim friends, to whom no offence is intended—that in my experience, in Ghana and Liberia, by order of the town councils, most towns were spruced up for the Christmas celebrations and Tinkersee was merely obeying the directive.

In 2002, I and several friends founded the Roman Ridge School in Accra, Ghana; the origins and details of which are contained herein and though I handed over the reins of the school to a very competent management team some years ago, I obviously maintain a lively interest in its affairs and progress and remain on the board of directors. As part of its Community Outreach Programme, the senior pupils have organised Ridge Readers, a literacy project where suitable books are handed to individual students, young and old, in schools and institutions, and the Roman Ridge pupils help them read the book, which stays in the possession of that student in the hope that it can be further used by his friends and family. Funds raised for this project are spent on books only and all other expenses are paid by the school.

During the past, I have written poetry and short stories, mainly to entertain the many students that I have had the privilege to teach in a variety of schools. Pupils and friends persuaded me to gather these into an anthology last year and thus *Sleeping with Lions, Sitting on Crocodiles* was issued as a fundraiser for Ridge Readers, which has been quite successful. Several stories and the poems it contains will feature in this book so be warned! It would appear, however, that there is now a quickening of interest in the details of my life and friends and pupils want more, hence the launch of an autobiography.

Some months have been spent researching facts, figures and geographical and historical details and I was very grateful for the help of national and local archivists. I must admit that I found the archive work very interesting indeed, though my friends and relatives must be now very bored with my narrations with the discoveries I made about the details of my past life. There are still several intriguing avenues that I would like to explore where I seem to be blocked by official non-release of files at national level or important documents have 'disappeared' and it has been impossible to maintain my writing at a localised and personal level as I have been drawn into national and international affairs.

I have also become aware of how much the world has changed in my lifetime, which now spans 70 years in the UK and abroad, and then further amazed by being able to contact old friends and colleagues who have given me powerful reminders of the past and invaluable details to assist me in my writing. To be frank, I have also been absolutely astounded by the past lives of many of these friends whose exploits and achievements surpass mine by far and of which I was totally ignorant.

Family and relatives are worried; evidently, autobiographies have a tendency to upset some family members. Former pupils who are now lawyers assure me that truth cannot be challenged but I understand that even the truth might prove difficult for several members and branches of my extensive family.

There has also been the realisation that outward normal appearances in childhood covered up a complicated pattern of relationships and events within the family and with many of which I had no conception.

I also encountered great problems with time scales and the chronology of various events, especially with the family happenings prior, or leading up to, my birth. Having established what I thought was a clear and concise timeline, details would be proved wrong by my research and reliable sources. Nostalgia was pleasant but also a handicap to my version of my life.

I was also determined to use only my personal photographs. It was a struggle but I think I have achieved this, even though many are very old or of poor quality.

My Early Years.

Chapter One
I Saw the Brabazon Fly

'When does one first begin to remember?' began the opening lines from Winston Churchill's first autobiography, My Early Life, and I recall this clearly as it was one of our course books in English Literature at secondary school. Winston Churchill's early life started in Ireland, but of course, he would not really have known that then because no one has memory at such a very early age. Churchill, like me, would either have been informed later by his family or transposed his early memories back into the past so that they caught up, as it were.

Following the end of the War, people expected a better life than those years before, with the Great Depression causing poverty and hardship worldwide. There were major shortages in jobs and housing for those returning from the War but there was also a big baby boom. I was part of that boom. Unlike Churchill, I would rate my early upbringing as being fairly normal for a post-war boom baby in Great Britain in the late forties and the early fifties, so what are my credentials for attempting an autobiography? The spur has been explained in the preface but I would suggest that you read on because events in my life become more exciting and eventful as we progress, especially bearing in mind that my childhood was normal and I liken it to breaking out of a traditional lifestyle into the void, but that is not to say that there were not interesting events in my early years.

I was born on 11 July 1946 at 78, St Michael's Hill, the Bristol Maternity Hospital, the son of Lawrence and Mary Allen, weighing 10 pounds and was adored and spoilt by all the nurses. I wish I had been aware of that at the time! Three months later, I was baptised at the church of St Thomas the Apostle at Eastville and my godparents are listed, rather formally I thought, as Mr and Mrs S Grant and Mr J Foord…namely, Uncle Sid and Aunty Gwen on my father's side and Uncle Jack on my mother's but I have a complaint. My godparents were admonished at the bottom of the Holy Baptism Certificate (my first certificate gained!): 'Ye are to take care that this child be brought to the bishop to be confirmed by him.' Well, I was confirmed on 13 May 1962 by the bishop of Bristol but I do not think I was prompted or guided by my godparents and they were not even there.

My earliest memory is quite mundane. There were stairs leading up to a glass door and my mother was struggling to bounce a pram up them and I was somehow watching from the bottom of them. These stairs still exist at 137 Fishponds Road when my parents were living in a flat above a shop—part of a small shopping row—though the whole building seems a little rundown and seedy now. With that memory, my mother assures me that every day, I was taken out around Eastville Park, virtually on the doorstep, and the process of getting the pram up the stairs while someone held me at the bottom was a daily routine.

This was the flat where my parents lived, for the most part, during the War and a little while afterwards. This was the flat where, one evening, my father, who had just cycled back from work in pouring rain and was drenched right through, climbed up the stairs, dripping all the way up, opened the door…and had me thrown bodily into his arms with my mother's comment, 'He's your baby as well!'

I had been teething rather painfully and loudly with my screams echoing around the flat all day and my mother was exhausted. My father reckoned that I could have screamed for England during that time and it got me into further trouble when we were asked to leave a café in Weston-super-Mare and a caravan site at Uphill. I might add that I have been asked to leave much more salubrious and high-class establishments than those in later life, but I suppose it was quite an achievement back then, and I occasionally added the fact to my various CVs—much to several future employers' amusement. But back to memories.

My next clear one is that of furiously peddling my much loved three-wheeled bike along the pavement at Ridgeway Road to get back to the house my parents had recently purchased as I was really desperate to go to the toilet. The image is crystal clear but alas, I did not make it. I would only have been about four years old at the time, and why was I out playing with friends in the streets at such a tender age? Because it was absolutely normal at that time. No cars, no traffic! Most children played in the streets but not main roads.

We moved to the small house in Ridgeway Road in 1949. It cost £700 and I believe my parents were very brave to commit to the mortgage. Prior to that year, various important events had taken place. The Berlin Blockade commenced, the Windrush docked, Gandhi was assassinated, the National Health Service was introduced and the Morris Minor hit the roads. It must have been quite a time but in 1949, the only event was the introduction of Noddy. All last year's events pale into insignificance compared with that.

The house was a straightforward, traditional, 'two-up and two-down'. Downstairs, the back room was used as a sitting room and had a dining room table for family meals while the front room remained pristine for Sundays, high days, holidays and special visitors. Front rooms remained traditional in my family for many years and it was only when my parents retired to Newquay in Cornwall that they adopted the idea of only one lounge.

The kitchen was attached to the back of the house in the form of a fairly solid lean-to and it was cold and dark. We cooked on gas and had a gas-meter under the stairs requiring top-ups with shilling coins. There was one outside toilet and there was no bathroom and we used potties for upstairs requirements all the time we lived there. Our washing and sanitary requirements would be considered as being very basic and inadequate nowadays. We had a splash wash, face and hands, in the morning and in the evening before going to bed, in an enamel bowl in the bedroom, and yes, we did have to break the ice on cold winter mornings because we only washed with cold water and there was no central heating or immersion heater.

Cold lino covered the floors and frost often appeared on the inside of the windows. Baths were taken once a week in a zinc tub in the kitchen in order of seniority. My father was first, then my mother, both of whom did it privately and alone, nudity was always suppressed as something slightly rude, then my mother would supervise my brother and I, taking particular care always to scrub our necks and behind our ears. Being the youngest, I was always the last and I do not think the present day environmental agency would have been too happy at the quality of the water in which I bathed, we survived.

My parents had the front bedroom with quite a large double bed donated by my grandparents with the traditional wardrobes, dressing table and cane chairs while my brother and I had two single beds in the bedroom at the rear with an old chest of drawers and a couple of boxes. There was no going out to furniture shops to buy 'new'. Most of our furniture was given to my parents or was purchased second-hand.

My brother and I had the misfortune to contract chickenpox, quite badly I understand, and were quarantined and confined to that room for two weeks. We were bouncing off the walls after only three days and my mother's patience was fully tried as she sought jigsaws, books and toys to keep us occupied. My father brought the radio up in the evenings to keep us entertained but, apart from Children's Hour, the content was mainly adult in nature. I used to adore sitting in my mother's arms in the afternoons to hear Listen with Mother and I think she used to like cuddling me and listening as well. My brother, on the other hand, would never be cuddled, and squirmed and wriggled until he was let go.

My mother had very firm views on what was correct and what was proper and was always slightly worried about what the neighbours might say so she was highly embarrassed when the doctor called during the second week (yes, doctors did call on their rounds in those days and held great respect) and discovered Roger and I pillow-fighting in the bedroom. He was amused—my mother was not.

I never remember my mother smacking me though she assures me that she did on several occasions. Discipline was quite tight but never excessive. We were told off as was necessary and a treat might be withdrawn or a threat to cancel an outing issued but that was about it. Her severest punishment was being made to sit 'on the stairs'. That was serious. I do not recall being really badly behaved at any time but I was quite adventurous and energetic so some problems arose. She was generally in a good mood and helpful and kind and had much patience.

My father was not so patient. He could become more easily exasperated on occasions and as we grew older, he took to giving us a clip around the ear, sometimes referred to as a clout or a cuff, if we had misbehaved. This was nothing abnormal in those days and was regularly used in schools, by parents and other figures in authority such as the police. From the outset, I hated this and I have struggled to fathom my reasons. I felt uncomfortable and unsure about it somehow. Though it was never used formally on me, I would have had no objection to corporal punishment being administered by my father, teachers or headmasters in the form of the cane or a slipper but I hated being hit around the head.

My father hit me quite rarely but as we grew older, my brother took to hitting me quite regularly, away from my parents' eyes, and this went on for at least six or seven years until one fateful day, when it suddenly stopped. I also recall the time when I stood up to a particularly vindictive and sadistic teacher but I digress from the main narrative.

We were not well off by any means but my mother was a very good housekeeper and cook and it was she who managed the family's affairs and kept a very firm rein on them all. My father was not very good with financial affairs and left everything to my mother. It must be remembered that rationing was a feature of our lives until the early fifties so the provision of food and the planning of menus was not an easy task for the first six years of my life. We always ate at the table—breakfast, lunch (at weekends) and tea. We had to sit up straight, make proper use of the cutlery and maintain reasonably good table manners according to age. We had to eat what was put on the plate. If we did not, it reappeared at the next meal until we ate it!

I am slightly bemused nowadays by the lack of table manners. Blame it on TV suppers or fast food or an even faster pace of life if you will, but the younger generation's eating habits are quite appalling and many young people cannot even handle a knife and fork. My young cousins are asked what they would like for a meal and their choices and preferences heeded and delivered, and please, please…why do we allow mobile phones at the table during a meal? I make a plea, please, for good table manners and etiquette. Bread and butter was provided at most meals to 'fill us up' and this tradition still exists in some families today. We were not allowed to 'get down' until everyone had finished and were normally required to thank our mother for the meal.

People will be quite amused nowadays at the type of food that we ate. The emphasis was on the food available, not the food of choice. I am not sure if it was because my father was a grocer but tinned spam and corned beef were regulars and we liked tinned peaches with the old standby…evaporated milk, which was very sweet to the taste. Bread and potatoes were a mainstay but we often had to substitute margarine for butter and then there were the greens, which most children did not like and was the cause of most friction at mealtimes. Boiled cabbage, peas and beans.

It was odd but I liked tinned peas and not garden peas. We nearly always had a roast Sunday dinner and meal types followed a weekly pattern. Monday was minced meat in the form of a cottage or shepherd's pie, Tuesday—soup or stew. Wednesday could be fish and chips, Thursday was normally liver and bacon and Friday was always fish (with parsley sauce…Yuk!) and thus it was so on most weeks.

My father also brought home a constant supply of broken biscuits from the shop—I do not think I ate a whole biscuit until much later in life.

As a young boy, I was dressed in a shirt, long-sleeved or short-sleeved, according to the season, worsted grey shorts, long grey socks and leather shoes. I also had a variety of ill-fitting homemade woollen pullovers knitted by my mother and other relations that were held up against me at regular intervals to check for size and fitting. The grey shorts were thickly worsted and so was our underwear-trunk-like pants and vests and…they itched!

We had Sunday clothes that normally were newer versions of the above but included a white shirt, red tie and a blazer with shiny black shoes. As we got older, we wore a long Burberry raincoat as did a whole generation of young boys. Pyjamas were warm winceyette and warm in the winter but caused overheating in the summer. We normally slept without the pyjama jacket in a heat wave. (Yes, we used to have heat waves)

Swimming trunks were made of wool, weighed a ton when immersed and drooped around the top of your legs like clammy limpets. Being a younger brother, I was also dressed in my older brother's hand-me-downs and I think my mother was aware of this because one winter, she made me a jerkin from one of my grandfather's old coats and I really liked it and wore it until it was outgrown.

My parents' dress was semi-formal as was usual for that period. My father wore a shirt, tie and grey flannels with a sports coat in summer and just added a short-sleeved pullover in winter accompanied by a warm, long overcoat that was remarkably heavy while my mother seemed always to be in a dress with stockings and small high heels and then added a cardigan if she needed warmth. For outdoors, she also wore a long coat and never went out without a hat. Whenever we were in town, she would complete her shopping and then she would head for a department store, which in those days, sold hats by the dozen…most ladies wore hats. There would be hundreds of them laid out right along the counters with mirrors placed at strategic points and the ladies would gradually move along the aisles trying hats on as they went. My grandmother would do the same and if the choice of a wedding hat was under consideration, I could be in there for quite some time.

Our back windows, upstairs and down, looked out over the Fishponds Coalfield and I think we lived very near Castle Pit. It was a blackened, dark landscape which was only relieved by the Midland Railway line passing along towards one side of our back garden and under the road bridge and a little steam engine working on the tip taking coal down to the main line that we called Puffing Billy. The original line was built to transport coal from Coalpit Heath down to central Bristol where all the main industries were situated right near the centre of the city. Northbound the trains travelled up to Staple Hill and Mangotsfield and then further afield from there.

Mangotsfield was quite a large station and had six platforms and a branch line bypassed the station completely. Arnold Ridley, author of the play, The Ghost Train, was allegedly stranded there late one night and heard what he thought was a ghost train because he could not see it. It was however just a train passing at speed on the branch line in the distance right out of sight. Spooky!

The view from the back windows has changed drastically. Where the coal tips once stood has now been levelled and grassed over to supply pitches and play areas to provide a wonderful facility for local residents. I would have loved to have played there as a child for we were forbidden to go anywhere near the area. Not because of the dirt, grime, pits and industrial dangers, as you might think, but because of the 'strange men' who could be lurking there. The strange men warnings were also used to keep us away from Purdown when I occasionally visited cousins who lived on Muller Road. I learned later, of course, that the 'strange men' were the residents of Purdown Hospital who were deemed unsuitable to be out in the community at that time, and some of whom, I was to meet a little later in life in rather unusual circumstances.

Across the road was the Fishponds Lido surrounded by a six-foot fence and this was out of bounds as well and I only got to swim there when I was older just on the one occasion. The railway route is now a cycle track and though I applaud the construction of all such initiatives and provision for cyclists and leisure users, I do not like using them. Too many of them, within the city areas, have dark uninteresting cuttings and most of them have no views. Quite frankly, they are boring but there are good ones such as the Avon Gorge Route and the ones around Leigh Woods.

My little world gradually expanded. Accompanied by my brother, I was allowed to attend the Saturday Morning Children's Cinema at the Vandyke and I remember the noisy queue and many of the cartoon, comic and cowboy films I saw there. No educational or instructional films involved. It was pure entertainment.

During this time, I had begun to suffer very badly from Bristol Catarrh which was quite unpleasant especially in the winter months and I must thank the family doctor who suggested that I get more sea air so day trips and holidays were arranged at Weston-super-Mare and Uphill mainly for my benefit. I think I have mentioned that I was quite adventurous at the time and one lovely sunny afternoon, I was playing on the sands and ventured too far out at low tide. I got stuck in the mud!

My father had to remove his shoes and socks and sports jacket, roll up his flannel trousers and squelch out to my rescue. There was no such thing as beach wear in those days! He was not amused. On another of our seaside trips, my brother and I were playing on the small area of shingle beach at Clevedon when I accidently dropped a large pebble on one of his fingers. He had to be treated by the St John's Ambulance volunteers up on the prom and I understand that it was very painful. I don't think he talked to me all the way home on the bus. A sign of things to come?

My brother had started school three years before me and by the time I started at Chester Park Infants School, he was at Doctor Bell's on the Fishponds main road. It was not until I started college some twelve years later that I realised this was THE Doctor Bell who founded the Dr Bell's Monitorial System whereby he would teach a selection of his senior pupils, or monitors, the particular lessons for that day and then they would disperse to teach classes of students the same topics. It was considered quite enlightened in its time.

I cannot recall any build up or preparation for my arrival at Chester Park Infants School and there was no trepidation on my part as my mother escorted me up Ridgeway Road to the school as she did every day from then on, faithfully collecting me at 3.30 pm every afternoon even though she was working in my grandfather's electrical shop on Warwick Road. I think I was looking forward to it because I did not even turn to wave as I entered through the gates, about which she was quite upset. The school was built on old traditional lines with a central hall lined on both sides by the classrooms. I regret I cannot remember the staff or any of my lessons or activities there but I was told that I did enjoy it so I have to take my mother's word. I understand that it has just recently closed.

1951 marked the Year of the Festival of Britain which opened in May of that year on a 27-acre site on the Southbank area of London. I suppose there must have been some sort of family preparation for Roger and I were woken up very early one morning and dressed in our Sunday clothes, had a quick breakfast and were then driven down to Bristol Temple Meads by my Uncle Jack in his old Standard car. I remember the make because a few weeks later, he was driving through Broadmead when the bottom fell out of it due to severe rust.

As a young five-year-old, I had conjured up the picture in my mind of him and his passenger propelling the car with their feet. We had been to Temple Meads before but I was always slightly in awe of the soaring edifice and the sweeping platforms. I found it quite spectacular. John Betjeman described it thus: Bristol is the only city in England where the railway station looks like a cathedral and the cathedral looks like a railway station.

We visited the festival site first that day. Its main feature was a 93ft high and 365ft in diameter dome which was, at the time, the largest dome in the world. Under it were housed various themed exhibitions celebrating British Industry, Arts and Science; all to inspire thoughts of a better Britain. The themes were Discovery, The New World, Polar Regions, The Sea, The Sky and Outer Space. The other main feature was the Skylon which can best be described as a huge cigar shaped elongated tower floating above the ground and supported by cables. It seemed to float in the air.

Pundits cheekily and playfully suggested that it represented the British economy at that time: No visible means of support! The Festival Hall was also built for this great event, which, by design or intent, opened almost exactly 100 years after the Great Exhibition. Now, my readers may be musing at this point, how, as a five year old, I could remember all this detail with such clarity. Was I a genius? Alas no. I do remember some of it but we purchased a big souvenir book of the event which was my constant companion for many years.

We 'did' London on that day. We saw the Horse Guards Parade and visited Trafalgar Square where we were allowed to feed the pigeons (taboo nowadays I am afraid) and we also visited The Tower of London. The evening was spent at the Battersea Fun Fair where I spent all my pocket money on rides within the first hour, my brother spent nothing. We caught the last train home back to Bristol and I was fast asleep for the whole of the journey. I was quite upset the next morning to hear that we had travelled back to Ridgway Road in a taxi. My first taxi ride and I had missed it. I do wonder in retrospect how my parents afforded that day. They must have saved up for many weeks. The entry fee for the Exhibition alone was 15 shillings per family, which was a big sum in those days.

Finally, I think I can say that I took part in a small piece of British History. You will understand that I had no personal part to play or specific role to perform but I did see the Brabazon fly. 'What was that?' my young cousins enquired.

The Bristol Type 167 Brabazon was the largest aircraft in the world when it first flew on 9th September 1949. It was so big that a special hanger had to be built for it and the Filton runway had to be lengthened to nearly two miles which entailed demolishing the small hamlet of Charlton at the far end. It had a wingspan of 70ft, a top speed of over 400 miles per hour and was powered by 8 Bristol Centaurus Radial Engines. Designed for the Trans-Atlantic Route, it had two decks but only room for 100 passengers because it had sleeping berths, a cinema, lounges, a dining room, a cocktail bar and well equipped dressing rooms for ladies and gentlemen. Well, they had to dress for dinner on board. It was unashamedly luxurious.

As it lifted into the air on its maiden flight, the chief pilot, Mr A J Pegg (I love the formality of that) exclaimed, 'It flies!'

And fly it did. It handled beautifully and was rated a wonderful aircraft but it was described as, The plane that flew too soon, or, by its critics, The prettiest piece of useless scrap ever! Though it survived for several years and was touted around international air shows, it attracted no buyers. It was commercially unviable and was scrapped. Only one was ever built. Its advanced technological features however, such as the fully pressurised cabin, hydraulic systems and electric controls, were incorporated into later aircraft, so a little of the Brabazon flies on.

I knew nothing of this only that my parents took us to one of the Filton Air shows that were held annually then and we found a place by a chain-link fence alongside the great runway with hundreds of other Bristolians and visitors. There was a hum of expectancy. I am sure that there were other aircraft in the display which I cannot recall. I wanted to see the Brabazon—the giant aircraft. Eventually, it slowly emerged from its gigantic hanger and my father lifted me on his shoulders to watch this huge bird creep on to the apron. It was awesome and obviously the biggest thing I had ever seen in my life. Then came a gradual ever-increasing cacophony of noise as each of the mighty engines were brought into life and it was truly deafening.

With all eight engines started, the Brabazon eased forward with a mighty roar and trundled down to the far end of the runway, turned and…stopped. And we waited…and we waited…and we waited, for a full half hour. Was it going to fly? Then, the engines were started up fully again and roared and screamed as the great aircraft, shimmering in its own exhaust haze, moved forward up the runway, accelerating as it made its approach towards us. I had to cover my ears but I still had a grandstand seat on my father's shoulders. This great aircraft lifted into the air directly opposite us and roared its way into the sky. I was shaking with excitement and my father had to put me down in case I fell, but I had seen the Brabazon fly and I had made a mark in history somehow.

War Years

Early 1940's

Liverpool 1943 – Father Centre Smoking Cigarette.

Chapter Two
My Parents

For many years, I lived abroad in Central and West Africa, and though I am sure there were many common factors of the various cultures and communities in which I lived, the overriding feature was that of the extended family system whereby families were much larger than the nuclear system we have in the UK. Extended families consist of grandparents, aunts, uncles and cousins all living in the same house or within a local area or village. This enables them to build friendships and networks and everyone helps everyone else.

Your relatives are on the spot to deal with the important events of your life and are your first resource when you need help and assistance. All extended families have great respect for their elders. I might add that not all relationships within the extended family are always harmonious, but disputes and disagreements are normally and quickly resolved within the family.

I often muse on the differences between family groupings at home and abroad. We rely heavily on the Welfare State which assumes many of our family responsibilities and I do wonder how much longer the UK can sustain it. We are used to the government supplying most of our needs through the NHS, social services, public services and local government and we are cared for from 'womb to tomb' but at an astronomical cost not all of it spent wisely or well. Many friends have enquired about such services…say, in Ghana. The answer is that there are none.

There is no free health service, no unemployment benefit, no old peoples' homes, very poor social services, a very limited police force and the last time the fire service was called out in an emergency, in my experience, they demanded money for fuel. There is no safety net for the very poor. Most problems are dealt with within the family and some charitable institutions. A Social Security Scheme does provide a pension to which all employers must contribute but this is very small and often considerably late in payment.

We should, perhaps, also recall that the NHS and National Insurance was not introduced in England until after the Second World War. There has been a radical increase and growth of the Welfare State since then. Now everything, however, is the government's fault and we expect it to come up with all the answers to all our woes. Our family units are not so close knit in these modern days and we abrogate many of our family responsibilities to the state. It has become increasingly obvious to me that family units were more closely knit when I was young. Perhaps this is due to my limited environment at that period but my upbringing and care was wholly within the confines of the family with no outside care or free nursery places or baby sitters.

My family is extensive in terms of sheer numbers. My mother had five siblings—three brothers and two sisters and, I learned much later, two half-sisters and one half-brother.

More about that conundrum later. She was born in Southampton on 1 August 1919 and was the daughter of William and Mary Foord. My grandfather was an electrician and worked on submarines in Southampton during the First World War. I find this difficult to envisage for two reasons. Firstly my grandmother met my grandfather on the Strand in London and she informed me that he was dressed like a Toff even sporting a silver cane. I understand that he did come from a well-to-do family, having attended a second-rate public school in Devon but I have no idea how he got into electrical engineering. Secondly, he was a big man… Big-boned and hardy handsome, as Gerald Manley Hopkins stated in one of his poems. How did he fit into tight little submarines?

The family then moved to Bristol in the early 1920s and lived in a big old rambling house at the bottom of Ashley Hill from where my grandfather launched himself into the electrical retail business. He was surfing the initial wave of the domestic introduction of electricity to Bristol. It was evidently tough going, so he opened a second-hand shop at Sussex Place to keep the wolf from the door, but was eventually successful and opened a retail electrical shop in the Arcade in Broadmead and a contracting company which still exists to this day.

It has been difficult to fully establish what my mother did between the wars. Her childhood was not easy and she relates stories of being inadequately clothed and shod, using Granddad's old army trench coat to keep her and her two sisters warm at night and running constant errands for her parents. She suffered from rheumatic fever at the age of 12 and spent many months in the Bristol Royal Infirmary. She was forced to lie still during that time and must have

had a fairly miserable time of it. My grandmother visited her but she never saw her brothers and sisters because children were not allowed to visit. Her grandfather saw her once when he visited the ward quite tipsy with a couple of his drinking pals and the Matron threw them out. My mother was highly embarrassed.

Her illness affected her schooling at St Simon's. (St Simon's Church has been taken over by the Greek Orthodox Church and is now much brighter and lighter.) She was obviously highly intelligent and won several prizes but could not matriculate because of her long absence and she was often late back in the afternoon because her father sent her on long errands in the lunch hour. She was the oldest daughter and shouldered many responsibilities which seemed most unfair.

Sometime before I was born, my grandparents moved to 18 Warwick Road and opened an electrical shop which had a large yard at the rear sufficient for the electrical contracting business, but I have no idea what prompted this move from Ashley Hill or the shop in the Arcade. My guess is that my grandfather drank the business away. My mother left school at fourteen and worked in the electrical shop for long hours until 7 pm. My grandfather would not allow her to leave early to attend book-keeping classes along Pennywell Road which started at 7 pm, so she had to run all the way there. She passed, however, with distinction and was top of the class. She then took over my grandfather's accounts.

An eventful Saturday night arrived and my mother was due to go out, probably dancing at the Berkeley at the top of Park Street with her brothers and sisters, but my grandfather stated that he could not pay her weekly wages because he was 'short', even though he would not fail to go out to the pub later. My mother stormed out and the next week she was taken on as a book-keeper by David Grieg's the grocers and this was where she met my father who worked there as a grocer's assistant.

Strange as it may seem, however, my grandfather sprang to my mother's aid late one Christmas Eve. She was forced to work until after 11 pm by orders of the manager, and my grandfather ordered her brother, Bill, to walk down to Grieg's and bring her back to Warwick Road, complaining at such late hours. My mother obviously decided that enough was enough, however, and applied successfully for the job of cashier in Bright's Restaurant on Whiteladies Road, which was considered to be a very good position indeed, and she much enjoyed the work.

She had a little glass booth in which she and the till were situated and the money arrived in screw-top canisters on overhead cables into which she placed a receipt and

My Father's Family.

Father Centre Front. Aged Two

Father extreme right – mid 1920's – Aged 14.

whizzed it back to the waitresses. Every evening she was escorted to the nearest bank safe deposit box just along the street. I understand that my father, her fiancé, did not approve of the gentlemanly escort and was quite jealous.

While working, my mother seems to have started to enjoy herself more in and around Bristol. There were impromptu parties in friends' houses, dancing at various venues, long cycle rides to Weston, Clevedon and Portishead and a very wet one to Cheddar Gorge. They would often put their bikes onto a Taunton Train and then cycle from there to Barnstaple—my father's home town. It seems strange to us nowadays but long cycle rides were then the norm, long before the universal introduction of the motor car.

During the War, without their bikes, and after they had been married, they had missed the last late night train from Taunton to Barnstaple and had to hitch-hike. A large staff car pulled up in the darkness and dad was reprimanded by a four-star American general who told him that he should not be hitch-hiking with a lady late at night. My father gave the classic reply.

'That's no lady. That's my wife, sir.'

The general ordered his driver to drop them both off in Barnstaple.

During the war, my mother found it difficult to cope. My brother, Roger, was born in 1943 but my mother had received no advice or help about having a baby and had no experience of childcare. She was thrown in at the deep end. I have also been trying to establish whether serving soldier's wives were treated as single mothers because she could not give birth at a normal maternity wing but had to go to a Salvation Army Maternity Home on Ashley Hill, which seems very peculiar.

She did have two good friends to help her—Auntie Peggy, who was a lodger, and Aunty Ethel. Auntie Peggy played no part in my upbringing but she liked looking after my brother, who, I understand, was a difficult baby and toddler and disliked socialising. His early claim to fame was that he buried all the teaspoons of the flat in the back garden one day. I think Aunty Ethel helped with me but I did get to know her when she moved back to Barnstaple as we regularly holidayed there in the early 50s. She was not directly related to us though there is a tenuous link between her and my paternal step-grandmother.

My mother was always short of money living on my father's army pay and experienced problems finding the rent for the small flat on Fishponds Road.

My father was born in Barnstaple on 28 March 1913 and had a brother and two sisters. He had led me to assume that he had come from a much larger family but this was not so. He was the son of Earnest Robert Allen and Elizabeth Belle Balman, the daughter of a Barnstaple butcher, who had a shop in Butcher's Row. They kept a Newsagents Shop at 39 High Street in Barnstaple and I reckon my father must have had a good early life there as the First World War barely seemed to touch them personally.

His brother, Reggie, was four years older than him and was quite a lively chap, getting up to all sorts of schoolboy pranks and mischief, whereas his two sisters, Gladys and Gwen, were nearly ten years older than him so must have played a major part in his upbringing. In 1919, my father's mother died, possibly of TB, and Mr Allen remarried. His new wife, my father's stepmother, was Mary J Hill. Her father was a woodturner as were many relations on my father's side and we seemed to have many French polishers in the family in the past based around Barnstaple.

Indeed, the Allens were very active in Barnstaple. Apart from their involvement with wood, various branches of the family were drapers, booksellers, builders and decorators, farmers, commercial travellers, milkmen and labourers. Two ancestors owned and ran the rather splendid Commercial Hotel, which is a very handsome building to this day on Taw Vale and another, the Stag and Hounds on Bear St, while Reggie appears to have owned the Rising Sun at some stage near the parish church.

Barnstaple Scenes.

Allen's Ice-creams – Me top right by my father.

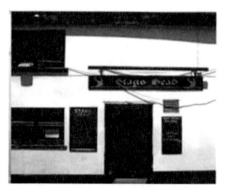

The Stag's Head owned by Alfred Allen

The Eldorado Ice-cream Parlour

Allen's Commercial Hotel.

In the early 1920s, the whole family moved to Bristol except for Reggie, who ran grocery shops at various sites in the centre of the town and played a leading part in Barnstaple life, especially the swimming club. My father was apprenticed at David Griegs in Stoke Croft in Bristol, when he was 14 and worked himself up to be an Assistant Master Grocer and that was where he met my mother.

Mr and Mrs Allen, my grandparents, opened a sweet shop and grocers at 126 City Road and resided there until 1941 at which time they returned to Barnstaple because of the Bristol Blitz. My paternal grandfather was a master carpenter, an excellent wood turner and French Polisher and held a reserved occupation during the early part of the war in a factory on Newfoundland Road.

My parents were married in October 1939 and thus commenced a very different type of life for them. War had been declared on 1 September 1939 and my father was called up, after the end of the Phoney War, on 7 August 1940 to join the RAF Defence Squadron. I have often wondered about my father's thoughts and feelings at this time, and indeed, the feelings of the thousands of conscripts and volunteers, being wrenched out of their civilian lives and plunged in to their training in preparation for war.

My father's world had been limited to North Devon and Bristol and, in reality, he had led a normal and fairly uninteresting existence until this time with no major excitement or events. It must have been quite nerve-wracking and slightly frightening. I know the RAF Defence Regiments had an initial requirement for 29,000 men, later expanding to 80,000 and this would have been multiplied many times over for the army, Navy and Air Force. My father was 24 when he was called up but we must remember that many conscripts were only lads of 18 and they must have been very nervous indeed. And what of the wives, parents and relations left behind? They must have been very worried and concerned as well. I know my mother felt very lonely and had difficulty coping alone especially after the birth of my brother.

I hope my father would have been blissfully unaware of the disputes and arguments that led to the formation of the RAF Regiment prior to his call-up, but it is quite obvious from various archives that there was a long interminable continuous argument waging at the High Command level as to whether the army or the Air Force should supply ground defence for airfields. This rumbled on throughout the whole of the Second World War and I believe that it was never satisfactorily settled or finalised. In the initial stages of the war, it was very much a case of, 'the emperor fiddling while Rome burned!'

Initially, it was the army's job to provide ground defence but they were very overstretched and were often called to other more urgent areas in the initial stages of the offensive. While the debate was on-going another farce was in the making. Britain had very few guns to the extent that very often museums and collections were ransacked for pistols, rifles and machine guns to equip the initial troops. Just the job for a diving Messerschmitt 109…I think.

Early in 1940, there were only 433 '303' machine guns and 72 assorted AA weapons from the army, with 835 gunners available for airfield defence. Documentation from 1932 proves that the high command was aware of this failing and that all the services were woefully ill-equipped. At the commencement of the outbreak of hostilities, some airfields were only protected by a couple of light Lewis machine guns and several old rifles.

There was an absolute scramble to strip Lewis machine guns from old aircraft and the Boffins were tasked with coming up with new ideas for airfield defence, two of which I like. It must have been an impressive feat of skill and stamina to erect, but, initially, many airfields had huge metal net curtains called Dannut Wires hung at the end of their runways to deter low level aerial attack and my father also reckoned that they assisted the gun aimers. The other idea I liked was the Pop-Up Pill Boxes (just like some modern loos) that could emerge and submerge into the ground. It is important to note that the RAF gradually invented and evolved much safer airfields as the war progressed to ensure their protection against ground and air attacks.

We tend to forget the many theatres of war. The RAF Regiment alone served in the UK, the Middle-East, North Africa, Sicily, Italy, Greece, the Balkans, India, Burma, Malaysia, Indo-China, the East-Indies, North-West Europe, West Africa, the Azores, Norway and the Cocoa Islands and other service units covered even more countries. Logistics must have been a nightmare and there must have been times when units were spread very thinly on the ground. We also forget the myriad of airfields in Britain at this time, all of which needed defending.

The High Command's position slowly began to change. In May of 1940, the Directorate of Air Defence was formed with the aim of providing 100 defence airmen per flying station. By December of that year, 29,000 airmen had gained their Gunner Badge including my father who gained his on the 9th October, but, overall, rewards of gallantry in the air were being negated by inadequate organisation for defence on the ground. There was a definite need for the enemy to be destroyed as they attacked bases.

Two major events seem to have brought about a more complete sea change and thus the formation of the RAF Regiment. On 10 May 1940, there was a shift in German tactics and strategy. An RAF airfield in France had most of its aircraft wiped out on the ground because of inadequate ground defences, and this was such a success for the Germans that they adopted this policy in all theatres of the war, much reducing the effectiveness of the RAF. On 20 May 1941, the Fall of Crete catapulted the problem into the limelight. Crete's ground defences were feeble and could not withstand the German onslaught from the air. Accounts and official records give a low-key report on these events but I reckon there was panic at a very high level with the RAF saying, 'We told you so!'

It was almost instantly decided that the RAF should be responsible for its own ground and anti-aircraft defence to protect against attack. An Air Commodore-in-Chief was appointed to be responsible for advising and planning the defence of RAF airfields and installations against ground and low-level air attack. At last the High Command had appreciated the absolute necessity for the establishment of an RAF self-contained permanent defence force. Thus the RAF regiment was officially formed on 1 February 1942 with its motto 'Per Ardea'—Through Adversity.

One cannot resist the jibe that most of the adversity suffered in the lead up to the formation of the Regiment was experienced at High Command level and not on the airfields. The Regiment's insignia is two crossed No.4 Lee Enfield rifles but I am sure these do not represent the only weapons that were used initially for the defence of our airfields.

My father rarely talked about the war and I wished that he had told me more stories now. Other members of my family who had served also seemed reticent and were keen to get on with life, so I never knew whether he had any knowledge of all the above. I suppose his first concern was to get on the train with his rail warrant and get himself up to Padgate, the National Training Centre for RAF recruits which had opened in 1939, near Warrington, where he was equipped with his uniform and all necessities, his Official Number 1110545, and shown his billet just like the other 1,500 men who arrived there every week. It must have required some organisation.

He was assigned to I Wing 3 Reserve Command and underwent three weeks basic training where he was drilled and trained as a Combat Infantryman, before he was shunted off to Roundhay on the Isle-of-Man for his IGDGS (Ground Defence and Gunnery School). This could have been a little more pleasant for him because accommodation was in the various boarding houses and hotels in the surrounding areas, though I understand the officers pinched the best accommodation.

The course, however, was tough. It lasted for ten days during which time they were trained on the Lewis and Browning gun then the Bofors. Westward Wallace Aircraft dragged drogues out at sea which the gunners used as targets, firing from St Michael's Isle (Fort Island) and Santon Head. Those poor pilots. They were airborne for most of the day going up and down…and up and down…it must have been excruciatingly boring. It was probably dangerous as well because the gunners were all novices, but I did not pick up any reports of accidents or fatalities.

My father seems to have done well and passed the course on 10 October 1940 with a comment of 'Very Good'. (Not like my school reports) I do find it rather odd, not that the airport on the Isle of Man stayed open for civilian flights all through the war, but that the gunnery practice was conducted in the same area. From the Isle of Man, my father moved back to Liverpool and we have a rather fine photograph of him walking jauntily down the prom with a cigarette in his mouth accompanied by a pal. He looks confident and happy.

RAF Liverpool was the Number 3 School of Technical Training. Squires' Gate provided technical courses for the RAF and several of my relations attended there for aviation and electrical engineering. My father was not in the least bit technical and I have dutifully followed in his footsteps, so I do wonder what he did there for a whole two weeks. Perhaps it was training on gun sights or care of the guns? He was lucky enough to be in good accommodation however again. He qualified as an Aircraftsman 1.

He was then posted to 4088 Anti-Aircraft Squadron at RAF Colerne. (Fighter and Bomber Command). I wonder how he got this posting? Was it the RAF in their kindness posting him close to my mother or just luck? I am not entirely sure of the timeline here, for his address on his army record is 21 Albert Park Redland Bristol and my mother is also listed as residing there as his next of kin. They lived there until March 1941 when bomb blasts blew out all their windows, then moved to the flat in Fishponds Road.

Anyway, remembering my former comments about long-distance bicycle rides completed by my parents in their youth, my father used to cycle from Bristol to Colerne and back on most days…a fifty-mile round trip. I suppose that he must have stayed in a billet for night duties. But what of Colerne?

In 1940, the grass runway was replaced by three fully tarred runways and a large variety of aircraft flew from there flown by Polish, Canadians and British pilots. For some reason, my father did not get on with the Polish but he would never explain his reasons. My mother thought it was because he thought the Poles had capitulated too easily at the start of the war.

As the war progressed, the airfield developed and expanded. Aircraft were moved to protected dispersal bays and 20mm Hispano light anti-aircraft guns were placed in sand-bagged pits with built-in dugouts. A Dannut Wire was built at the end of the runway and strategically placed strongpoints were built around the field while the all the other buildings were upgraded and better protected. By 1941, more light machine guns were provided and Bofors guns arrived.

He never told me that Colerne was weakly defended, but, by one of those odd quirks of circumstance, and when I was a slightly older teenager, I was made an honorary member of the Officer's Mess at Colerne and got to know a little of its wartime history. During my father's first few months at the station it would never have survived a sustained attack. He was trained on Bofors but they did not have any then. While at Colerne my father was promoted from Aircraftsman 1 to Leading Aircraftsman.

My father appears to have been at RAF Colerne for some two years. His service record is almost illegible but my Mother is adamant that he manned guns during a brief period of the London Blitz. This may be possible as there is an entry on his record for such a movement. He seems to have fired guns all night and then helped dig the poor people out of the rubble in the East End next morning. Evidently the sights that he saw upset him for some time later in his life.

His fairly quiet war at Colerne was suddenly to be disturbed and this may have been another shock for him, especially as he had never been abroad during his life.

On 25 May 1943, he was shipped out on a troop ship to 2860 Squadron in the Middle East. Some of the following is guess work and has been obtained by reference to various squadron records of that time, my problem being that my father's service record becomes even more illegible so some of this may be inaccurate, especially bearing in mind that many of the RAF Squadrons at that time consisted of seven officers, one hundred and seventy eight airman, organised into three rifle flights, an armoured car flight and a support weapons flight and there were independent anti-aircraft flights of one officer and sixty airman, and I guess my father was part of the latter. The difficulty is that these sections were often split and operated independently or with other units so tracing actual movements has been quite a nightmare.

2860 Flight was formed at Leuchars in 1942 and then reformed in May of 1943 at La Marsa in Northern Tunisia. It seems that my father was part of that reformed group having landed in Bizerta, then make his way down to La Marsala which is the national airport of Tunisia nowadays and was the hub for all troop movements. His previous 4088 Squadron combined with the 2860. They moved south to Hammamet and then on to Sousse (used by Hannibal during his conquests) before returning to Bizerta. Their role presumably involved airfield defence.

My father was using the Bofors guns on which he had been trained. These 40mm guns were very powerful and though they were a multi-purpose weapon the RAF Regiment used them primarily for anti-aircraft work. They were an auto-cannon designed by the Swedish manufacturer—Vickers, and could shoot 120 rounds per minute. If you want more technical details, they weighed 1,981 kilos and had a range of just over 7,000 metres. They were also transportable and helped to win the fast-moving war in the Western Desert.

At this point, an interesting item makes its appearance in my father's service record. On 30 October 1944, he was stopped one day's pay for abuse, I would be very interested to know the circumstances especially as he was awarded a Good Conduct Badge prior to that in 1943.

Then, stated quite clearly for once, in my father's service record is the simple phrase 'Transferred to Army'. There is no date but I do know that somewhere in Tunisia, his units formed part of the X Corps 8th Army with the Dragoon Guards and Household Cavalry, and he took part in Operation Torch. He had badges to prove that and he was immensely proud of this. But I am not sure if he was aware of the battle yet to come, for the invasion of Italy was in hand. My father seems to have been moved swiftly between 2828 and 2746 Squadrons when he was attached to the Invasion Forces. He was deployed to cover airfields in Palermo, Baglioni, Cerignola and Anzio and later moved on to Naples.

In May 1944, his squadron, then 2746, was paraded somewhere near Naples. All married men were ordered to take one pace forward and told that they were going back to the UK. All the single blokes were shipped off to continue the war. He was discharged on 13 March 1945 with the simple statement 'Services no longer required'. And that was his war over.

As a rider to my father's war record, I met, by chance, in Brecon, some years ago, a former war veteran, in the Welsh Borderers Museum, who, upon hearing that my father was in the RAF Regiment referred to him as a 'Rock Ape.' I was bemused by this and enquired the origin of the phrase. It seems that Goebbels had hatched a plan to kill all the apes on the Rock of Gibraltar because legend has it that if the apes left the Rock, England's rule over the colony would end. Churchill ordered that the apes be guarded and protected and a unit of the RAF Regiment was detached to ensure the apes were safe. Hence the nickname for the members of the RAF Regiment.

My mother meanwhile had stayed back in Bristol. I use this phrase advisedly as quite a few residents had left due to the Blitz and I am sure there had been discussions about moving somewhere safer, especially after my brother Roger was born. The flat on Fishponds Road was away from the main bombing targets but there was always the chance of a stray bomb. She was working at Warwick Road near Newfoundland Road and that area suffered badly as many vital factories were situated there. My father's parents were on City Road and they lived with the constant danger right on their doorstep and of course, my paternal grandfather worked in one of the factories.

Certainly, during the night raids the noise and explosions must have been heard in Fishponds and there were anti-aircraft guns ringing the city, some 200 in various forms. Less than a mile away, at the top of Purdown, was the legendry Big Bertha, whose huge bang could be heard throughout the north side of Bristol on many a night, except…there was no such gun. Big Bertha was a myth! My mother thought it existed because she heard the huge bangs it made night after night and was quite upset when I told her the facts.

Atop Purdown was a battery of four 3.7" calibre guns each in a different gun pit (the pits are still there) facing in various directions. All four guns were electrically linked and a flick of a switch fired the four guns simultaneously hence the tremendous noise, probably aided by the echoes rumbling up the Frome Valley. All Bristol's guns were fired regularly but I am afraid that this tremendous effort through many a night only resulted in two downed planes throughout the whole of the war, I am quite sure though that many were damaged or put off by the bombardment. It may be of interest to note that Lundy, in the Bristol Channel, claimed the same figure without firing a shot. Sneak German Raiders used to make low-level attacks on Bristol at sea level. Two of them forgot Lundy was there and flew straight into the cliffs on the west of the island. A few remains can still be seen.

Bristol was a prime target because of its docks, factories and the Bristol Aircraft Corporation and was easy to locate on a clear night. All the bombers had to do was follow the River Avon up through the gorge and they were right in the centre of Bristol. With such a clear navigational aid I have pondered on two other seemingly ambiguous facts. The two islands of Flatholm and Steepholm lie out in the Bristol Channel and would have been on the flight path of incoming bombers. Flatholm had guns, 7 inch 7 ton guns mounted on the famous Moncrieff Carriages. They were termed 'Pop-Up', because they could be loaded in the relative safety of the gun emplacement with the barrel in a vertical position.

To fire, the barrel would swing up into a horizontal position and the recoil would pop it down and back again. Clever, eh? Steepholm had no guns but a series of concrete look-out posts built at the base of its cliffs, some of them quite hazardous to approach especially on a dark and wet stormy night. One has a stairway that seems to soar out away from the cliffs and was called, 'the Stairway of the Gods'. It is rumoured that inspecting officers would never brave these steps. Why were no guns put on Steepholm? Were the ones on the end of Brean Down deemed sufficient to cover attacking aircraft? On my one and only visit to Steepholm, I was somewhat surprised to note that a Bofors gun has now been emplaced outside the refreshment hut; a bit late, don't you think?

A decoy night-time airfield was situated high on the Mendips on Blackdown, even having night lights, but apparently, the Germans were not fooled and no bombs fell on it, possibly because of the River Avon navigation aid that geology had thoughtfully and so clearly provided millions of years ago?

My parents were quite reticent about their war experiences, but I know that my mother would have had a particularly bad war, for Bristol and its inhabitants had a particularly rough time of it. Everyone was fooled into a false sense of security during the Phoney War period from September of 1939 until June of 1940 and one would have hoped that the relevant authorities might have taken the opportunity to bolster their defences. Be that as it may, Bristol was woefully unprepared.

The fire service was under strength and undermanned, only 50 ambulances were available from a requirement of 200 and only 3500 places were available in bomb shelters when it was estimated that 35,000 were required. There had been air raid rehearsals, most gas masks and blackout curtains had been issued and some homes had received the famous Anderson Shelters but supplies were sporadic and there was a great shortage of them. There were also food shortages resulting in queues and the whole period was filled with rumours. I suppose the worst effect for many was that weaker beer had to be made by George's Brewery and I am of the opinion that they continued this long after the war because I could never drink it.

Bristol was a prime target because of the Bristol Aeroplane Company at Filton, Parnell's, which switched to war work from shopfitting, Gardiner's, the iron foundry, Sheldon Bush which made ammunition, Strachen-Henshaw—manufacturers of submarine equipment and Butler Oil which was a main provider of tar for runways including RAF Colerne where my father was stationed. Then of course there were the docks supplying the immediate hinterland of the Bristol area and much of the Midlands which included Avonmouth and central Bristol. Bristol was for it, especially as it was so weakly defended.

And so it was that, on 25 September 1940, 58 Heinkel bombers protected by 40 Messerschmitt 109 Fighters flew, unmolested, straight over Bristol heading for the BAC. It does seem remarkable that this formation was not harassed and attacked and there was an enquiry about this incident. The air raid alarms did sound but there had been many false alarms leading up to this one so the workers packed and locked up their tools and ambled to the shelters thinking they would be back at work in a short time but all hell was let loose. 350 bombs landed on Filton and the Works and many workers were killed and injured.

The workers described being lifted three or four inches off the floors of the shelters from the impacts. There were many unpleasant and ugly sights as the rescue workers dealt with the casualties and the dead. Much damage was inflicted and, of course, the loss of so many skilled technicians was a great loss to the war effort. Many relatives and families grieved that day. Thus commenced the Bristol Blitz.

There were now regular raids on the Bristol targets. On 24/25 November, 134 long-range bombers adopted new tactics, for, instead of arriving all in one formation, the experienced crews were sent in first to drop the flares followed by the incendiary bombs in their thousands—13,000 on this particular raid, then followed the HEs—the High Explosive Bombs, not as saturation bombing but as a steady trickle over several hours which must have been a terrifying experience when you were on the ground. The less experienced pilots came in last guided by all the fires that had already started so a raid could now go on for many hours. Bristol became a city of flames.

Incendiary bombs were not that difficult to deal with and were easily extinguished either by using water or beating them with a coat or cloth or even stamped out but though there were hundreds of Fire Watchers all over the city they could not cope with the thousands that were dropped in any one raid. My Uncle Jack, while on a weekend pass from his RAF station, was due to meet an attractive young lady at the Regent Cinema in Old Market one Friday night, but just as they met, 'Moaning Minnie' went off and my uncle spent most of the night putting out incendiary bombs on the cinema roof. Some date!

Further raids were inflicted on the city on 2 and 3 December 1940, the 3 January 1941 and on 16 January as well. Businesses, houses and the commercial heart of the city were reduced to vast piles of rubble. On March 16th, the target was Avonmouth and the Bristol Docks but the bombers were misled somehow, so bombs rained down on Fishponds, Whitehall, Easton, St Paul's Montpelier, Kingsdown, Cotham and Redland. The seven-hour raid dumped 33,840 incendiary bombs and 164 tonnes of HEs on the residential areas of Bristol. I have recorded the fact that my parents were living in Redland at that time and though they survived the raid all their windows were blown in and the house was damaged so this is when they moved to the flat at Fishponds.

Morale in Bristol at this stage of the war was very low and Bristolians suffered physical and psychological damage. Severe shortages ensued, power cuts, hunger, sickness and poor housing were effecting everyone. The winter of 1940/1 had also been one of the worst in living memory with frozen weather and deep snow. I cannot estimate which of these effected my parents or relations but I am very impressed that they managed to get through it all and Thank God not one of my extensive family was injured or killed at home or on active service. The entire family came through the war relatively unscathed bearing in mind that Bristol was the 4th most bombed city in England. It had suffered 548 raids, 1,229 deaths, 3,305 injured and over 3,000 homes had been destroyed. The Reverend Paul Shipley described Bristol as, 'Writhing in Agony. Tested in Gold', because there were many acts of bravery and heroism as well as kindness and help.

I came across a sobering item of news recently as a Rider to the Bristol Blitz. Several bombs have been unearthed recently in the city and three in the Bristol Channel opposite Hinkley Power Station. It has been estimated that over 6,000 High Explosive bombs were dropped on Bristol during the war. These were designed to penetrate the ground at high speed but a further estimation suggests that 10% of them did not explode. Bristolians could well be sitting on 600 unexploded bombs. Bristol also has the distinction of having received the biggest bomb ever dropped during the war. It was huge but did not explode. It landed in Beckington Road in Knowle on 3 January 1941. It was not removed, however, until 1943.

My mother and my brother had spent their air raids in a cupboard in the flat at Fishponds, with the mice, the official shelters being at some distance and not in good condition. She was adamant that Bristol's air raid shelters fell well below any acceptable standards of decency and hygiene. She detested them. This seems to have applied to many family members at that time and some even stayed in their houses preferring to, 'die in their beds'. This was interesting and so I researched a little more and the findings were quite startling.

The overall picture was that Bristol had insufficient, inadequate and inferior air raid shelters. Most of them were cold, stank of urine and musty bedding and were not well made so people were often injured while in the shelters during raids mainly from collapsing roofs. Bristolians were absolutely convinced of the fact that deeper was safer, so, rather than use these wretched facilities, they adapted a whole range of Bristol's underground world in which to seek a safe haven. It gave a new meaning to the Underground Movement. Redcliffe Caves, Clifton Rocks Railway, The cellars at Blaise Castle and many railway tunnels were used for shelter as well as church crypts, the caves in the Avon Gorge and even the old Ice House on the banks of the River Frome and the area where the Frome goes underground at Eastville.

There was a long, ongoing dispute involving the use of the Portway Tunnel that lingered on for several months but the simple fact was that Bristolians preferred the dangers of moving trains, being entombed in church crypts and being buried deep underground to the awful shelters that were being provided: not that there were enough spaces anyway.

Thus commenced another War Phenomenon and I refer to the Trekkers, often called the Yellow Convoy. Quite simply thousands of Bristolians used to trek out of Bristol every night to avoid the bombing. For the more affluent this entailed packing themselves and their family into their car or commercial vehicle and heading out to a rented room or property in the outlying districts for which they paid large sums. The less well-off might have paid for transport on a bus or lorry and then found shelter in a barn, or, under a hedge. Some just walked out and found shelter as they could.

Most spare accommodation—houses, outhouses, barns, single rooms, churches and church halls, sheds and old vehicles were rented out and the residents in these areas made quite an income I suspect. Places like Burrington Combe looked like a crowded encampment with all the people sheltering there. It must have given a whole new meaning to the hymn. 'Rock of Ages cleft for me. Let me hide myself in thee.' The next morning the flood of people would make their way back into the city. It is estimated that 10% of Bristol's population joined in this 'Trekking' and it led to ill feeling for it left fewer citizens for fire watching and civic duties and an added burden to those who remained. Those who walked, however, showed some stamina walking up to fifteen miles out at night then another fifteen miles back in the morning. Some endurance. I am aware that in other cities, special trains were arranged to take businessmen out of the city centres in the evening and these were called Funk Expresses.

No members of my family were involved with this. It seems they mainly found protection in their own homes during raids but they must have been frightened and scared for a lot of the time. One piece of advice was to shelter under a stout table but I am not sure what protection this would have been against a High Explosive bomb. Having dealt with my parents in the events leading up to my birth, I am aware that mention must be made of both their immediate families because they also had a great effect on my development especially my mother's family and it was my grandparents with whom I had most contact, especially my grandfather.

In retrospect, I now appreciate that he did not raise his family well and there were many tensions within the family involving his offspring. When I was very small, I had no idea of these tensions at all and it is now obvious that I was his favourite grandson and he devoted much time to me.

My father's medals:

1939-45 Star—Africa Star with Clasp—Italy Star—Defence Medal—War Medal

My Parents and Grandmother

Which one is my Uncle Jack?

My Family and Me.

Chapter Three
The Extensive and Extended Family

My grandfather, William Foord, was a big man. He stood about 6'4" and had quite an imposing presence. He was born on 11 January 1884 and was the head of our family. He was always in a good mood with me but he could be cantankerous and difficult. In modern-day terms, he would be described as an alcoholic but my mother called him a drunk for he did drink in quite large quantities. He demanded, and somehow got, all the best things in life even during the austerity of war. It was he who had egg and bacon for breakfast while his children had one egg, once a year, at Easter.

He wore well-tailored suits and a waistcoat with a gold watch and chain, and expensive underwear and expensive shoes, all from the best shops in Bristol. However, his children were poorly clothed and shod, sleeping under his old trench coat for warmth in bed at night, yet he would walk into The Three Blackbirds on Stapleton Road on a night and buy drinks all around. He allowed no ambitions for his children, turning down a grammar school place for his eldest son, Bill, and his main idea seemed to be getting his children out working as soon as possible.

He had no aspirations for his daughters, and his sons were expected to join the business. He ran electrical shops, or retails outlets, as we would call then now, and at one stage, he had six electrical shops in Bristol and a contracting company, which, by the time I was born, had dwindled down to one shop on Warwick Road and the contracting side. It is most likely that this was due to excessive expenditure on alcohol but he also seemed to maintain the belief that stock was money and he did not seem to sell much through the shop which contained elderly brass and Bakelite electrical lamps and switches and a jumble of electrical accessories which I often tidied. That same static stock nowadays, of course, would be worth a fortune. He did make money on domestic repairs which came in over the counter and always turned over a tidy sum at Christmas repairing fairy lights.

I have already mentioned the arrival of the Windrush and my early childhood marked the beginnings of the West Indian community in the St Paul's area. When I was a little older my Uncles Jack and Dennis, the two younger sons of my grandfather, used to do electrical work for this community on the 'Lump' during evenings and weekends because my grandfather never paid them a decent wage.

My mother assures me, because she kept the accounts, that my grandfather rarely paid any tax. He was ever ignoring tax demands and many was the time that my mother had to go down to the tax office to pay his five-shilling fines which probably seems a good way out of paying tax for us in modern times. On the lighter side of things, I bet he gave the tax office staff nervous breakdowns with his antics. My grandfather seemed to enjoy beating systems and even gained some of his stock by illicit means.

A selection of travellers (salesmen) from the electrical wholesalers would regularly visit him and surreptitiously unwrap electrical items, for which my grandfather paid cash. These travellers were always a little smarmy with Brylcreemed hair and looked more like horse racing touts than representatives. My grandfather was not a good payer and was often behind with the settlement of his invoices and bills and would not always stump up the electricians' wages on time, which was very hard for them.

Later, Foord and Sons acquired vans but in the early fifties, it was the apprentices' job to load up handcarts with all the items and gear required for the particular job that day, including any ladders, and push it to the place of work while the electricians rode their bicycles to the job with their bags of tools hanging from the handlebars.

My grandfather controlled everything from his upright easy chair at the far end of the back room facing the big table, where we had our meals, and next to the fire. Either a coal fire or a gas fire was lit and I thought this very strange for an electrical company but my grandfather took no notice of the irony. It was a good position to keep an eye on his safe as well, which was next to a huge mahogany desk alongside a wall and was full of hidden treasures such as cigarette cards, stamps, some medals and a selection of naughty postcards that he seemed to collect, which, we young children much enjoyed. He also had a secret wall safe upstairs in the back bedroom that we all knew about and probably gave a new twist to the old phrase, 'Up in nanny's room'—behind the wallpaper.

When in a good mood, he would place a sixpenny piece on his forehead and sing:

I've got sixpence. Jolly, jolly sixpence. I've got sixpence to last me all the day.
I've got tuppence to lend, and tuppence to spend and tuppence to last me through the day!

At which juncture, he would nod his head forward and catch the coin as it flew off his brow. If he was in a really good mood, and we had been good, he might then give you the sixpence!

I only ever had one run-in with him one day when I had gotten myself into a paddy about something at the bottom of the stairs in the shop at Warwick Road. My grandmother tried to control me, then my grandfather tried and I kicked him on the shin. I am ashamed of it now but I did. To my surprise my grandfather, this great tall, strong man, almost dissolved telling my grandmother, 'May, he kicked me, May, he kicked me!' And I was not sure whether he was astonished or hurt. My uncles were quite amused by the whole thing and thought it was a David and Goliath moment.

Later on, when I was about twelve, my grandfather was hospitalised at Manor Park in Fishponds and I accompanied my grandmother to help him settle in but it was he who threw a tantrum then. Manor Park Hospital had been a workhouse many years ago and my grandfather could not forget it. He pleaded pitifully to my grandmother that she should not leave him there and then started screaming and shouting. I was embarrassed and upset for him. His tantrum was stopped by a furious matron for whom he was no match. He had to admit defeat in the end and they did treat him very well indeed.

From an early age, my mother was still working at the 'shop' on Warwick Road and most people worked Saturdays. This applied to all the adults in the family whatever their occupation but I must now admit to a developing mystery which is applicable throughout my boyhood. It would appear that my brother has disappeared. Every Saturday, I would be taken down to the shop at Warwick Road and I would go out with my grandfather. This developed into a regular event but Roger never came down to the shop.

He would only have been eight or nine then but I am sure my mother would not have left him to play at home or play out in the street with no supervision and he did not spend time with our cousins as he was unsociable and did not really like their company. Anyway, with my grandfather proudly clutching my hand and the grandson proudly clutching his, we would set out and walk the short distance along Warwick Road to Stapleton Road to catch a bus. Obviously, my grandfather knew all the routes but his knowledge was unerring for on every bus we boarded, he would tell the conductor, 'One and a half return to the terminus please!'

We would always make for the upstairs front seat and sit grandly there while we trundled off to the extremes of Bristol…without fail ending up at a pub at the terminus where he would buy a pint of bitter for him and a lemonade shandy for me. In summer we would sit outside while in winter, I would be allowed to sit in a quiet corner of the saloon bar—but never the public bar. Drinks finished, we would then catch the next bus back to Stapleton Road.

We talked all the way there and all the way back and he always answered all my questions wisely and well. I learned a lot about Bristol from those outings; a knowledge I still hold to this day and which came in handy when I was a little older. Perhaps it was this experience that led to my travelling on buses alone at quite a young age. When I was only eight, and we were living out at Kingswood then, I was allowed to catch the Number 8 from Kingswood to Lawrence Hill, walk through to the bottom of Stapleton Road and then catch the Number 2 bus up to Warwick Road and I regularly travelled into town with the customary warning about strange men.

My well-dressed grandfather always carried a smart stick and he was quite a ladies' man. His favourite expression was, 'Don't forget the corners, my dear,' as we passed housewives busily cleaning their windows and he was not above giving them a little tap on their bottoms as he passed. I don't think they objected. I am sure he would not get away with it nowadays. There would be strong protests from the Feminist Lobby.

As I grew older, the trips became extended. On one or two Saturdays during the summer, my Uncle Jack or Uncle Dennis would drop us very early down at Hotwells, near the old Rock Railway, where we happily boarded one of the Campbell's Paddle Steamers for a trip around the Bristol Channel and set out down the Gorge. I am not sure if I can remember the full itinerary now but I think we stopped at Portishead, Clevedon and Weston after which we crossed to Cardiff, Barry, Penarth and Swansea, then back across the channel to Watchet, Blue Anchor and Ilfracombe. We then completed the whole trip backwards.

It was a long, long day starting at 7 am and arriving back in Bristol after 9 pm My grandfather spent most of the trip in the bar and I was given free run of the paddle steamer all day and got everywhere from the bridge right down to the engine room, and the nicest thing was all the crew used to share their sandwiches with me and my grandfather could never understand why I was not hungry, though he did buy me copious amounts of lemonade.

Many years later, I completed a nostalgic trip on the Waverley (which is not a paddle steamer) leaving Bristol sailing down the gorge, which I always enjoy, and visiting many of the ports mentioned. Again it was a long trip and because of high winds, we could not be dropped off at Clevedon to board the coach to bring us back to Bristol. Instead, the Waverley's captain skilfully brought us into the marina at Portishead at 11 pm at night in total darkness. The channel is quite narrow and there is a tight dog-leg into the dock there. I was impressed.

My parents could not afford a car and we rarely travelled in one. Uncles Jack and Dennis had old cars which they used but we relied almost entirely on buses to travel around. For me there was a game-changer when my grandfather bought a big second-hand Humber from my Uncle Ray on my father's side of the family. It was the big 'square' one with three seats in front and it could take at least six passengers in the back. It had a very long front bonnet.

I think my grandfather bought it solely on the basis that it had interior picnic tables and a small cocktail bar and from then on, my grandmother had to supply picnics suitable for such fittings. I do not think she was too pleased. My grandparents had maintained their associations in Southampton so I often accompanied them down there for the day in the Humber, stopping at a pub on the way down and back, the highlight being a picnic in the New Forest. In those days, we would be the only vehicle under the trees and traffic was non-existent. Some years later, I drove my horse lorry down to the Rhinefield Polo Ground for a weekend tournament set deep in the forest and was amazed at the amount of traffic and congestion. We also visited Swanage on several occasions and did local trips to Cheddar and Weston. I loved that big car and could almost ride my bike around inside it.

My mother had been born in Southampton and I am sure that my grandfather had quite a few relations there but the only two I ever met were Auntie Pat and her second husband, Uncle Bert, and they lived in a beautiful bungalow in Shirley, having moved there from the centre of the New Forest, just after the war. It adjoined several orchards and they had a very mischievous son whose sole aim was to get me into trouble accusing me of all sorts of misdemeanours and petty crimes. It was not a good few days. Rather oddly there was a shed at the bottom of the bungalow's main garden in which an old man resided. Auntie Pat would deliver food to the shed on a regular basis but I caught very few glimpses of the man.

My grandmother informed me that the man was Auntie Pat's first husband who had taken to being a tramp, but occasionally reappeared for short periods. Auntie Pat's other son, Aaron, was a steward on The queen Mary Liner and he took my grandfather and I down to view it from the dock. It was a beautiful ship, with graceful lines and towered over our heads as I gazed up in awe. He promised to get us on board sometime. Alas, this invitation never arrived though my grandfather and I did go on the Isle of Wight ferry the next morning, visited a pub at the terminal and came straight back. My grandfather true to form as ever.

My grandmother was Mary Foord, her family name being Sweet, and she was born on 10 December 1896 so she was twelve years younger than my grandfather. I wish I could now add their wedding date to bring everything into line as it were but there is a problem. My mother was one of six children and their parents were William and Mary Foord. That is a fact. I knew all my uncles and aunts but I could never understand the presence of an adult called Big Jack who had two sisters. Auntie Dolly ran the Oakfield Hotel up near Clifton (whom I never met) and an Auntie Kitty who lived with her family on the banks of the Severn at Bullo who I had met several times. Big Jack visited my grandfather quite regularly and sometimes 'subbed' himself out to my grandfather to do electrical work on various jobs otherwise he seemed to do a lot of hanging around and I think he did some driving for his sister's hotel. I could not really work out who they were.

My grandfather and I did go over to Auntie Kitty's one day with Uncle Big Jack driving the Humber. Auntie Kitty lived in a cottage close to the River Severn and had two boys older than me. It was suggested that we go and play on the river together and the two boys promptly stripped naked and I was told to do the same. I thought, initially, that it was a bit odd but rather enjoyed running down to the river with no clothes and the people we met seemed to regard the sight of three naked boys running past them as absolutely normal.

I felt free and eagerly followed the other two out on to the mud where we mud-larked for well over two hours and were joined by several other boys as we jumped and skidded through the mud, rolling and diving as we frolicked along the shore line. It was great and we were totally covered from head to toe in thick Severn mud which dried and caked on us as we ran back to the house where we were hosed down and cleaned off, laying down on the grass to dry off. I would have liked to do it again but the chance never arose and my mother was not too pleased in case someone might have 'seen us'. I did not let on that all the neighbours had.

Though it entails jumping ahead on the time scale, I never really got an answer about the mystery surrounding these relations until I was about 11 years old, when my mother whispered, in very hushed tones, that they were my grandfather's children, the inference being that they were illegitimate. This was a great scandal to have in any family.

Fast forward to 1961, the year my grandfather died. Social Services, in trying to establish the pension rights of my grandmother, obviously wanted to see documentation. Birth, death, marriage…marriage? She looked very young and I am sure they did not believe her age. Oh yes. There was a marriage certificate but it was dated 9 February 1955 long after my mother and all her siblings were born. The official explanation was that my grandfather had to await the conclusion of long, drawn out divorce proceedings before he could marry my grandmother.

To all intents and purposes, my mother and her siblings were illegitimate and with another burst of realisation was probably the reason why my mother could not receive her matriculation papers. I was 15 years old when all this came out and thought it was quite funny but my mother did not. She gave me a very sound dressing down.

My grandmother had three sisters and four brothers but I only knew Auntie Kate and Auntie Mabel. Auntie Kate lived just off Stapleton Road and was married to Uncle Walter who had sailed the seven seas on the old fashioned sailing boats and had lots of stories and mementos to show but he did go on a bit. I would occasionally go around for afternoon tea with my grandmother and even had 'boughten' dainty iced cakes. Her house furnishings were almost Victorian with Aspidistras, chintz and lace everywhere.

Auntie Kate died when I was quite young and Uncle Walter then became slightly morbid trying to insist that my mother and grandmother visit his wife's grave every week. My grandmother was having none of it and refused so she was cut out of his will. Auntie Mabel lived in Wandsworth and I stayed with her on two occasions, but we get ahead of ourselves again.

My grandmother, whom we called Nana, stayed in my life for many years and we got on very well together. She, like my mother, was always kind and helpful and provided strong support for me. She was always interested in what I was doing. She was generally always practically dressed and always wore a headscarf during the day but she also enjoyed dressing herself up on special occasions and she was quite adept at threatening my grandfather if he did not stump up for a new hat she wanted. She drank Guinness all her life and I do not think it was for medicinal reasons.

Her domain was a rather antiquated kitchen that had been tacked on to the back room of the shop with a small oven and a couple of gas rings and an old fashion kitchen cabinet. It is truly amazing what excellent food she produced from such limited facilities and she usually turned out up to six or seven 'Dinners' on a lunchtime for which everyone except my grandfather and me had to pay one shilling and thruppence and she quelled all protests very firmly when it went up to one shilling and sixpence. Again, however, all meals were served at table with the habitual bread and butter and she insisted on good manners from everyone.

One major offender was my Uncle Jack who often got a rap with the flat edge of her bread knife. Her reactions were swift and accurate. And there was always a pudding which I remember with much satisfaction and they were all to die for. Bread and butter pudding, treacle sponge and custard, spotted dick, jam roly poly (Grandma's Leg as it was known), banana custard, apple and blackberry crumble and apple pie. There was nearly always rice pudding as an extra if you wanted it—'to fill you up'.

When the table was clear, a game of Cribbage was played for about half an hour or, sometimes, my Uncles Jack and Dennis and Big Jack would reach for the 'Meger' and see who could stand the biggest electric shock. Do not try this at home. My Uncle Jack normally won but my nana thought he had no sense—no feeling.

I was always in slight trouble with my mother because I preferred Nana's cakes to hers. They were moister than my mother's and I do not think she was too pleased about my preference though I did prefer my mother's jam tarts. I often did the errands for Nana, mainly shopping on Stapleton Road. I never visited the butchers but Flooks the greengrocer, David Greig's the grocer, the post office (where my Auntie Linda worked—Uncle Jack's wife) and the cleaners were all on my round and I often sneaked down to Bennett's the toy shop on the corner just to window shop.

I sometimes did a round trip to take in Claremont Street on the way back as this was where the 'ladies of the night' operated. I thought I was quite daring but have to admit that I never saw anything untoward. I never widened my education in that street.

In retrospect, again I know now that Nana must have had a hard life. My grandfather was very difficult to cope with and she was forever kept short of house-keeping money and found it extremely difficult to raise six children. My Grandfather treated me very well indeed and I know he was reasonably kind to all the grandchildren. He neglected his family very badly indeed and his drinking, and the cost that it entailed, was nearly the ruination of his family. From his fine house in Park Row where he was brought up as a boy and sent to private school he slowly made his way down, 'the primrose path to the everlasting bonfire and nearly ended up bankrupt, leaving little provision for his family and business.

I wonder—did he realise what he was about? Surely he knew what hardship his family was suffering? It is not a pleasant note to add but most of his family did not mourn his death. Most called him, 'The Devil Incarnate' and Nana told me in no uncertain terms that she was glad that he had 'gone' and now she could get on enjoying her own life. Even to mention his name sends my 98-year-old mother into a tirade against his lack of care for he never kissed her once in the whole of her lifetime. She is very bitter indeed and when one of my cousins inadvertently mentioned that she was going to attempt to locate his grave at Arnos Vale Cemetery in Bristol, she told her not to disturb the old devil.

One person did shed tears. Me. And I know several of his grandchildren mourned his passing. He really had been exceptionally kind to me throughout my childhood and I liked him. I was due to set off on a school trip to the South of France shortly after his death and asked my mother whether I should cancel to attend his funeral. This autobiography cannot record her answer. The book would not pass the censors.

My knowledge, relationships and interactions with the family were ever growing and I spent quite a lot of time with my Uncle Jack who was always full of life and liked fun. Even at quite a young age he would take me off to jobs on his bicycle as he had a little saddle on the crossbar on which I perched. He was not a good cyclist and we crashed on the cobbles of a street in Easton one morning while attempting to transport me, his tools…and a ladder. The cobbles hurt!

It was normally him who helped my parents if they needed a car and we met up at family events. He figured much in my life during various times and I likened him to a banshee rogue jumping bean bouncing around in the gene pool of our family. Nana reckoned that he had been 'touched' by the African sun when he was deployed to Egypt during the war. He had a daughter, Christine, of whom I did not see a great deal of in my early years but we have made up much time since.

His war service was with the Royal Gloucesters and he served as an Aircraft Technician in Alexandra presumably as an Electrical Engineer. His war was much disturbed by football, however, because he played centre forward for the Regimental Team and was thus not allowed to work at any of the forward operational airfields in case he missed a match. Centre forward—Service at the Front. Same thing is it not? My Uncle Bill was also serving in Alexandra at the time laying mines and he and my Uncle Jack were in fairly regular contact.

Uncle Jack received a message one night to say that Uncle Bill had been severely injured so rushed to his bed to comfort him and kept an all-night vigil until the morning only to discover that the patient was not Uncle Bill. He was alive and kicking and dealing with his mines as normal. Uncle Jack always related an experience he had while being transported back to England on a troop train which was jam packed. Finding no space he slid under one of the seats but awoke with a panic attack and a feeling of being trapped and that fright stayed with him for years causing many nightmares. His brothers reckoned that his travelling companions have probably shoved him under the seat to keep him quiet.

My Uncle Dennis was a much more sedate and sombre character and was married to my Auntie Violet. The term 'madcap' could not be used of him and he was very shrewd and astute. Being my mother's younger brother he had just missed the war but he did do National Service in the Navy, which, I cannot help thinking was rather odd as all our family was in the RAF. Perhaps the Navy needed an electrician at the time? His daughter, my cousin, was disabled and again, I did not see much of her in my early years and only occasionally see her now.

The eldest brother, my Uncle Bill, was also a qualified electrician and when I first knew him he had married my Auntie Mary, had a daughter, Pat, my cousin, and they were living in South Road Kingswood. On his return from North Africa, he had worked with my grandfather's electrical contracting company but my grandfather paid below the rate and did not treat his electricians well including his sons. Like my mother, my grandfather said he could not pay Uncle Bill's wages one week so he walked out and found an electrician's job at Wick Quarry, where he stayed for many years.

My mother's two sisters married—one twice. Auntie Ruby married Uncle Bob, a painter and decorator and had two sons—Michael and John. John was more my age and we did get into trouble occasionally together. They lived off Stapleton Road initially and I stayed over once or twice then they moved to Muller Road close to Purdown. They were Catholics and during one of my visits there was a knock at the front door. Peering through a slit in the front curtains, Auntie Ruby exclaimed, 'It's Father…!' (I've forgotten the name) and there was a frantic clearing up and a plumping of cushions before he was allowed into the front room.

He had come to enquire about their non-attendance at Mass.

John and I also did apple scrumping together. My Uncle Bob had an unusual war. I was led to believe that he served in the Merchant Navy. Wrong. He was trained as a Gunner in the RAF, and much to his surprise, he was posted to ships on the North Atlantic Convoys. His ship was torpedoed and they were picked up by the nearest ship, but, just as they were climbing on its decks, that one was torpedoed as well and they had to man the Lifeboats and were soon picked up. He reckoned he was lucky, if they had been below decks when torpedoed for the second time, they might not have survived.

My Auntie Doris married my second Uncle Jack (she was divorced from her first husband at a time when divorce was not approved of) who worked for the BAC and they had four children Gregory, Stephen, Janet and Caroline who have popped in and out of my life occasionally at home and abroad. They lived in Beaumont Street which was very close to my grandfather's shop on Warwick Road and also spent a little time living in Mexico where they fried eggs on the pavements!

At the age of five, I had my grandparents, ten uncles and aunts and twelve cousins just on the maternal side of the family and I liked them all though I saw some more than others but I did come into contact with them quite regularly.

I did not receive that level of contact on the paternal branch. My paternal grandparents were Earnest Robert Allen, a newsagent, and my step grandmother was Mary J Hill daughter of a wood turner and a widow both based in Barnstaple. My real grandmother was Elizabeth Belle Balman, a butcher's daughter from Bear St in the centre of the town, and I suspect she died of TB towards the end of the First World War. My father was born in Barnstaple and had two older sisters, Gladys and Gwendoline, and an older brother Reggie. They all lived in Barnstaple until 1919 when they moved to Bristol and set up a provisions and newspaper shop at 126 City Road, except for Reggie who stayed in Barnstaple and became quite an entrepreneur opening several grocery shops and the Rising Sun pub near the parish church.

My father then was employed, as mentioned by David Grieg's on Gloucester Road. My Auntie Gwen and Auntie Gladys were married between the wars. Sid Grant was Auntie Gwen's partner and Ray Sweet married Auntie Gladys. My Uncle Sid was employed in the office at Gardiner's I think and was a leading light and secretary of Bristol Rugby Club. He lived very close to the Rugby Ground, while Uncle Ray ran a coal and removal business on Fishponds Road with his three sons—Clifford, Roy and Graham. The big arch where the vehicles entered the garage is still in situ.

As previously mentioned, there was a huge bombing raid on Bristol in March 1941. The bombers missed all the main targets and bombs rained down on the residential areas of the city including City Road so my paternal grandparents returned to the safety of Barnstaple where they opened an ice cream business.

Allen's Ice-creams. We wish to inform patrons we are in the happy position to supply our pure ice-cream throughout the season at no 3 Barum Arcade. We are members of the ice-cream alliance having won several diplomas for our delicious ices with fair and reasonable prices. Allen's ice-cream may be obtained for dances and parties, etc. at short notice. Eat Allen's pure ice-cream for health and beauty,

24/9/1946. The North Devon Journal.

Number Three at the Barum Arcade was named the Eldorado Ice-cream Parlour and the Allens held the Summer Ice-cream Selling Concessions for all the North Devon Coast from Ilfracombe right around to Westward Ho! My paternal step-grandmother (we always knew her as Mrs Allen) was an astute businesswoman. The Allens started with bicycle ice-cream sales to begin with then moved on to the traditional ice-cream vans with the long bonnets and a stand up section to serve ice-creams at the rear.

I had only just entered the world at this stage so knew nothing of these developments.

As a toddler, I was taken on short holidays to Barnstaple and really only saw my Barnstaple relations for a couple of short weeks. Mrs Allen was a terrible driver to the extent that most of the Barnstaple residents kept clear of her at all times. In those days, you were allowed to freewheel to save on petrol and Mrs Allen had it down to a fine art. She could freewheel for miles.

Mr Allen, my paternal grandfather, reckoned she could freewheel uphill. One brave policeman stepped out into the road one day to attempt to bring her to a halt. She drove straight over his foot! My most vivid memory was when she took my brother and I over to Saunton Sands in one of the ice-cream vans and we were being thrown around in the back because of her erratic steering and speed. As she swooped madly down the rough ramp to the beach I was screaming to be let out! I was pacified with an ice-cream however.

On a previous rundown to the Sands, it is alleged that she hit one of the donkeys being taken down to the beach for rides and it was so badly injured that it had to be put down. It is also alleged that she berated the donkey owner for not controlling his donkeys on the road! In fairness to Mrs Allen, I could find no record of this in the local papers of that time. For Barnstaple, it would have been major news surely?

The Allens lived in a brand new split-level house called Woodlands in Higher Raleigh Road where they made the ice-cream just above and between the fire station and the hospital (it is all sheltered housing now). I was impressed even though I was very young. I had never been in a big new house before. Earnest Robert Allen, my grandfather (presumably my middle name Robert originated from him), was a quiet man and it was obvious that Mrs Allen was the power behind the family.

He was a skilled carpenter and he liked nothing better than to escape into his garage and find peace and quiet with his wood. There was a fine lathe there and I used to watch fascinated as the bowls and chair legs were turned. He did not pass on his innate skills to me, however, for at my first attempt on the lathe during a woodwork lesson at my secondary school, a fine fruit bowl dwindled to a dish and then an ashtray. No woodworking genes seem to have been passed on to me.

In the early fifties, the Allens emigrated to New Zealand probably under an assisted passage scheme and Uncle Reggie and his wife, my Auntie Joyce, neither of whom I really knew which came as no surprise to my mother because ever since my parents had been married Mrs Allen had tried to sell her businesses on to them. First offer was the Provisions and Newsagents in City Road, Bristol in an area that had substantial war damage, then the ice-cream company in Barnstaple which my parents could not afford (and they did not want to abandon the family in Bristol anyway) and, finally, the offer to go to New Zealand with them.

This was a simple no as my parents did not have the occupational qualifications that New Zealand required. They were not looking for grocers or book keepers trained or otherwise. And so the Allens departed our lives and I only ever saw Mrs Allen on one occasion again when she made a return visit to Barnstaple in the early 1960s and she was staying with my father's Auntie Annie, with whom my mother and father often stayed during the war on brief visits. She was not well.

I could only have been in my early teens then and, at her request, I went alone because she said that she wanted to know more about me. Mrs Allen was Mrs Allen. For two hours, she never stopped talking, mainly about herself, and I think I only got about four or five words in for the whole of the time so I have no idea what she found out about me. The voyage to New Zealand was interrupted by a shipping strike in New Zealand

and the family were held up in Australia for a while.

Eventually, they had to fly to New Zealand and their possessions were air-freighted as well. This was no small operation because Mrs Allen had opted to take her grand piano and Mr Allen…his lathe. As a young six-year-old, I had this wonderful vision of them all sitting on top of the grand piano as they flew to Christchurch for their new life.

Having had so little contact now with the paternal side of the family, I am not sure of its true size but Parish records indicate that there were a large number of Allens in Barnstaple just after the war, many of whom were distantly related and, of course, my father had his two sisters in Bristol, one of whom I hardly ever saw and the other whom I saw regularly.

Auntie Gwen and Uncle Sid, my godparents, we saw but occasionally. They would make formal visits to our house and were always invited into the front room. Auntie Gwen was, small and thin, and not really all that cheerful. That is to say, she moaned quite a lot. She could be quite sharp with some of her comments and she was a fastidious housekeeper. We really had to be on our best behaviour when invited to tea at her house. Uncle Sid was quite formal and could be gruff on occasions but was essentially benign and kind. He was a great rugby devotee and seemed always most disappointed that I had not taken up the game but he did own a Morris Minor and I always felt a bit disgruntled because he never offered his godson a ride in it.

Auntie Gwen and Auntie Gladys were as different as chalk and cheese. Auntie Gladys was warm, kind, round and friendly, perhaps a little bit disorganised on occasions, but always pleased to see you. Her husband, Uncle Ray, seemed to be in the background somehow, and was a little terse in manner—not a great conversationalist but he was not unkind. He ran a very successful coal and removal business on Fishponds Road and was helped in this enterprise by his three sons, three of my much older cousins—Clifford, Roy and Graham.

Roy was always smartly dressed and ran the office and did the accounts. There was seemingly tension between him and Clifford who ran and was directly involved in the on-going operations as Roy liked to keep all the finances under his direct control releasing few details. Clifford was always busy driving, moving the coal, lifting furniture and repairing the fleet of vans and lorries. He was a good mechanic and was not afraid to get dirty. Graham had served long-term in the Navy and joined the business later but he was a hard worker and gave much support to Clifford. Auntie Gladys also had two daughters—Jeannie and Carol.

Jeannie did not enter much into my life though we have maintained contact since. She married an American and lived in the US working on airlines and in the travel industry. She attended dancing classes and was a good dancer. Carol was much younger and we met up quite often on my mother's visits to see Auntie Gladys so we knew each other quite well. Being the youngest and with all her other children grown up, Carol had all her parent's care.

It must be admitted that Carol was a fussy eater when she was a child and I think Auntie Gladys despaired of getting her to eat properly and well. Brought up as Roger and I were on strict food economy and being made to eat what was put in front of us, we were aghast when Carol used to eat just the icing off a cake then leave the cake itself, and she ate basically what she wanted. Many years later, Mum kindly agreed to let Carol stay with her when Auntie Gladys was on holiday and found it quite a trial to feed her properly.

When we were very young, we played dressing up games together because Jeannie had left many of her costumes back at home. We dressed up in one costume or another then ran down the stairs to let the adults guess what or who we were. It was fun. But my mother was worried. On the bus on the way back home, she leaned over to me very confidentially and asked, 'You didn't take your pants off, did you?'

I wondered what would have happened if I had replied in the affirmative?

Auntie Gladys and Uncle Ray loved cruising—not the sort of cheap versions that are on offer today but proper cruising on liners and big ships on long voyages and visited many exciting parts of the World. Perhaps it was just an impression but Auntie Gladys never appeared to be interested in the scenery the history and great buildings. She loved shopping instead. When she visited India one could ask about the Taj Mahal and she would brush the query aside and tell you about the wonderful department stores in Bombay.

She collected all the freebies on route, all the teabags, all the coffee and shampoo sachets, all the little envelopes of sugar, all the free biscuits. etc. etc. I am never sure what she did with them all. Was she eventually hoping to open a café? One of my Ghanaian friend's wives worked as an air hostess for Ghana Airways and she began to notice that all the waste coffees, teas, milks and sugars were being collected by one of the junior male stewards. Why? His wife had opened a small 'chop bar' (restaurant) in Osu in Accra and used them for her customers. Recycling at its best, I think, but then the Ghanaians never let anything go to waste. They recycle or reuse everything.

By tradition, and we came to recognise it as a very long tradition, Auntie Gladys had a Boxing Day party for the Allens and the Sweets. This was not always easy because many people like my father had to work on Boxing Day morning and had to work the following day as well if it was a Saturday or weekday. I do realise, that, these days, the party would also be difficult to organise as many of the staff in retail work all Boxing Day.

By tradition, the men would walk up to the Memorial Ground to watch Bristol play rugby always against a Welsh side and then walk back to Fishponds through Eastville Park. Many big tables had been laid out prior to their departure and upon their return it would be laden with a huge variety of food. Auntie Gladys did not stint in her provision of a Boxing Day high tea and over 30 of us would sit down for the feast which could last for an hour and a half. Food and tables would then be cleared and party games would be played, mostly organised by my parents, and they also performed a mind reading act upon conclusion of which, there was rapturous applause. My mother played the piano well and she would play while we all sang along. We had a piano for most of our childhood and first, Roger took lessons, then me…but the piano teacher died and that was that.

My standard joke at this event was that my mother smoked regularly—half a cigarette on Boxing Day every year. Many other of my relations smoked regularly throughout the year. I am not sure if this was because of the stress of the war, the availability of cheap cigarettes or the fact that Bristol was the home of W.H.O. Wills, the tobacco factory at Bedminster. Strange to relate, however, that very few of the Allens or Sweets smoked.

Uncle Ray liked his pipe and cigars were also handed around for the men on this Boxing Day event which they enjoyed, especially my father. He smoked for most of his life, firstly cigarettes, then small Pantella cigars in packets of fives. My mother, who always alleged that my father was tight with money thought it a huge waste. Smoking at that time was not banned anywhere so there would be a haze of cigarette smoke at any events. I am still reminded of this in Ghana where smoking is still allowed in restaurants and public places. Africa is still a growing market for the sale of cigarettes. My father never smoked in any of the houses that I owned. He would dutifully go outside to the garden to puff away even if it were raining. I smoked.

As boys we would often break up dried leaves in our den and roll them into scraps of newspaper then puff away and cough and choke! Occasionally we bought Dominoes or Woodbines in open packets of five which gave some relief to our lungs I suppose rather than the homemade variety. I did then smoke from the age of about fifteen to sixteen mainly due to peer pressure and for effect but I never inhaled. I dislike cigarette smoke intensely and its smell lingering in your clothes.

Unfortunately, the ban on smoking in clubs, bars and restaurants means that non-smokers cannot eat outside anymore and I find it difficult to wade through crowds of smokers littering our pavements and paths and why do smokers get more breaks than non-smokers? It is an unfair world.

Games and party tricks completed, amazingly, all the tables were wrestled in again, more food arrived, and supper was served by tradition including a chocolate Yule Log which no one seemed to eat, so, by tradition, the appearance of the Yule Log was greeted with great cheers and everyone asked if it was the same one left over from last year? When we lived at Ridgeway Road it was easy to walk back but when we lived at Kingswood we very often had to rush to catch the last bus or Clifford would very kindly offer to give us a lift home. Clifford, Roy, Graham, Jeannie and Carol all married and had children giving me even more relations and their children have also started families as well providing another big contribution to my extended family here in the UK.

At the age of six, little did I appreciate another big change in my life looming on the horizon; a change that would open all sorts of new experiences, events and excitement. It was going to be another life changer, a much better environment and there would be many challenges to face.

Kingswood and Boyhood.

Chapter Four
Kingswood

In essence, biography merges into autobiography as my memories come much sharper into focus without the need to rely on family memories and research. My memory now clicks into a much higher gear within my direct experience with all events under my control, well, 'kind of', as my young cousins would say. Now I can write about my genuine thoughts and feelings and there may even be a few revelations to encounter on the way.

It was, firstly, an auspicious year because, on 6th February, the king died. George VI had been ill and had suffered from lung cancer and Queen Elizabeth ascended to the throne on 8th February in his place. I could not truly say that I recalled the actual events but I did watch a Pathé newsreel at the Saturday morning children's show at the Vandyke Cinema in Fishponds, which showed the queen leaving Treetops in Kenya and flying back to the UK.

At the age of five years and eleven months, we moved to the lawless forest of the Kingswood, a former Saxon and Norman hunting estate under the aegis of the Constable of Bristol Castle. He was also the Chief Ranger and the whole area operated under Forest Law where deer, boar and wolf were the king's prerogative while the tenants had right to the smaller game, or, more simply put, we moved out to the suburb of Kingswood, on the eastern border of the City of Bristol some three and a half miles out from the city centre. Kingswood had been an important coal mining area, and like Fishponds, provided coal for the ever voracious industries of Bristol.

The miners were regarded as being less civilised than savages by one George Whitfield, one of the founders of Methodism which was very strong in the district. In 1952 it was the home of the famous Douglas Motor Bikes (Duggies), Kleeneze (which started in a small cupboard), Carson's Chocolates and a thriving boot and shoe industry most of which closed down before or just after my arrival. The shoe making skills were transferred to...corset making.

Now, here I have to confess that I became quite knowledgeable about corsets. How? My mother wore corsets and as I was too young to be left at home by myself, I would be included in her visits to Mrs Bamford, her corset maker. I would sit on a high stool while corsets were discussed and my mother measured but of course, I had to wait outside in the hall while they were being fitted. Kingswood was also the home of DAPS—Dunlop Advanced Plimsolls worn by generations of Bristol school children.

I held no awareness of the preparations for this move but it was the family removal company, R A Sweet and Sons, that arrived one early morning at Ridgway Road to transport the family and our furniture to 50 Church Road Kingswood and my father, brother and I travelled in the back of the furniture van with two of the removal men while my mother primly sat in the front cab. Tut! Tut! Travelling in the back of a moving commercial vehicle. The Health and Safety Executive would not approve of that! All normal again I am afraid—common practice. How much more would they have disapproved of the sight of 36 scouts and four leaders loading their equipment and kit into the same sort of vehicle then everyone piling in on top as they set off for camp?

And one would imagine that the health and safety inspectors would have an apoplectic fit if they visited Ghana and saw the overloaded open-top lorries and pickups jam packed with passengers. It is not, however, a form of travel that I would recommend. When I lived at Harlyn Bay in North Cornwall, there were fatalities and serious injuries when a speeding pickup overturned on a bend while transporting a party of teenagers down to the beach.

The upper doors of the removal van were pinned back along the sides so we were able to mark our progress as we trundled up Lodge Causeway, slogged up the steep hill to Cossham Hospital then made our way to the top of New Cheltenham Road, at the bottom of which we turned right to drive halfway up Church Road. We could not approach our new house from the top of Church Road as there was an extremely sharp and tight corner at the top which our vehicle would have struggled to negotiate. We lurched to a stop outside our new house and I was helped out over the tailboard. There it was. 50 Church Road. A large pebble-dashed semi-detached house with bay windows situated on a corner plot with a big garden (which my father came to dislike because there was also a large garden behind).

Collingwood Avenue ran the length of the front, side and back gardens on the un-tarred road leading to a row of fairly elderly houses. (later my parents had to extend their mortgage to pay for the tarring of the avenue). The house looked very tall and impressive, almost imposing, bearing in mind that we had just moved from a little two up and two down. I was allowed to explore the house while the unloading commenced on the basis that I would not get in anyone's way, I think, which I constantly did. There were basically three downstairs rooms. The obligatory front room that looked huge without our best three-piece suite, a living room where we were to eat and sit (one of the most used rooms we had) having French windows leading out onto a small patio and a fair-sized kitchen, which led to the back door and a cold larder.

There was also a cupboard under the stairs. The stairs, that were big and wide, led to a landing and three bedrooms. I was scheduled to have the small one at the front while my parents took the bay-windowed one also at the front while Roger was to have the large bedroom at the back…and…and…and…a BATHROOM complete with toilet, sink and bath.

At the rear of the property, in line with the kitchen and the larder, was a stone built shed which we were to use for the garden implements, mowers and bikes. The garden was big running almost fifty yards down to a rough back lane that led up behind the row of houses of which we formed the end. It was a wonderful house though I felt like a minnow in a big pond. It was light and airy and gave the impression of tremendous space and I wondered if our meagre furniture was not going to look a little lost rattling around inside it.

Not everything had been connected and the electric stove stood idle for that evening awaiting one of my uncles from the electric contracting business to come up the next day. In later life I found this to be a 'constant' rule. Builders, plasterers, plumbers, electricians and carpenters almost invariably neglect their house and family jobs and push them on to an eternal waiting list much to the exasperation of their wives and relatives. One of my uncles, who was the part owner of his electrical business, had a dangling light switch hanging from the wall of his lounge for many years.

A five-minute job consigned to an everlasting delay. The gas was 'on' however, and after we had eaten a picnic tea, 'O frabjous day, Callooh! Callay!' I chortled in my glee, I luxuriated in the delights of my first proper bath ever that evening. Of course, economy decreed that we still all had to use the same water, but as I was the youngest and thus going to bed earliest, I had the first bath, I was in seventh heaven and did not want to get out.

And thus, I was tucked up in my own small bed, in my own small room at the front of the house overlooking Church Road with all my own possessions and toys. I was not tucked in alone, however, because I would always sleep with my Golliwog from whom I was never parted at night. After a period of 65 years, the term Golliwog has been erased from our vocabulary and dictionaries, of course, but they were very much part of the scene when I was young, available in many shops and mentioned and portrayed in books and comics of that time including Enid Blyton's Noddy series. It was a pity because Golliwogs were friendly and kind and I loved mine. To my mind, he was a friend.

I was an early reader and was reading all of the Enid Blyton books at that time, even The Famous Five! Fixed in my mind was this perfect picture of a lovely and homely lady author writing away in her picturesque cottage who loved all children and was amiable and nice. Luckily, my idyllic picture was not shattered until some twenty-five years later when, to fill in an unemployment gap, I whiled away my time as a postman on a semi-rural round out on Frenchay Common on the outskirts of Bristol.

The postman on the round next to mine had come down from Sussex, I think, and used to deliver to Enid Blyton's Cottage. I did my round in a brand new white Triumph GT 6 and often gave him a lift. He assured me that she was quite rude, unpleasant in manner, her dogs were out of control and she used to berate him on most days to the extent that the Royal Mail refused to continue her deliveries. He was an interesting chap anyway as he had moved down to Bristol to send his only son to his father's old school… Colston's out at Stapleton. The fees were not within normal reach then and he used to complete up to three rounds a day on overtime rates to maintain his son at the school. Some effort, I would say. I fell asleep tired, happy and very content indeed.

The next morning, I was allowed to explore.

Our garden was as big as an allotment and had three lawns, fruit bushes, a rockery and a huge vegetable patch as we were right on the corner of Collingwood Avenue so were blessed with an extra strip of land running from the front to the back. My father would have probably erased the word 'blessed' and inserted 'cursed' instead. I could safely play on the lawns and it was even big enough to play chase and tag as well as cowboys and Indians and other exciting games.

The next four houses up along our rank all had children so I had plenty of playmates of various ages and we were allowed freedom to roam. We could play football in the cul-de-sac where there was only one row of four houses then and joy of joy, a big area of waste land behind the houses that provided us with endless possibilities for adventure. We were very happy. We played games, built dens, lit fires, smoked the most revolting home-made cigarettes consisting of newspaper and leaves, had the occasional run in with neighbouring gangs that were all light hearted and fairly harmless mainly involving name calling and gestures.

We always built a huge bonfire for 5th November, which had to be guarded and protected from Bonfire Raiders in the run up to the event and in which we made more dens. We burned, cut, stung and grazed ourselves (my knees still bear the scars), fell out of trees and into streams and off our 'trikes' and raced our Dandies down the hilly parts of Church Road. That was exciting, but, though there was little traffic about, we always kept lookouts at the top and bottom of the run to warn of approaching vehicles. One of the older boys failed to heed our warnings one day and ended up entangled with the milkman's horse and cart. (In one of those odd quirks of life, I met the self-same milkman last year again after 65 years.)

I played in the summer evenings and at weekends. Never on a Sunday. I had set bedtimes and my father always called me in just before. I was often dilatory in heeding his calls. (Well, all children are the same—are they not?) I would often receive a clip about my head as I ducked in through the back kitchen door but that form of a punishment was building up a head of steam for the future. We had the freedom to roam and play and we had the freedom to be bored.

Very often, we would just lay down on the grass and do nothing. We would never be stuck indoors even during the winter months. We would wrap up warm and go out in most weathers. It was raining; we stuck our noses to the windows and watched the rain trickling down until we could go out again. There was, of course, no TV or any of the electronic devices that children devote most of their attention to nowadays. The only form of entertainment was the radio and then this only had a limited choice of programmes for children. I had a wonderful first summer and had never had so much freedom in my life. In later life, my mother seemed to regret moving to Kingswood but I did not.

So heady was my pleasure and fun that it came as quite a shock when she mentioned school again. School? At the tender age of six now, I had forgotten all about that. I had been registered at New Cheltenham Road Infant's School on New Cheltenham Road, a ten-minute walk from our house through the lane that was a short cut between Collingwood Avenue and Alma Road, down to the bottom of the hill and there was the school. My mother always walked me down and brought me back at 3.30 pm which left her very little time to catch the number 8 then the number 2 bus down to my grandfather's shop on Warwick Road to work, then back again in the afternoon. She must have been very fit.

New Cheltenham Road Infant's School was much more modern than the Victorian Chester Park and we had spacious classrooms with lots of windows and individual desks. It was a rather grey one storey building built with concrete and asbestos I think. I fell in love, not with any of the pretty girls that were in my class, but with my teacher, Miss Brewer. She was young, lively, caring, helpful, bright, kind and cheerful. I adored her. I was like a young puppy always anxious to please, always keen to do my best in lessons and if she praised me, I got that warm cosy feeling and tried even harder.

She was a good teacher and she became my staunch friend when my mother objected to me taking a Dinky Toy to school every day but she sided with me and said she liked her pupils to bring a toy into the classroom, especially during registration. She was my hero. I thrived and she was the first of many good and caring teachers by whom I have been taught right up to the age of eleven. Alas, the school was demolished and there is a block of flats in its place.

My brother, in the meantime, had been registered at High Street Junior School, next to the park where he stayed for two years and must have been taught by the same teachers that I was to be with one year later. He was in Class Four when I was in Class One but I never remember him being there, though in fairness the Senior Classrooms were on a lower level down by the canteen and I certainly never remember him helping or guiding me in any way. We never walked to school together and never played together. I think he had older friends but he was very much a loner.

He could be very pleasant but he had a nasty temper and I was forever wary of my contacts with him especially as he used to physically hit me, generally around the head in his more aggressive moments. My mother always told me that he was a difficult baby and could never be cuddled. She gave up taking him to play with other children as a toddler as he had embarrassed her by his aggression towards them and he was not very amiable towards his cousins either.

He had something of Jekyll and Hyde in his character which led to many difficulties on his life's path and he seemed to grow away from contact with us and strange to relate, my parents cannot recall ever receiving a school report from either his primary or secondary schools. Fault must lie with my parents here and I have often pondered about this. Why did they not approach the schools and enquire about their non-appearance? My mother did once hint that they were nervous about Roger's temper and angry reaction and I do believe that they were quite frightened of this.

Me? I was in a little cocoon of happiness, delighted with everything at Kingswood which had opened up a whole new world of outdoor pleasures and I do not think I had a care in the world. Aged seven I could not have been happier and better was to come. Two major events occurred in 1953 and readers will be amused to hear that I consider them to be of equal importance. The first was the Coronation of Queen Elizabeth the Second which took place on the 2nd June 1953 and was the first Coronation ever to be shown in full on TV. Readers might well now pinpoint a problem. We did not have a TV nor did most of our neighbours…except one who was kind enough to invite all of us in.

The Guinness Book of Records could well have been interested in the number of adults and children crammed into the Pillanger's front room that day for we were crammed in like sardines though we smaller children ran in and out to have play breaks for most of the day and everyone had brought plates of snacks and a huge variety of sandwiches and lemonade and alcohol to feed and water the 'five thousand' who demolished the lot. Twenty-seven million people watched the Coronation that day out of a population of thirty-six million and they calculate that the remaining nine million listened to it on the radio. We watched from early morning until well into the afternoon. We saw the Gold State Coach with its eight matching grey geldings—Cunningham, Tovey, Noah, Tedder, Eisenhower, Snow White, Tipperary and McCreery—carry the queen and the Duke of Edinburgh out through the gates of Buckingham Palace having been giving a rousing send off by all the palace staff.

I have a passion for grey horses for they are mostly steady and reliable in character and indeed, my first two polo ponies were greys—Geronimo and Prince. The processions consisted of over 30,000 men and women and 8,251 guests were seated in Westminster Abbey. Crowds lined the route and many people had camped out all night to get a good spot, especially on the Mall. As children, we became slightly bored by the three-hour long ceremony as the Recognition merged into the Oath and the Anointing but we were all there to see the actual crowning before the Enthronement and the Homage. This was another historical milestone in my life.

On that day, it was announced to the world that Hilary and Tenzing had conquered Everest but I later learned that this was not scheduled or planned. It was not supposed to be so. The wrong chaps had reached the summit first. Hilary was considered a bit of an outsider by the other climbers in the expedition. He had not been to the right school and was considered to be a nice chap but a bit too serious as a climber. Everest obviously had no consideration for class or country, however, as the initial planned summit attempts were thwarted for various reasons so Hilary and Tenzing were able to try. The rest is history.

Everest and Sir John Hunt, the Everest expedition leader, popped back into my life some years later when I was teaching at St Bede's School in Eastbourne, situated at the far end of the prom close to the famous white cliffs of Beachy Head which were used by the members of the 1953 Everest expedition for practice as, evidently, the chalk held the same consistency as ice, I am told.

This is not permitted in today's world of conservation. I had built a climbing wall on the back of new tennis courts overlooking the sea near the top of a small section of cliff. While walking down the rough track to the entrance of the climbing wall, one afternoon, draped in ropes and wearing a climbing harness with slings and karabiners attached, I met a gentleman coming up from the beach dressed in walking gear.

'Goodness gracious me!' he exclaimed. 'Are you going to climb on the cliffs?'

I explained that this was no longer allowed, not that I would have attempted anything so difficult anyway and proceeded to unlock the gate to the climbing wall. He was fascinated and looked keenly at all the artificial holds and cracks and chimneys unable to resist gripping and feeling one or two of the lower holds.

'I don't suppose I could have a go?' asked the gentleman cautiously.

I started to make all the usual comments about insurance and I think I even mentioned the need for some experience, etc., etc.

'Oh! I have not introduced myself. I am John. John Hunt!' and he shook my hand.

'Go right ahead,' I replied.

He took off his walking boots and climbed up and around the wall like a monkey though he did have a slight problem with the narrow chimney. When the children came down I introduced him and he stayed and watched even offering advice and help. They were well impressed that I had managed to find such a famous instructor for the afternoon but the staff thought I was pulling their leg when I explained what had happened. Sir John was visiting Nick Escort, a young professional climber, whose home was in Eastbourne and whom he had mentored for some time.

Unfortunately, Nick was killed in an avalanche while attempting the West Ridge Route on K2, not long after and it was a sad loss for the climbing world. Some of our pupils had met him when he presented a slide show at the National Mountaineering Centre at Plas y Brenin in North Wales where I took a group every year just before Christmas. The Everest Summit Bid had not been easy so I was rather amused by a photograph that another friend of mine, Nick Banks—who had been senior instructor at Plas y Brenin and later operated as an international mountain guide—had shown me. He was dressed only in shorts and wearing sunglasses sunbathing on the South Col of Everest. He was acting as a guide to a wealthy client and his wife.

The next day, in perfect conditions, the client and Nick successfully reached the top. The day after, in perfect conditions, the man's wife and a Sherpa set out for the summit and were never seen again. Everest appears not to take prisoners, because on another attempt on the East Side, Nick became very seriously ill at well over 20,000 feet and was lucky to get down. He climbed Everest several times and was the 100th man to conquer the mountain.

Stephen Venables was also acquainted with several schools at which I have taught. He is a brilliant mountain photographer and his pictures are stunning but Everest nearly did for him. He made a successful summit bid, without oxygen, via the dangerous Kangshun Face from the Tibetan side approaching the South Col from a different side. He was benighted just below the summit and was forced to Bivouac there at 8,600 metres. It was the highest bivouac recorded and the record still stands, but he lost some of his toes in that attempt. He took a small school group out climbing on the Dewerstone in Devon one Sunday not long after his Everest trip and his balance was not 100% even then and he fell into the stream.

I am not sure if it was before or after the Coronation but a huge street party was organised for our section of Church Road and Collingwood Avenue and the preparations were long in the making. We children tried to help but probably got in everyone else's way as we were wildly excited. The avenue had been decorated on the evening before and there was a splendid array of bunting, flags and flowers in abundance everywhere. Each house brought out a table and chairs of which there was an amazing assortment but it did not seem to matter that they were all of different height as they were quickly covered with large rolls of paper. (One of my friend's fathers worked in a paper mill.)

Then crockery and cutlery was brought out from the kitchens or sideboards…and… finally, out came the food. Cakes, sandwiches, pickles, sausage rolls, small pasties, jellies, custard, trifles, chocolate and cream. The tables were groaning with the weight. This was all to be washed down with squash and lemonade for the children and beer and wine for the adults. All the mothers had competed against each other to provide the most and the best and I can vouch that it was all delicious, but, before we sat down we all sang, 'God Save the queen' lustily and loudly, followed by grace.

At each child's place, there was a commemorative coin and a special pack of chocolates, funded by the nation I think. Everyone then tucked in and the eating continued for over two hours until nothing was left. This was followed by a few speeches before all the cutlery and crockery was taken back to the houses and amid much heaving and hilarity all the tables were lifted back into kitchens and living rooms and the chairs put into a big circle leaving a big space in the middle. A whole host of games were then organised for the children with prizes and balloons and this was followed by dancing and singing well into the night. I fell asleep during the dancing and woke up the next morning in bed. But it was a wow of a party.

The second important event which I rated just as important as the Coronation, though I am not sure if her majesty would agree, was that I joined the Cubs. On July 11th I had reached the grand age of seven and was thus eligible to join one of the two cub packs of the 111th Holy Trinity Church which conveniently met at the top of Church Road in one of the big halls set deep in the graveyard behind the austere and plain looking church. I was kindly received by Akela, Bagheera and Baloo all of whom were veterans of managing excited young boys and I was delighted to fulfil a wish that I had longed for during the last twelve months.

For those of you not in the know, the whole of the Cub Programme is based on Rudyard Kipling's Jungle Book with the adventures of Mowgli and his animal pals. I appreciate that there is still some criticism of the 'jungle' theme but it has survived and no one else can come up with a suitable alternative. It has colour and atmosphere with many examples of loyalty, team spirit, and obedience to a just law. It is a very suitable conveyance for the imagination and practical training of young boys, and girls nowadays. The Cub Law and Promise, the noise and symbolism of the Grand Howl and the training programme are still valid and valuable in this day and age and above all…it is fun!

I particularly liked the games. There were relays, training quizzes, sleeping pirate, whipping the gap, obstacle courses, bridges and tunnels and British bulldog. I was quickly into uniform—just a green sweater, neckerchief (yellow and black), a Cub cap and green garter tabs. In later years I ran Beaver, Cub and Scout packs and troops in Ghana and they were utterly taken with the whole programme. They really did try to follow the various Laws and Promises and loved all aspects of the programme. I was a full and active participant in the pack for four years and gained my two stars and my Leaping Wolf Badge. The stars were put into your cap: one star—one eye open; two stars—both eyes open. I was carried away with utter enthusiasm for everything cubbing.

As boys, we were often noisy, unpredictable, positively overflowing with energy and always full of beans, keen to try everything and cubs catered for our every need. The Cubs were, and are, a wonderful programme for boys and girls aged 7-11. Long may they continue.

Once a month, there was a Church Parade when we were expected to attend a service in the church with the Scout Troop, Senior Scout Unit and the Rovers. It was quite formal and we had to take extra care with our uniforms and turn out smart and clean. All four units paraded their flags and another 'Colour Party' presented the 'Union Jack.' These were processed down through the church at the beginning of the service and presented at the altar, then collected again at the end. The Matins Service congregation was swollen some tenfold with our presence even in those days. The units and flags were also paraded with even more formality at the Armistice Parade when we were joined by the British Legion, guides, brownies, boys' and church lads' brigades.

The Armistice Parade always sticks in my memory because we were not allowed to wear jackets or coats. As Cubs, we had fairly thick pullovers but as Scouts, we paraded in short-sleeved shirts and shorts. Brrrrrrrrrrrr! Kingswood also held a Whitsun Parade which I think was a possible left over from the Temperance Movement. All the organisations and all the churches, chapels Sunday Schools and associations paraded or were driven along on floats and the route went straight down the main street and back again. Whitsun can be cold too, believe you me.

As Cubs, we were taken to Arrow Park Sutton Coalfields to visit the Jamboree. There were 35,000 Scouts on site. Our Cub leaders must have been brave souls escorting us on to trains and around a very crowded Jamboree site for the whole day.

Then, there were Cub camps. These were invariably held at Dyrham Park (where I was to set up a special relationship in my scout troop days) up on the A46 and we arrived and left by furniture lorry which also carried all our equipment. The Scouts helped in putting up the tents and we always had a campfire singsong fairly late into the night to make us tired…it never worked. We chatted away until the early hours but slept like logs on the second night.

Cooking was centrally done on open fires and we all ate together normally sitting on the grass and the Scouts had dug latrines and set up bowls for washing so we were fairly well catered for The house was not open then but a good programme was provided and we did have time to roam and look for antlers and spinal bones that we painted and made into woggles to fasten our scarves. An inspection was held each morning and we and our kit would be presented properly and checked before any activities commenced. We were expected to help with the light chores such as fetching water and washing up. It was all good training.

Seven was a magical number as I was now also due to attend High Street Junior School for Boys. Instead of going down the hill I now walked up the hill of Church Road, along the main road and entered the school by the old-fashioned small side gate. This building had a history because it was opened in May 1892 along very traditional lines. It was built of Pennant Stone with Bath Stone dressings and it cost £6,000 and another £1,350 to equip. I was arriving sixty one years later and the buildings, fixtures and fittings were still original apart from the new classroom block on the lower level running on from the canteen.

Generations of schoolboys had been educated there including sixty-eight of them who called up and were killed in the First World War. The original buildings next to the main road still exist with alternate uses but the Park School now has a wonderful complex of new buildings where once we used to play football and compete in athletics. I have a very clear picture of my mother escorting me to school on the first day and waving me goodbye as she turned back to catch her bus down to Lawrence Hill and then on to Warwick Road. This became the daily pattern but she also very wisely and gradually cut down on her son's escort work and I was soon allowed to walk alone to and from school with the absolute proviso that I could only cross the main road using the Lollipop Lady.

Another of my secondary school set books was Cider With Rosie, and I vividly recall Laurie Lee's first day at school and his entry into the playground at his Slade School where he described the yells and calls of the children as they whizzed, swooped and screamed around running here and there. It was bedlam. High Street was no exception and some of the bigger and older boys looked frighteningly large from my small stature. My mother's instructions to my brother to look out for me and make sure I was alright had been ignored. I moved to a wall on the nearside and waited.

A handbell clanged near one of the entrances. Suddenly, everyone stopped in absolute silence and then quietly walked into class lines. It was amazing. It was if someone had flicked a switch. I was helped into a line with the other new pupils and we were escorted into a classroom that was directly next to the playground by a kindly, somewhat elderly teacher who introduced herself as Mrs Handel. She was kind, helpful and patient throughout the whole of that first year and she was an excellent teacher.

There were no classroom assistants in those days and Mrs Handel 'Handeled' our class of 35 with skill and good humour. She was always firm and she had us wrapped around her little finger. I never heard her shout or even raise her voice and I particularly enjoyed the stories that she read aloud to us. She obviously sensed my exasperation, however, when reading exercises involved the whole class. I suppose it was a little unkind for I could never wait for the slow readers as they stumbled and stuttered through the text, and was always pages ahead. Instead of punishing me, she called me to her desk and gave me a hard text to read from a geography book.

I sailed through it and it was obvious that I was way ahead of the class in reading. She was sensible enough to let me have my head so I read quietly in one corner while the rest of the class proceeded at a slower pace. I do remember her utter astonishment when, towards the end of my first term, I checked out the last book I had read and she told me to get another one from the shelf. I had to inform her that I had read them all!

She escorted me to see Mr Dobbs, the headmaster, who was a very kindly man. He regarded the situation as most amusing and said it was a first at the school. He personally organised a small library of a dozen books to be kept in the classroom for my use only and it became quite a regular thing for me to take the finished books to Mr Dobbs and ask for another supply, so much so that the school had to make arrangements with the local library to guarantee my reading material. Before I attended Teacher Training College, I did a two-week teaching experience with Mr Dobbs at the school in various classes and at the first assembly, I was introduced as the only boy who had read every book in the school and the children's section of the Kingswood Library.

During my first year, we received very sound teaching in reading, writing and arithmetic and I wish to record that I was very good in all aspects of the three R's. I say this with some feeling because later in my schooling, my Math declined miserably because of very bad teaching but not at High Street. I much enjoyed geography and history—an interest that I maintain to this day—and we also had Nature Study Scripture and PT. (When was PT changed to PE? And what was the difference anyway?)

All subjects were taught by Mrs Handel who also played the piano and we really enjoyed her music lessons, especially as this was the first time we used triangles, tambourines, castanets and drums in accompaniment. I think the cacophony of sound produced might not be described as truly musical and was much embellished by our enthusiasm. We also learned the recorder which we were allowed to take home to practice. I wish I could say that my parents were delighted!

The radio was used for Musical Movement and there were regular programmes for schools in other formats and subjects as we progressed through the school. In those good old days, the sanitary arrangements could hardly be described as hygienic or adequate according to modern standards. Our toilets were situated on one side of the playground: the side furthest away from the classrooms. They were covered by a leaking roof and contained three or four urinals at two cubicles. These were deemed sufficient for over 200 boys.

In summer, they were rather 'high' and housed a large selection of flies and insects while in winter they were freezing cold and often iced up with icicles dangling from the inadequate roof. In inclement weather, i.e., rain, wind, snow and hail, you emerged from your classroom and then made a quick dash to the toilet and an even quicker dash back. We all wore shorts so there was no attempt to use buttoned flies. We quickly raised the leg of the shorts, did our business and got out again as quickly as we could.

These were still the days of school milk which was dutifully delivered every morning and collected by the 'milk monitors'. We had to drink it. There were no ifs or buts—drink it we must. The one third of a pint was good for our health, so they said; so, under the careful eyes of our teacher, we drank the milk with the use of straws, which was another responsibility of the Milk Monitors. Some days it was OK. On hot days, it was warm and horrible and we hated it. On cold days, there would be lumps of ice in it and difficult to drink. We really enjoyed the days when it was completely frozen and could not be drunk.

Our canteen was down on the lower level where the new school is now sited and it was in here that morning assembly was held every morning of the week with a hymn and the Lord's Prayer followed by notices. No one seemed to mind and we all enjoyed the singing but often the first odours would be emulating from the kitchens in which we took a sharp interest and would start guessing what that day's meal would be. The worst smell was boiled cabbage and I still cannot understand to this day why they seemed to cook it for three and a half hours and all meals seemed to be boiled.

Boiled fish. Boiled meat. Boiled stewed vegetables. Chips were never provided and there was no fried food. We also hated semolina and tapioca pudding and the custard was lumpy and horrible. These were still the days when what was put before you had to be eaten and there would always be a small contingent of boys left in the dining room as they held their noses and tried to swallow the food they hated—myself included. Perhaps the ingredients were good but the cooking was not! Where was Jamie Oliver in my school days? He should have been born much earlier.

I have racked my brains in an attempt to remember who taught me in my second year at High St and I have failed. Though I was definitely in one of the very old Victorian classrooms right next to The Cupboard. This was a long bench on which naughty boys were placed to incur the wrath of the headmaster but in truth, it was rarely used. I and my class sat on it once but only while we were awaiting an injection from the school doctor. Nor can I recall ever being hit or shouted at while I was in the school and I never knew Mr Dobbs to use the cane, yet firm and fair discipline was maintained.

The school retained two sets of buildings. The old section was on the upper levels and was little different from the old Victorian buildings I had used at Chester Park Infants' School while the newer classrooms and the canteen were down on the lower level. Having sampled new desks and equipment at New Cheltenham Road Infants' School I now had to revert to old classrooms and old desks which sat two pupils at a time.

Moving up into the third and fourth years we were back to more modern desks and furniture and I was taught, for the first time, by male teachers. Mr Knight taught the third formers and Mr Burgess the fourth and final year in preparation for the 11+ Exam in the lower-level classrooms, that were quite modern, next to the big games field. Mr Knight was a quiet unassuming man of medium stature and like the other staff-maintained discipline in a quiet manner. He had the 'look' that I was to acquire in later years and I am not referring to fashion.

If you started talking out of turn, or became a little noisy in an activity, dropped your ruler or pen…he would just look at you. No admonishment or harsh words, just…a 'look'. And that would be sufficient. I had it down to a fine art later in my teaching career and I could even do it without looking at the pupil. If I heard the wrong sort of noise, I would stiffen and half turn around, and that was generally enough. If a pupil was well out of order, I would look and narrow my eyes then they knew it was serious.

Mr Knight taught all subjects and he was good. He raised my standards of English, arithmetic, including mental tests, and massively improved my creative writing. Verbal reasoning tests were then much in vogue and he offered sound advice and direction on those in preparation for the final year.

All pupils in Bristol at that time faced the 11+ exam and there was no debate about it. Everyone sat the exam. It is still part of the mosaic of parts of our education system to this day from some fifty years ago but now invites fierce controversy as 'Selection' is seemingly frowned upon. One thing is for sure. None of the pupils were coached by outside tutors for this and any other exam. There is today a vast industry in tutoring children for SATs. Tests, 11+ and GCSE's and A Levels and the tutoring is mostly unnecessary and variable in quality and delivery.

I have direct experience of poor tutoring ruining the good teaching in an approach to an exam resulting in a fail or low pass. One school that I visited in Devon had a Class Six of high ability and the children were well en-route for a good 11+ pass, but earlier on, one parent employed a tutor, then another, and the whole process snowballed into every child being tutored purely due to peer pressure. Why? It was all unnecessary.

Head teachers of selective secondary schools have recognised this coaching phenomenon and now prefer to invite the children in to the school for the day and conduct in-house testing and activities for the candidates with no exterior input. Parents do not like this but it is a good system. Regrettably, the SAT Tests are becoming increasingly unreliable as results are either manipulated or doctored and direct help is often given by teachers in the actual tests. It is only a valid test if it is administered honestly and marked and graded accurately. Do I believe in testing? Yes. Do I believe in the use of the 11+? No.

I would be looking for a much fairer selection procedure probably akin to the one mentioned above but I also firmly believe that we do need to nurture and encourage the more able children if we need skilled talent in the future. Some Comprehensives do this well but many do not and have no real aspirations for their pupils satisfied with a low overall mediocre standard for all rather than encouraging the more able to reach their upper limits. Pupils are individuals and cannot be lumped together in a 'one size suits all' mentality and 'Dumbing Down' must be avoided.

High St School was fairly keen on sport. We had PT and we played organised football, but ever the individual, I developed my athletic prowess in…sack racing, though I did claim an honourable third in the potato race one year—galloping backwards and forwards loading potatoes into a bucket. Imagine explaining that to an alien!

Sack racing required technique. To start, you had to lie down inside the sack with your head resting on the starting line. When the whistle blew, you jumped up and raced about forty yards to the finish. That was the essence of it but there was no hopping and jumping along the track for me like a drunken kangaroo. Firstly it was important that you did not wear your Daps. (Daps is the Bristol name for Plimsolls.) You needed stockinged feet and you placed them in each corner of the sack so you could actually run down the course, not hop.

Now, all the faster sack racers did that but I also had another technique. While resting my head on the line ready for the start, I would look at the starter so I could see when he was about to blow the whistle; thus, I always got off to a fast start. I was the champion for two years and I never revealed the secret of my fast starts!

Mr Jefferies was in charge of sport and I found two photographs of him recently with the school's soccer teams. In one, he looked very young and in the next, he looks a little more tired and haggard. I suppose that's what teaching does for you. Like the other staff, he was always kind, helpful and encouraging.

Mr Burgess took me in the fourth and final year. He was tall, handsome and always smartly dressed in a sports jacket and flannels and my mother reckoned that he must have been in the RAF as he sported a wonderful moustache. He was calm, placid and spoke clearly and well and organised his top class very effectively. We all had class jobs ranging from Milk and Ink Monitors, pupils in charge of text books, pupils in charge of exercise books, pupils in charge of pens and pencils, etc., etc. which were changed every week according to an enormous chart he had posted on the wall. Everyone had a job!

He had the ability to get the very best out of his pupils and we responded to his methods and ideas. The classroom was full of enthusiasm and we still managed to have lots of fun and study interesting topics despite the necessary practice ready for our exams. In this final year we received homework which generally entailed about one hour of Arithmetic and English and I never bothered my parents for help. As a headmaster much later in life, the school was programmed for the children to complete their homework/prep in school which had two advantages.

Firstly, the parents could not interfere in the prep in any way. This could be by helping or doing the set task for their offspring or, in some cases, hiring tutors to do it for them. This could lead to pupils gaining full marks for a piece of work about which they knew nothing. The junior pupils did theirs in class which had the added advantage of the teacher being able to identify problems and difficulties immediately.

Secondly, all books stayed in school so there were no problems with forgetfulness, 'exercise books chewed by dog' or late nights. There is an educational idea, mainly circulated by worried parents, that the more work you do in preparation for an exam, the better will be your results. This is not true as the pupil will become overtired and overwrought which will adversely affect his or her result. Better to adopt a far more measured approach with regular works, activities and rest. Mr Burgess did just that and we all took our 11+ fresh, well prepared and with no histrionics or worries. Did I pass?

My life in Kingswood proceeded at pace. Home life was stable and reasonably secure. I played, I went to school, I argued about my bedtimes and I was reasonably healthy but there were two scares. One night I awoke with the most appalling earache and it was so painful my mother heard my sobs thinking that I had been experiencing a bad dream. Between my sobs, I managed to tell her about the piercing pain in my left ear. She organised a hot water bottle and warmed a flannel on which I rested my left ear. It did little to ease the pain and I cried for the rest of the night until I was totally distressed and exhausted.

The doctor was called early the next morning and my mother told him that I was just whacked and would not be able to get to the surgery. He agreed to put me on his list of calls first. He arrived just after 10 am and gave each ear a thorough examination with his otoscope then pronounced that I had a perforated eardrum. My mother was upset when he suggested that she had cleaned my ears too thoroughly and roughly and she had not and she remained grumpy for the rest of the day. He prescribed drops and stated that it would heal itself which it did but the left ear still produces much more wax than the right. How odd! My father was very sympathetic as he suffered from constant ear problems having been a gunner in the war. Indeed, my mother urged him to apply for a war pension to which he would have been certainly entitled but he stubbornly refused to do so.

When I was about nine, my brother persuaded me that we should share the back bedroom together so that he could set up his model railway in the small front bedroom. I cannot think now why I agreed to this as, upon reflection, there was no advantage to me. Firstly we had to share a double bed, secondly I was forbidden to touch any of his possessions and thirdly, I was never allowed to play with his train set. Upon reflection, I believe my parents pressurised me to just placate Roger who was becoming increasingly belligerent in all aspects of family life.

One early morning we were pillow fighting and suddenly I experienced a stabbing and severe pain in my neck. I screamed loudly and it was very loud, probably due to all that practice I had had when teething as a baby. Every time I tried to move my neck, the stabbing pain returned. Both parents came running. My father started to berate Roger but it was not his fault. My mother was very anxious about me. I was carried down to the warmth of the fire in the living room where the slightest movement of my neck resulted in more loud screams. My mother was ever true to form, however, because her first thought was what the neighbours might think!

Meanwhile, I was in total agony and frightened to move, curled up in the back of an armchair. It was 8 am on a Sunday morning. My mother called the emergency number at the doctor's surgery. There were doctors on call in the good old days and he arrived very quickly, sensing the urgency and concern in my mother's voice. He was very jocular but kind as he examined my neck with me still yelling and whimpering. I had sprained a muscle in my neck quite badly it seems and he added that he must now add pillow fighting to his list of high-risk sports.

A prescription was written for a special ointment and bandages were also required. It fell to my father's lot to cycle all the way down Two-Mile Hill to St George's to the duty chemist on his very sturdy Raleigh Rudge bicycle to obtain same and, what was worse, to cycle all the way back up Two Mile Hill to Kingswood again. I was anointed with the ointment and then my neck was bandaged but that was a bad Sunday because I was in pain for most of the day.

The only consolation was that I had a week off school and I was waited on hand and foot for several days supplied with a bountiful supply of comics, books and goodies. My brother was not happy because he took over my chores for the week. Did I 'milk the situation'? What do you think? Mrs Crew from next door baked some jam tarts for me, glad that I had not been murdered on that Sunday morning. Later, her daughter, Carol, told me that she nearly came running in that morning as she was dying to know what was happening.

Our neighbours were always kind and friendly. Peter Langridge, two doors up, was of my age whereas the other children were slightly older. Mr Langridge worked as a signal man for the railway over near Shortwood and we visited him in his 'box' several times while Mrs Langridge had a part time job at Carson's Chocolate Factory. She used to cycle to work most days but we were surprised to see her right arm in plaster one day because she had collided with a cow on Syston Common. Now I suppose it could be considered unkind but Mrs Langridge was a lady of somewhat large proportions and I did wonder how the cow had fared in the encounter? Neighbours could be inquisitive in those days, never nosey, but would always help in a crisis and were mostly good natured and helpful.

Good news or bad news? Builders moved on to the wasteland at the back of our house to fill Collingwood Avenue with new houses. We lost the wasteland but we had tremendous fun playing around on and in the building site though we were not supposed to and I became quite interested in the construction. The builders became our friends and were also kind enough to give us waste wood and scraps for our bonfire.

Roger and I both had chores to complete upon which our pocket money depended. No chores—no pay. My sixpence a week would not appear. I generally saved my pocket money for Dinky cars that we purchased at Blanns on the High Street. I understand that most of the ones I purchased are now with my brother who collects them. My early chores were cleaning the family shoes and drying the dishes. When I first started, I spread more polish on myself than the shoes so my mother gave me one of my father's very old shirts to wear which almost reached to the ground. I was very amused. I completed the drying up without ever breaking a dish. Aged nine, I was promoted to dish washing and lighting the coal fire in the living room.

I was good at this and I was a 'One Match Kid'. I had to clean out the ashes and take them out to the bin, then lightly screw up pages of yesterday's paper and place sticks on them, then strike the match to light the fire. It always caught well and I waited for it to burn up brightly before spreading the coal over the sticks. I returned after five minutes to ensure all was well. In winter, my mother always came home to a warm fire after work.

We had no central heating and there were only small electrical convector heaters in the bedrooms while hot water was supplied by a gas immersion heater. During the winter, the house could get pretty cold so we all mostly lived in the living room. Hot water bottles would be placed in the beds during the evening. At bedtime, I often used to put my pyjamas near the living room fire (not too near) and then change into them before grabbing my clothes, racing upstairs and diving into a warm bed. Needless to say, duvets had not yet materialised.

My Uncle Jack would often employ this technique when visitors were reluctant to go up to bed or leave to go home. He would put his pyjamas by the Rayburn then threaten to change into them there and then. His guests would disappear at speed at the thought of him in his long johns. It would have been quite a sight. I was always allowed half an hour's reading time but lights were turned out strictly on time according to age… and negotiation. The system could always be overcome by the aid of a torch under the blankets.

In the house, apart from board games, jigsaw and playing with our toys, the radio provided the bulk of our entertainment through the Light Programme, the Home Service and the Third Programme. The Light Programme sometimes provided us with background music during the morning. Who mentioned background music? Unequivocally, I can state that I hate, loath and detest background music and have done so for many years. When I was a student in Exeter, it was the 'done thing' to go to Bobby's for afternoon tea because I think they were one of the first departmental stores to have 'Muzak', which was a continuous tape machine that played all day, every day, throughout the year.

It was played at very low volume and not intrusive but we most enjoyed and eagerly awaited the times it malfunctioned so it would then play Christmas carols in July and we would all join in. Since then, every restaurant, pub, café, coffee bar, newsagents, butchers, shopping mall, waiting room, hospital, clinic, departmental store and supermarket plays background music. Even when you make a phone call, you are bombarded with it and it has slid insidiously into the background talks and commentaries of TV and radio. Why? Why do we need constant and often loud background music?

On one occasion, a surgeon, who was performing a minor 'op' on me with local anaesthetic asked which music I would like. I do not think he liked the answer, 'None.' If I am in a restaurant, I want to talk with my friends and hear them. In one restaurant I visited recently, the sound system was malfunctioning and all we could here was the constant thumping of the bass. When I requested for it to be turned off, they stated that they could not because it was centrally controlled by one of their establishments in Cardiff. And we were in Bristol. I excused myself to go to the loo and whipped the plug out of the machine and, amazingly, no one noticed at all.

The Third Programme was very high-brow and not suitable for family listening so we relied on the Home Service. When very young, I had always listened to Listen with Mother cuddled up in my mother's arms and we later listened to Children's Hour at 5 pm, I think. The Archers, Mrs Dale's Diary and Woman's Hour had all started but we never regularly listened to these. Two of them are still going.

Sunday lunchtime, the Forces Family Favourites, Educating Arhie and, wait for it… Wakey! Wakey! The Billy Cotton Bandshow were regulars. There were good programmes during the week. Dick Barton and Journey Into Space were good and I remember being frightened to death by Quatermass. The comedy shows were great and many of them can still be heard on Radio Four Extra. Beyond our Ken, Life with the Lyons, The Clitheroe Kid (who was a girl), Hancock's Half Hour, Much Binding in the March, The Glums, The Goonshow and The Navy Lark. My parents always listened to Friday Night is Music Night. I also enjoyed Just a Minute and that is still on air.

I still prefer the radio. I have forgotten who first said, 'The pictures are better on the radio!' My TV viewing can be rated at three or four hours a week just watching selective programmes. I am never quite sure of my radio listening time because I am one of those peculiar characters who keeps Radio Four on all night as it drifts though Sailing By, the Shipping Forecast and the National Anthem into the One o'clock News and World Service programmes until 5.20 am, when it reverts to the shipping forecast and news and weather.

I do sleep but have become adept at waking up for the more interesting items. My brain has the capacity to sort out the relevant from the irrelevant while I am sleeping and can be immediately alert when required. In this respect, I have woken up suddenly hearing the broadcast reports on two air crashes.

One September night, in 1992, I awoke with a start. A Pakistan International Airline flight had crashed on its approach to Kathmandu. Ten miles out from landing, the pilot had reported his height and position. His height had been clearly stated as 8,200 ft— he was 1,300 ft lower than he should have been to avoid the jagged mountain ridges surrounding the airport but Air Traffic Control gave no warning. (Why?) 42 seconds later, the plane crashed. It flew straight into a mountain and everyone on board was killed.

Instructors and friends of mine had told me that they were planning a trekking and climbing trip when I last visited Plas y Brenin, the National Mountaineering Centre in Capel Curig, North Wales where I had attended several courses and often took school parties. The schoolboys and I had thus become acquainted with many of the young leading climbers of the day, who in turn often visited the school at which I taught for exciting lectures and an outdoor climbing session on Dartmoor. One of those had been Mick Hardwick and we also knew his wife, Sue. When I last phoned Mick he stated that they were going for two month's trekking in Nepal and were going to attempt to climb the South Face of Annapurna. I had no idea of his schedule or programme. That night I had a strong premonition that they were on that flight and could not get back to sleep at all. I phoned Plas y Brenin the next morning. Four instructors from the centre, Mick and Sue Hardwick, Dave Harris and Alison Cope, were on that flight and had died instantly.

Eight years later, in the early hours of a January morning, I was suddenly alert and heard that a Kenyan Airways Airbus 310 had crashed on take-off from Abidjan. It may sound odd but any one of my Ghanaian friends could have been on that aircraft as many international airlines stopped over on West Coast Flights heading to and from Accra. The flight had been due to land in Lagos but could not do so because of the thick dust haze caused by the Harmattan and I have to assume that Kotoka, Ghana's International Airport, had been closed as well, so the pilot, one Paul Muthee with 11,500 flying hours logged, flew along the coast to Abidjan and proceeded to wait there for three hours, after which he was cleared to fly back to his destination—Lagos.

Just after take-off, an errant stall warning sounded, the correct reaction procedure for which is to put the nose down and dive, which, despite the objections of the co-pilot, Paul Muthee, he did. They were then only at 400ft and the plane dived straight into the sea. Of the 179 people on board, only ten survived. But why was I so worried and concerned? I had this nagging premonition of doom again. I phoned Accra early in the morning and my friends there assured me that everyone was well and safe. I just knew that it was not.

At lunchtime, I took a call from Accra to be informed that a polo playing friend, Simon Kuseyo, had been on that plane and had not survived. Simon was the highest ranked polo player in Kenya and the top indigenous African player in the world. I had known him as a young boy at Gilgil from the times I had played polo in Kenya and then met him many times in West Africa and the UK. He was a very good player indeed and occasionally asked if he could use Singsing, a very fast polo pony I had acquired in Taunton, for his matches there. He had been hired to play at a High Goal Tournament in Lagos hence his reason for being on that ill-fated flight, but for the vagaries of the Harmattan, he might still be alive.

The World Service had been my constant companion in all my foreign travels in Southern, Central, Eastern and West Africa. It has an excellent reputation for accurate and reliable news and is a Flagship Ambassador for the UK. I rate it highly and in times of crisis, danger, disaster or political upheaval the local inhabitants trust and rely on its broadcasts as well rather than trust possible local radio's version of features and stories which can be biased on occasion.

We had tried the usual pestering of our parents to get a dog but they insisted that it would not be fair to leave it alone all day and now I realise the validity of that decision and as a consolation for the refusal, we obtained two rabbits that were kept in two hutches because mine was a Buck and Roger's was a Doe and never the twain could meet. Roger's was white and very docile—mine was black and white, called 'Buck' and hated being picked up. I used to get scratched all over my chest. They were given to neighbours when I started at secondary school. There is a sequel to this part of the narrative. After I left for college I came home one weekend and discovered that my parents had a dog—a black poodle called Pepi that one of my cousins was unable to keep any longer. I pulled their leg about this for some months.

Every fortnight, my mother, my Auntie Gladys and a friend, Mrs Upton (who worked for the BAC and had been involved in that first bombing strike at Filton during the war), had a theatre night out either at the Little Theatre attached to the Colston Hall or the Theatre Royal in King's Street. They liked the old traditional plays which they much enjoyed and hated modern drama but they still went regularly. This was an opportunity for my father and me to go to the Odeon Cinema at Kingswood to watch a western, war or adventure film.

I think the Odeon originally opened as the Ambassador Cinema and was built in the then favoured Art Deco Style. It was big. It had 1314 stall seats and another 480 in the balcony area. Hitler gave it a small present in 1941 when a bomb dropped at the back of the cinema and blew one of the salesgirls into the foyer. Alas, 'they changed our local pally into a bowling alley' after the cinema was closed in 1961 and today, it has been replaced by a supermarket. I was not keen on ten pin bowling and I think I only ever bowled there once. It was pleasant to talk to my father on these outings and we sometimes chatted about problems and plans.

He had been steadily employed by a Mr Palmer as an assistant manager in his grocers shop on Two Mile Hill. The shop is still there but not as a grocers. I think my mother had been gently nagging my father to get a better job for some time and he openly discussed this with me one Tuesday evening as were returning from the cinema together. I felt proud that he had confided in me and I said he should try. He did.

A couple of weeks later, he proudly announced that he was to be trained as a manager for the Maypole Group of Grocery shops and this would entail going away to their training centre in Worcestershire for two weeks up in the Malvern Hills. He was worried about leaving the family for so long so my Uncle Jack kindly took us all up to see him at the weekend. It is odd, as a couple of months before that, my mother was admitted to Cossham Hospital to remove a lump on her breast.

This was quite a magnificent building built in the queen Anne Style in 1909 kindly bequeathed by Handel Cossham who was one of the local 'King Coals' and industrialist of Kingswood. We visited my mother on the evening before the operation. Strictly speaking children were not allowed but I was smuggled in and out quickly by an understanding matron. On the way home, to my astonishment, my father started weeping and I became a bit worried. Was the operation more serious that I had been told?

He explained it was just that since he had returned from the war in 1945, my mother and he had never slept apart for eight years. He was alright. The operation was a success with no complications and I was allowed to see her again the next evening for another fleeting visit. The matron and sister at Cossham eventually became my mates when I was a little older for I now realise that I led the way with the introduction of mountain biking to Kingswood which was not supposed to exist at the time. Another activity I had introduced ahead of its time?

I was forever challenging the hard terrain of Syston Common and some of the grass-covered coal heaps while attempting to ride through streams and down perilous tracks so I was on the receiving end of bangs, bumps, cuts, bruises and grazes from my mad escapades, many of which were patched up by the kind staff at Cossham, with good humour and much patience. One of the doctors even gave me a lift home as I had bent my front wheel.

My father had a bike, my mother had a prim lady's bike with the usual shopping basket on the front handlebars and my brother had an adult bike. Aged nine, I was given a small, maroon-coloured bike with brakes but no gears. My parents had visions of long cycle rides around the outskirts of Bristol with hot sunny days and picnics but first, I had to learn to ride. My father did his paternal duty, that is to say, he ran up and down Southey Avenue holding on to my saddle while I wobbled, swerved, fell off and banged my legs on the pedals in my desperate attempts to keep my balance on two wheels.

Suddenly, I did it. Dad sensed this and let go and there I was pedalling up and down the road. My riding was still slightly erratic and I had yet to master turning but I was well on my way. Whoopee! We could now venture out as a family on our bicycles. Shall we say that practical and technical difficulties arose in relation to the differences in wheel size?

My parents and Roger would pedal slowly and majestically along admiring the route and the scenery while I was lagging far behind, pedalling like fury. I needed about 100 revolutions to their ten. I just could not keep up. My father would often come back and push me along for a while which was a great help but I could not go any great distances. When we came back to Kingswood from visiting the Iron Acton Show one weekend, I crawled onto the sofa and fell into a sound sleep and did not wake up until the following morning in my bed where my mother had undressed me the night before and I had known nothing.

We did go out on the bikes again but only for short journeys. We tried walks instead and often went out in company with my Uncle Bill and Auntie Mary and my cousin Pat, not forgetting the dog, Paddy. Uncle Bill worked at Wick Quarry and knew the surrounding area like the back of his hand and most of its inhabitants. It is said that he had rewired every house in the Wick and Marshfield area to earn extra cash. We would go out using Sunday Bus Routes and walk through the St Catherine's Valley picking primroses on the way or walk along the River Avon at Saltford.

Once and only once, he made a mistake with the timetable and we were stranded in Marshfield but he knew the local taxi driver and we managed to get back to Kingswood.

I joined in as many of the Cub activities as I could including the sports and district pack meetings and there was also the tradition of the Gang Show, which was produced annually and in which we were all expected to take a full part. I was a snowdrop one year and we danced in tutus, vests and had yellow petals and leaves on our heads and the second year, I was a decorator.

Little bits of putty. Little bits of paint.
Make a row of 'ouses look like what they ain't. Talk about magicians when they come along.
Miracles can happen all day long.
Down our street. Down our street. There's a lot of people you really ought to meet.

It was all a bit daft and silly perhaps, and I think very few Gang Shows are produced nowadays. It was fun at the time and it was a good attempt to make our own entertainment. While in the cubs and scouts, I learned many campfire songs (another good form of entertainment) and still retain many of them in my head. They have been a valuable asset in my teaching and instruction and have provided much fun and amusement. I seem to have had some facility for learning text and poetry and hold a vast store of quotations now including the scripts for the Art Show that I tour around the South West. I think it is probably because of my love of poetry and English literature.

We were not well off as a family but my parents always took us away for a holiday every year bearing in mind that my father only had two weeks a year off, thus we visited caravan sites at Uphill, near Weston (the site is still running) and a marvellous site at Westward Ho! right near the beach. At Uphill, I learned to ride a horse. His name was Crackers and I was given lessons and walked and trotted around a rough field adjoining the site. My brother, ever antisocial, would spend all day looking for golf balls on the local course, which, presumably, he sold. We never knew.

It was at Uphill that my mother was involved in a freak accident. We were walking along the road back to the site one afternoon and she was struck on the head by a hard object. One of the white china insulators had dislodged from a telegraph pole somehow and she had been unfortunate enough to be directly underneath it. It bled profusely. An ambulance was called and my mother came back to the site some two hours later with ten stitches and a white bandage covering her head. We thought she looked like a Maharaja. She was later given £30 as compensation, which was a sizeable sum then.

Mostly we visited Uncle Fred and Auntie Ethel in Barnstaple. We would catch the mainline train from Bristol Temple Meads to Taunton, alight from the Express and then just walk ten yards to a small side platform where the Barnstaple train would be waiting. It was the days of steam and I have vivid memories of the saddle tank engine that was waiting for us. It would be gently simmering like a boiling kettle with the driver fussing around the wheels and pistons with his oiling can. We would clamber aboard with our suitcases having ensured that we had used the toilet first because there were no corridors in the three coaches, and with a toot of the whistle and a gathering momentum of a burst of puffs, we slid out of the platform.

This was the Devon and Somerset Railway—44 miles of beautiful and varied countryside, sometimes displaying stunning landscapes then diving in to cuttings and woodland. There was one very steep gradient 1:58 which the engine huffed and puffed up and we always enjoyed going through the Castle Hill Tunnel. The train stopped at every station—Norton Fitzwarren, Milverton, Wiveliscombe, Venn Cross, Morebath, Dulverton, East Anstey, Bishop's Nympton, South Molton, Filleigh, Swimbridge finally pulling into the town station at Barnstaple, so the journey took some time but this did not deter the driver and firemen who would regularly stop the train in a shady wooded area and invite everyone to get out to pick primroses which Auntie Ethel received with much pleasure.

Alas, that hooligan Dr Beeching put the line on his dreaded list and it was closed in 1966 and it is now largely covered by the North Devon Link Road. As you drive towards Barnstaple, however, the signposts still point to the towns and villages in the list above. Some of the old magic remains.

Little was I to know that I would have future links with some of these names as later I used to attend the Boxing Day meet with the foxhounds at Wiveliscombe, normally with my favourite polo pony, Ace. For those of you who are anti-hunting, be assured that we never caught a fox at the three meets I attended, for Sir John just seemed to enjoy a good gallop over open country. He did not appear to have any interest in the fox at all!

I was always slightly amused with the hunting horses and their control or, in some cases, lack of control. Polo ponies are obviously very agile and you control them on a loose rein with your left hand. This means that you can spin on a sixpence, stop quickly and go into an immediate gallop. Hunters are not so easy to handle but I do not suppose that Ace could have survived a full day's hunting in bad weather. We only hunted for a couple of hours then hacked back to the pub for the more important business of the day.

Dulverton also has a polo connection and is now the home of the Somerset Polo Club which is at the back of The Craven Arms. Few people are aware that the whole area, particularly Dunster, became alive with polo in the Twenties. It was traditional for the Maharajas and Princes to come over from India to warm up and prepare for the Hurlingham Season in the summer near London. The first match was played at Dunster in 1926 and after that it must have required some feat of organisation to bring ponies, grooms, household items and servants and family from India by ship down to Somerset and the British section was achieved by use of the railway. It must have been quite an impressive sight for the local residents.

Uncle Fred and Auntie Ethel lived above Barnstaple in Chaddiford Lane in a big house with a garage for Uncle Fred had a car, an Austin Cambridge, on which he lavished much care and attention. Now I much enjoyed my visits to Barnstaple but I always had one ordeal to get through and that was always prompted by the question posed by Auntie Ethel, 'How are you getting on at school, Geoffrey?' I was generally doing well in my academic pursuits but I could not be described as a scholar in any shape and form and I was never top of the class.

I was comfortable with hanging around the 10th position in a class of 35. Joyce, their daughter, however, was very intelligent and came top in everything, I think. She would gain high percentages in most subjects and Auntie Ethel was very proud of that. I must add immediately that Joyce was no weird geek or swot. She was a very normal and pleasant girl and we got on well together. The topic of school and ability never arose between us. The academic comparison showed a big deficit on my intellectual ability.

For many years, I lost contact with Joyce which has now been re-established. I knew she had attended the Girls' Grammar School in Barnstaple and then chose an academic route studying Medieval English at Leeds University. She proceeded to gain her masters and doctorate and then lectured at Leeds later becoming head of department and finally ending up as pro chancellor. She still lives in Leeds and is now an emeritus professor and a leading authority on Medieval English. I note that nowadays school leavers seem never to be encouraged to pursue an academic route. I wonder why?

Uncle Fred had a pleasant and placid character maintaining a very dry form of wit. One would have assumed, though I suppose assumptions are dangerous, that he had lived out his life quietly engaged in his accountancy work in the town of Barnstaple for which he had much affection. Like thousands of other men in England, however, his pleasant world had been turned upside down by the advent of war. Like my father he had suddenly been shaken from the comfort and security of his job and family and thrust into another world of danger, fear and excitement.

At the outbreak of war he had been an articled accountant in Cheltenham and he and his friend very wisely decided to volunteer before the call up so that he could choose his unit and possibly extend his professional qualifications as well. He signed up for the RAOC (Royal Army Ordnance Corps), completed his training at Tidworth Camp and was posted abroad. He was in France when it fell where he was attached to the Second Army Division, which was motorised and thus not allowed to go out via Dunkirk that being reserved for the infantry. They raced for the port of Brest with the Germans in close pursuit where they were ordered to smash the lorry engines, throw their motorbikes into the sea and leave all their kit and ammunition behind. They commandeered a small collier with its cargo still on board, which inched its way painfully and slowly around the Cherbourg Peninsula and limped across to Falmouth.

From here, he sent a telegram to Auntie Ethel stating that he was safe. He returned to army life in England but he was warned that he was going to be posted somewhere tropical so he obtained a special marriage licence from the bishop of Exeter and married Auntie Ethel immediately during his embarkation leave. He reported to the troop ship and waited on board for the two days it was loading.

Prior to its imminent departure he was called over the Tannoy to visit the dock office where he was ordered to report to North Wales for Officer Training (he had been promoted through all the NCO ranks). If he had sailed with that troop ship, he would most probably have been captured by the Japanese and become a POW. In 1942, he served in Egypt in a big supply, transit and prisoner camp where he was promoted to captain and then major. He did key work in the repatriation of prisoners and was commended for the same before returning to Barnstaple in 1946. What a war? There was nothing ordinary about Uncle Fred.

From Barnstaple we visited all the North Devon Resorts from Westward Ho! right around to Ilfracombe, including Saunton Sands and Broughton Burrows and we travelled by train and bus. Distant relatives were visited and we always liked the town with its Butcher's Row and Market. I liked Bideford as well and I later revisited the town many times to board the Oldenburg en-route to Lundy which is a special island. Roger came on these holidays when he was young but eventually drifted off to scout camps and other activities so we could then go out in Uncle Fred's lovely car. Men in the front x 3—ladies in the back x 3. There were no seatbelts in those days!

On one such trip, we were driving in lanes near South Molton and Uncle Fred had to stop because there was a large parcel in the road. I reckon that he knew what was about to happen because I was delegated to go and pick it up. As soon as I bent down to do so, it was whisked away and through the hedge to much laughter from the other side. Uncle Fred was laughing and stated that he used to play the same prank when he was a boy? It must have been good entertainment.

We also had two caravan holidays in Newquay in Cornwall, which was, at that time, a more family-orientated resort; not like the sun, sin and sex city it has become nowadays. We travelled by train as usual but it was a much longer journey and we had to change at Par. For some unfathomable reason, we had left Bristol very early so arrived in Par at 7 am and the connection to Newquay did not depart until 9 am. The on-duty station master took pity on us and plied us with tea and biscuits and would not take any money in return. That was a good turn.

We surfed on body boards and even my parents had a go, though my mother could not swim, played crazy golf, swam in the rock pool under the Island and walked along the cliffs to the Huer's Hut where watchers would announce the arrival of the Pilchard Shoals, and…I have no idea what tempted my parents to do it, we rode horses along the Gannel from the stables by the boating lake.

My father was on a sturdy cob that was obviously placid and bomb proof. He had ridden once before—on a camel in Tunisia. The camel owner was charging for my father and his friends to be photographed sitting on the beast but had also demanded another fee to make the camel sit down again so that they could get off! My mother was mounted on a grey horse called Airborne and was ever such a horse so misnamed. It would not walk on and I had to constantly turn back and give it a slap on the rump. Having crossed the main road, you then have to ford the Gannel River. Airborne walked halfway through and stopped dead.

I am afraid we laughed but not unkindly at the sight of her stuck in midstream trying to kick the horse on. I had to go to the rescue. It was very brave of them to try riding. That was their first and last attempt. I was quite a competent rider at that stage and used to ride out most mornings. Whilst on that site we awoke one morning to such a banging on the roof and shaking of the caravan that we reasoned the world had probably come to an end. It was seagulls fighting over scraps of food picked out of the site's dustbins. We visited Newquay twice and took the same old caravan. It had calor gas lights and cooking facilities and was three berth. It had a wonderful smell that I always liked, unlike the modern vans that are very hygienic and sterile.

If someone had asked me when I had first started canoeing, I would have probably answered that it was when I first started college at the age of eighteen but it was not. I learned to paddle and row on Trennance Lake in Newquay and was really quite happy to take out a bigger boat and row my parents around for hours.

When I was very young we holidayed in Goodrington. I think I was probably only five and I can only remember two parts of it. We loved the beach and spent many glorious days bathing, building sandcastles and eating ice creams while my parents lounged on deckchairs. Two doors down from our Bed and Breakfast accommodation lived an enormous Great Dane under which its owners' children used to play. It was very friendly and loved attention.

I think my last childhood holiday with my parents was Brighton. Again, we travelled by train and we had obviously decided to take pot luck with the Bed and Breakfast accommodation. We found a large house opposite a twenty foot wall and I suppose, in retrospect, we should have enquired what was behind the wall. We found out that night that it was a Goods Train marshalling yard. We were kept awake all night by the noisy sounds of shunting, clanging, bumping and screeching of wheels. We had breakfast, packed and found other accommodation as far away from the yard as possible.

I am wondering now, however, whether its owners were German sympathisers because the accommodation seemed to be full of German war memorabilia and there was a very large Swastika pinned to the back of the toilet door. It was the first time I had visited the South Coast and I was not overly keen on the flat coastline. I was very impressed by the Devil's Dyke and Beachy Head and enjoyed the splendour and magnificence of the Royal Pavilion. I had fun on the pier and we did a show in the town and I swam in the sea but did not like the pebble beach. There is a rather funny photograph of me with a parrot on my shoulder with my parents in the background and my father in a strange pose.

I enjoyed all the annual holidays provided and know they came at no little cost. I was very grateful for all the experiences.

As I approached the age of eleven, I sought ways to increase my income. I would regularly help in my grandfather's electric shop at Warwick Road and received pocket money for same and I would often get a sixpence or a shilling from one of my uncles or aunts. The shop at Warwick Road was the hub of family life and I mostly met the various branches of the family there. Amazingly my father gave me my first chance of regular employment. He was working for the Maypole now in the shop halfway up Blackboy Hill at the top of Whiteladies Road. Many people in Bristol believe that the hill was so called because the affluent and rich ladies of Clifton had African boys as servants to carry their shopping.

I am afraid that this was not so and the White Ladies were the Carmelite nuns based at St Michael's Hill. There were remarkably few black slaves in Bristol during the history of the terrible trade because they were a very valuable commodity fetching good prices in the plantations of the colonies. The clinching fact is that Blackboy Hill was not named as a road until after the abolition of the Slave Trade anyway. Its present rather splendid road and buildings are a far cry from the muddy track with sewage running down it of the past for it was the main route to bring farm produce into the city.

I understand that there was a pub at the top of the hill called the Blackboy Inn where these farmers stopped to refresh themselves and were often robbed of their takings. The inn acted as a courtroom where trials and hangings took place and the tarred corpses were left as an example to the criminals of the district. Tarred corpses: Blackboys?

The Maypole employed one van driver, called Ron, who rotated around the Maypole shops of Bristol on a weekly schedule to deliver groceries. Six shops/six days. Ron was a chain smoker and always had a cigarette in his mouth. His habit had already seen the removal of one of his lungs and thus he was experiencing extreme difficulty with his breathing completing the Clifton Round as many customers lived in flats involving several flights of stairs or even more. I was to be Ron's legs and helped him. I did—at the grand rate of seven shillings and sixpence a day.

I would cycle, yes cycle, from Kingswood to the Centre then all the way up Park St and Whiteladies Road to the shop, though occasionally I would use Park Row and, in the late afternoon I would cycle all the way back. In bad weather I caught the Number Eight bus. I would help Ron load the van and we would set off with both the front sliding doors open to begin our deliveries. I was still in shorts, of course, but my father gave me a small white coat which made me look quite smart. The customers were enamoured of me.

They thought I was wonderful and I must say I was particularly polite and helpful carrying their grocery orders into their kitchens and asking if they wanted help to unpack it. My father was told constantly when his customers came into the shop what a polite, cheerful and helpful young chap I was. He must have thought that another boy was doing the round. I was often plied with sweets, chocolate, biscuits and squash because it was assumed that all children liked squash in those days.

Meanwhile, Ron would be waiting patiently for me in the van. He rarely enquired about the obvious small delays. If he did, I just told him that the customer wanted help with the unpacking or that there four flights of stairs to go up to the attic flat. He may have guessed but what I never told him about was the large number of tips I was receiving. I would often get a sixpence and the richer ladies popped a shilling into my hand. I never let on to my father either until many years later. On good days, I could double my wages. Fifteen shillings a day. I was rich.

My father's shop was situated on a corner and I witnessed two events while I was in the shop which had two plate glass windows and, unusual for Bristol then, a plate glass door into which the customers constantly bumped. I was helping out on a weekday morning during a school holiday and was refilling the glass-topped biscuit barrels in one corner when there was an explosive bang. We all jumped. The plate glass door had shattered and no one knew why. It happened two more times and it was finally traced to the Bristol Omnibus Company. Most of the buses changed down into second gear in front of the shop and the vibrations it caused shattered the glass.

I was returning from the bank one morning getting the change for the tills. As I turned the corner, someone shouted, 'Watch out!' and I jumped to one side as a car turned at speed and crashed into the side window. The car was half in the shop. My father walked across to the lady, opened her door and asked her what groceries she would like. She had pressed the accelerator instead of the brake.

My great friend among my father's staff was Mrs Rudge. Everyone called her Rudge except me. I called her Mrs Rudge. She spent most of her time upstairs in a small back space slicing up the ham, spam, corned beef, bacon, tongue and cheese with the aid of the bacon slicer and the cheese wire. I often helped her prepare the meats and I was forbidden to touch the bacon slicer but she often let me have a go until my father's voice would boom up the stairs.

'RUDGE! You're not letting that boy use the bacon slicer, are you?'

She would pat me on the shoulder to move into my place and shout down, 'Oh no, Mr Allen. He's just helping with the meats.'

She would then turn and give me a big wink.

She and I took our lunch break together and we would go up to the nearest pub where she would have a Guinness and I would have a cider. We chatted a great deal and liked each other's company. I once visited her in her flat as she had not been well. She lived in a small flat on the Clifton Triangle at the top of Lloyds Bank. I thought it was wonderful as it gave you a view of the Victoria Rooms and the Fountains and Magg's and Bright's—two fashionable department stores on Whiteladies Road.

The one abiding memory I have of my father in the shop was the speed of his adding up. He always kept a pencil behind his ear and as he served the customers, he would jot down the price on the corner of a piece of greaseproof paper. Upon completion, he would just run his pencil down each column and get the total in a matter of seconds.

At the same time, during the holidays, I would go into Bristol to draw pen and ink sketches of various landmarks such as the Lord Mayor's Chapel, the Naval Volunteer, St Stephen's Church, etc. Perched on a pavement or a seat on College Green I would studiously draw, feigning complete disinterest in the world around me and was bound to attract interest from passers-by. A small boy drawing was an object of interest.

In truth, I was not that good at drawing outdoors so I always carried a small portfolio of sketches of many of the Bristol landmarks that I had completed much more carefully at home. Invariably, someone would ask if the drawing was for sale. I always explained that I could not sell it as it was a commission for one of my relatives but let them browse through the portfolio and made many sales at five shillings a picture. Was I heading for the Super Tax bracket?

My teeth and eyes now enter the scene. I had been diagnosed with a lazy right eye and had attended several appointments at the eye hospital but there was no cure even though it may have been the cause of my debilitating migraines in my youth and my right eye is a problem with which I still live so I am ever anxious to protect my left eye otherwise I would be partially sighted. The consistency of my teeth was also suspect and I was referred to a Mr Blythe, an orthodontist at Clifton in Bristol.

Later, dentists have impolitely suggested that he was simply an 'Awful Dentist' for it was he who decided that I had too many teeth in my head that led to four teeth being extracted by gas, which I hated. The smell and the tugging and the blood set me on edge every time. He then fitted me with a brace for my front teeth, which, of course, I hated and also insisted that I wear a strap device around my head at night to correct something or other. The first time my Padstow dentist looked into my mouth, he was horrified and I explained the history.

'The man was a butcher!' he declared.

On one memorable day, I had attended the Bristol Eye Hospital in the early morning and a lunchtime appointment with the 'Butcher of Clifton' who made horrible impressions of my teeth and, finally I had a dental appointment for a filling in the afternoon at a surgery on the Causeway. The dentist there was a big brute of a man and stood no nonsense.

My mother explained that I had already experienced a bad day. He was having none of it and quite roughly pushed me into the chair and pulled my mouth open. I gagged and he hit me over the head and I screamed. It must have frightened the waiting patients to death. He then held his hands over my mouth to stop me screaming and I thought he was going to choke me so I bit his hand. He started to go berserk but by that time my mother had run in and she flew at the dentist and really gave him a telling off, took me gently out of the chair and stalked out, never to visit that dentist again. I had never seen my mother so angry and I was impressed. She had been fighting for, and protecting, me.

It may be of some interest to learn that during the course of my writing, I returned to the environs of 50 Church Road. Initially, I could not make it out. There was no side garden and no real corner. I then realised that one of its owners had hit on the idea of building another house on the side plot on the building which was now 50 A and B—flats I think. A shrewd move and the gardens have thus been compressed into smaller areas. My Father would have been pleased. Furthermore my parents purchased the house in 1952 for £2,300 and sold eleven years later for £15,000. The house sold for £200,000 the last time it was sold but now seems to be valued at £300,000. My mother was amazed. I was amazed at the huge number of cars parked outside, and in Collingwood Avenue. It was chock-a-block. When I lived there, it was rare to see cars at all.

But how did I fare in the 11+ exam?

Chapter Five
Rodway

I passed the 11+ examination. Previously, I had been offered three choices. If I had failed, I would have gone to Kingswood Secondary Modern School, though I understand you could still have chosen to go there if you passed. I was very lucky and my educational guardian angel must have been keeping a very close eye on me. Normally, all pupils who passed the exam had to go to Kingswood Grammar School.

This was an old-fashioned establishment where the teachers wore gowns and mortar boards and lessons were all based on traditional academic subjects, including Latin. The buildings were rather austere, old and many lessons were taught in Nissan Huts at the rear of the school. I must digress. Five years ago, I revisited the site. The school is now known as Forest Oak and I was mightily impressed with the new façade and many of the modern classrooms and specialists subject areas until I happened to glimpse a view of the rear of the buildings.

The Nissan huts were still there and in full use. They must have been well over seventy years old. I was not very keen to attend that school, especially as they studied Latin, which I falsely assumed I would not like. Latin scholars will be pleased to know that I felt the need to teach myself rudimentary Latin to cope with the many references to it in English Literature in my 6th form studies. Touché.

Rodway Technical High School was built in 1956 and opened in September 1957. I had timed it perfectly. It was my first choice. Here was a brand new campus with modern buildings, splendid facilities, acres of game fields and an all-weather running track—the only one in South Gloucestershire. Its 17 acre site was situated just outside Mangotsfield on the outskirts of Bristol right on the edge of Rodway Common. The derivation of the name Mangotsfield came from The Battle of Dyrham where Man 'got his field'. That is to say, he won.

The Parish of Mangotsfield was evidently huge at that time. Rodway cost £172,000 to build and £24,000 to equip. It held the status of a grammar school but offered a technical emphasis with a wider curriculum. Academic subjects were all on offer (except for Latin) and there was also Woodwork, Metalwork, Technical Drawing, Commerce and more Arts-based subjects. It was a splendid school and that was my choice especially as it had a big, up to date library with thousands of books. There was a limited entry. There were to be three classes for the First Form and three for the second, accepting pupils from other schools, I think, then a smaller intake for the third and fourth forms.

My parents were delighted and so was I. Roger was not. My success soured a somewhat shaky relationship with him and his bullying, aggression and temper became worse that year. When he first heard the news, he shot me a very hateful glance and strode off. His attitude to me worsened and I tried to avoid him as much as possible. One afternoon we were in the house alone and he started bullying and taunting me. I tried to walk out of the front room and then he hit me on the head. I saw red. I turned and went for him like a mad animal.

I screamed and shouted, hit and kicked him. I did not let up and I drew blood and I think I really hurt him. It had consequences. Mrs Crew, the next door neighbour, came in to enquire about all the noise and, of course, Roger blamed me.

Mrs Crew was waiting for my mother when she came home from work and told her 'allegedly' what I had done. I was not in the mood for taking the blame and told her that I was just fed up with him hitting me. I was told to sit on the stairs and wait for my father. On his arrival, he started to berate me and though I was not rude, I just told him that no one in the house was going to hit me again. I told him enough was enough.

I was sent to bed with no tea and I think my parents must have talked it over because Dad never hit me again and I think Roger was too scared to do so in case it prompted another attack. It was a difficult time within the family but I stood my ground. I was pleased but I never boasted about my pass and it was never directly mentioned again. My parents' pleasure fell to earth with a bump, however, when we received the list of uniform required all to be supplied by Horne's the tailors, in Broadmead, which was a very expensive shop indeed. Everything was strictly proscribed and deviations not permitted. You could not buy cheaper options from other school uniform suppliers.

The uniform was a grey shirt and grey pullover with cherry red piping around the neck, a grey and cherry red tie (to be worn at all times—summer and winter), grey worsted shorts (again), knee-length grey socks with cherry red bands at the top, a cherry red blazer and we boys had to wear a grey and cherry red cap with the school badge, Reeds for Rodway. Derivation 'Reedway' and the Book of Knowledge. Everyone wore black shoes (still supplied from Horne's).

For rain and any form of bad weather, we had to wear a grey Burberry raincoat (amazingly with no cherry red on it). And it did not end there. We required white and black aprons for woodwork and metalwork, cherry red socks and grey shirts with black shorts were worn for games (we were allowed to buy football boots elsewhere) and white shorts for PE.

Mother on Airborne

My brother Roger

Bristol Scenes

My Father and Mother – not the Family car!

My father wondered why we did not have to wear cherry red underpants! I know that my parents struggled to meet the cost and further expense was incurred with my stationery requirements—a fountain pen, pencils, rulers, rubbers and a geometry set.

There was near uproar when all this was insisted upon by the new headmaster at the first parents' meeting in June before the September Entry. The main source of dispute, however, was shorts. Boys who had already completed one year of their secondary schooling had changed into long trousers aged 11. At Rodway you had to wear shorts until you were thirteen. All objections were overruled by Mr Hughes and he stood his ground in everything. All regulations concerning the school uniform were to be obeyed. School rules were insisted upon throughout his tenure.

The wearing of shorts never really bothered me as I always wore them in summer anyway and I still do and for a rather odd medical reason. When I was young, we wore short shorts and this was particularly so even when we were Cubs and Scouts. Indeed, I remember Senior and Rover Scouts wearing extremely tight short shorts which I was not to see again until I worked with the Settler Farmers in Zambia.

Up to this year, men generally wore long shorts, that is to say, below the knee, but, hey, I notice now that shorts are getting shorter again. Perhaps that could be described as a short intermission?

For the first two years, I was very happy at Rodway. I enjoyed all the subjects and we were taught well and thoroughly by mainly young and enthusiastic staff during that period. I would happily cycle off in the morning and enjoy my school day before returning home and completing my homework up in my bedroom before tea. We always had homework, which had to be completed on time and handed in promptly. Failure to do so resulted in punishments. I had very few but I did find completion of homework quite a strain on occasions.

Some subjects were quite hard. I enjoyed school life and joined the choir and also took part in small plays. We had organised games and cross country and athletics in the summer. Athletics was my favourite and I was quite a good long jumper. My godfather, Uncle Sid, was quite annoyed that I had not chosen the Grammar School because they played 'rugger' there, a game that was destined to put me in the Exeter police cells.

Upon reflection, the PE and games master, Jimmy Munn, a diminutive Scotsman, was a little odd and we all became quite wary of him. I suppose in today's environment of personal safety, he would have been reported. He never resorted to sexual abuse but he had 'strange' ideas.

All boys did gym just in a pair of white shorts, under which you were not allowed to wear pants, and you had to be barefooted. I suppose this was alright for us smaller boys but the bigger boys were more developed and had difficulty holding everything up and in. As we filed out of the changing rooms, he would pull back the front of our shorts, checking for underwear.

When in the gym, he would always control the lesson where we had to go upside down on the ropes and wall bars. Even more strange was if you had forgotten your shorts, you did gym naked. This was obviously embarrassing for the individual concerned but made much worse by the fact that when filing into the gym, we would pass the girls going in for their dance lessons the other way. I always kept my PE shorts in my locker. Upon completion of the lesson, we always had to take a hot shower which consisted of a spray race system with cold showers at the end (to close the pores).

He would stand at the end to ensure that everyone used them and would administer a couple of swipes with a cricket bat on our rumps if we did not. After football we had to turn around naked in front of him to ensure that we had washed all the mud off. On one famous occasion, he made us run the cross-country course with no shirts (how odd was that) and it was freezing cold. Parents complained and it never happened again.

By the time I entered the Third Form, I was being taught by older and more mature staff and I quickly realised that there were two distinct groups. There were the 'Goodies' and the 'Baddies'. The 'Goodies' were helpful and kind. The 'Baddies' were not. My favourite teacher was Ken Peace who taught geography and it is to him I owe my present love and interest of that subject and I did 'O' and 'A' Levels with him.

Mrs Goodfield coped with my math and she made me sit right in front of her so she could assist with my algebra. Mr Gay, metalwork, realised that I was not going to set the metalworking industry alight but was patient and helpful while Mr Woodman taught us music and scripture and also ran the choir and I responded to him well. 'Noddy' Lock was my chemistry teacher but he was one of those unfortunates who could not keep a class in order. He was a charming man and incredibly kind, much too nice to be thrown into the teaching arena with us lions!

There was a twilight world between the two groups. Mr Renault, who taught history, was a strict disciplinarian and quite boring as his lessons always followed the same format and then there was Mr Neesham, head of science, who shouted and ranted a lot.

1. My father and his staff
2. Mum in Tunisia
3. Family Outing
4. My father on Holiday

He could be quite scary on occasions but never resorted to abuse or violence. I had an unusual encounter with him at his house in Pucklechurch because one of his boys was in the Cub Pack that I ran there when I was in the sixth form. I was delivering details about a Cub camp and he recognised me. He invited me in for a cup of tea and then suddenly asked, 'Tell me, Geoffrey, what do the pupils think of me?' I told him.

'I thought so,' he responded.

Now I have no idea if my comments were heeded because he did not teach me in the sixth form but perhaps he moderated his outbursts.

The 'Baddies' were led by the Deputy Master, Jack Petit, a big, burly man who took enormous delight in plying his pupils with sarcastic derogatory remarks and belittling them as much as he could. He was a regular 'Hitter' about the head and I felt his wrath in nearly every lesson. I was so scared of him I could not learn. I began to hate and loathe his lessons especially as he was so unpredictable. If you did not understand, he hit you. If you got low marks, he hit you. If you gave a wrong answer, he hit you. He was aligned with 'Duggie' Board, our English teacher who had a very old gown, the frayed ends of which he had knotted, and he hit all slow learners and pupils supplying wrong answers. He used to swoop down on you like a demented crow and whack you hard over the head. It hurt.

He was also an expert with the use of chalk and blackboard rubbers and he aimed to hit…not miss. My namesake, Mr Allen (no relation, thank God), taught French and was quite evil and spiteful as well as using a high degree of sarcasm. He was the world's expert at the put-down which he used in an attempt to gain laughs from the more fortunate members of the class. He would pick you up and physically shake you roughly and would beat boys (only) with the edge of a blackboard ruler. He would never touch the girls. Mr Springett, the woodwork teacher, was also short in patience and would hit you for even minor reasons. It was permissible to use corporal punishment in schools in those days but this was to be applied only by the head at Rodway.

I think I have explained previously that if I had been involved in a serious misdemeanour while at school, I would have expected to receive six of the best, but seriously, I was not expecting to be physically attacked by a gang of teacher thugs who had obviously received their training at Dotheboy's Hall, were downright bullies and a disgrace to the teaching profession. I was very surprised that it was happening in a new school with high standards. I was not a bad pupil and was generally conscientious and well behaved. It is not as if I was disruptive or rude.

Things came to a head in the fourth form. I had not been feeling well in the morning and had visited the school nurse during break to get some aspirin. The fifth lesson of the morning was French with Mr Allen and I was slow in responding to a vocabulary question that he asked. Without warning he dragged me from my seat, up to the front of the class, shook me and bent me over his desk, then started to hit me with the edge of his blackboard ruler. I snapped. I stood up and pushed him away.

'Keep your bloody hands off me! Don't you dare hit me again!'

He went berserk and dragged me out of the classroom, cuffing me all the time and pushed me up against the wall, banging my head in the process. I was sent to the headmaster hurt and sobbing almost uncontrollably. His secretary took charge, despatching me to the school nurse who examined the bump on my head and washed my face. She gave me a drink of water and after I had calmed down, I was ushered in to see Mr Hughes.

Mr Hughes was a 'Goodie' and I held him in enormous esteem. His manner and methods later supplied me with exemplars for handling my own school and I commenced by apologising to him and he quite simply said, 'Tell me what happened please.' I recounted the immediate events and by careful questions, he extracted the fact that Mr Allen often hit boys and was quite spiteful and malicious. He asked if other members of staff had also hit me and for what offences.

He seemed a little bemused and it was then that I realised that he had no idea that this was going on in the school. Later, as a headmaster and school principal, I was always aware what was happening in my school as I was often involved with the pupils and physically patrolled the campus. I now wondered if he ever knew how weird Jimmy Munn was as well? At the end of the interview, I asked what my punishment would be but he simply thanked me and told me I could go, adding that if I wanted to go home for the day, I could, which seemed odd at that time.

I stayed at school, otherwise I would have some serious explaining to do to my father but I was quite the hero. I did not repeat the story but my classmates did and all the other pupils treated me with a great deal of respect.

Two points occur to me here. I had told my parents about the teachers but it was not usual for parents to complain about treatment in schools in those times. My parents never complained or commented to the school. Secondly, in relating this story to friends some of them took the view that it always happened in the good old days. 'School were the best days of your life.' A good clip around the ear and a good beating were all part and parcel of your education. The incident had an immediate effect throughout the school.

One can only guess what went on behind the scenes. Perhaps an urgent staff meeting? A quiet word with the teacher thugs? Who knows? As far as I recall all hitting and physical abuse ceased from that date. It was too late to save my Math and French but I could now get back to enjoying school where I thrived.

Despite all this, I rarely missed school and my attendance was good. I would have to make the occasional dash home if I started to suffer from a migraine before I was violently sick but I was quite ill once. I had always suffered from catarrh while in Bristol and in the winter of 1960, I contacted Sinusitis. It was painful and dragged on for two weeks most of which I spent in bed, utterly exhausted, feeling fatigued and washed out.

During the second week, I did nothing but sleep and my mother was very anxious when she had difficulty waking me up. I carried a Doctor's Note back to school to put me off games for at least four weeks. I nervously handed it to Jimmy Munn, the PE teacher, who, to my amazement, sympathised with me and told me to go up to the library. (Was there ever a better instruction?) I was now mainly reading books on travel and expeditions. Fuchs had crossed Antarctica two years before and I was fascinated by the account and I read all of Richard Attenborough's books as well as the humour of Gerald Durrell. The year before Richard Leakey had found the earliest man in the Olduvai Gorge in Tanzania. I was also beginning to take an interest in the news and was spellbound by the account of the escape of the Dalai Lama from Tibet.

As a lesson to others, I mention that I was so scared of getting low marks in maths that I got help and cribbed from my best friend Richard Bane who lived on Warmley Hill. He must have considered me a bit of a nuisance but he never refused. This led to no understanding of the maths in hand, my results plummeted and to this day I still cannot get my head around algebra. Let that be a warning. I should add, however, that in my later work, I was involved with the world of international finance and banking even at a governmental level and understand corporate accounts, balance sheets, cash flows and company figures. It was the algebra that was my downfall of which I never felt the need. My young cousins and niece can run rings around me in that branch of mathematics now.

Here is an intermission and a very important one. I had cycled back from Rodway when I was in the first year, as usual, and carefully hung up my very expensive cherry red blazer by the door, as usual, then slung my satchel over the banister ready to go upstairs to do my homework, as usual, and walked into the living room to greet my mother for she did not work on certain days, as usual, and gave her a kiss, as usual. She asked about school, as usual, and I told her it was OK, as usual, then turned to leave the room and go upstairs. She burst into tears…not usual.

I was upset to see her upset and asked what was wrong. She just pointed to the corner of the lounge. There, in the corner was a brand new working TV which she had planned as a surprise and I had not even noticed it. I had to make up an excuse about my thoughts being on a particularly difficult piece of homework and gave her a hug and another kiss before disappearing upstairs.

I do wonder if my first encounter with a TV set was to set the pattern for my relationship with TV for the rest of my life because I am not an avid watcher even now and only view two or three selected programmes per week. Family viewing became the norm in the evenings. My father adored all Westerns so Bonanza and Rawhide were his favourites. At the young age of eleven, I watched The Lone Ranger, Lassie, Flipper and then Mr Ed, The Adams Family and The Munsters.

My brother liked Dragnet, Mission Impossible, Hawaii Five-O and the Avengers while my mother preferred The Dick Van Dyke Show and I Love Lucy. We generally all sat down to watch The Black and White Minstrel Show (not PC nowadays, of course) and Sunday Night at the London Palladium. My father also liked his football, wrestling and boxing.

Richard was well impressed with our TV, however, and often came over to watch some programmes. He and I were great friends and as we entered our middle teens, we were almost given complete freedom to roam. Our normal limit when we were younger was to go train spotting at Mangotsfield, just below Rodway Common, but that became uninteresting because the same engines plied the line, especially Barfleur, which we must have seen a thousand times! We played tennis and completed long cycle rides to Weston, Wells, the Mendips, Portishead and Clevedon and, one weekend, my grandmother let us use her caravan at Brean for a night. It was a small dark green elderly caravan that had seen better days and it was sited in a field of long grass right under Brean Down. The only facility was an Elsan bucket in an old tin hut. We were happy enough, however, but we had to cycle back to Bristol on the Sunday in pouring rain and gale force winds.

Shortly after I and another cousin travelled up to London with my grandmother to stay with her sister, Auntie Mabel, in Wandsworth. Auntie Mabel lived in a block of flats on the third floor. Nana took us around London on the first day and I remember seeing the Changing of the Guard but on the second day, I had free rein. I was up before dawn and caught the bus to Convent Garden. I mean the real Convent Garden, when it was a fruit, flower and veg market and I wandered around it for well over two hours fascinated by the hustle and bustle and the cries of the porters who were carrying stacks of loaded baskets on their heads.

The next morning, I was off to Smithfield to see the huge meat market which was just as interesting, then on my last day I visited Billingsgate. The smell of fish was overpowering and what fish! There must have been well over several hundred species on sale as well as the seafood. I had not been exploring long before I realised that I had lost my wallet. Careful thought deduced that it must have been on the bus on the upstairs front seat. I returned to the bus stop and stopped two or three returning buses whose conductors let me have a quick search. I gave up. The next bus that came along, I jumped on, showed my return ticket to the Clippy and climbed dejectedly up the stairs to the front seat. Staring at me was my wallet and it must have travelled quite a few miles around London before it found me again.

Another friend, Stephen Gay, lived in a big detached house in Alma Road and introduced us to the game of 'King'. No! No! Not the Game of Kings. That is polo. The king of Games. 'King' took place on Stephen's back lawn. Each competitor had a wooden tent peg as a stick and the idea was to hit a tennis ball through your opponent's small goal. It was akin to mini hockey but gave us hours of fun.

Methinks that I have galloped ahead on the scholastic developments of my life. Eleven was another milestone in my life in several respects. I gained much freedom from parental control. I do not mean that I was allowed to run wild. I was fairly sensible and responsible and we did live in, what might be considered, a much safer environment than that of the immensely complicated and risky world which we now inhabit and there was far less traffic speeding on our highways. I could be left alone for short periods if necessary so I often had the house to myself after finishing school and I enjoyed the emptiness and silence for a short time.

I am quite a gregarious person now but I do enjoy solitude as well so I gain the best of both worlds. After I had lit the fire in the living room I always got on with my homework which was two subjects and generally took about an hour and a half to complete. By that time my mother and Roger would have come in and my father arrived at about 6 pm when we all sat down to an evening meal and recounted our day.

My brother was always slightly reticent about his activities and had never taken a full part in family life. He was not withdrawn as such but I always thought he was wary and secretive. He could be absolutely charming, friendly and helpful and he could also be unkind, spiteful and malicious. His mood changes were quite sudden and his bad moods could go on for days.

He had at that time some contact with the Reverend Sidebottom whose big house backed on to the hilly part of Church Road and I think Roger helped him in his garden. The reverend was a Bristol City supporter and often took Roger to matches. Roger was ill one Saturday, however, so my mother told me to pop up to the Reverend's to tell him that he could not go to football. To my amazement, the reverend asked me to go instead.

In truth, I did not enjoy it much for I am not really a 'spectator' type but when he was driving back at the end of the match, he suddenly asked if I was mistreated at home. I clearly replied in the negative but he then went on to ask about any possibility of maltreatment to Roger. I gave another definitive no. I nearly said that he constantly mistreated me but decided discretion was the better part of valour. I wondered what stories Roger had told?

Sometimes he would be helpful and friendly and it must have been during one of those times that we agreed to a bike swap. He wanted my red and white Raleigh and offered his racing bike in return. My parents brokered the deal and it was agreed. Two days later he changed his mind and wanted his racer back. In the face of his mounting anger, again I was forced to capitulate for the sake of my parents.

In the August after my eleventh birthday, I 'flew up' from the Cubs into the Scouts in a formal ceremony with the Cub Pack on one side and the Scout Troop on the other. There is a tendency for the modern generation to mock such ceremonies but they were well respected at that time and the formality of them injected a high level of seriousness and respect. I 'flew up' from the Cubs as a Senior Sixer (Top Dog) and a Leaping Wolf and became…a lowly Tenderfoot Scout.

From being a big fish in a small pond, I became a small fish in a big pond. Hence began the fun of being in the Scout troop. I quickly passed my Tenderfoot Badge and the right to wear a Scout uniform with the presentation of the black and yellow troop scarf and I was thus duly invested into the Scouts having learned the Law and Promise with the instruction to do a Good Deed every day. Old Hat nowadays, I think, and yet, even the updated modernised version of the Scout Law and Promise has a relevance to modern life and the world would be a much better place if more people kept to its simple aims.

Many years later when I started a school in Ghana we formed Beaver, Cub and Scout Units catering for some 200 pupils—girls and boys. There is a charming sincerity about Ghanaian children. Some of them believe in Father Christmas to a much later age than young sceptics in the UK and every one of them actually tried to keep their respective Law and Promise to the extent that there was a tremendous improvement in overall attitude, helpfulness and kindness throughout the school. I was approached by two of the Junior School staff shortly after the Investitures who cautiously asked me what I had done to the children.

'Why?' I asked.

'It is like magic. They have become much more kind and helpful,' they replied.

'Ah! That is Scouting for you!' was my response.

There is an addendum to this. One of the parents, Voncujovi, was a magician of international repute and had performed all over the world including 'The Strip' in Las Vegas. He very kindly agreed to perform for the children and as I watched the pupils' reactions in the audience, I realised that they did not think these were tricks—they believed it was magic!

The school has a very special atmosphere. The children are well behaved, fun and a pleasure to teach and we involve the parents in their child's education so the atmosphere is one of care and friendliness backed up by the helpful and competent staff. One leaver clearly identified this quality in her end-of-term remarks.

'I think Mr Voncujovi left some of his magic here when he left'

I attended troop meetings, held in the 'Scout Hut' regularly on a Wednesday evening which always started with Flag Break and finished with Prayers and had a good mixture of games, training and fun. We mostly worked in out patrols with our Patrol Leader. I was in the Wolf Patrol and was promoted to Patrol Leader sometime later. I attended all Church Parades and activities and 'tried to do my best!' The Group Scoutmaster was run by Don West; Skip and a David Macarthy ran the Scout troop. All leaders gave up their time voluntarily to run such troops and I am very grateful that they did. Scouting is still a very popular movement today throughout the world and I have always rated it very highly.

The troop was attached to the Holy Trinity Church which still stands in its original position at the top of Church Road. It was a 'Waterloo' Church, built by funds voted by Parliament to mark Napoleon's defeat at Waterloo (whom I later met…strange but true!).

It was one of the first, completed in 1821, having been built as a matter of some urgency as Kingswood was a 'Hotbed' of non-conformists. The church has a tall tower ninety feet high with pinnacles and a clock and two bells and it was designed deliberately, so I am reliably informed, to directly 'look down' on the Wesleyan, Whitfield and Moravian Chapels that are nearby. It was built in the Early English Style consisting of a Chancel, Nave and a North Porch and could seat a big congregation in a rather plain interior. It was set in a huge graveyard that ran alongside Kingswood Park and two church halls and the Scout hut were deep inside it as well. It was considered good sport to hide up the trees to frighten Akela and Skip on dark winter nights fuelled by ghost stories that they had related.

In charge of the church was Canon Radford who lived in the rectory to one side. He was a fairly serious man always dressed in a black cassock and cloak, and he was a little daunting but he never came down to cub or troop meetings anyway so we only encountered him at Church Parades once a month. That was about to change in a big way. One troop meeting selected younger Scouts were called in to the small room at the back of the Scout Hut one by one. I could not think what it was all about.

There had been problems with the previous scout leader who had been hurriedly asked to leave concerning unconfirmed allegations about naked romps with the more senior Scouts in visits to various local woods but I thought that was 'All water under the bridge' now and it was not during my time anyway. Skip was very serious looking and I quickly scanned through any misdemeanours, perceived or unperceived, that I might have committed in the last few weeks. It was spring so there had been no graveyard scares in the night, I had passed my Second Class Knotting at last and I had not annoyed my Patrol Leader recently. What could it be?

There was a conspiracy afoot and I am ashamed to say that my mother was involved. The plot thickens. The Press Gang had been let loose and Skip was part and parcel of the treachery. Holy Trinity Church was short of choirboys and the Very Reverend Canon Radford was the gang leader. Surely, he had reasoned, the young Scouts would be willing to come to the aid of the church of which they were members to make up numbers and he had brought Skip and my mother around to his way of thinking. My mother had mentioned nothing to me which made me a little upset at the time. I made my excuses. Homework? My Saturday job? Scout activities? All my arguments were systematically demolished by Skip but then I played what I thought would be my ace card.

'Skip! I can't sing!'

'Mr Parker, the choirmaster, has heard your singing at church parades and rates your soprano as one of the best he has heard for a long while. You would be his first choice. He has asked for you in particular and you have also sang well in the gang shows.'

This was a three-line whip. I was sunk. I was, after all, only eleven and could not resist so much adult pressure. I joined the choir. To this day, if anyone ever asks me about the choir, I maintain that I was coshed over the head and press-ganged—'Shanghai-ed Aboard a Trawler!' And there was no hanging around.

The next evening, two other scouts and I presented ourselves for 6.30 pm choir practice with Mr Parker looking as pleased as Punch now with a total of eight choir boys. He was a very small man and he had to conduct the practice by swinging around backwards and forwards on the bench seat of the organ and we often wondered how the feet on his short legs ever reached the organ pedals. He could only see our heads and shoulders in the organ mirror above his head so our hands and feet were free to misbehave. We were no angels I am afraid and used to lark around quite a lot. He never became annoyed. The constant instruction was to: 'Settle down. Ready?'

The following Sunday we were asked to come to the church vestry early at 10.30 am to prepare for the Matins Service at 11 am. We were fitted with our black cassocks with white surplices over the top and the standard choirboy's white lace ruff, which had to be cleaned every week, then I was paired up with one of the regular boys to show me the ropes. By this time the ladies and men of the choir had arrived and donned their robes. At 11 am, a prayer was said with a cross carried before us, we processed, singing, down the aisle, boys leading, ladies next, then the men, followed by Canon Radford, and filed into the choir stalls.

The service order sheets had already been placed in the stalls by the church warden and so we knew what was coming and the service would begin. Holy Trinity Church was not a High Church but the service was traditional from the opening, 'Dearly beloved Brethren, the scriptures move us in sundry places to…' and the confession, the Lord's Prayer, psalms, lessons, the Venite Te Deum or Benedictus or Jubilate Deo, the Apostles' Creed, collects, the sermon and the thanksgiving and the blessing at the end.

There was always a processional hymn when the choir would file back out to the vestry and the choir received a special prayer in the West Porch. I can still recite the Matins Service nearly all the way through and, of course, I know many hymns and psalms by heart. The sermons were rated as being of no interest to we choirboys so we were allowed to draw on the back of the service sheet to keep ourselves out of mischief.

At 5.45 pm, we had to present ourselves for Evensong which was slightly less formal. My favourite Processional hymn at the end of this service was, 'The day thou gavest Lord is ended'—I always enjoyed it and still do. The men of the choir were always keen to disrobe quickly after the evening service. There was a great tradition at Holy Trinity for many of the men of the choir and congregation to take a sharp left turn on coming out of the churchyard straight in to the pub next door.

I am, perhaps, one of those rare people who still enjoy the formality of the format for Matins and Evensong and find it comforting and secure though I also enjoy modern church services of all sorts and am also quite willing to visit those of other denominations and faiths. The school that I helped found in Ghana had a Christian ethos but we welcome all children of all faiths and the school still has regular Morning Assemblies which most secondary schools do not seem to hold in the UK and more. I think it is a valuable time for the school community to join together and share a common experience religious or otherwise.

We were paid! As choirboys we were paid seven shillings and sixpence a quarter for twenty four services and twelve practices but as I was then only on half a crown a week pocket money it was a welcome extra income. We could also be paid five shillings for a wedding. FIVE WHOLE SHILLINGS FOR AN HOUR'S WORK! which only happened two or three times a year BUT you were paid seven shillings and sixpence if you sang at a funeral!

I sang at two, once with the other choirboys and once as a solo—Psalm 23. The Lord is my Shepherd. I tell you there was not a dry eye in the church. Canon Radford thanked me especially, the first thanks I had received from him ever!

He was not so pleased with me several weeks later when I played a prank in the vestry. It has already been mentioned that the tower contained two bells which had been rung in the past but were now condemned to everlasting silence because of cracks in the tower. The call to the faithful was thus delivered courtesy of a vinyl record of the bells of Salisbury Cathedral from a radiogram come record player in the vestry and out through huge amplifiers in the tower. As choirboys, we were occasionally trusted to put on the record while the church warden, who was very deaf, went about his duties.

One Sunday morning, for a dare, I swapped the 'bells' record for Rock Around the Clock which boomed out of the tower all over the High Street and the immediate area. It was actually nearing the end of the record when Canon Radford rushed into the vestry, took one look at the record and quickly turned off the turntable. Now Canon Radford had what he considered to be an errant son, called Andy, who was very much into pop groups and pop music and automatically assumed that it was he who had swapped the record and sent for him immediately.

I could not let Andy take the rap and owned up. I was suspended from the choir for two weeks with no pay but had to sit in the congregation for that time. All members of the choir and the congregation had been highly amused by the prank, however, and a 'Free Geoffrey Campaign' was launched immediately. By the end of Matins that morning, I was a hero and Canon Radford let me back into the choir with immediate effect. Some parishioners even suggested that, perhaps, Rock Around the Clock should be played every Sunday to make churchgoing more attractive. Canon Radford was not amused.

I was in the church early one Sunday morning helping in the preparations for the Mother's Day Service and suddenly, I heard Canon Radford almost shout. Andy had climbed up into the pulpit and Canon Radford berated him about his behaviour and general lack of morals which did not entitle him to even climb the steps, in front of many members of the congregation.

Catching up with Andy's life much later, I think it was probably he who had the last laugh. On leaving school, Andy had worked with a builder then started his own building company, absolutely determined not to go the way of the church. He had his own pop group still and this developed into a forty-strong Pop Gospel Group. He then trained as an architect and surveyor during which time he must have received the Word. In 1972, he studied theology, took Holy Orders and became a curate in Shirehampton in 1972. He had worked with local radio before but he became a popular and innovative speaker and was excellent at Broadcasting. He became Communication's Officer for the Diocese of Gloucester.

In 1998, the Right Reverend Andrew Radford was appointed suffragan bishop of Taunton and he was there for two years before he unfortunately died of cancer. I reckon his father might have let him climb into the pulpit before he died.

I was very young, of course, but I was aware of whispers alleging that Canon Radford was not mindful of his weaker and less fortunate parishioners. I could not doubt his calling or methods. Suddenly, a very kind and affable curate appeared at the church. He was a big man and an ex-submariner I think. He wore a cassock and cloak and big black boots but their appearance was shabby. He walked all around the parish visiting the sick, disabled, the old, the frail and anyone in need of solace and comfort. He lived in cheap digs and gave most of his salary to charity. He would stay up all night with parishioners often giving carers a well-earned break and he always had a kind word for everyone. He was an exemplar for all priests.

During this time, I learned to swim at Speedwell Baths using snorkel, mask and flippers. A friend had let me use them and I was swimming by myself in twenty minutes without them. Speedwell Baths had a very heavy chlorine content and we always used to emerge with our eyes streaming. I never looked back and swimming is one of my main sports, now usually early morning at the adult length sessions. I also much enjoy swimming in the sea and in rivers and streams. I am a wild, wild swimmer and like the solitude and the freedom to roam and swim. My star sign is Cancer and my mother has suggested that I have an affinity with water. At least I am clean for most of the time.

What with school, scouts, choir and my jobs I led a very full life indeed. I was lucky. It is really quite difficult for young people to get a part time job nowadays but I never seemed to be out of work from the age of eleven onwards. I had always ran errands, helped in my grandfather's shop and did chores but I think my other jobs and incomes provided just a little help to my parents. I provided my own pocket money from about thirteen onwards and thus it never became an issue between us.

My Uncle Jack was amused that I was now being paid for my singing. Ron, the Maypole vanman died, alas, so my very lucrative Saturday job came to a halt. I was unemployed for one week and then managed to get a job on a Second Hand Car Lot next to the Tizer factory in Kingswood washing and cleaning all the cars. It paid well at three pounds for a fairly short day and the owner was very pleased with my conscientious approach and I was a good draw for customers.

Job done!

Bob-a-Job in the Cubs and Scouts was an annual occurrence where you personally raised five shillings by doing jobs for your annual Capitation fee for the National Scout Association and the balance went to the troop. My poor long suffering relations. They were always persuaded to give me jobs so I would get approximately ten shillings from them just leaving them as 'poor' relations! After that I normally knocked on doors with a fellow Scout to get a big variety of odd jobs. We were caught out one day, however, when we decided to go mobile and cycled out to Syston Common where we called at a farm thinking we could make a few bob in one go.

The farmer was delighted. He took us around the back of the farmhouse and pointed to two 60ft sheds. He opened the door of one and it had hundreds of chickens in it. The other was empty. We had to transfer about five hundred chickens from one to the other. We always had to wear uniform and the farmer said it would be better to remove our scarves and shirts. The chickens had to be moved by catching their legs and taking about five at a time to the other shed, so, with the two of us, this entailed ten chickens per trip.

Fifty trips in all backwards and forwards. It took about three hours. The farmer signed our cards in the job done section and then gave us…a shilling each! We could not complain. 'A Scout smiles and whistles under all difficulties.' His wife objected, however, and gave us another shilling each and a glass of squash. We washed off under a pump and cycled wearily home. Chickens have never been my favourite birds ever since. The other Scouts in the troop thought it highly amusing and we did win a small prize for the most unusual job done. Some consolation, eh? A Mars Bar each.

Having worked in many schools, the Mars Bar was the in-house currency of the time for prizes, helpfulness, bribery to complete a job, topping a climb, completing a difficult squeeze in a cave, kindness, competition prizes and thanks. In the good old days, the Mars Bar was quite large but over the years it has increased in price and decreased in size. Do you like Mars Bars? Do you think it is the best chocolate bar in the world?

A friend of mine applied for a managerial post at the Mars Bar Company. The whole factory runs democratically, offering equality to everyone. All the managers and workers eat in the same canteen and have equal opportunities and my friend was really quite impressed with that and the modern production methods. In his interview at the end of the day, they posed the two questions above. My friend hesitated. Mistake number one. Then he prevaricated saying that he quite liked it. Interview terminated. No job.

I attended many weekend camps and two full week camps. The first was at Sidmouth and it did not have a good start. Skip had not matched up the approach lane with the width of the lorry and we ended up having to lug the gear one mile up the track. The site was also on a slope and the wood that the farmer had supplied at a price was completely damp and rotten leading to cooking issues all week. The Senior Scouts and Rovers always came with us and it was their job to set up the store tents, dig the latrines and help service the camp but we cooked in patrols and set up our own camp kitchens and other camp gadgets.

The patrol system was in full swing and it taught us independence and how to get on with each other. The beach was at the bottom of a cliff path and we swam but it was a steep shingle bank entry so lifeguards had to be posted. One of the senior Scouts was appointed to this task one afternoon but as I was about to enter the water, I suddenly remembered that he could not swim. Oh, the joys of Scouting.

My parents had wangled a lift down on the Saturday. I was with my patrol in the town of Sidmouth. I was informed of this by a senior Scout so I ran all the way back along the shingle beach, and panted my way up the cliff path to find that they had left ten minutes before.

At the termination of each camp, the latrines had to be tidied up and filled in. Our toilet facilities were slit trenches with stones in the bottom as a urinal and Elsanol Buckets for Number Twos. I actually preferred just a slit trench for poo and pee which you covered up and filled as you used it because big pits had to be dug to take the Elsanols' waste. Once you had helped with this job, you were declared an Honorary Member of the Royal Ancient Order of St Elsanol. An honour indeed. The solemn act of pouring the contents of the Elsans was always conducted as a mock funeral.

We would blow a whistle three times and everyone in the camp would stop and stand to attention. Upon completion of the burial, the whistle would blow again and the 'Filling of the Hole' would begin. By tradition, it was the 'Second' of each patrol who was assigned to help in this task. When the pit was 'Stoppered', the turf would be laid 'proud' on top, and then jumped on by one of the Seconds.

Alas, at Sidmouth, the Senior Scouts had not dug a deep enough pit and the Elsans were very full. The turf appeared to be wobbly and when a Second jumped up and down on it…he sank to above his knees in…let us not go into detail!

The second week's camp was at Hope Cove which was a long journey down by furniture lorry and this was run along the lines described above. The weather was hot and there was a smashing beach so we were able to spend a lot of time swimming and sunbathing though the latter did not qualify for any scout test. It was here that I completed my First Class Hike but a violent storm shredded the small hike tent we were sleeping in and we had to take shelter on the veranda of the Thurlstone Golf Club.

What we did not put in our logbooks is that the steward arrived early next morning to check the storm damage and cooked us a slap up full English breakfast. The hike had a planned ending in Kingsbridge and we caught the scheduled bus back to Hope Cove. Even then, Hope Cove had a one-way system and the small bus rattled its way down to the cove. It was near our stop and the driver told us to get ready. Suddenly, he braked hard and we nearly overbalanced.

A car in front had collided with a van coming up the wrong way. There was only us and an old gentleman on the bus so the driver and my friend and I rushed out to assist. The van driver seemed OK and the bus driver started to check over the car driver while my friend checked the lady in the back seat. The lady in the front seat appeared to be dead! Her head and shoulders were covered in blood and there was grey matter probably coming from her brain dribbling down on to her lap. I gingerly reached for her pulse but she let out a scream and sat bolt upright. I assured her that everything was OK—very sure that it was not.

I had never seen such a terrible injury and I thought it was well beyond my first aid skills. It was then that I noticed a broken china bowl on her lap. It materialised that she had been carrying a bowl of tripe on her lap and the impact had pushed her face right into the bowl. She was cut badly but not at death's door. I had my leg pulled for the rest of the camp with everyone asking if I would like tripe for tea?

I became a film star at an early stage. Michael Stephens, who was well known at the BBC, roped my mother, several members of my patrol (I was a Second then) and me, to make a film about a deaf Scout. Even our house in Church Road was filmed and in one of the later scenes, we appeared on the banks of a stream in Hambrook (the present site of which I have been unable to locate), with the scenario that the 'lost' deaf Scout was on the other side. I was picked to vault the stream with the obvious conclusion that I would not make it and received a ducking. I have tried to locate the film without success.

Aged fourteen, I was promoted to Patrol Leader of the Wolf Patrol. My singing career had come to quite a sudden end before that as my voice broke and could thus no longer sing in the choir. I did not want to become an altar boy and I could now concentrate on my other activities…and school, of course. You will recall that I was a very strong believer in the patrol system whereby you did most of your Scout activities as a unit of six boys of varying ages and I was determined to make it work.

I could now also have a say in the running of the troop as the patrol leaders met once a month in a Court of Honour with the Scouters to generally run the troop. I firstly organised the patrol with the three more senior boys helping the younger ones on a Buddy system so that they could learn their Scouting skills much faster and help with any problems and if they were free, we would meet in one of the Patrol's house for a specific activity at weekends probably twice a month. This had a wonderful side effect because the mothers would compete with the cakes, drinks and goodies on offer.

At a troop camp at Dyrham Park, prior to my promotion, I had wandered down near the house area. We were allowed to roam. The house was out of bounds but one could go right up to the gates. Masons had been working on some of the statues that had been taken down and propped against the railings and I was fascinated by some particularly hideous gargoyles. I started to make faces back. Suddenly, I heard a noise and, almost in front of me, was an old lady dressed in black, with white lace cuffs and neckpiece, laughing at me.

I smiled and she invited me into the house introducing herself as Lady Blaithwaite the previous owner of the house now living there by grace and favour, I believe. She took me into the kitchen where a House Keeper gave me squash and biscuits. I was very enthusiastic talking to her about the park and the deer and I could see that she was pleased with my comments. Thus began a very peculiar friendship. She never seemed to leave the house and told me to visit whenever I was camping. I was a little bit late getting back to the campsite because of our conversation, so apologised.

'But where have you been?' asked Skip.

'Having tea with Lady Blaithwaite,' was my reply. They all laughed thinking I was joking.

I approached Skip at the next troop meeting to ask if I could run a patrol camp at Dyrham and he said that no adults could be available.

'Skip, I want to run a patrol camp. That means no adults!'

This had never been requested before but I became quite the barrack room lawyer quoting Baden Powell's views on the patrol system and the gaining of responsibility and independence. Finally, transport became the only real objection and I put it to the patrol.

'We have a trek cart in the troop. If we resurrect it, we could use that. Agreed?'

'Agreed.'

We worked hard on the wheels and tyres and one of the fathers gave us paint. It looked quite smart in a dark blue livery. One of the mothers had even made a sign,

'WOLVES', to go on the front. I had written to Lady Blaithwaite for permission to camp and it had been graciously granted through the estate manager and so it was that one June evening, we all dashed home from school, ate a hurried tea, changed into our uniforms and met outside the garage of one of my patrol's parents at the bottom of New Cheltenham Road where we had stored the trek cart with all the camping equipment.

We threw our kitbags in and the six of us set off pushing and pulling and puffing, over Syston Common, up through Pucklechurh and on to Doynton and Dyrham. It was hard work but the trek cart did have pneumatic tyres and that really helped. We arrived at the bottom gate of the Dyrham Estate at 9 pm and it was still light. We faced our biggest hurdle. Those of you who know Dyrham House will appreciate that the front of the house is approached from the A46 then down a long steep drive to the house at the very bottom of the hill.

We now faced going up that hill with a fully loaded trek cart. It appeared worse that the North Wall of the Eiger at that stage of the evening. The Estate Manager took pity on us. He fetched a small tractor to which we lashed our trek cart and walked up the hill behind it. By 10.30 pm the tent was up and we were organised. I quickly said prayers. By 10.45 pm we were all asleep. I can recommend pushing a fully loaded trek cart for twelve miles as a cure for insomnia!

That first patrol camp was wonderful and the sheer freedom it gave us was appreciated. We cooked over open fires in those days so we all scavenged for wood first thing in the morning after a quick splash wash under the tap. By 8.30 am we had cooked our fried breakfast and washed up, ready for a formal inspection about which no one complained, for this was an important necessity with all camping and best carried out in a certain way. It ensures that Scouts are clean and their gear and clothing.

The morning had been earmarked for various scout tests. In camp was the best place for the practical skills associated with Scout craft such as fire lighting, use of axe and saw, tree recognition, camp cooking, and airing the tents. We quickly gobbled down sandwiches made with tinned corned beef and a cake that one of the mothers had kindly supplied after which we practised map and compass work for a couple of hours. I had not let on about our surprise tea.

I just explained that we were going out and needed to be dressed smartly so everyone bathed under the tap, put on full uniform, buffed up their shoes and slicked down their hair. Little did they realise that the whole process would have to be repeated tomorrow again. We walked down the long drive and I wish that I could have photographed the faces of the Wolf Patrol as I stepped up to the front door and rang the bell. We were ushered in by the housekeeper, and I told them to take berets off and put them under their epaulettes, before we were taken up the grand staircase to greet Lady Blaithwaite in a splendid salon-type room with rich tapestries and portraits on the wall.

I formally introduced all the members of my patrol and she insisted on using the left handshake used between scouts world-wide and wanted to know about the weekend so far. She had seen us come in late last night and had been concerned for us. She was very practical and had organised a big tea in an adjoining room at a big table. She had immediately realised that the younger members of the patrol could not have coped with balancing cups and plates on their laps. She and the housekeeper were very kind to us all and I think we ate nearly all the sandwiches and cakes. It was a wonderful spread.

We offered our profuse thanks and filed out. I stayed for a couple of minutes just to confirm that we would be attending church next door tomorrow and that we had brought the troop flag for presentation at the altar and yes, according to flag protocol, it had been blessed by Canon Radford himself.

The patrol were gobsmacked. How had I arranged that tea? How was I so friendly with Lady Blaithwaite? I had to explain that I was actually a member of the British aristocracy directly descended from William the Conqueror who had bequeathed land just beyond the Badminton estate on the far side of the estate and I was also indeed related to the Duke of Beaufort and was thinking of dropping in on him tomorrow for Sunday lunch. Would the patrol like to come? They realised that I was winding them up and started to whack me with their berets all in good fun. I told them the real story over a campfire that night having sizzled sausages on sticks washed down with cocoa. It had been a good day.

The next morning, we let the three younger members of the patrol cook breakfast. It was hard not to intervene as they struggled with the fire but practice makes perfect and if you don't make mistakes, you don't make anything and the results were dubious but just about edible. We all helped with the washing up and then I insisted on another full wash and kit inspection in full uniform. This raised suspicions as camp inspections were generally from the knees up to prevent your socks and shoes getting soaking wet.

I then informed them that we were going to church and though there were only six of us, I chose the three most senior scouts for the Colour Party and they had a quick rehearsal. The younger two and I were going to distribute hymn books at the door. Thus we walked down to the church which is situated very close to the house. The Blaithwaites have a private passage which leads down from the house. The church at Dyrham is a delightful little one built of stone in the Decorated and Perpendicular Style and its unusual feature is an embattled tower with a splendid clock and six bells that are actually rung.

When the full troop was in camp, it was overflowing at church parades and for the church to be full was very rare I think. There were about a dozen parishioners and they were very pleased to see us. The Colour Party was lined up at the door and the young patrol members were charming the members of the congregation as they handed out the hymn books. One gentleman asked where our leaders were.

'I am he,' I simply replied.

'No adults at all?' he asked.

I shook my head. He proceeded down the aisle. 'Wonderful. Wonderful.'

Lady Blaithwaite arrived last with perfect timing and insisted that I and the two young Scouts sit in her pew. The processional hymn started, the colours presented and the colour party returned to the other front pew and the morning service began. It was the normal Matins Service that I had said and sung for three years. I had briefed the patrol on the basics of the service and helped the two young Scouts find the correct pages in the hymn and prayer books to which I did not refer at all. The Scouts sang up and out. The organist was very pleased. All the while, however, Lady Blaithwaite kept on glancing towards me.

The service was delightful and my patrol members were stars. The vicar rose to deliver his sermon but announced that the one he had written last night could wait and he proceeded to tell the congregation about the Wolf Patrol. There were the usual jokes about Dyrham now having wolves patrolling through the park and concerns about the deer etc., and he asked the audience to look around to spot the leaders.

Of course, everyone turned around. No leaders. He went on to describe what splendid chaps we were and what a magnificent example we were to the youth of the day and even described the manner of our arrival. He had the congregation applauding us in the end. He asked Lady Blaithwaite if she had anything to add and, to my amazement, she turned around to the congregation and asked.

'I know many of you attend this church regularly and you and I all know the service very well indeed. But I ask you. Is there anyone here who did not refer to his or her prayer or hymn book for the whole of the service?'

No one responded.

'Well, I have to tell you that I have been sitting next to a remarkable young man here who did just that. He knew every hymn, prayer, psalm and response by heart and never once referred to a book. Geoffrey. How?'

I had to come clean.

'Lady Blaithwaite, I was in the Holy Trinity Church Choir in Kingswood as a chorister for three years and attended over one hundred and fifty Matins Services!'

The congregation laughed and applauded again. When filing out of the church, we were profusely thanked by the vicar for enlivening the service and I was told that Lady Blaithwaite would like a word. She wanted to know what time we were leaving and wanted our promise to return. I stated that we would probably be at the bottom gates by 2 pm that afternoon and that we had every intention of making a return.

We packed up camp, making sure that we left the site spotless and clean, and then sat on our equipment eating a late sandwich lunch when the main gates opened and a large pickup drove slowly over the grass towards us. It was one of the estate staff who was under very strict instructions from Lady Blaithwaite to load our trek cart, equipment and us on to the pickup and drive us carefully back to Kingswood. What luck! We loaded in about two minutes flat and then all jumped up on the back. We sang all the way home. We returned several times to the park and Lady Blaithwaite always entertained us with wonderful hospitality.

There were two outcomes from that first camp. At the next troop meeting, after Flag Break, Skip informed the troop that he had received a letter and he had to read it to everyone. It was from Lady Blaithwaite extolling our behaviour, manners, courtesy, politeness, deportment and our love of fun and enjoyment. Secondly, 'I had a cunning plan.' My patrol's Second was due to go off to boarding school so I had to accept another Tenderfoot into the patrol.

New patrol members were normally allocated to patrols by the Patrol Leaders Court of Honour before they came up from the Cubs. One young Cub had been quite difficult and the other four patrol leaders I knew were not keen to have him in their patrols. He was full of energy—a bit like a jumping bean as you might say but he was not deliberately naughty. If the Court of Honour could not decide then we would have to draw straws.

I took my new Second to one side and discussed my plan. He agreed. There was a stony silence in the next Court of Honour when this Cub's name was mentioned. Skip was just reaching for the straws when I quietly agreed to take him into the Wolf patrol. Stunned silence. Skip narrowed his eyes and looked at me keenly.

'Why?' he asked. I returned his stare.

'Skip. A Scout is a friend to all and a brother to every other Scout no matter to what class or creed the other may belong.' (Scout Law)

We took the new recruit into the patrol and he turned out to be an excellent member but why was I so keen to have him? His father ran a building company and had a fleet of vans, lorries and pickups. We would never need to push the trek cart again. His father became such a fan of the Wolf Patrol that we called him Old Wolf, he enjoyed scouting more than his son.

I took my 'O' Levels at the end of my fifth year and passed in six subjects. As a treat, my parents took me and my friend Richard out on a Campbell Steamer to Cardiff. In those days one had to walk through Tiger Bay to get into the city and I mean the real Tiger Bay, not the gentrified rebuilt version of today's model. The area was very rundown with dilapidated warehouses and slum houses. It was not the best approach to Cardiff!

We visited the castle and then took the organised bus trip out to St Fagan's the Welsh Folk Museum which displays old traditional houses which have been carefully rebuilt on site. It offers a fascinating insight into the history of Wales. Amazingly, I 'hang' in the castle at Cardiff. No this is not a mistake. I was not 'hung' in the castle. I hang there as part of a mural. I modelled for an artist friend who was painting a commission for the castle and wanted my face. I am recognisable.

Perhaps this is a rather odd interjection but it was at this point, in writing this account, I suddenly realised that I had reached the age of sixteen and I suppose it was normal for those days, but no one had told me, or instructed me, or advised me—about sex. I was told nothing by my parents or school, not even in biology lessons and yet I had gleaned the basic facts while I was passing through my teenage years. As far as my parents and family were concerned, it was a taboo subject and sex and nudity were considered rude and improper, let alone masturbation, consensual sex at a young age or activities involving striptease or other erotica.

Even boys showering together was considered improper by my mother though I do not know if she was aware that thirty of us did so at school after games and PE lessons. I did have several girlfriends but I only ever reached the stage where we kissed, had a quick fumble with their bra and nothing more. The only erotic material I ever saw was old copies of 'Titbits' left around by my uncles or workmen and the world of 'porn' was unknown to me.

I suppose I did tell dirty jokes as a young teenager when with my friends but I was really too busy with all my activities to worry about it all. The question of problems surrounding nudity have never bothered me and I have no hang ups about it at all as long as it is normal and not flaunted in any way. Most autobiographies contain lurid details about sexual encounters and sordid confessions. I am sorry to disappoint.

Now I need to finish this chapter on a high note, and all the way through it I have been evading mention of that, which really, perhaps, I should have had the courage to mention right at the outset. You have read about my schooling, my scouting, my parents, my brother, my relations, my holidays, my activities and even my illnesses, but, no mention has been made of…my dancing!

It is one of those anomalies of life, and believe you me, I have racked my brains in attempting to do so, but I cannot fathom how and why I started dancing lessons! Suddenly, I was there at Miss Pierce's Dancing Academy, off Fishpond's Road, with my cousin Carol, immediately after school, once a week, every Tuesday, and I attended for over two years. I was taught Victorian and sequence dancing including the Veleta, the Old fashioned waltz, The St Bernard Waltz, The military two step, and the Gay gordons, and then the Quickstep, Waltz and Foxtrot in the modern ballroom section, followed by the Tango, Rumba and Cha-cha-cha in the Latin American category.

I was awarded the Junior Bronze, Silver and Gold Medals of the Imperial Society of Teachers of Dancing Incorporated for all three categories and I retain the certificates (personally signed by Victor Silvester himself), and medals to this day. I even did a demonstration Pasa Doble dance in one of Miss Pierce's dance shows at the Victoria Rooms and appeared as a pageboy in the pantomime Cinderella on the stage of the Hippodrome in Bristol. It was not easy.

I was the only boy for a start and I laboured under a distinct disadvantage. I would normally dance with Miss Pierce who had a rather fulsome figure with an even fuller bosom which was at my eye height. When we danced, I could not see a thing to left and right or up and down as my head was locked between her breasts. Now please do not get the wrong idea but I preferred to dance with her husband, who was known as Mr Punter. He was a very tall man and looked not unlike a boxer as he had a broken nose. He would often call across the studio.

'Geoffrey—you cum and dansh wiv me!'

I did not mind truly as he would take the lady's part perfectly for he could move like a feather drifting across the floor. He was very light on his feet and a superb dancer but it must have been quite comical to see little me and big him doing the quickstep.

There are more confessions to come however. Wait for it. I also did ballet lessons. This was not in any vain attempt to join the Corps of The Royal School of Ballet but merely to improve my poise and posture for ballroom dancing competitions. I did wear tights (real men wear tights) and my PE vest and borrowed a pair of ballet shoes to join the girls in their lesson. Lucky me. So you see, all in all, I was far, far ahead of the pack. 'Strictly Dancing' is playing catch up as far as I am concerned but I still do not know why I started ballroom dancing.

Chapter Six
The Grand Hotel

Were things better in the past or is it just a trick of our imaginations, our memories kidnapped by a benign afterglow of delusion and rose tinted glasses? From my perspective, that is to say, studying the present from the past, I have no doubt or hesitation in stating that I prefer the past to the present. Unequivocally I can state that I would not swop my childhood for a modern one. I desire no part exchange into the life of this social media-led childhood. The present headlong rush into the ever expanding world of technology is committing future generations into a barren land.

Technology is good for mankind but only as a servant—not as a master and there-in lies the rub. I have occasion to drive into London quite regularly now and a sure sign of this mindless technology is displayed on my route over the Chiswick and Hammersmith flyovers, where, from the elevated position, you can see hundreds of workers at their desks staring at their screens, and there must be millions doing the same all over the country. It is mind bogglingly mindless! Is this the future for our children? We give them fourteen years of education so that they can sit behind a screen for their working lives?

Our children are now living in a mechanised and electronic world where everything is instant and they do not even have to think anymore. Evidently, already it is recognised that our youth is under great pressure and stressed out by attempting to cope with the digital morass that has developed. This cyber bullying, body shaming and 'fear of missing out' culture is of our making and if our children are in trouble with the minefield of social media and interacting online then perhaps it is our fault? We bred and fed the beast and now it has bolted out of our control, the problem being that very few of us oldies are technically competent to apply controls. We have no reins.

Things were better in the past. There was not so much traffic on the road. Everything was built to last. We had the freedom to go out to play at an early age, and unaccompanied. People had much more respect for each other and had good manners and patience. Not EVERYTHING was broadcast on the Internet. No throw away culture and no goods with built in obsolescence. People took time to talk to each other and families ate around a table. There were Matrons controlling hospital wards and Bobbies on the Beat. Free college and university education. Fresh produce with no chemical additives.

People spent more time outdoors. Community spirit. Classic sitcoms and light entertainment and songs with recognizable tunes and lyrics you could understand. Good neighbours. Clothes were better made. The morning milkman and milk bottles. Bus conductors. Affordable rail travel. Good education.

I was more than fortunate. I had a free education until I was twenty-one. The NHS was started just after I was born and I missed doing National Service. (Well, I had been in the Scouts) During a short period of unemployment, I had been very well looked after and I am now able to draw a full state pension. I am part of the 'lucky' generation that has done well but I do worry about our legacy. The young face a very difficult road ahead.

I should imagine that it is quite daunting growing up in the world that we oldies have created. I am now much involved with older students so know something of their thoughts and feelings. The 'young' were visibly upset at the nation's decision to Brexit as it was the older voters who swung the vote but future generations wanted to stay in the EEC. Unfortunately, however, this young generation has grown up in a 'Nanny' state and expect everything from the government of the day.

The normal throes of adolescence passed me by. I did put on growth spurts but I did not have acne and I think I was reasonably pleasant and amiable with no mood swings, rebellious behaviour, bouts of anger and indignation or grunting! I did argue with my father but that was normally about politics and current affairs. We also argued about my right to argue! I had relationships with several girls but they never advanced to promiscuity or sex.

I actually gave up my first girlfriend because I could not afford her. Why? I was saving up for a bike! I told her formally and I think she was a bit upset at my preference. It was a very nice bike, however, a red and white Raleigh, and I paid it off at half a crown a week. We did have a social group of girl and boy friends who met up for dances, events and the occasional party and had plenty of fun. I maintained the same friendships with my school friends some of whom I met socially as well.

Unfortunately the throes of adolescence did not pass by my brother. They hit him full on and he became quite unbearable and unpleasant to everyone and they developed into quite serious problems for the family. He had black moods and was better avoided at such times. I was leading my own life and tried to keep out of his.

Grand Hotel Times – teenage years.

Our paths crossed at times but caused me no real problems until Skip asked him to run the summer camp because our scouter was ill, I think. This was at the time that I was patrol leader of the Wolf Patrol. I had persuaded all my patrol to attend the camp and we were looking forward to it though I did have some misgivings about my brother as leader.

A short period before the camp, he insisted that he was breaking up the patrols and making four new ones for the camp. I objected, politely, to Roger and Skip, saying that we were a good patrol and wanted to stay as one unit adding that they were destroying the patrol spirit within the troop. My brother was angry and insisted that the patrols be shuffled. I was insistent that we remained as a unit. I understand that my brother threatened to walk out and not run the camp which put Skip in a difficult position. He had to back Roger. The Wolf Patrol held a meeting and I explained the problem.

Every member was adamant that they wished to stay as a patrol so I simply explained this to my brother and Skip. Roger was unmoving in his stance. The nest troop meeting, having spoken to all the parents of my patrol members, I informed Skip that none of us would be attending the camp. He tried to argue the decision. We held firm but I know Skip received several angry letters of complaint from parents. Of course, my brother had been giving me hell at home all week and now he flew into one of his rages. I would not join in any discussion and that was that.

As Roger's involvement with the troop increased, so mine decreased, I am afraid, and to save any more disputes, arguments and rows, I resigned from the troop. I deeply regretted having to do so but it did lead me on to pastures new. I was now preparing for my 'O' levels and that obviously took up a little more of my time. Spare time was still available so I joined the YMCA whose building was only about ten minutes from our house not too far from Regent Street in Kingswood.

It was a fairly large brick built, two-storey building with a large lounge and coffee bar, a secretary's office and a large hall with a stage and changing rooms and toilets with showers spread out over the ground floor and two very large rooms on the upper floor for table tennis and snooker and billiards. I just enjoyed the facilities. It was pleasant to relax with friends, and though it might seem odd, the 'In Thing' for our age group, was coffee bars. Young Men's Christian Association it might have been but I can never remember anyone mentioning religion to us.

As time progressed, I did become more involved and I was asked if I could run a Boys' Club on a Saturday morning which I had free anyway having lost my job as a car washer, for the requirement was now for someone full time and that counted me out. The club proved very successful. The programme was quite limited but there was an outdoor court for games and football and the hall was used in winter for badminton and games. These were the main attraction as I had brought all the games we played at Scouts and they were wildly popular. The secretary was pleased as we even finished with a prayer as a nod to Christianity.

The YMCA was famous for its Night Hikes. You would be taken out by car, blindfolded, and then dropped in a lonely spot about fifteen miles out. The aim? To get back to the YMCA. They were fun and it was a bit nerve-racking walking through the long, lonely hours of the night. It always concluded with a slap up full English Breakfast. The hikes were stopped, however, as one group, trying to get back first, took a shortcut… through the Box Railway Tunnel on the London Main Line!

I also helped start a gardening scheme for residents who were unable to cope with their gardens. It lasted some time but many residents were taking advantage of us and demands were coming in for landscaping and allotments. We let it die a quiet death. During my time at the YMCA, there were two secretaries and I got on very well with the first but my relations with his successor were not quite the same. I liked the idea of running a YMCA as a career and experienced a lot of fun at two youth conferences at their holiday centres in Rhyl and Barry Island which prompted me to ask about such a career.

I was despatched to the Central YMCA in London on the Tottenham Court Road for an interview. I was interested and they were interested and asked me to reapply when I was 21, their age of entry. That appeared to me as being a remote spot in the distance. Six years was a long time and my interests led me elsewhere. I did write to them, however, to inform them of my teaching career and they responded with a charming reply wishing me luck.

One auspicious day, I was informed that my brother had been appointed to run the girls' Duke of Edinburgh's Award Scheme at the YMCA. The inevitable had happened and there was a big row in the Scout troop and he had walked out. Was he following me? Wait a minute! Am I not writing about the Young MEN'S Christian Association? During my time, they had admitted girls as well. An uneasy peace was established but he was ever critical about all my efforts with the Boys' Club.

Under the second serving secretary, behaviour and discipline were more relaxed and overall standards dropped probably more than was wise. It made running the Boys' Club difficult as I had no real back up from him in any way or any other adult. There was a lot of high jinks and practical jokes now and one Saturday morning someone let off a stink bomb on the premises. For some obscure reason, the secretary accused me and got the older members to strip me naked in the changing rooms and put me under a cold shower as a punishment for an offence which I had not committed. I was very upset about the whole incident. It was not only the embarrassment but the fact that he thought I had been childish and silly enough to have done it. I told him so but he just laughed it off thinking it was just good fun.

I resigned from the Boys' Club, there and then, and never visited the YMCA again. I think the secretary had not expected that. I was contacted, however, by Gordon Farnsworth, the Chairman of the YMCA, who asked to see me. He was a good chairman and also editor of the Bristol Evening Post and had encouraged my interest in the news when on his regular visits to the building, local and national, and I actually met him in his office, which, of course, made me think, perhaps, of being a journalist. He was very kind and understanding as we discussed what had happened and I think he was hoping that I would change my mind, but I strongly felt that I could not work under the Present Secretary and I also explained about my brother.

In the meantime, I had passed my 'O' Levels in English language, English literature, geography, history and commerce thus gaining entry to the sixth form at Rodway to do 'A' Levels in English literature, geography and economics. This was a good time despite the hard academic work. I loved English literature and geography and found these two subjects easy and enjoyable. Economics was more difficult not helped by the fact that the teacher was in poor health with many absences so we had to be content with the RSA Economics qualification instead.

Our 'A' Level teachers were brilliant and all four were Goodies. I have already mentioned Ken Peace, the geographer, with whom we learned well and thoroughly. There were only five of us in his class so lessons were informal and pleasant and Ken was always available to help you. He moved on to Makarere College in Uganda after I left and on one safari I had made arrangements to meet up but there was civil unrest and we decided not to enter Uganda. Mr Bridges and another member of staff whose name, I regret, I have forgotten, taught English. If it had been the dreaded Duggie Board, I would have left school at sixteen!

The latter member of staff looked like a railway ticket clerk. Quite short, with a bald head, pince-nez glasses and a slight stoop. I thought he looked rather like Mr Polly, a character who appeared in one of our course books, 'The History of Mr Polly.' He also taught the Romantic Poets with which we struggled slightly but he got us through if safely. He owned two Dachshunds called Romeo and Juliet and we used to pull his leg about calling them in at night. 'Romeo! Romeo! Wherefore art thou?'

Mr Bridges taught the more modern literature such as the War Poets and Gerald Manley Hopkins and T S Eliot and was amazed to discover my true appreciation of them all. He was also gobsmacked by my ability to learn quotations, all in my head still to this day. I totally, utterly and wholeheartedly latched on to the language and verse of Gerald Manley Hopkins and he remains a favourite. My classmates did not as they found his poetic construction and language too hard to comprehend and there would be me expounding the fact that it would be better to close their eyes and allow the words to flow over them. They obviously thought I was nuts.

We studied Macbeth in cold blood in the cold light of day in the classroom so I asked if we could go down to the hall and act it through. It came instantly alive and we did this whenever possible. I estimate that I know at least a third of the play by heart.

I was writing some poetry at this time. Chris Levinson, one of the English staff, encouraged me in this activity and we would have regular discussions about my work. I think he was a minor poet and I much enjoyed his interest on my behalf. I was persuaded some years ago to enter several of my poems in a competition. I was surprised that they only chose the ones that I had written during that time and not those from my adult life.

I appeared in two school plays. La Malade Imaginaire—The Imaginary Invalid and Hobson's Choice. I appeared as the Apothecary in the former and was everyone impressed? I am not sure but two of the school cleaners reckoned I had a lovely pair of calves with silk stockings! Hardly a rave review! I was appointed as a Senior Prefect in my final year. I understand that I had been put forward for Head Boy but I had blotted my copybook the previous term. With the Headmaster's permission we had formed a CND Group as long as we did not bring the school into disrepute.

As teenagers, we had all the answers to correcting the wrongs of the world, nuclear weapons included. All went well until one of our members wrote a letter to the Evening Post and signed off with the school name. I asked to see Mr Hughes first thing next morning and I told him the circumstances apologising for the blunder and stating that I accepted full responsibility. He dismissed me with a curt nod and we never heard another thing about it. He had a dry sense of humour as well. He taught us technical drawing in a classroom which to reach, you had to walk through the library. I often took him off in reasonably good impersonations, especially as he used very precise and strictly grammatical constructions. He would refer to two angles for instance and say, 'Let us propose that these are they.'

I was in full launch at the front of the class one morning but our lookout was not looking out properly. I was just in the middle of, 'And let us propose that these are they,' and pointed with a stick to the whiteboard when 'Rex' walked in. He gave a wry smile.

'I suppose, Allen, that if I were to fall ill for a day or two, you could fill in for me. Would that be acceptable? Please sit down. Oh! Were you aware that you have the two angles the wrong way round? You will never make a draughtsman but I think you could do well as an impersonator on the stage.'

About this time, as well, we received career advice from a career's officer, whom we had never met. In a five-minute interview, he thought he knew the correct career path for every senior pupil. Without even greeting me, he launched into his advice.

'Banking is the path for you, old chap. Yes, banking. You would like that wouldn't you? Eh?'

'No. I would not,' I responded. 'I have thought of going in for teaching.'

'Oh! No. No. No. That would not suit you at all. You will never make a teacher. Banking. Yes. Banking. You have to have the right temperament for teaching, you know?'

I almost asked if he'd said temperament or temper, bearing in mind the gang of teacher thugs that had taught me earlier in my time at Rodway but merely replied, 'I would be bored to tears in a bank.'

'Oh no. The work can be very interesting.'

'Not according to my friend's mother, sir. She thinks it is very boring.'

'And what would she know about banking?'

'She is the Assistant Manager of the National Westminster Bank in Fishponds, sir!' The interview came to an abrupt end.

If only. Our world is full of 'If onlys', but if only I could meet that pompous career's advisor and hopefully give him a copy of this book to see how much of my life had been dedicated to teaching and education and what I have achieved in my life. Banking indeed!

In the summer holidays, between my lower and upper sixth years, a friend and I visited Torquay for a week with the blessing of our parents. My friend, John Small, was very keen and even persuaded his father to drive us down to the bed and breakfast accommodation which had been recommended by my parents who had stayed there the year before. I appreciated, of course, that before we left, my mother had already phoned Mrs Hearder to ask her to 'keep an eye' on us.

Upon arrival, she insisted that she take John and I around Torquay to show us the town in her little red Mini. John sat in the back and I took the front seat. She was a horrendous driver, far worse that my paternal step grandmother in Barnstaple. She drove like a maniac, taking on buses and lorries and berating every other driver on the road. She thought that all the other road users were bullying her on the road because she was a woman!

We came face to face with a double decker bus and she made the driver reverse! I think her driving instructor must have been a failed Kamikaze pilot. We returned, somewhat shaken, to our accommodation and politely turned down all further offers of lifts, preferring to use bus and train instead. It was a good week with fine weather so we explored Torquay and surrounds and devoted much time to the beach. We both returned home with sun tans.

If you recall, in my Preface, I did admit to having problems with my timelines and this is one such occasion, for a person came into my life, upon which he exerted much influence, and I am not sure of the how, why and when. His name was Jim Pickup and he also lived in Kingswood, with his parents, though he was some ten years older than me He was a member, of course, of the Pickup family based somewhere in the North Country, and one of his uncles was Harry Pickup. Take away the letters r and y and then k, u and p. What do you end up with? HARPIC. The toilet cleaner. 'Clean around the bend.'

This simple toilet cleaner got us out of a spot of bother with customs when Jim and I were returning from a French trip sometime later. We came in through Southampton, which, at that time, had a huge custom's shed, almost as big as the Brabazon's Hanger, and there were four lines of cars that had disembarked from the ferry. Jim's little Mini looked totally overwhelmed by the mass of cars around us.

Jim was somewhat of a connoisseur of wine and he had used my allowance, his and much more to bring in a big selection of wines. We were way over the allowance. Something was up. Customs were checking every car thoroughly and we crawled at a snail's pace towards the check point. Jim began to get worried. The custom's officer looked officious and stern as we drew up alongside him.

'Passport!'

Jim handed his passport to the officer who scrutinised it carefully then looked back through it again. 'Pickup? Pickup? Eh. You must be young Jimmie—nephew of Harry, right?'

Jim confirmed that he was he, if you see what I mean, and the custom's officer, obviously a distant family relative, waved us on with a broad beam on his face and we drove out of the shed with alacrity. Jim had been a pupil at Bristol Grammar School and had gone on to University at Southampton to take a degree in French, and then, returned to teach French…at Bristol Grammar School. Which was where I caught up with him sometime, somehow.

My brain-wracking elicits the following clues. Jim had a connection with Miss Tidder who was the Chair of Governors of Rodway School and she was also a county councillor. She also ran a home for disabled children on Frenchay Common near the hospital. She and Jim were trying to start a swimming club for the disabled and had possibly asked Rodway for help? Or. There was a Scouting connection. Jim had become involved with scouting and helped with a troop while at Southampton University and he was, when I first knew him, the assistant district commissioner for Scouts for South Gloucestershire. This was strange! Why?

From the time I first met him, he kept this fact from his parents and I never understood why. He was a secret ADC! The plot thickens because it was through Jim, who was obviously aware of my scouting connections, that I helped run the Pucklechurch Cub Pack and Scout Troop for two years while I was in the Rodway sixth form. In truth, I had missed scouting enormously and was happy to do so.

In the background to all this, I was offered a Saturday night job as a glass runner in the ballroom of the Grand Hotel in Broad Street. While at Rodway, a friend, Ian Cooling, and I walked to school and back together. Ian lived at the back of Kingswood Park and he would call for me at about 7.45 am every morning. We had a set route. Down Church Road and through paths and back alleys to Station Road, all to avoid climbing Lees Hill, then up on to Rodway Common.

We became very friendly and our mothers seemed to know each other though I do not think Ian was in the Scouts. Ian's mother worked in the Grand Hotel as the resident florist and had obtained the glass runner job for Ian who worked with another lad until he did not turn up for a couple of weeks. Mrs Cooling and Ian kindly recommended me for the job. There was no interview, I merely accompanied Ian dressed in black trousers and a white shirt and the hotel gave us bow ties and a short white coat. In essence, the job was straightforward. We only worked in the ballroom.

At that time, Saturday nights at the Grand were very popular and there could be over five hundred dancers in at any one time. The ballroom was big and spacious with tables around the perimeter. On the far side there was an enormously long bar at least one hundred feet long that was manned by well over a dozen bar staff serving mainly Pimms, which then had up to twelve different types, One to Twelve.

As the evening progressed, Ian and I would patrol the tables collecting the dirty glasses and then take them through to the Still Room where an operator put them through a gigantic glass washer. We would collect the clean glasses and scoot them over to the bar. Easy? Not when the ballroom was packed and busy.

It was difficult to manoeuvre around the tables then and the dishwasher could not always keep up and we had to fight our way back around the dancers to deliver the clean glasses to the impatient bar staff who were forever running out of the same. It could become very hectic, almost frantic, and we were always glad when we could knock off at some time after midnight to be taxied home.

The above provides quite a jumble of events. Certainly, when I was only just 16, I was studying for my 'A' Levels at Rodway during the week, running the Cubs and Scouts at Pucklechurch on a Friday evening, and then the YMCA Boys' Club on a Saturday morning, reporting for work at the Grand Hotel that night, then helping with the disabled swimming club on a Sunday evening. Often as well I would be needed for Sunday Scout activities at Pucklechurch including church parades. I shed the Boys' Club fairly early for reasons already explained but all the other activities were maintained for two years.

Ian and I did take one break on a weekend. We decided to hitch-hike up to Earl's Court in London to see the Motor Show so got out to the A4 and started thumbing. Lifts were slow and it started snowing and we took shelter in The Golden Arrow Café at Silbury. Who was stuck with us? The Temperance Seven. They took out their instruments and played for about an hour, then, as the snow was easing, one driver said he was going and could offer us a lift right into London.

We accepted not realising that he was driving a coal lorry but we were dropped off in Knightsbridge and then walked down to the Embankment where we got a couple of hours sleep before visiting the Motor Show. Ian wanted to see an ailing aunt somewhere near Reigate so we hitched down to there. His parents were by her bedside and had come up on a whim so we had a very comfortable lift home.

Constantly, as a background to all this, family life continued with us visiting relatives and relatives visiting us. Nana Foord used to enjoy visiting Church Road and took and active part in picking the black and red currants and the gooseberries and strawberries that were all mostly preserved for Jam. I do not think, for one moment, that she was lonely without my grandfather. She was really enjoying her freedom without him. She could have survived anywhere really as long as she had her supply of Guinness and Babycham (Black Velvet) and Harvey's Bristol Cream Sherry.

My brother's behaviour and attitude had deteriorated to the extent that my parents became quite frightened of him. It all came to a head one day when there was one enormous row and he became so violent that my father phoned the police who promptly arrived and talked him down but he was incredibly rude to them. This was the period when he was a Leader for the Girls' D&E Award and he was actually courting one of the girls involved (try that nowadays). The outcome was that he moved into a small caravan under a lean-to roof at his girlfriend's house.

Much later I learned, through the girl's brother, that Roger had turned the whole thing around and told everyone that he had been badly treated and physically abused, putting all the blame on my parents. This upset my mother deeply. Eventually, the relationship collapsed in another wave of anger and malice.

This had merely conformed to his usual pattern of behaviour as the same breakdown had occurred with his first girlfriend who was distantly related to one of my uncles and it was to be a pattern that was to dog him right into the future, for he is still alive without a friend in the world. I realise now, of course, that he was suffering from a mental illness and have much sympathy for him. His main problem was that he never recognised that fact for he could easily have been treated. He was, and is, ill.

When he left school at 15, he was taken on as an apprentice electrician with the family's contracting business. I understand that he was quite good at his job and performed well at his Day Release Course at Ashley Hill. He was quite highly regarded for a couple of years then started the usual pattern of deteriorating attitude and anger. His downfall was brought about by two things…well, three I suppose, if you count his temper.

Firstly, he was cheating my uncles on time. He would leave a job at three in the afternoon and book until five. Secondly, it is alleged that he stole hundreds of pounds worth of stock and materials. Several years prior to this, my father, for some obscure reason, had decided to build a garage. My immediate reaction was—why? We did not have a car! I think it was to obtain a small income but be that as may, he, who was the most impractical of men, started digging in preparation for the concrete slab on which the garage would be erected.

I know because I helped him and amazingly, he laid the concrete. I was impressed. Now he did have help with the erection of the garage; a prefabricated asbestos building, which, seemingly just bolted together. It was to be let for ten shillings a week. Roger, my brother, offered to rent it, as by that time he had purchased a Wartburg car. My father agreed but I know there were constant ongoing problems with the payment of rent.

A little while later, I realised that his Wartburg was never parked in the garage but alongside our side garden in Collingwood Avenue. When I innocently asked him about this he blew his top and told me to keep my nose out of his business. Returning from a friend's house one afternoon for my tea the garage door had been left open and Roger was nowhere to be seen so I could see what was inside. Without going in I could see that it was full of electrical goods—cable, conduit, light bulbs, switch plugs, lamp holders, light fittings and boxes of screws, clips, nails and starter lamps. An electrical Aladdin's Cave. There was no room for his car!

He could never have afforded the goods in the garage. They had been 'knocked off' by him for use in 'jobs on the side'. This was not old stuff taken from current jobs; it was all brand new. I am not sure if an attitude that existed in those days is still prevalent. Many workmen appeared to consider that it was normal to steal from their bosses. If challenged, the standard reply was to say that it 'fell off the back of a lorry'.

My brother may have thought this but he was stealing from the family firm and my uncles who had been very good to him. In consideration for my mother, Roger was never formally accused of this theft but he was caught out by my Uncle Dennis who visited the site at which Roger was working every day for a week at 3.10 pm exactly. No Roger.

He was challenged on presentation of his time sheets at the end of the week, tried to blag his way out with lies then completely lost his temper. He was 'fired'. He was told to collect his cards and go. My uncles had had enough, he was taken on by Porters and ended up wiring council houses for the rest of his apprenticeship. Boring. Amazingly, he was allowed to have his old room back in the house and that was a brave decision as he was what he was, and his attitude, behaviour and anger did not improve. I had my own life to lead and really let him get on with his but I would take no bullying or verbal abuse from him and that was clear.

I can clearly identify four aspects of my life during this time; namely, my life at school, my scouting, my charity work and my work. I have explained that they were all irrevocably linked and intertwined, weaved one within the other into a coat of many colours. It was a good, happy and stable time in my development and the anchor was Jim, my older friend. There is a realisation now that I was just as happy mingling with adults as I was with friends of my own age. Nowadays of course my relationship would be regarded with some suspicion especially as there was a ten year gap in our ages but nothing ever untoward happened between us despite being in close contact in tents and hotels. He was a very good friend and mentor for me.

In school, I had my prefect duties and 'A' Level Studies and I was diligent in both. A short time ago, I was decluttering and gave all my school memorabilia to High St/ The Park and Rodway schools. I was given a tour of Rodway, which, of course, is not so modern though I must say that they have some tremendous new buildings and facilities. The original 'modern' buildings are now sixty years old.

Parts I recognised as I was interested to see how well the new buildings had been carefully aligned and joined with the old as I was shown around. I was taken into the admin offices and there met a former pupil. She had been in Form One when I was in the Upper Sixth and said that she clearly remembered me. How? I used to let the juniors in during breaks and lunch hours to warm themselves by the radiators remembering how cold my bare legs got in the freezing cold atop Rodway Common.

I attended two field trips in the Sixth Form. There was a one day visit to Burrington Coombe to study features of the limestone scenery and we also visited Cheddar Gorge. On this particular trip I was able to travel there on my father's old NSU which he had sold to me for five pounds. It gave me a new sense of freedom and I used it to go to school on most days. It was great for visiting the coffee bars of Bristol including the two Balis and the Dunelm and I also met friends at Aunt Jemima's Pancake Kitchen in Broadmead.

I remember the day clearly because I was stopped by the police when I was returning home on Fishponds Road for…speeding. I was doing top speed. Thirty-four miles an hour! They just gave me a warning. Little was I to know that some twenty years later I would get to know the underground caves of those areas better than the ground above and I would take successive teachers of the year from my school in Ghana to show them the exact spot where the Reverend Augustus Montague Toplady sheltered from a storm inspiring him to write the famous hymn, Rock of Ages. They were well impressed.

The second field trip was to Fort Dale in Pembrokeshire where we studied Coastal Features and Weathering and Erosion. My notebook for that trip is in the archive at Rodway. Naturally it was all serious work. Not. I can almost pinpoint that time because Gerry and the Pacemakers were Number One with I'm in Pieces, Bits and Pieces. We had eventually to stop singing it as everyone stamped on the floor and Fort Dale's floors were rather old at that time. I cannot declare that we formally gained a world record on that trip but we did get thirteen students into a Ford Anglia Estate car to drive us back from the pub up a winding narrow road to the field centre which was right out on a headland. I thought the trip was great. The geography was OK as well.

In my final year, I travelled to St Luke's College at Exeter which was my choice of teacher training college for an interview. I had not been able to attend the initial scheduled one as I was attending the Gloucestershire Scout Jamboree at Miserden Park as aide-de-camp to the Chief Scout, Sir Charles Maclean, whom I met again at the National Scout Headquarters at Gilwell in Epping Forest on a Chief Scouts Woodbadge Course. He had a phenomenal memory. I was browsing in the Burlington Arcade one lunch hour where I had been working in the adjoining Royal Academy on a film project (name dropper) and there was a slight tap on my shoulder. It was Sir Charles who had recognised me from some forty years ago.

Mr Hughes had kindly written to St Luke's asking for another date if possible and they had agreed, so I proceeded to Exeter, by train, for the interview. It was all slightly shambolic. Fred Smith, head of English, who was to become my house tutor, and was quite forgetful, forgot that I was coming and had to be dug out of the staffroom, started the interview, which was immediately interrupted by a small group of third-year students who were scheduled to have a tutorial at that time, so I was left out on the landing to complete an intelligence test that he had shoved into my hands with the injunction to just do what I could.

I am not very keen on intelligence tests and the last one I had taken was for my 11+ some six years prior. It appears that my guardian angel was hovering overhead, however. A few weeks ago, one of my school friends had applied for officer training to go to Sandhurst and they had sent him practice copies of the type of written tests he would be facing. The intelligence test was fairly daunting and it took the combined efforts of three of us to accurately complete it. The test that Fred Smith had given me was the exact same one. What luck. I completed it in record time so that when he re-emerged from his study, ushering the students out, I was sitting and waiting quietly.

'How far did you get?' he enquired.

'I managed to finish, sir,' I replied.

'Really? That's a first!'

He produced some marking matrix, skimmed through my answers and then looked at me intently. 'I don't suppose anyone helped you with this outside?' he asked cautiously.

'No sir,' I honestly replied.

'It's incredible. You only have one wrong answer. You're at Mensa Standard! Remarkable. Why are you coming here? You should be applying for university?'

By a strange coincidence, these were the exact words uttered by my external examiner in my final interview at the college some three years later.

Interviews were odd at that time because I had also applied to VSO thinking I might have a year abroad. The foreign bug had already begun to bite. My only experience of going abroad to date had been a school trip to the South of France to Avignon and the Camargue which I had much enjoyed and for the ghouls among my readers we also visited the very room where Van Gough cut off his ear. Unusually the German teacher who was leading the trip also gave us the option of attending a bullfight. I was interested but it was a very poor show and quite scruffy so we did not enjoy it. It was nothing like the great arenas and crowds of Seville and wait until I recount the time when a famous bull fighter fell in love with my mother!

I attended the VSO interview in London and as I recall, the very jolly interviewer asked none of the questions one would normally expect. He showed me a postcard from his son at a well-known public school in which he had written about a 'row in the tank'.

'Can't fathom that out. Can you?'

'Yes sir,' I replied.' Your son has practised rowing in a specially designed tank for that purpose.' And then I added slyly, 'I do not think that it refers to him arguing inside a battle tank.'

'Oh droll. Very droll.'

We then proceeded to discuss the poetry of Gerard Manley Hopkins for ten or fifteen minutes and he was impressed by my knowledge of his verse quoting ad infinitum from many of his works. The interview ended and when I was wandering down Piccadilly I was wondering what the whole interview was about. Had any of it anything to do with VSO? The result? Another deferred acceptance. They would take me after my teacher training and that was too far ahead to even consider at present.

Every Friday afternoon, I would ride out of the gates of Rodway at three o'clock and cycle all the way to Pucklechurch. I believe I had asked permission to do this but no-one ever stopped me anyway so I supposed it was alright. It was quite a long ride, great in the spring and summer, but it could be very cold wet and dark during the winter making my way back to Kingswood. For those interested in statistics I calculated that I did that ride over 130 times on my faithful red and white bicycle. I would be given a tea by the scout leader and his wife and then change into my Scout uniform. Cubs and Scouts met in the same 'hut' which still seems to be there.

It was all a little rundown and though the Akela/Scout Leader was a very kind and pleasant man, he did not have full control. He cajoled instead of telling and thus had slight problems with discipline. There was no enthusiasm with the programme and not a lotta fun! The boys (no girls allowed then) all lived in the village and were either local or from the RAF Married quarters at the top of the village.

By a strange twist of fate, the married quarters were for RAF Colerne where my father had been stationed in the war. I think the Scouter welcomed my help and we managed to formalise the Cub and Scout meetings so that they became quite orderly and well disciplined. My experience as a patrol leader stood me in good stead and it must be recalled that I was only sixteen at that time. I took an interest in all members of the group and got to know them individually. We injected a faster pace and more fun into the meetings. The units remained small, only about sixteen in each but Pucklechurch was a small village then anyway.

We held several fundraising events for new equipment as the camping gear had well passed its sell-by date and Jim very kindly drove me over to Colerne in the Christmas Holiday in his father's Austin Westminster which he was allowed to use occasionally (I loved that car!) to meet the base squadron leader. He was expecting Jim to take the lead but he kept in the background and let me do the talking and I explained that the Cub and Scout Units were now revived but needed help and support from the base and the RAF.

I told him that the boys in the village had absolutely no other activities to attend apart from Cubs and Scouts. He responded by informing us that there was a Community Fund on the base and we could apply to that for specific equipment and activities and he would give us all support. I also handed him a list of general items of which we were short such as trestle tables, benches and stacking chairs, ammo. boxes, wooden boxes, rope ends, etc., which he handed to one of his admin staff.

'Pucklechurch Scouts would like to borrow items like these on a long-term loan. Could you please scour the airfield and see what you can find?'

The next day, all the items were delivered to the Scout hut with an invitation for Jim and I to attend Christmas dinner in the officer's mess in a few days' time. We did and in Scout uniform. We were formally introduced with a brief resume of what we were about and formally presented with passes for the base and the officer's mess. I have to admit that I did tease my father about the pass. As an airman, he had never set foot inside the officer's mess and so I took him in as a guest sometime after, where he was warmly welcomed with much respect. They had arranged for the old siren to be sounded while he was there and everyone joked that he should be standing by his gun.

I took camps with the Pucklechurch Scouts and we were allowed to camp at Dyrham but Lady Blaithwaite was then in a nursing home somewhere up near London. I bet she hated it and I missed her. Jim kindly invited the Pucklechurch Scouts to a Summer Camp in Brecon with his former Southampton Scouts and that is when I discovered The Login.

One weekend, Jim invited me to go with him to Brecon with the Southampton Scouters to look for the Summer campsite. One of the Scouters was a wonderful character called Brian Musket and another Richard (Dick) Little. Brian was a road surveyor and was promoted to be the road surveyor for Berkshire and then became the road surveyor for East Cornwall. As a bachelor he lived in a small flat in Hungerford where he kept all his dishes and crockery dirty and only did the washing-up before the meal not after, but when he got married he moved into one of those posh country mansions divided up into flats. I met him years later by the side of the Launceston by-pass!

Dick Little was a scientist and his speciality was rust on wheat. He was also into fungi and was a professional recorder player. I always had considered recorders as 'An ill wind that nobody blows any good'. Dick played beautiful music and it was a pleasure to listen to him. Jim and I, in his Mini, and Brian and Dick, in Brian's Hillman Imp, drove up early one Saturday morning. Jim never packed a bag—he packed the Mini, as it had many door compartments and other nooks and crannies. It was fairly new, so much so that when he wanted to check the battery, he could not find it. It was hidden under the boot boards.

We looked at one or two sites around the one-way lanes to the South of Brecon and then were directed to Cwm Llwch directly leading up to the track for Pen y Fan through an area known as The Login. I liked it at first sight. It was sheltered with a mountain stream, a plentiful wood supply and a wonderful area to explore. You could see the Beacons rising high above us. I must have used that site well over a dozen times since. The owner was a Mr Powell (very appropriate for Scouts we thought) and we agreed fees and dates. He and his father and his whole family became very firm friends.

We popped back into Brecon to check the local grocers Hogarths, that was there for many years, then drove back to the Aust Ferry to cross the Severn (no bridge then). There was the usual queue so we carefully checked with the ferrymen about number of trips and times. It was tight but we would be the last two cars to get on. We waited in the queue and were not far from the ferry when a driver in an MG sports car tried jumping the queue. He was arrogant and stood his ground and all the other motorists in the queue argued with him but he would not budge. There was nothing for it. As Scouts, we had not yet done our good deed for the day so we let his tyres down amid loud cheers from the other motorists and that was that.

The camp was a big success and the Southampton Scouts were very kind to the Pucklechurch boys. It was all new to my chaps and they struggled with the hiking and more adventurous activities but were determined to have a go. Programmes in those days were very much Scout orientated with the completion of tests and badges and First and Second Class hikes. Adventure activities had not really hit the Scouting programme but I adopted them with a big bang a little later.

We had one mysterious incident at the camp. Food was disappearing from the store tent. We obviously thought it was the Scouts but everyone flatly denied it and we believed them. On the third night, we arranged store tent guards in shifts. Brian and I had one of the big patrol cast iron frying pans on the first shift. Around midnight, a shadowy figure crept into the tent. Brian warned him to stay where he was but the figure attempted to run. I tripped him and as he went down Brian gave him one almighty bang over the head with his frying pan. He was out cold for some while and we tied him up and called the other Scouters.

Jim drove down to the nearest call box and phoned the police, and by the time they arrived, the man had come around and admitted that he was out on a survival exercise from the SAS base in Hereford. The police called in the military police who shoved the SAS soldier roughly into the back of their jeep and drove off.

Not twenty minutes later, a senior SAS officer arrived, took our statement, sincerely apologised and asked if it would be OK to come around again in the morning. A major arrived at 8 am in the morning with a seven-and-a-half-ton lorry full of supplies that he was adamant we should accept. It was unnecessary but very generous. And the Scouts ate like lords for the rest of the week because they had included sweets and lemonade in the delivery. They were very happy. We took all the Scouts to the cinema as a treat on the money we had saved. It is the only time that the boy Scouts defeated the SAS but I would not like to say that to an SAS soldier.

Two patrols of the Pucklechurch Troop attended the rally at Miserden Great Park but they were now under new management. I was concerned that when I left for college the troop might decline so Jim and I worked very hard to persuade two parents to take over, and one of their wives even agreed to take the Cubs. It is no easy task to take over in these circumstances when you have no Scout knowledge and they achieved good results in no time so I could bail out with no worries.

The Soundwell Dolphins was a swimming group for the disabled and met at Soundwell Baths every Sunday evening. I think it was the brain child of Miss Tidder and Jim Pickup and they also involved me and several other members of the Rodway Sixth Form. It catered for well over thirty disabled adults and children with a wide variety of disabilities and every one of them had a swimming carer. It was a friendly club and everyone was kind and helpful. I understand that it is still running to this day so we did leave a 55-year legacy.

The Grand Hotel job paid well. I was paid at the rate of six shillings and four pence per hour and I always worked six hours on a Saturday night, sometimes seven. When I saw my parents the next morning and told them, my mother swore me to secrecy. The electricians only received six shillings and two pence. Roger was very annoyed because his hourly rate was much lower which led to more bad feeling. We were under the control of the general manager and/or the banqueting manager, who were both Italians. They were fairly strict and I was quite nervous of them to begin with.

There was an initiation trial, which for us, consisted of taking four pints of beer slops out of the lower area of the Still Room and back through the upper area of the Still Room. I knew something was up—as you returned through the upper door, the other staff would throw water all over you. That was to be my fate but as I approached the upper door, I stood back to allow the general manager to go through and he got the water all over him. Sorry. It was called Nurdelling for some reason.

On the third night I was working, I was crossing the edge of the ballroom returning clean glasses to the bar when someone called my name. I could not think who she was but she reminded me that she was Sandra and she had attended ballroom dancing lessons with me some four years before.

'Come and have a dance,' she said.

I refused saying that I would not be allowed to as I was on duty. Ten minutes later, the general manager called me over.

'Allen. Can you dance?'

I explained that I had taken lessons for a little while.

'Umph! I am told that you have medals, certificates and trophies, true?'

'Yes sir.'

'Well, the next time that young lady asks you to dance—you dance. Understood? It will be very good for business.'

Five minutes later, Sandra was back for a quickstep and as we stepped onto the floor, Victor Silvester (I kid you not) announced our dance and we did a demonstration quickstep to cheers and applause. Every Saturday after that, I had to do two demonstration dances with Sandra or a couple of other partners and we danced to the live bands of Victor Silvester, Jo Loss, Monty Sunshine, Acker Bilk, Edmuno Ross, etc. All the times I danced, I wore my white waiter's jacket. I was a bit worried. Being only three weeks into the job, I felt that I was making Ian do my work. By mutual arrangement, it was agreed that I would go halves on any tips that I received for dancing and some of these were quite large. Ian was well pleased.

Towards the end of the night, Ian and I would be allowed to go down to the kitchens to have a supper but it was normally restaurant leftovers and not very nice. The Chef was Stefano (another Italian), and he was one of the first TV chefs. On screen he was always smiling, assisting the ladies with him and a real charmer. In his kitchen, he was a tyrant. He would swear, curse, rampage and roar and we really should have worn crash helmets to protect ourselves from pots and pans.

Ian was ahead of me one evening as we walked into the kitchen. He saw something coming towards him and ducked just in time. It was a meat cleaver and it stuck in the door jamb just to my right. That was it. We informed management and they arranged for us to have a plate of sandwiches up in the banqueting kitchens. Between midnight and one o'clock, we would be taxied home but Ian warned me that shenanigans could be abroad.

Many of the waitresses that served at banquets were moonlighting from the various hospitals on Purdown and on Saturday night there were normally parties in the offing. We would go up to the nurses' quarters and social areas and enjoy the party atmosphere and that was where I met Napoleon one night. A quiet young man sidled up to me dressed as Napoleon almost in perfect detail and proceeded to tell me about his life as a gunner and living in Paris, what it was like to be an Emperor and the boredom of life on Elba.

He told me everything in total historical accuracy. He could have been a leading world authority on Napoleon. He seemed so sane and sensible. Why was he a patient? I had no idea. I met many other interesting characters all of whom seemed to be quite sensible and totally coherent. The taxis would take us back home at about three in the morning and if my mother made any comment, I said that I had to work overtime. Little did she know that I was extending my educational knowledge and otherwise in the psychiatric units of Bristol during the hours of darkness.

The job grew. Ian and I would often be called in to help with the preparations for banquets during the holidays and we were both trained to Silver Service so that we could lay up if they were short of staff. A younger cousin of mine had also been trained to Silver Service at Carwardine's at Clifton and I have always joked with her that I was trained in an establishment that was far superior. I was then moved on to be a wine waiter at banquets where you supplied the drinks for two tables of twelve people. I was now being paid seven shillings and sixpence an hour and making one hell of a lot of money on tips.

I still looked fairly young and the lady guests in particular would mother me and call me cute, demanding that their partners give me a big tip. I could earn fifty shillings in wages and another fifty shillings in tips. The biggest tippers were the workers of the Bristol Omnibus and Transport Company who held their annual dinner and dance at the Grand every year and it was at that function I heard Russ Conway, Winifred Atwell and Mrs Mills play the piano. I would come away from that banquet with over ten pounds in my pocket.

Alas, in 1963, the company was imposing an employment colour bar on immigrants in the city and the hotel staff stated that they would not serve them at the banquet that they had already booked. I agreed with their stance but I lost out heavily on tips. Occasionally, we were transported up to the Grand Spa in Clifton when they were short-staffed as both hotels were in the same group. It was a much smaller establishment and the staff did not seem so well trained or professional.

During one Easter holiday in my second year at the Grand, I was summoned to the restaurant upstairs to be informed that the wine waiter had been arrested (don't ask me why) and I would have to take over immediately until a replacement could be found. I knew about some of the wines I had served at the banquets but these had been invariably in the cheaper brackets. The restaurant manager gave me a crash course and informed me that Tommy the cocktail barman, from whom I got the wine, would help me all he could. By midday I was ready, or I thought I was. Who were the first customers? A coach party of French tourists. There was some confusion as they came in about seating so I moved in with my schoolboy French.

'Asseyez vous ici sil vous plais. Pour vous monsieur. Madame ici.'

Eventually, they were all seated. I helped them select from the menu as well and by now the restaurant manager, Dick Court, was looking stunned. I moved in on the wine announcing to everyone that I was a Maitre de Vin Nouveau, 'aidez moi sil vous plait.' And they did help. They were wonderful to me and thought that my fractured French was very amusing. It was a wonderful lunch which I enjoyed immensely and the tips? Ah! That is a secret but it was my biggest haul ever.

Dick Court was delighted and delightful and I stayed with him for two weeks, working lunches and evenings until quite late. The Grand Hotel was 'the' Hotel in those days and received many famous guests. The writer William Thackeray ate and supped there as did Samuel Taylor Coleridge and Samuel Plimsoll was also a guest as was Winston Churchill on his inspection of the bomb damage during the Second World War. Dick Court insisted that all the male diners should wear a proper jacket and tie and so adamant was he about this rule that he turned the scruffy Rolling Stones away and refused to serve them.

At the end of the second week, I was called into the general manager's office downstairs who thanked me profusely for my efforts and gave me twenty pounds as a bonus. He asked if I was interested in hotel work as a career but I politely declined. He was pleased to see me back occasionally and always asked about my college work and progress. All through my student life, I knew that I could be employed at the Grand and they helped me enormously. Tommy became a firm friend and was always pleased to see me. And I also maintained contact with Dick Court. They always looked after my interest. One evening I was helping in the cocktail bar just stacking, collecting, washing dirty glasses and keeping the ashtrays clean, when in walked Hughie Green and Michael Miles. Tommy called me over.

'Steer clear of those two and whatever you do, don't help them carry anything to their rooms. Understood?' I understood. They did make two approaches to me but Tommy deftly intervened.

We celebrated my 21st birthday at the Grand with my parents and a group of friends. Tommy was still there and greeted me warmly. We sat down and ordered and were then approached by a hesitant young man who was the wine waiter. He was obviously new to the job so I helped him along in a friendly way. I ordered two bottles of Volnay 1956 to get us started. When he returned, he plonked the bottles on the table, none too gently, opened both and then poured a little in the bottom of my glass. I glanced at the opened bottles and noted that they were Volnay 1957. I tasted it. Looked puzzled. Took another sip then asked, 'Did I not order Volnay 1956? This is 1957!'

Tommy came rushing over and sorted it all out. He let us have the 1957 on the house. I just quietly let him know about the faux pas with the serving, that is to say, no presentation of the unopened bottle first. Tommy mentioned that mistakes were happening all the time now as he no longer issued the wine. This was now the responsibility of the main bar.

Jim and I were very firm friends and both us were very good friends of Mr and Mrs Minns who lived out at Wick in a big house near to the main road that had a proper name but was always called The Minnery. Mr Minns was the District Commissioner for Scouts and Mrs Minns helped out with various charities. Mr Minns ran an engineering company in Kingswood and I was very surprised to see his logo on a van that passed me on the M5 motorway about a month ago. They had two standard poodles—Mandy and Poo. Poo was named for obvious reasons and it was best not to be around him too much.

Jim was a very staunch Tory and Mr Minns was a fierce Labour man and they had many a political argument in the evenings when we visited. Mr Minns had a soft spot for me because I had taken over the Pucklechurch Scout Troop which had been problematical for him in the past. It was one of those houses where you could always just drop in and our visits there were always very pleasant and often very amusing. I taught Mr Minns some campfire songs that he did not know and he kindly acted as one of my references for my college application. Mr Minns would often ask Jim and me to go on summer camp inspections which was a requisite for all out-of-district troops camping in the area. Upon termination of the visit, we would normally adjourn to a local pub, still in Scout uniform.

I think it must have been Jim who shaped my drinking habits. From the age of sixteen, Jim, friends and I would visit pubs to drink. Presumably because I was with adults, no landlord ever challenged my age. Prior to this my drinking had been confined to coffee bars and I had never thought to go out drinking with a group of lads and get the worse for wear like many of the young generation today.

My parents had always allowed me a small glass of wine or a mini sherry from the age of ten onwards but they did not drink much at all, which is surprising, because many family members drank quite heavily, some to excess. My grandfather's shop at 14 Warwick Road was just along from the Warwick Arms and my grandparents and uncles used it as their regular especially as my uncles played darts. When I was first taken in there it had sawdust on the floor and that must have been in the early 50s. The beer was horrible. A pint of 'Bitter' tasted very bitter to me when my grandfather let me have a sip. I would screw up my face and the regulars would laugh. My grandfather also drank at the Three Blackbirds round on Stapleton Road but, strange to relate, I never stepped foot inside the door of that pub.

It may be that all the family's excessive drinking put the brakes on any excessive drinking for me for my imbibing was always under control. When Jim and I went out with friends, we would only have two pints and then retire back to a house for coffee. We also went out for meals and drank mostly wine in good restaurants. The Restaurant de Gourmet was one of our favourites near the BBC on Whiteladies Road and they were famous for their sweets. One was a vertical upright wafer stuck between two scoops of ice-cream with a small candle on top. It was called The Virgin's Prayer!

The Berni Inns were also popular then, especially the Rummer in St Nicholas's

Market and we even visited Thornbury Castle once or twice a year. You will recall that I was earning good money at that time and was really quite solvent. I do not think that I could match today's youth, however. I often help students with their budgeting and after they have noted all the usual expenses there is invariably a short fall; a kind of, sort of, Black Hole. You know what I mean? Many students, while complaining about Student Loans and the high costs they endure will quite happily go out and spend up to £100 on a Saturday night. Strewth!

Having completed my 'A' levels and having received my college acceptance, Jim and I planned a three week trip to the South of France but Mr Wilson was in power and it was the days of the £50 Travel Allowance which was very strict. Our plan was to use the Southampton Channel Crossing, then drive in Jim's Mini right down through France to a small resort called Canet Plage and stay in a cheap hotel for two weeks. Jim was anxious to relax on the beach for most of the time but I was not so sure.

We started with our £100 allowance, then Jim scrounged some US dollars from an uncle and an aunt gave me a wad of Lira as a residue from one of her Italian holidays and this turned out to be much more that we thought. We would be alright. All went well. We had a good trip down, though the hotels en-route were pretty ropey, and our hotel room with veranda, in Canet Plage, had a good view over the main street. We ate our breakfast there every morning and watched the town awaken. I was fascinated by the tradition of French parking.

Going in or out, they would just bump the cars with their cars to move them backwards and forwards. Jim was very worried about the Mini but the hotel owner allowed him to park in his space at the back. We relaxed on the beach, about which I wrote a poem, and swam lazily in the Med. We visited a wine festival somewhere inland but the cheap wine on offer was awful so we left early. We went to Carcassonne to visit the fortified town which was the highlight of the trip for me being a history buff but somehow got mixed up in right-wing political demonstration and had to beat a hasty retreat.

Late one night we went skinny-dipping at one end of the beach which I suppose was the second such experience for me after wandering over the River Severn mud banks. We were nearly arrested one night when, after a meal of bouillabaisse, Jim lost one of his contact lenses in the gutter outside and as we were grovelling around on the pavement, two gendarmes thought we were drunk.

We visited La Ballet, a famous French restaurant for which we had saved up some of our allowance. We were the only customers, the restaurant was very cold and we were disappointed with the hors d'oeuvres and the main course which we could not eat. We thought we had been conned. Jim, as a French teacher, was able to rip into the owner and left him in no doubt about our disappointment. The owner was full of apologies and said we have nothing to pay. He brought out a bottle of red wine and a good cheese board and we talked at some length. Upon enquiring where we were from, he exclaimed loudly, 'Mon Dieu. Graham Bell. Thornbury Castle. He best French chef in the world!'

Well, there you are. Better to stay in Bristol if you want good French food. I was nearly arrested, I think. I had hired a canoe when we were on a small rocky beach south of Canet Plage and proceeded to paddle lazily in and out of the small bays. It was quite delightful and in one little bay there was a few people sunbathing and an ice-cream seller. I beached the canoe and walked up the beach and asked for an ice-cream in French. He was not French, he was Spanish and became quite agitated with me. An English couple suggested I scarper quickly, which I did. When Jim asked me where I had been, I replied, 'All the way to Spain and back.'

Would I have been arrested as an illegal immigrant?

That night Jim was feeling unwell so I walked up the beach to the far end where a party of young ladies were skinny-dipping. I turned to go back but they saw me and invited me to join them. They would not take no for an answer so I stripped off and spent a very pleasant half an hour with them. I had no towel so had to use my shirt to dry myself and walked back to the hotel.

'What have you been up to?' Jim asked, noticing my damp shirt.

'Jim. If I told you that I have just spent a wonderful half hour skinny dipping with a bevy of beautiful naked maidens, would you believe me?'

' No!' he said and turned over and went to sleep.

The next morning, we were sunbathing and relaxing on the beach. We 'people-watched' for a lot of the time and we were fascinated by one English family that always came down to the beach at 11 am exactly and set up their space with military precision under the direction of the father who always wore a long-sleeved shirt and tie and shorts probably from the First World War, and black socks and black shoes. Normal beach wear for the British? We called him Poona. We were disturbed by four bikini-clad girls coming towards us.

'Geffroi! Geffroi! Savat Bien ce matin?'

Jim was flummoxed. I explained that these were some of the ladies with whom I had been swimming last night. They were fascinated that Jim could speak such perfect French and stayed with us until lunchtime. Jim and I agreed to meet them for a drink at about 9 pm…and of course, we took our towels in case anyone spilt some wine. It was a good holiday and I did return to Canet Plage many years later.

Chapter Seven
Virtue Et Industria

Climbing to the top of Cabot Tower atop Brandon Hill on a fine day, the City of Bristol is spread out in front of you like a gigantic three-dimensional map and the panorama it presents contains much of the geography and history of the city in one easy visual presentation to beat any interactive PowerPoint presentation. It is clear to see why the city chose this site when the castle was built within the protective sweep of the River Avon with the advantage of a bridging point alongside.

The castle was the centre of Bristol then, of course, but is not far removed from the present centre which lies at the heart of Bristol, for, indeed, it has long been known for its port and the then smaller ships could come right up into the city and dock near the Hippodrome and even up as far as the former SWEB Building at the far end, not that either of those edifices were there during that period of Bristol's history. Cabot Tower, built to celebrate one of the famous sons of the city, who discovered America in 1497, on the good ship Matthew, the replica of which is now often to be seen 'sailing' up and down the harbour on its interesting tours.

Indeed, I know it well, as it often moors directly opposite my cabin cruiser at the far end of the harbour to allow its passengers to have a meal and perhaps to visit the various pubs in the vicinity. To all intents and purposes, her exterior is an exact replica of Cabot's vessel but she is not all that she seems. Below decks, she is quite modern with all facilities for the crew and she must also conform to modern maritime regulations. She also has an engine, I should imagine that Cabot would be very impressed if he could see her now. She is a valuable asset to the city providing a vivid insight of many aspects of Bristol's history.

Throughout my childhood and teenage years, I maintained a very lively interest in the history of the city and have always recommended visitors and friends that they visit St Mary Redcliffe Church, 'the fairest, goodliest and most famous church in England,' according to Queen Elizabeth the First, for it has stood on its present site for 800 years and from it one can obtain an overall picture of the history of Bristol and national historical events. Be impressed by the beautiful height and space created by the skilled masons of the fifteenth century as one's eyes are automatically drawn upwards by the huge columns to the elegant vaulting to well over a thousand wonderful gilded golden bosses, intricately interwoven with geometric patterns but a more detailed exploration is a must. Entry to the church is normally by way of the North Porch with its richly carved exterior and you find yourself in an unusual hexagonal-shaped foyer.

It was a tradition that sailors would come into this area to pray before a statue of the Virgin Mary when they were due to embark on long voyages. Above the porch is the Muniment Room and this was not only the strong room for parish records and church treasures, all stored in huge chests, but also where Thomas Chatterton wrote his poetry under the assumed name of a medieval priest called Rowley. When the truth emerged Chatterton was discredited and shunned and took his own life at the age of 17 but his verse probably provided the spur to the Romantic Period and his death attracted other poets such as Shelley, Wordsworth Coleridge and the rather portly Ben Johnson, who got stuck in the rather narrow staircase that leads up to the room, to visit the church in search of inspiration. Unusually, the room contains a medieval toilet!

The church provides an insight into one of the darker periods of Bristol's history namely the Slave Trade. Many Bristolians are unaware that a Slave Trade was flourishing in Bristol from Saxon times and was said to be thriving by the time of the Norman Conquest for the city was the centre for collecting, selling and exporting slaves. Young women were the main commodity and in Saxon times a father was allowed to sell his son into bondage for seven years though evidently there was no legal instrument or provision for his release at the end of that period.

Slavery was common practice everywhere. Bristol was a huge hub and the merchants and dealers did very well indeed from the proceeds of the trade. The slaves were exported all over Europe but the main market was Ireland and there were ships shuttling backwards and forwards from Bristol to Ireland with an outward cargo of women, men and children, returning with Irish goods in exchange. There was no interference with this trade as everyone was receiving a cut including good old King William.

Wulfran, Bishop of Worcestershire (Bristol was then in his diocese) and who had St Mary Redcliffe under his authority, was appalled by the barbarity and inhumanity of the trade, and fought a long, hard campaign to have it abolished. It is said that he gradually brought the people of Bristol around to his way of thinking. Lanfranc, Archbishop of Canterbury, pressurised the king to forgo the slave tax so William renounced and forbade it. Even in 1172, however, white slaves were common in Ireland as traders were still buying children from their parents in Bristol and exporting them to Ireland.

SS. Great Britain

Park Street and University Tower

St. Mary Redcliffe.

The lure and the attraction of the slave trade could not be resisted by the merchants of Bristol. New colonies in North America and the West Indies were growing fast and needed labour. There was a perfect chance. In August 1648 Cromwell's troops defeated the king's army in Lancashire and took thousands of prisoners so application was made from Bristol to transport 500 of them to the plantations of Jamaica and Virginia—many owned by Bristolians.

Successive defeats in England and Ireland provided a rich supply of prisoner slaves, some 30,000 in all, and when this dried up, an illegal business in kidnapping commenced. Though strictly against the law, the members of the 'Bench' were probably involved in this shady trade and gave light sentences to offenders. It also became common for judges, 'at a price' from a merchant or trader, to reduce a death or a long prison sentence to transportation for life and before he knew it, the prisoner was shipped off to the colonies.

There appears to be a modern day debate about the conditions under which these slaves were treated and regarded. Some historians have insisted that they were treated like the African slaves were to be, that is to say, inhumanely, as chattel slaves, and that they entered a harsh life as a possession of the slave owner. It is argued, however, that the slaves sent to the colonies prior to that time were treated as indentured servants and were not the property of their buyer, suggesting, somehow that they were better treated. This seems a somewhat abstract argument to me. All slaves ripped from their homes and families must have led a rotten life in terrible conditions.

There can be no doubt that the most prosperous times were about to come for the city of Bristol with its involvement in the transatlantic slave trade which began in the eighteenth century. Bristol was a major player in the Triangular Trade. Bristol ships would leave the port, normally from the area around the M-Shed, and sail to the West Coast of Africa full of trade goods—muskets, brandy, beads and copper—which would be bartered for slaves that were then jam packed in the ship, in appalling conditions, for the Middle Passage, across the Atlantic to the West Indies or East Coast of America, where they would be sold. The ship would then load sugar, molasses, rum and tobacco or any other profitable cargo and recross the Atlantic to Bristol.

It is estimated that six million slaves were transported as part of this terrible trade and two million of them were transported by British ships mainly from Bristol and Liverpool. It was a cruel atrocity and all in the name of profit. It was eventually abolished due to the initial efforts of William Wilberforce and the Abolitionist Movement. St Mary Redcliffe had no active part in the Slave Trade as such though it may be that the slave ship crews came to gain protection from the Virgin Mary before they set sail but many of the individuals who gained from all aspects of the slave trade gave money for the improvement of the Church. The improvements were at the instigation of a large number of individual merchants who grew rich on the inhumane trade.

Entering upon stormy waters, I fully recognise the sheer wickedness and evil of such a trade and what is more, I think that present day Bristol has recognised that its history was neither acceptable or right, whatever the arguments for justification, but, I am wary of any attempts to either alter or eradicate history in any form, and thus have doubts concerning the renaming of Colston Hall or trying to expunge his name from the history of this city.

Few Bristolians appreciate that Colston Hall has a history all of its own for its construction was solely due to Victorian concerns about keeping Bristolians away from the bottle, i.e., alcohol and a political dingdong between the Liberals and the Torys and another pop at the snobs who lived in Clifton, Clifton had the rather splendid Victoria

Rooms opened in 1842 which was built by the Torys. Colston Hall had its origins in the very much working class Bristol Temperance Progressive Society, which, though not large, had union backing, for it was reckoned that the bosses in the city were deliberately keeping the workers stupefied with booze! To alleviate this, they organised events involving arts, music and culture.

One newspaper report on one of their concerts poked fun at the amount of toxic tea they drank. This perhaps an unfair jibe for, even today, it is reckoned that we drink 165 million cups of tea in the UK every day. The PTPS proposed the building of a 3,000 seat public hall with function rooms to be named St John's Hall (no Colston). But the company formed to build the same was underfunded so enter the wealthy Liberals who were only too keen to get one over on the 'Tory' Victoria Rooms but the project lacked a site until Colston School very conveniently moved out to Stapleton and the Hall was eventually built though at a much higher cost than originally budgeted for. No expense was spared, however, as this was a prestige project. Bristol was anxious to be ranked with the new industrial cities of the North and the Midlands.

The Lord Mayor's Chapel drawn by the Author aged eleven.

As you sit sipping your gin and tonic in the bar of the new atrium at Colston Hall, perhaps one should muse on the fact that for many years it had no bar and no drinks, furthermore, if you are a Bristolian, it was the venue for a fortnight of Gospel Temperance Mission Rallies in 1882 and 21,000 Bristolians took the pledge. It would also appear that the name of Colston was only decided upon as a sign of respect for the school but the people of Bristol were not involved with this decision. It had been made for them by the more affluent, influential and rich citizens.

History should be left alone as it provides us with valuable lessons for the present and future, but, my main concern is to ask where will the whitewashing stop? Would there be an attempt to expunge all bad characters and evil from our past? It would lead to enormous complications especially as it would reverberate out into other areas on to the national picture…and even the international scene. Once that snowball starts moving it will start rolling faster and faster, collecting more and more snow until…? We could find ourselves under one mighty avalanche. In a hypothesis, I expound later I identify what I think to be a historical problem with the control and governance of our city from time immemorial for the last one thousand years and the wrongs of this large chunk of history would be impossible to expunge.

Many of the gifts to St Mary Redcliffe came, therefore, from the merchants and traders such as the Canynges, who made a substantial contribution to its rebuilding and has two near identical monuments in the church. How odd is that? Most of the beautiful stain glass, produced locally, came as gifts and is of very high quality. Henry VIII ordered the use of English in churches during the sixteenth century and English bibles had to be brought from Geneva. St Mary Redcliffe has a copy of one in the North Ambulatory.

The pulpit is interesting leading straight into another facet of Bristol's history. The pulpit was carved in the nineteenth century with the figures of the Twelve Apostles each holding a symbol by which they are identified. Rather gruesomely they are mostly the instruments used in their executions. St Bartholomew holds a knife because he was flayed alive and now he is the patron saints of butchers and surgeons. It is beautifully and intricately carved…by a Quaker whose religious beliefs would be of simplicity and plainness according to the dictates of their religion. The Quakers were a big and influential group in most aspects of Bristol's varied past.

Perhaps more of an anomaly is the unusual cope chest used to store the beautifully embroidered copes worn by the clergy for different religious festivals—the contents of which were lovingly stitched by Mrs Fry, another Quaker, and member of the chocolate family 'Fry' If you ever have time visit the Quaker Cemetery just across the road from the church which includes an old hermit's cell. They all died at a ripe old age. Why? The Quakers looked after their elderly with much care and affection and they lived longer.

Due to a gift from good Queen Anne, the congregation was able to lash out on the interior of the building in the eighteenth century hence its somewhat flamboyant baroque style including a great three piece painting by Hogarth, who was paid the princely sum of £525, and wrought iron gates, made by William Edney in 1710. The ironwork is rated as some of the best in England and I find it absolutely stunning but can you find the mouse? The American connection is in the church for all to see.

There is a model of the Matthew which the crew donated to the church before embarking on the replica's voyage to Newfoundland and two whalebones. The first is the original, brought back by Cabot and the second is original brought back from the modern journey. A legend is abroad, however, that Cabot may not have been the first person to land in Newfoundland. It would appear that several Bristol ships had been fishing on the

Newfoundland Banks for years before he set sail. What do you think?

The church's other famous son was Admiral William Penn, father of the Quaker William Penn, a contemporary of Samuel Pepys with whom he often quarrelled and argued in the Admiralty. He had a distinguished career. He had been a Parliamentarian in the Civil War commanding the naval forces of Oliver Cromwell. Upon Cromwell's death he shrewdly offered the navy to Charles II and built up a very close relationship with him. Penn had captured Jamaica from the Spanish and as its governor, he stripped the island bare for which he was not popular. He was able to loan huge sums to the king.

On his death, his son requested (I do not think you could 'demand' from a monarch) the return of the money. The king had an empty purse at that time so offered land in America on condition he named it after his father, hence, the State of Pennsylvania. There is a further link with the tomb of Joanna, daughter of Richard Ameryck—America?

Of special interest is the Harris and Byfield Organ which is allegedly one of the finest they ever made. It has been completely restored but Handel and Samuel Wesley and his son Sebastian certainly played on the original instrument and there is a stain glass window recording this fact.

The 'Naval Volunteer', King Street.

The church contains two more modern items of interest. The Chaotic Pendulum is thought provoking and fascinating. It is part of the church's Journey into Science Project and challenges watchers to think about the universe and then there is the Tramline. Outside, at the back of the church, alongside the railings of Colston Parade (Will we have to change that name?) is the end of a tramline embedded deeply into the earth at an odd angle. It was a present to the church from Hitler during the air raid on Good Friday 1941. There was much destruction all around the church from the bombs and the stain glassed windows were badly damaged. A bomb landed on the tramway on Redcliffe Parade and the force of the blast blew the rail over the church landing in its present position.

Finally, it is worthwhile finding the grave of Tom the Church Cat who died in 1927. He is said to have attended more church services than some of the clergy

which reminds me of a verse that I can only just remember from The Faber Book— Grave Shades from the Graveyard—involving a suggested epitaph by a yet to be deceased bishop: 'Tell my priests that when I am dead o'er me to shed no tears. For I am no deader now, than they have been for years!'

Beneath the church, spreading from the harbour and over towards Bedminster, are extensive caves quarried out for the red sand for the Bristol glass industry and the network is big so one can easily get lost unless one is accompanied by a guide. You can see part of the cave in the Ostrich Pub and legends and myths have evolved around them of smugglers, slaves and hermits. The caves have been likened to a petrified forest because of the great stone tree trunks that are actually pillars left to support the roof.

Some parts may have been formed naturally, and indeed, one legend has King Alfred hiding in one of these to escape the Danes and that is why the wharf on their frontage was once called Alfred's Quay. It is also part of myth and legend that one of the pillars in the church was hollow so that you could enter by that route. One section actually leads to the Hermitage of St John that is in the Friends' Cemetery. The caves were, for the most part excavated for their fine sand to make bottles for the export of the medicinal water from the Hotwell and fifteen glass factories needed the sand which had to be mined, not quarried because of all the buildings above. They were used extensively as waterside storage and probably the remoter caves had contraband in them.

Many of them are now lost due to infilling and building works. William Watts' famous shot tower was nearby and it dumped much ash and waste into the caves so now, only the caves on this side of the river can be visited and even these were used as bomb shelters during the war.

I have explored much of Bristol Underground though I have yet to gain access to Pen Park Hole or the caves of the Avon Gorge (except Giant's Cave underneath the Camera Obscura), and again, the later were used for air raid shelters. Due to 'security alerts', I have also not visited the tunnels under Temple Meads Station. Sally Watson, who has written a book about Bristol's underground world, likens Bristol to a 'Swiss cheese full of secret holes'.

Daniel Defoe thought Bristolians were a money-grabbing lot with 'souls engrossed with lucre'. I have no idea what Bristolians had done to upset the poor author but many of these hardworking businessmen seemed to have a romantic side to their souls for some of them had underground grottos and I have visited two. One is set in the gardens of Goldney House in Clifton which is famous world-wide and the other is hidden away in Warmley.

I had obviously lived only two miles away from it while I resided in Kingswood but had no idea of its existence. I was highly amused to read that hermits were in great demand to live in the grottos of England despite the fact that they could not speak or cut their nails or hair so this was somewhat of a barrier to long-term employment.

Thomas Goldney built an 'underground extravaganza' in the grounds of his Clifton mansion which is a delight to visit though it is only open at certain times of the year. No one can truly establish the reasons for it being built and there was some mysterious connection with Daniel Defoe but the real enigma to me, again, was that Goldney was a Quaker.

The Warmley Grotto is shrouded in mystery and it beggars belief that I did not know of its existence, but, further enquiries, have revealed that it had been badly neglected for years and it is only in recent decades that local historians have revealed its secrets, hence my lack of knowledge. It is a huge grotto, built by William Champion, inside a large mound, containing tunnels and chambers some 20ft high and there is a complicated system of waterfalls, channels and pools. The chamber walls are rough in texture coated with waste material from Champion's industrial works nearby and there is little or no decoration which is another mystery. It has been suggested that Champion, who was a very capable engineer, designed it with lighting and water effects and even pyrotechnics.

The focal point of the grounds in which the grotto stands was a thirteen acre lake. It is now dry and a caravan site occupies it but there is a huge 24' statue of Neptune still standing and which was the focal point for the lake. He looks a bit rough and battered yet he exudes power. All the locals think he is the biggest garden gnome in Europe. Thomas Champion was an inventive industrialist and invented integrated brass manufacture, which, initially was very successful. His over active mind absolutely seethed with all sorts of plans and schemes and he over reached himself. His business failed. One grand idea he proposed in 1767 was the creation of a floating harbour in the centre of Bristol. What a silly idea, ridiculous?

My grandfather and I had waited on many a morning in the 1960s for one of the Campbell Steamers on the pier opposite the bottom of the Clifton Rock Railway at Hotwells and again, as child you could not gain access to it then so it was some years later when I managed to arrange a family outing to view inside. It is one of the few Rock Railways that is constructed inside cliffs and it was, by all accounts, a very tough job. From Hotwells, a 500ft tunnel rises on a 1:22 slope to appear at the St Vincent Rocks beside the Avon Gorge Hotel that I knew as the Grand Spa.

By arrangement, you can make your way down the handsome wrought ironwork protecting the staircase and into the tunnel from the Clifton top station. Now, evidently, the people of Clifton were dead set against the construction of this funicular railway. The Hotwells area of Bristol had become very dingy and was full of Riffraff and rough sailors. The Avon stank like the open sewer it was and the whole area was highly unpleasant. Clifton residents did want any of that up in their somewhat isolated eyrie!

Against Clifton's objections, construction started and was completed in two years instead of the twelve months originally planned and went three times over the estimated budget. The funicular consisted of four cable cars working in tandem wending their way up and down the cliff by the 'water balance' technique. As a safety measure, each car had three sets of independent brakes. The railway opened on 11 March 1893 and on that day 6,220 people made the return journey. An average of 11,000 passengers used it each week and the fare was a penny to go up and ha'penny to go down but it was never a practical mode of transport—just a curiosity. In 1934, it closed but its story did not end. It came to life gain during the Second World War when it was pressed into use as a civilian air raid shelter and part of it was used to repair barrage balloons but it must have a difficult job to get the deflated balloons in and out.

The BBC was also looking for a safe studio outside of Bristol and their first choice was the old Port and Pier Railway Tunnel. It was well suited to their needs but they dithered. The Director General seemed in no haste and amazingly, he decided to send down a 100-strong symphony orchestra, under the baton of Sir Adrian Boult, to test the tunnel's acoustics. What a hoot! He was too late. The people of Bristol had adopted the tunnel as an air raid shelter in spite of the fact that it was dark, wet, stank of urine and thoroughly unpleasant. 'It deserved full marks for having everything that a shelter should not possess.'

There was confrontation on New Year's Day 1941. The BBC had lined up all their vehicles and equipment to move in but were blocked by a sit-in of nearly a thousand people. The BBC politely withdrew and took up residence at the bottom of the ramp of the Rock's Railway. It was set up to continue broadcasting in any event but was never used. Much of their equipment is still in situ and my visit was an interesting insight into Bristol's past.

'When I was a boy—'

'Oh no, he's off again!' I can hear my young cousins respond.

When I was a boy, I often went out with the electricians on their various jobs and one morning my Uncle Jack asked if I would like to go over to Bath with him to a job at 'Macfisheries' in the centre of the town. I liked looking around Bath. It was a nice city and I have retained an interest in everything Georgian ever since. I like the simple elegance and architecture of that time. Macfisheries had a big cellar under the shop where much of the preparation of their fish and fowl took place and on this occasion, I noticed that part of the wall was down.

I looked behind the wall and there was a tunnel. I asked one of the apprentices if I could borrow a torch and proceeded to explore. The tunnel was slightly damp but led downwards towards the river and there were other tunnels built on a grid pattern most of which ended at the river. Evidently they are part of a sewerage system and some of them could well have been Roman.

I 'came up' about two hours later and I gave two of the Macfisheries' apprentices quite a shock when I came out from behind the wall. My uncle asked if I had had a good time exploring Bath and I replied in the affirmative but I never told him until much later that I was underground and not overground.

But enough of this troglodytism. Bristol is far more interesting above ground. We need to climb Cabot Tower again, for the view it offers now is very different from the view I first saw some sixty-five years ago!

'When I was a boy' Bristol still had many bomb sites and I still vividly remember Castle Street. To me, Woolworth's and the Co-op buildings stood alone amid the dereliction and opposite them was the Star Restaurant, standing in solitary loneliness on the opposite side of the street, where you could get a three course meal for one shilling and sixpence, which we could only afford occasionally anyway. On walks with my grandfather, we would pass many bomb sites in the Easton area and of course, the Hartcliffe and Knowle Bus Terminals were surrounded by houses on new estates built to absorb people who had lost their homes during the war.

Somehow the aftermath of the war was all around me when I was in the city and it could be slightly depressing. But Brandon Hill and Cabot Tower however always seemed to inject a breath of fresh air into us as the gardens and surrounds were well kept then and we boys always had tremendous fun playing Roly-poly down its steep slopes. Our girl cousins and friends would never commit, for it would be unseemly for them with their skirts and dresses to show even the smallest sight of their knickers because 'it would be rude'.

I returned to Brandon Hill some five or six years ago after returning from abroad and was very disappointed to find it unkempt, littered and scruffy and the Tower barred and shut up. It had become a home for vagrants and the homeless and judging by the activity in nooks and corners, many promiscuous girls had no hesitation in displaying their underwear for all to see. I am assured that the whole area has been cleaned up again and the tower reopened—so to the top again.

In the late fifties and early sixties, one would have been able to pick out St Mary Redcliffe and the Baltic Wharf with its graceful cranes. The cathedral could be partially seen with its squat towers adjoining College Green and the Will's Tower at the university and some of the houses sprawling down to the Floating Harbour. (Not such a daft idea after all, Mr Champion. Eh?) Downriver, the squat tobacco warehouses sat foursquare and solid alongside the Cumberland Basin and then across the River Avon (New Cut) you had the vista of Bedminster and its Down and glimpses of Clifton and Long Ashton.

My father used to joke that the Floating Harbour was barely afloat at that time and the whole of the area around the waterfront looked quite seedy and run down. Perhaps this was just 'post-war blues' on our part but we were never eager to go down and explore the waterfront in those days. What a difference now.

Much of the area around St Mary Redcliffe still looks rundown but new buildings, mainly offices and hotels, abound in the Victoria Street area and at long last the wharfs opposite Queen Key are being renovated and renewed. The king Street area has been sympathetically renewed and Welsh Back is quite trendy while Queen Square has lost none of its original magnificence though very few Bristolians live there now.

The Centre? At time of writing, this looks like a modern bombsite with workman, diggers and concrete mixers scurrying all over in an attempt at pedestrianisation and easing the traffic flow. Why? It seemed alright as it was and I am not sure if there was a public consultation about the project anyway? Many Bristolians to whom I have spoken would have dearly loved to see the whole area revert to water again. I think Bristol lost a major opportunity to develop the Centre into a major marina attraction.

The Baltic Wharf Building is now the M Shed, an industrial museum, and this has become a major attraction to Bristolians and visitors alike. I particularly like seeing dozens of school parties there and the crowds of children descending on it during the holidays. I often take my 98 year old mother there and she loves visiting the Bristol Exhibitions though she is not too enamoured of the high prices in the café! I am particularly impressed by the kind and helpful staff at all levels.

Opposite the Baltic Wharf is Canon's Marsh which has been completely rebuilt apart from the buildings on the Bordeaux Quay. The Lloyds Building is most impressive and its forecourt is a major events venue for the harbour now hosts many such events. This area always seems busy from the Watershed Centre right around to the Millennium Square. The @Bristol Building is great for children and the Planetarium is most impressive and it is great fun to wander around Millennium Square looking at the statues and the water features.

I have a confession. I have never visited the aquarium and I suppose this represents a big hole in my experience of Bristol? It is quite something to visit this area at night when it buzzes with hundreds of revellers which seem to spread right up Park Street and Whiteladies Road spilling out on to the pavements and roads. The rest of Canon's Marsh is flats. Thousands of them I should think, spreading down almost to Poole's Wharf where smart town houses have replaced the sand wharf that was once very busy there. The gem of the waterfront must be the SS Great Britain.

At her launch on 19 July 1843, she was 322' long and was designed to carry 252 first and second class passengers on the transatlantic route. Designed by Isambard Kingdom Brunel she was unable to operate out of Bristol (why?) and her home port became Liverpool. She could cross the Atlantic in about fourteen days but her initial voyages were dogged by damage due to poor navigation. She was eventually sold to the Eagle Line and sailed back and forth from Australia for 24 years but she then fell afoul of new insurance regulations, had her engines stripped out and became a pure sailing vessel plying between Liverpool and San Francisco as a 'Down Easter' ship that rounded the Horn.

In 1886, she attempted to 'round the horn' for thirty days but had to limp back to Port Stanley where her hull was found to be sound but her rigging was ruined and too expensive to repair. She then became a floating warehouse hulk used as a wool and coal store. The rest is history and familiar to most Bristolians. She was placed on an ocean going barge and hauled back across the Atlantic to Avonmouth where she was floated off the barge and towed up the Avon with some difficulty arriving in the Cumberland Basin on 5 July 1970 and then towed in the Great Western Dock on 17th July, exactly 127 years after she had left it.

The Prince Consort had seen her out and HRH Prince Philip welcomed her back. It is interesting to note that on her first trip down the Gorge in 1844, the suspension bridge had not been completed. I had been back in Bristol some little time after her return and visited the ship in her original condition not long after her return. It was many years later when I made a return visit and I was absolutely amazed how brilliantly she had been restored to her 1844 state and often take visitors around her who are suitably impressed.

Amazingly, I was also able to take my elderly mother around her because the ship is now 'wheelchair friendly'. I visit her regularly even now but not to go on board. I often canoe in the Floating Harbour early in the morning and paddle past her getting a water-level view but the SS Great Britain is impressive from any angle.

Brunel left two more legacies for Bristol. He designed the original plans for the Clifton Suspension Bridge and under his supervision the two towers were built (quite recently the brick buttresses on each side have been discovered to be hollow and I think you can now venture inside), but not the roadway, as the work was interrupted by the Bristol Riots. Brunel only ever crossed the gorge, which is 331' above the Avon, depending on tide, by means of the basket like device that was pulled backwards and forwards by a system of ropes. It must have been quite hair-raising.

His original design was altered quite extensively before completion and the chains were brought down from another bridge in Hungerford to span the gorge. If you visit the newly opened Visitor Centre it contains photographs of the bridge workers jauntily walking along the chains about their work. The Toll Bridge was opened in 1864 and its length was 1352'. Unfortunately this was five years after Brunel's death. Amazingly, the original wooden supports for the roadway are still in place.

On 8 December 2014, it celebrated its 150th anniversary and the bridge is still used regularly by over 4,000,000 cars a year. I have travelled over the bridge many times by car and foot and I still enjoy the experience but several friends and relations will not set foot on it. It is believed that the sport of Bungee Jumping was started off the Bridge by Oxford University students. At a short distance from the bridge is Bristol's Camera

Obscura where you can gain a photographic panoramic view of Bristol and also visit the Giant's Cave which leads down to an observation platform jutting out into the gorge. I have an even larger group of friends and relations who will not do that.

Nearby is 'The Slide'. There is a fifty-foot rocky ramp that finishes near the pavement down which generations of Bristol children have slid so it is now smooth and fast. I must have greatly added to its lustre. If you were going to ask, the answer is…no. Not many girls used it in my boyhood because they mostly wore dresses and…got it? But girls of this generation do.

Brunel was also responsible for building the GWR railway line between Bristol and London though he does seem to have slipped up by deciding on a 7' Broad Gauge model which had to be reduced to standard gauge not that long after. It was nicknamed 'God's Wonderful Railway' and I still much admire Bristol Temple Meads Station which is magnificent both inside and out. The exterior is stunning and the interior is almost as good as a cathedral.

With the advent of Bristol Parkway, of course, there is not much point in the residents of South Gloucestershire using Temple Meads which is a disappointment for me. He built the Box Tunnel of course and what I did not know until very late in my adulthood is, that, on the western approach to the tunnel is a branch line that goes into a whole system of other tunnels and this is where ammunition was stored during the war.

What few people realise is that, prior to the above, another great ship was launched in Bristol. Designed by Brunel, The Great Western was built and then launched on 8 April 1838 just below the M Shed. She was a wooden paddle steamer providing luxurious accommodation on a regular two week schedule across the Atlantic and was a huge commercial success. Very few Bristolians even know about her.

The Cathedral has been mentioned. I hope God does not get to read this but I have never been a devotee of Bristol Cathedral. 'When I was a boy', its exterior was blackened by the city grime and I do admit that it does look much more presentable now it has been cleaned but it just does not tick my box. It lacks the air, aura and ambience of a cathedral, and may I say it is dull? I know it was founded over 1,000 years ago, has a medieval past and it has a sound gothic style but I cannot feel that. It communicates nothing to me.

I have filmed recently in several of the London palaces in which I was able to feel and sense the history and I can get shivers down my spine when I look at a great painting or work of art. I will admit that the Romanesque Chapter House and the Elder Lady Chapel are of interest and almost touch a nerve. Just across the College Green is the Lord Mayor's Chapel which I much prefer.

As a boy, I was taken to the zoo which I enjoyed, and I was awed by the collection of animals contained therein. I rode on 'Rosie' the elephant and I am sure that 'Alfred' the gorilla was taken out of his cage for walks with his keeper? I place the question mark there in case my mind is playing tricks but several of my cousins think they saw this as well.

Poor old Alfred is now tucked away in an upstairs exhibition of other stuffed beasts in a dark corner of Bristol Museum which I think is a bit unkind. Surely he deserves a more interesting spot? The only excitement he has ever had was when he was kidnapped by university students during one of their Rag Weeks. He was such a popular chap that I think he should be taken out more. Perhaps we should start a Free Alfred Campaign?

I have to admit that I am not a great lover of zoos. I do accept the need for captive breeding programmes and the need to raise money for survival projects but wild animals need to be in the wild. To be frank I am also not very keen on the keeping of pets in our modern western society and the whole world of commercial marketing that surrounds them seems slightly odd to me. I am just uncomfortable with the whole humanisation of animals and I do truly believe that this over pampering and spoiling of pets is cruel and unkind.

One could mention dogs that need much exercise being only taken out for a short morning and evening walk, or big dogs confined indoors all day or dogs that are not under proper control or poorly and cruelly treated. One also now hears horror stories about the breeding of animals for the pedigree market. England is now probably over populated with dogs and one cannot escape them or their mess. I came across one famous personality who was adamant that all dogs in England should be put down on the sole basis that they were ruining the bluebell woods of our fair land. Now that is a bit extreme but I know what she was driving at.

Wherever I walk, or cycle, or canoe, that acrid smell associated with dog poo, even though most of it had been picked up according to law, remains. Many of our commons and footpaths stink! The equestrian world has also reduced many horses to the level of pets and many ponies and horses are inadequately cared for or exercised and I do not like to see birds in cages. The most healthy animals are those that work or serve a definite purpose.

One lady was upset about Dartmoor Ponies being sold for meat abroad so purchased one at the Tavistock Sales. She took it home and put it out on the veranda of her fifth-floor flat. It had to be rescued by crane. The donkeys at Clovelly used to work carrying light loads and baggage up and down the steps and were hardly overstretched or mistreated or beaten. That was alleged to be cruel and I was horrified to see that they had been moved to a small paddock where they spent their days giving rides to children going round and round and round. Cruel? I would hope for a much more balanced approach to the keeping of all pets and to be controversial…licensing fees?

There is a story that does the rounds involving Bristol Zoo and like all good stories, it has long been repeated, embellished and passed into myth and legend. Many Bristolians state emphatically that it is true but all that I can say is that I have only heard the account in story form on Radio Four Extra. A fairly young man noted one day, on a walk around the sea wall of the Downs, that an increasing number of cars were parked. This must have set him thinking. He came back the next week dressed in a fairly smart blazer, peaked cap, white shirt and tie and he had also acquired a second-hand ticket machine.

I think he was only charging a 'bob' a ticket but he walked along the sea wall collecting a shilling from each car and built up a tidy sum. This continued for a couple of weeks. As he was walking through the carpark outside the zoo returning to his home, he noticed a young mother struggling to dismantle a pushchair with a toddler in tow. He stopped to help her and she said that she had not realised that she had to pay for parking at the zoo and he duly issued her with one 'bob' ticket.

The next morning, he turned up at the gate of the zoo and introduced himself as being the parking attendant from the council. No one asked to see any form of proof so he did a very good business in selling his parking tickets. The whole enterprise grew. The zoo staff were pleased to see him in the knowledge that someone was now in charge of the parking and he was on to a very nice number. He started by putting up a few parking signs, brought a chair and an umbrella then a little table, and after that, even a small hut in which he could shelter and make his tea.

It is alleged that he worked there for fifteen years, and then, one day, he did not turn up. One of the zoo staff phoned the council offices to enquire about his absence but they stated that they knew of no such man. Legend has it that he had made enough money, with no real overheads, National Insurance or tax to retire comfortably to a nice villa in Spain.

I played on the Downs as a child and sometimes we would take the bus out to Blaze Castle and I was fascinated by the legend of the giant Goram. Then you could also look inside a real Gypsy Caravan that was in the grounds and that is the one on display in Bristol Museum—quite close to Alfred. The museum was always an attraction to me and I spent many happy hours there. When I returned from abroad, I also gave them, on permanent loan, several Ghanaian artefacts including carved doors and colonial figures.

I would often lead family friends and visitors around Bristol and the Georgian House and The Red Lodge were both on the list. I liked the Georgian House because of its simplicity, elegance and style and the surprisingly huge bath in the basement. It has now been cleverly presented as the Slave Owners house of John Pinney, a Bristol Merchant who prospered in the trade.

The Red Lodge was the home of the Bristol Savages a group of artists that still meet in the Wigwam adjoining the Lodge. The mixture of Tudor and Georgian interests me and The Grand Oak Room is wonderful to see. The lodge also has what are claimed to be the three oldest rooms in Bristol. It was also the first Reform School for Girls started by Mary Carpenter with funds supplied by Lady Byron.

I have always used Christmas Steps as a short cut into town and to this day I can still be seen sitting in the small alcoves below the Foster's Alms Houses, people watching. Attempts have been made to surround it with an Arts Quarter. Park Row fulfils this aim and has many interesting shops including galleries and antique shops but Christmas Steps sees almost determined not to join in and the lower part seems to have lost its way and soul somehow.

The musical instrument and stamp shop are still there as is the bow maker though he is nothing to do with Robin Hood. He is a world renowned bow maker for stringed instruments and they are not cheap. 'When I was a boy', I sold my stamp collection to the stamp shop at the top of the 'Steps' for the princely sum of ten shillings and sixpence and then treated my mother for lunch at what was then, a restaurant at the very bottom.

I roamed Bristol between the ages of eight and thirteen and there were few places that I did not know. I particularly liked King's Street, with its Theatre Royal and the Llandoger Trow included in 'Treasure Island' and the Merchants' Almshouse for retired seamen. Marsh Street was a disreputable place historically inhabited by Irish immigrants and unemployed sailors and the slave ship captains often found their crew members here. Corn Street was the commercial hub and business centre of the city and banks and insurance offices were located here as well as the coffee houses where much trade was completed.

Until quite recently you could go into Lloyds Bank in the street, lie on the floor and admire the beautiful plaster ceiling. I once prompted six people to do that one morning. The National Westminster Bank, the Commercial Rooms, the Exchange, All Saint's Church and 56 Corn Street (Coffee House) are all difficult to visit nowadays but a small boy was always admitted and then were the Nails. These are the four Mushroom like pedestals where transactions were completed, hence the term, 'paying on the nail'.

The Guildhall and the Courts were also situated nearby in Broad St that led down under St John's Arch. The St Nicholas Market then was a market selling fruit and veg and was a very lively place to visit first think in the morning though you could wander through it for most of the day. Old Market, as its name suggests, was a place for several markets and it was controlled by the Ancient Court of Pie Poudre which met in the Stag and Hounds pub by the present day roundabout. It was responsible for sorting out market disputes and though it was last active in 1870 it was not legally abolished until 1971.

I was away from Bristol for long periods but on one of my return trips I needed some camping equipment so stopped off in Old Market to visit a camping shop called Casey's. I walked up and down several times but I could not find it. I was stopped by a lady who asked me if I was lost. I told her what I was about and she informed me that Casey's had gone out of business many years ago. She then advised me that it was not very wise to walk up and down the pavements of Old Market as it was now the Gay Village and people might get the wrong idea.

A similar thing happened in Plymouth one lunch hour. Another member of staff and I had dropped pupils off for cricket coaching in the clearly marked school minibus and we had made our way down to Union Street so that I could pick up some camping gas cylinders at a camping shop. We could not find a parking spot so my colleague said that he would just drive up and down between the two roundabouts until I emerged from the shop. When I came out a little while later, there was the minibus stopped by the police who thought it had been 'cruising' and we were the butt of many jokes in the staffroom after that.

I have been aware from an early age of the outline history of Bristol and the more I have discovered and uncovered about its historic events, the more one can establish there has been a very definite failing in its governance, regulation and control. Our city fathers hesitated, prevaricated, deviated and procrastinated when faced with any problem or decision. They were ditherers and cunning in their use of evasion. They were also hypocrites often making laws that they by-passed or ignored.

The major problem was that the mayor was chosen by the aldermen. The second problem was that the mayor chose the aldermen. So when William Wilberforce denounced the white slave trade, they employed their cunning evasion tactics petitioning the king to let them have the defeated Scottish troops in their place. When the war came to an end, they resorted to kidnapping and then treated kidnappers leniently when they came up in front of the 'Bench', which consisted of aldermen, of course, and they resorted to frightening prisoners into accepting transportation rather than long prison terms.

For transportation, read slavery. All legal and above board—all controlled by the city fathers. They had no thought for the working Bristolians who had no hand or profit in the business. It was a closed shop for the super-rich! And it was very hard indeed for the petty criminals and thieves hoisted off to slavery in the colonies.

It is often stated that the people of Bristol were wishy-washy as to their loyalties in the Civil War but I suggest it had very little to do with the common people. The war commenced in 1642 and the aldermen agreed on one thing. They were going to sit on the fence to see which way the wind blew trying to be on both sides at the same time. Then they moved to being slightly in favour of Parliament but declared themselves not against the king. This was prevaricating diplomacy at its best! Then they hit upon the device of sending both parties a letter recommending reconciliation!

They would have been very happy if the war had just gone away. They hesitated again but it was too late because both parties fully appreciated the importance of Bristol. Charles instructed the council to admit no troops and to defend the city for his cause while the Parliamentarians issued requests for money. The Marquis of Hertford came west on orders of the king with a request to bring his cavalry into the Bristol. The mayor repeated the instructions of the king.

If Bristol was not for him then it was against him so he advanced but the strong men of Somerset intercepted them and put them to rout. The city then took upon itself to adopt a policy of armed neutrality and set about improving their defences but with no sense of urgency whatsoever and even extended their summer holiday. Parliament asked for more cash but the corporation turned the request down but did finance, on a business basis, the embarkation of 2,000 soldiers for Ireland. The Parliamentarians lost patience with Bristol's prevarication and inactivity and sent 1,000 troops to attack the city. Even then the corporation dithered, hesitated and played off both sides but now there was an intervention by the people, for the people?

The mayor's wife and one hundred women stormed the council chamber and the aldermen agreed to let the Roundheads in. Essex came into the city at the head of 2,000 men. This proved very expensive and thousands of pounds had to be spent on the garrison and many thousands more on a great defensive ditch that took over three years to build. Why? The great castle of Bristol was very vulnerable to cannon fire as it was now well past its sell-by date!

Under the command of Colonel Essex, who was a drunk and made slow progress on the ditch and then the Hon Nathaniel Fiennes who made very fast but very expensive progress with the digging, the ditch took three years to build. Its value for the defence of Bristol was questioned when a local uprising, led by one, Robert Yeamans, to win the city back for the Royalists, failed but some of Prince Rupert's men had easily breached the incomplete defences. Was the corporation worried? No. They were complacent as ever. Had the fine forts not been completed? Yes, they had. But what about the sections in between? They were still hacking at the rocks around Brandon Hill.

Then the struggle moved in favour of the Royalist. The Battle of Lansdown on 5 July 1643 was inconclusive but the Parliamentarians were smashed at Roundway in Wiltshire one week later. Prince Rupert attacked Bristol. Rupert made three attempts to take the city but Fiennes repelled each one which made the Cavaliers grumpy as they hated sieges. Prince Rupert ordered an all-out attack involving all his men the next day and by a stroke of pure luck, the sharp-minded Captain Washington (a direct descendant of George Washington) was ordered to attack between the Brandon Hill Fort and The Windmill Fort (near the site of the present museum) between which there lay a hollow where the rampart had never been finished!

Fiennes had left the back door open. Washington led a fire-pike charge and totally routed the defenders and then ordered his men to fill the ditch with earth which enabled the cavalry to charge in. The Royalists took heavy casualties but Fiennes asked for a parley and then surrendered. Bristol was now in the hands of the Royalists. By the terms of the treaty, the defeated Parliamentarians would be allowed to leave the city and be escorted to Warminster with their arms and horses and the sick and wounded.

The Bristolians had their rights and property protected and the corporation's functions and liberties maintained. All well and good. Some hope! Fiennes and his men were stripped and beaten out of the city, lucky to escape with their lives let alone take anything with them and then Rupert's men looted and plundered and mistreated the citizens. The corporation offered Rupert £1,400 to stop the looting which Rupert accepted. The looting continued. Bristol was made to pay for taking the Parliamentarian's side.

The Corporation had complained about the Roundheads' insatiable demands for cash but now the Cavaliers bled Bristol dry. By the end of that year, they had been forced to give £33,000 to the Royalists and also support the garrison at £400 per week.

Rupert took over as governor and the Corporation gave him and the king large supplies of wine and other gifts, as well as a hundredweight of sugar. In 1644 Charles graciously granted Bristol a pardon for having opposed him and he seemed to much enjoy mistreating the members of the Corporation and insisting that all aldermen and the Mayor and Sheriff should be staunch Royalists.

Charles was very suspicious about the city's speedy switch of loyalty to him and the worse the war went for him, the more he demanded money until the city was bankrupt, but still, the Royalists demanded more. The city was being brought to its knees and eventually could supply no more funds. Rupert then hit on the idea of using forced labour in lieu of payment. Amazingly the fortifications and defences of Bristol had been greatly improved to a high standard during this time. Considering the amount of money that Bristolians had given the king they should have been of a phenomenally high standard. In June 1645 Rupert's troops lost the Battle of Naseby and this meant the capture of Bristol was the next target.

Sir Thomas Fairfax charged through the West Country scattering the Royalists at Langport, Bridgwater, Bath and Sherborne. By the 20th August he and Cromwell were at Chew Magna and the Siege of Bristol commenced on the 23rd August. There were skirmishes, raids and fights for several days but the now enhanced fortifications held firm though the Roundheads made several classic blunders. The siege ladders that they had constructed were too short and they adopted the Death and Glory Approach of attacking the forts instead of the ramparts.

On the fifth day, more by chance than anything else, Rupert ordered a night attack (how ungentlemanly) and the soldiers at Lawford Gate got over the ramparts and opened the gates for the cavalry to enter. The game was up and the fierce opposition of the Royalists so angered the Roundheads that they gave no quarter. This in turn may have frightened the Royalists into surrendering as they still held many parts of the city and had ample supplies and ammunition. Rupert lost his nerve and surrendered. Or was there another reason far more sinister in its implications?

Charles I, in contemplating Rupert's surrender, stated that, 'This great error of judgement proceeded not from change of affection, but by having his judgement seduced by some rotten-hearted villains.' Enter the Corporation again. One such rotten-hearted villain was Alderman Hooke, Mayor of Bristol in the previous year and an obsequious and cunning opportunist who had mastered the art of switching loyalties to a fine degree.

The historical facts are difficult to ascertain. Hooke had something on Rupert and we will probably never know what but it must have been something pretty big. Blackmail? Bribery? Bluff? We cannot say but it was such a powerful 'hold' that Rupert surrendered immediately. Hooke's name does reappear several years later when Cromwell managed to extract him from legal problems stating that Hooke had done, 'something considerable' for the Puritans which must be concealed. When the Royalists had held Bristol Hooke had avidly adhered to their cause and could well have been in a position to indulge in spying or treachery.

Fairfax treated the Royalists well and let them leave the city unmolested but Bristol was in a very sad state and was dirty and neglected.

And can you believe it? Whom did the Corporation now back wholeheartedly and with much fervour. They became Roundheads again. They presented gifts to Fairfax and raised £6,000 for them. How? The Corporation's handling of the situation was true to form. Prevarication, hesitation and cunning evasion. Derek Robinson, a well-known Bristol author to whom I am indebted for the discovery of many facts put the Corporation's position very succinctly. 'When there was only one side left, Bristol knew which side it was on.'

Black Slavery brought immense wealth to Bristol. For nearly one hundred years between 1698 and 1807 all the mayors, sheriffs, aldermen, councillors, merchants and businessmen were involved in the ghastly business of the 'Triangular Trade' which brought them huge triple profit. The statistics are truly staggering and probably ten million slaves were transported during that time and there can be no escaping that Bristol was a major player in the whole enterprise, indeed, in the 1750s, Bristol was the leading port with nearly one hundred 'Slavers' operating out of it.

There can be no excuse offered in mitigation of this trade and I find it totally hypocritical that these eighteenth century 'Fat Cats' professed to be Christians. They knew the conditions under which the slaves were transported and cruelties which they endured in their bondage for many of them owned estates in the West Indies and America. What is perhaps even more shocking is that the priests and clergy also defended the trade. When one reads the various advice concerning the 'stowing' of the slaves and the fact that insurance companies considered them as cargo, the whole trade becomes a crying example of 'man's inhumanity to man'.

Though Wilberforce was the leading abolitionist on the national front, it was Thomas Clarkson who brought about its downfall in Bristol when, in 1787, he started investigating the awful trade, but unbelievably, it was the mistreatment of the crews on the slave ships that attracted sympathy. Clarkson very cleverly used this sympathy as a portal to reveal all the horrors of the trade and the Abolitionists Movement gained ground.

It would be pleasant to report that the members of the Corporation appreciated the errors of their ways but they fought the abolition of the trade 'Tooth and Nail' insisting that it would be the ruin of Bristol and indeed, added that, they could well be, 'left starving in the gutter,' themselves. Their lobby was strong and the initial response was to control the number of slaves that ships could carry. Wilberforce and Clarkson continued their long campaign and slaving was formally abolished in 1833. Interestingly Bristol had withdrawn from the trade over the previous twenty years. Had their conscience pricked them? Were they ashamed? No. Slaving was risky and the merchants could make just as much money on cargoes by sailing direct to the West Indies and many of them had gradually sold their ships and invested in the West Indies Estates and most of them of course made another huge profit from the compensation paid out by the government!

I find the Bristol Riots of 1831 fascinating. Its main cause has always been mentioned as the reform of Parliament, the basic proposal that MPs should be elected by their constituents but as Bristol, at that time, had two Pro Reform MPs so the Reform Bill of 1932 made very little difference to the actual number of voters in the city. Bristol had more voters than other cities.

The economy of the city was not good. The slave trade had all but gone and high port dues led to a decrease in trade. This hit the poor most. There was much unemployment and prices were soaring. There was much ill feeling too and nervousness about the French Revolution and a fear of a recent outbreak of Cholera in the North so when the Government threw out the Reform Bill in October the restlessness and ill feeling erupted.

There were riots elsewhere but in Bristol they were much, much worse. You will recall that in Bristol the mayor was elected by the council, the aldermen by the mayor and other aldermen. The councillors were elected for life by the council who also elected the sheriffs and chamberlain. It was a 'Closed Shop' and had no respect for the ratepayers and the ordinary working people whom they treated abominably so when the magistrates, who were all councillors, turned to the citizens for help in the riots, none was forthcoming.

The trigger for the whole event was one Sir Charles Wetherell MP whose constancy was that he was always against everything. He was the Attorney General and Recorder for Bristol and initially had much support in the city but he was under the misapprehension that the people of Bristol agreed with everything that he did or said. He made over one hundred speeches against Reform which upset Bristolians as he had no right to speak on their behalf. He was not their MP.

At the next Assize Sessions, the Corporation requested troops for his protection. The aldermen tried to get help from seamen in the port but they responded with, 'they would not allow themselves to be made a cat's paw by the Corporation' Thugs and toughs had to be hired who were to act as 'bludgeon men' in the event of any trouble and trouble there was aplenty. On 29 October 1831, a mob accosted him on the Bath Road and screamed abuse at him. His transfer to the Civic Coach at Totterdown led to more riotous outbursts and there was an even bigger mob lying in wait at the Guildhall who created such a furore that the court was adjourned and Wetherell then ran the gauntlet to the Mansion House in Queen's Square with the mob trying to overturn the Civic Coach.

Wetherell, the mayor and the magistrates took shelter inside and most of the crowd dispersed. The Specials on Protection Duty had been eager to punish the mob and now saw their chance. They lay into anyone that they could catch, many of whom were just innocent bystanders. The brawling and fighting went on for hours, then, amazingly, most of the Specials disappeared for the end of their Duty Shift. The mob redoubled their efforts and John Pinny, the mayor, read the Riot Act three times. This was ignored by the mob and they renewed the attack and smashed their way into the Mansion House destroying everything in their path then prepared to fire the building and burn all within. Wetherell, however, had escaped in the guise of a coachman and actually walked through the mob before dashing off to Newport.

Troops led by Lieutenant-Colonel Brereton rode into the square. The magistrates and mayor had hesitated and delayed in calling them in and total disaster was only averted because the rioters could not find a light. It is argued that if the mayor and magistrates had called in the troops immediately the riot would never have started. The rioters cleared the Mansion House upon arrival of the troops and the mayor ordered Brereton to clear the square using all necessary force. Now Brereton hesitated insisting that the rioters were, and I quote, 'a good humoured mob,' so ordered his troops to just patrol the square in a misplaced attempt to jolly the mob. It failed.

Brereton ordered Captain Gage's troop of the 14th Dragoons to charge using the flat of their blades. The mob was driven into the side streets and then went off to make trouble elsewhere and Gage made several more charges, this time using the sabre edge. The streets were cleared. Damaged windows and doors were repaired, the streets put to order and Brereton made another huge mistake. Assuming all was well, he cleared all the pickets…and by 8 am in the morning, the mob had discovered this and came back to inflict an orgy of destruction upon the Mansion House and the houses in the square. The mayor only just escaped with his life and sought out Brereton.

The cavalry were saddled up and returned to Queen's Square where one major change had taken place. The rioters had discovered the cellars and were now all roaring drunk and the news of free drink had spread throughout the city and crowds of people including layabouts, thieves, the unemployed and thugs turned up to share the orgy, many of them looking for mischief and spoiling for a fight. The Mansion House was recaptured and an alderman read The Riot Act again but Brereton refused to fire on the crowd in case it infuriated them. If he had the Riot might have ended there.

Brereton foolishly ordered his men back to their quarters at this juncture and the mob followed them with a hail of stones and other missiles that came to hand. One rioter was shot for attacking a Dragoon which made Brereton very angry and he insisted that the 14th Dragoons should leave the city immediately. Brereton went to Queen's Square and the mob hailed him as a hero and he promised no more firing. The rioters continued drinking and the good folk of Bristol went about their Sunday devotions. The rest of the city was peaceful and quiet.

There was another appeal from the Corporation for help which was generally ignored as most Bristolians dislike of the Corporation had now turned to total contempt. The rioters were eventually convinced that Wetherell had gone but they did not disperse. Instead, they attacked the Bridewell Prison and released all its prisoners setting the prison alight. An urgent message for help was declared too late by the aldermen and that day they also turned down an offer of help from 250 veteran army pensioners.

When volunteers eventually arrived at the Mansion House in the afternoon, they received no instructions but a meeting was called in which the mayor was asked about his plans and he had none. The town clerk gave a boring political speech and Brereton refused to recall the 14th or use his troops, the 3rd Dragoons, pleading that they were tired. Again in a monotonous repeat of their past record, they agreed to do nothing.

The mob was meeting no resistance and was making mayhem everywhere. They fired the New Goal and the prison at Lawford's Gate and were terrifying the citizens of Bristol. The Corporation and Brereton did nothing. Then the mob moved in on the bishop's Palace on College Green. A group of Specials were rushed to its defence and the 3rd Dragoons notified, but they were too late for the mob was busy destroying everything inside and starting fires. When the Dragoons arrived, the rioters ran off. The Dragoons formed up at the front door of the Palace and...did nothing.

Out they rode amid great cheers from the mob who then raced back inside to complete the havoc they had started and the Palace burned to the ground including the Chapter House. An attempt to fire the Cathedral was thwarted by the bravery of the Sub-sacrist so the rioters returned to the Mansion House and finally burnt that as well. The Dragoons left on guard at the Mansion House had been drawn away to protect the bishop's Palace and had returned to find the Mansion House ablaze. They waited for orders. All the Specials were nowhere to be seen and Brereton seemed completely withdrawn. They gave up and went back to their quarters.

In the evening, the Dodington Troop of the Gloucestershire Yeomanry arrived in response to the magistrates summons. Captain Codrington could not find any of the magistrates (I wonder why?) but did bump into Brereton who insisted that he could do nothing without an order from the magistrates. They had made a round trip of forty miles for nothing and went home. Colonel Brereton went home to bed. The mob was now having a whale of a time drinking and looting and it set fire to nine houses in the square. The fires spread to warehouses along King Street but then the mob decided to burn all the houses on the west side of the square.

By the middle of the night, half of Queen's Square, much of Prince Street, King Street and the bishop's Palace were ablaze. The looting continued on a grand scale though it is estimated that the hardcore of looters only numbered about one hundred but thousands were spectating. Very few Bristolians were abed that terrible night, oh, except Colonel Brereton, of course.

Samuel Goldney searched everywhere for the mayor and eventually tracked him down in a house in Berkeley Square. What passed between the two seems not to be recorded but Pinny did give Goldney a letter addressed to Brereton directing him, 'to take the most vigorous effective and decisive (that was a word not often used by Bristol mayors) measures in your power to quell the existing riot.'

Goldney rushed to the cavalry stables and encountered Captain Warrington who refused to open the letter at first but said that he could not turn the troop out without instructions from a magistrate. Another plea from an alderman arrived and Warrington then declared that he could only mount up on the direct orders of Brereton. They located him at his lodgings in Unity Street but he tried a catalogue of excuses not to turn his men out.

5 am. The Dragoons reached Prince Street as the clock was striking. The mob was around another warehouse fire that they had just started. The Dragoons charged them and galloped into the square where they met Major Mackworth who noted that Brereton was patrolling in his usual casual way. He took over.

'Colonel Brereton, we must instantly charge. CHARGE MEN AND CHARGE HOME!'

That was the beginning of the end of the riots. They cut the mob to pieces and killed or wounded 130 people. Mackworth galloped off to Keynsham to recall the 14th Dragoons and this troop was quickly followed by that of Major Beckworth from Gloucester. He found the mayor, aldermen and the town clerk at the council house bewildered and terrified. None of them would venture out but gave him full authority to do as he saw fit. He also found Brereton in a confused state so took effective control of the situation. The mob began to cautiously reappear but were immediately cut down by the decisive action of Beckworth's Troopers.

For over two hours, the rioters were hounded, charged down and routed. Other units arrived, frigates moved into the Bristol Channel and 4,000 Specials were put on duty. Trials took place and sentences handed down including transportation and the death penalty and Brereton was court martialled for his stupendous failings in his line of duty. He had actually caused much more damage than all the rioters put together. Before being sentenced, Brereton went home one night and shot himself in the heart. I can find no record, however, of any censure of the mayor, Corporation, aldermen and councillors whose indecision had also resulted in the extended length of the riots nor any documents relating to any one of their suicides. The phrase 'bungling and indecisive fools' comes to mind and did this incompetence cease?

Surely successive Corporations and mayors had learned their lessons by now? We all learn from our historical mistakes. If you don't make mistakes then you don't make anything. Right? Yes, but not the Bristol City Council. In 1775 Bristol was the second biggest port in the country and many people thought that was how it would remain totally ignoring its growing problems of overcrowding, difficult navigation, mud at low tide, inadequate wharfs and moorings and the constant danger of fires spreading from ship to ship.

What did the Corporation do? Nothing. In the meantime, Liverpool was beginning to overtake Bristol with the construction of a 'Floating Dock' and they did not have the Avon to navigate. Ships had then to be towed up the Avon by rowing boats and then back out again. Surely Liverpool was providing a clear example of what should be done.

In 1757 the Corporation set up another committee to look at the docks to discuss the problems about which they were perfectly aware. It never reported back It must have been made up of members of the corporation? A half-hearted attempt was made to build a wet dock which did not come to fruition, it being voted down by the council and a private initiative to build one was scotched. To save money the council agreed to extend the lease of the Merchant Adventurers who leased the rights to collect dues if they agreed to improve the quays and wharves at their expense. The deal took six years to complete and the Merchant Adventurers did very well indeed out of the arrangement as did the Corporation for many merchants were members of both.

The problems remained and were now extenuated by a general increase in the size of ships and for well over thirty-five years the authorities did…nothing, though this was when Champion put forward his scheme which was deemed too expensive. In 1787, in rapid succession, schemes were proposed by a Mr Nickalls, then Jessop and then Smeaton (of lighthouse fame). The Venturers passed it all over to the Corporation who formed a committee and after four years nothing happened again. In 1791 the committee eventually endorsed the Venturer's Scheme, whichever that was, and in another two years they asked the Corporation to promote an Act of Parliament. At last—action?

Nothing happened. Some eight years later, on 8 July 1801, the Merchant Venturers made a polite enquiry about progress. The Corporation seemed to shake itself awake and Jessop set to work on new plans, an Act of Parliament was passed authorising the work and on 1 May 1804, work commenced well over 75 years later than Liverpool. The whole dock area was to become a Floating Harbour, the New Cut would be the revised course for the Avon and a new sewer was to be built. Cost £300,000.

The French helped. Their prisoners of war based at Frenchay helped with the excavations but like all grand projects difficulties and snags arose and the cost spiralled to £580,000. This high cost was to dog the Bristol Dock Company which paid no dividends at all for fifteen years. Harbour dues had to be raised and this scared trade away. The Floating Harbour was a success in itself but the port of Bristol only stood still rather than moving forward. Whose fault was it? The dock company was incompetent, the corporation ever money grasping and extremely greedy and the merchant adventurers smug and self-satisfied and who served on all these boards and authorities? While Bristol slept, Gloucester, Newport and Chepstow all moved in and Liverpool, Hull and Whitehaven easily overtook us. Greed had raised harbour dues and taxes to an exorbitant amount, at least twice the rate of competitive ports.

A newly formed Chamber of Commerce petitioned the Corporation to cut its charges and made a full attack on the Corporation who stated that the dues were theirs by right but the Chamber reacted by stating that none of the inhabitants benefited from the dues which were never spent on the improvement of the city. A government commission came down to investigate and the Corporation had to back down and reduce its rates. The dock company was also taken to court as the 'float' was stinking and one could almost walk on the water. Everyone complained and the dock company did…nothing until forced to do so by order of the court, having to build a new sewer outfall into the New Cut. With all this activity, or rather, lack of activity, Bristol was losing business fast and it was rapidly being shunned by skippers and owners alike. Enter the cavalry.

In 1833 the House of Commons decided to reform the country's municipal corporations and a delegation was sent to Bristol. The Bristol Corporation acted as if highly offended, declaring that the government had no right to inspect local government and the Merchant Adventurers and the Bristol Dock Company boycotted the enquiry. The Corporation did not and the report blasted it to pieces with charges of mismanagement and sheet extravagance and it was recorded that a quarter of all expenditure went on eating, drinking and pomp and ceremony. This would all be well and good if the

Corporation had handled the affairs and economy of the city well but they had made a rotten job of local government finances and particularly criticised the crippling harbour dues and taxes.

In conclusion, they stated that there was 'something essentially bad' in the way that Bristol had been governed. That was the end of the old Corporation but the Dock Company and Merchant Adventurers continued to function and this was a continuing handicap to the port. It now also became obvious that Bristol's port was in the wrong place; eight miles up the twisting Avon and through the gorge was another handicap.

The Dock Company did not help matters by resisting the introduction of steam tugs. A solution had to be found. Suggestions were already in the public domain about building a dock at the mouth of the Avon and a road through the gorge. So Bristol waited again hoping that ships would still come to them but they needed to go to the ships. The Chamber of Commerce approached the Merchant Adventurers with a proposal to build a pier at Avonmouth which they snootily rejected but the Chamber of Commerce kept up their pressure and eventually they agreed to consider the proposal if the Dock Company joined in. This was 1839 and the riots put an end to all schemes at that time.

With the development of the railways, people began to envisage the revival of the port with Brunel's Great Western on the Transatlantic route from Bristol to America but she had to anchor off Avonmouth and was charged heavy dues. Bristol drove the Great Western away and her home port became Liverpool. The Dock Company, the Merchant Adventurers and the Corporation met to discuss a solution but could not arrive at any decision. At the root of all this was the Dock Company's almost fanatical payment of large dividends and this was bringing Bristol to its knees. The city was in the crazy situation of having built and financed these beautiful ships and then waving them goodbye as they departed for Liverpool.

Thus commenced an all-out attack on the Dock Company by a combination of businesses under the name of the Free Port Association. The Dock Company capitulated two years later and was handed to the reformed Corporation. The new company slashed all charges and the port instantly began to recover making a 50% increase in its revenues. The difficulties of the physical approach was still a problem, highlighted by the running aground of the Demerara but it was to take another twenty-six years.

Many schemes were put forward and cast aside and the Corporation even opposed a private development of Avonmouth but this company persisted and in 1864 the piers and railway were opened quickly followed by the Bristol and Portishead Pier and Railway Company which operated in fierce competition with each other and then some people still preferred Bristol. The Cardiff tug Black Eagle blew up in the gorge, the Kron Prinz ran aground and was stuck for three weeks and a steamer called Gypsy also ran aground and Bristolians realised that the fate of Bristol was sealed. Avonmouth now handles all of the shipping cargo and is now also served by a fine motorway network. At first the Corporation abused, harangued and insulted the new companies but then bought back Avonmouth and Portishead Docks.

Bristol was even hesitant about honouring our dead war heroes. Most of the major towns and cities unveiled their War memorials and Cenotaphs only a few years after the First World War while Bristol's Council prevaricated and argued over the cost and the design, and the site! It was a full fourteen years before it was finally unveiled on Colston Avenue.

And what of today? What of the governance of the city especially now we have full electoral rights at local and national level? I think we can dismiss cunning evasion but do I still detect a hint of hesitation and prevarication. When will the arena at Temple Meads be built? It has been the subject of many delays. When will Bristol see an ice rink in the city again? When will the Bristol/Portishead Railway be reopened to passengers and the station built at the Portway Park and Ride? When will Bristol fully address the problem of homelessness on our streets? When will the new Bus Expressway services actually start? Running over budget by some thirteen million pounds we poor Bristolians now have to wait until next year. Why? The ticket machines are not ready. I am not sure if this is hesitation or downright incompetence! It appears to me that it is not a wisely conceived project anyway. Bristol needs to change its Latin Motto to:

PROPAGATI NASCERENTUR IN OMNIBUS
PREVARICATION BEFORE ALL

Chapter Eight
San Luci Domini

San Luci Domini- College Scenes.

Preparations for college entry must have started in my final year at Rodway. I do clearly remember filling out the application for entry form because it was very vigorously checked by Mr Hughes, my headmaster, who insisted on dotting every 'i' and crossing every 't' before it could be posted and of course, applications in those good old days were handwritten in ink using a proper fountain pen. Am I such a relic? I still use a fountain pen whenever I can.

The decision to apply to St Luke's in Exeter was obviously upon advice and a very good decision it turned out to be. Readers may be surprised to know that my place was unconditional. The actual basic requirement was for five 'O' Levels which I had already achieved and the fact that I passed 'A' Level English Literature and Geography and RSA Advanced Economics could have been regarded as irrelevant but it did place me upon the Advanced English Course, which was a bonus.

St Luke's College was founded in 1838 in one of the houses in Cathedral Close which I believe now houses the present choir school and moved out to a more permanent site on Heavitree Road, opposite the police station, in 1854. The buildings still stand and are now part of the university campus. Conditions in 1838 were rather more strict than during my time. On Sundays all students wore frock coats, a black silk top hat and a white cravat (until 1913). Junior students waited at table, answered the door, cleaned the gas lamps and tended fires. No newspapers were permitted.

Rising bell was at 6 am prompt with a service in the cathedral and bedtime and lights out was at 9 pm. Obedience was exacted by senior monitors and corporal punishment was allowed. The food allowance was just under ten pence per day per student. Every day all students walked out with the Principal for two hours in the summer and one hour in the winter. The aim was to train student teachers primarily as good Christians and secondarily as Schoolmastery Christian gentlemen. All students had to sign a declaration that they were prepared to spend their lives in the service of the poor. I think they were aiming for muscular Christianity.

Though slightly adapted, this vigorous regime continued until the First World War when the college volunteers were formed as part of the Devon Regiment. The 3rd and 4ths were lucky enough to be sent to India for the duration, but, unfortunately the 2nd Devons were first 'over the top' in the Battle of the Somme and were wiped out. Sixty-seven names appear on the College War Memorial. It is fascinating to rummage through the college archives at the South West Heritage Trust in Exeter, for after the First World War, the regime was austere but not overly strict and the college, under the auspices of the Church of England, was producing good sound teachers. As a note of history, it was the first teacher training college that was ever established.

The Second World War. At 1.30 am during the night of 4 May 1942, the college was bombed during a raid on Exeter, extensively damaged and set on fire due to an 'Incendiary' but there was no loss of life as most of the staff and students were across the road helping the hospital that had also been set alight and damaged. A claim was made by the college in accordance with the War Damage Act and Ecclesiastical Insurance covered the damage to the chapel. The two claims are quite vivid in their detail.

Come 1945, James Smeall was appointed principal. His CV was impressive. Born in Middlesbrough in 1907 and the son of a Yorkshire doctor he was educated at Highgate School then studied English at Queen's College, Cambridge where he played Wing Three-Quarter for the first team and then did one year at the Sorbonne. His teaching career began at Merchiston after which he taught at the Royal Britannia Naval College at Dartmouth.

In 1934, he became a housemaster at Bradfield College, then a senior master at Epsom and then served as headmaster at Chesterfield Grammar School in 1939. In 1945, he took over a random rubble ruin—St Luke's. He was charismatic and quietly forceful and he knew exactly what he wanted. When I arrived some twenty years later, the college was a close knit community with superb grounds, a tradition of corporate worship and a highly professional staff all of which he had developed and nurtured.

'Jimmy'—for that was how we always referred to him—brought up the reputation of the college through sport, particularly rugby. He despatched 'Scouts' to West Country schools to recruit the best student players and succeeded. He often appeared on the rugger Pitch to coach and referee and in the 1953-4 season the first fifteen scored over 1,000 points and the college could boast over 36 rugger internationals.

St Luke's was developing an excellent reputation, especially as Jimmy was also recruiting highly qualified and professional staff in all faculties and he gave them their heads. St Luke's was unified and possessed common purpose to achieve the best in all areas. It was this unity that enabled the college to bypass much of the national student unrest in the early sixties and Jimmy poured oil on troubled waters by declaring that, 'the college does not belong to the staff or students—it belongs to the children the students are preparing to teach.'

Jimmy not only had charisma but he was also an intellectual with a wicked sense of humour for, he was indeed, the author of the book, 'English Satire, Parody and Burlesque,' published in 1952. Of one member of staff he commented that, 'he had an impeccable taste in enemies,' and he also described a bank's new premises on the High Street, as, 'An Egyptian Urinal.' He was a local figure of some repute and served as mayor in the period 1965-6 and also did good work in the Civic Society. The council meetings were somewhat enlivened by his wit and dry humour.

Upon receiving an honorary degree from Exeter University, when St Luke's amalgamated with their campus, he commented, 'I am happy to receive this honour in recognition that I am still alive.' Under his leadership, the college increased tenfold, the students came to be highly regarded in the teaching world and I was to be one of them. I came to realise much later in my career that Jimmy was an outstanding individual but he was very typical of head teachers in the independent school sector who offered their own brands of charisma, loyalty and energy to the schools to which they had been appointed.

Furthermore it would be fair to say that St Luke's was organised and run on independent school lines though we were all treated as young gentlemen in a very adult way and of course…there was no corporal punishment.

There was a strict dress code, however, which involved the wearing of a collar and tie, a sports jacket and flannel trousers with sensible shoes. Jeans and casual clothes were not permitted at all and there was no leeway or discretion offered. So, suitably and properly attired, I presented myself at Rowancroft House on the Heavitree main road in the early September of 1964. My trunk had been sent down by road some days earlier and I was accommodated in the first big room on the left with beautiful bay windows looking out over a football pitch.

My roommate was Wills Patterson from Carlisle and there were only about twenty-four students in the whole of the house though it was surrounded by Rowancroft Court which provided single-room accommodation for students in modern surroundings. I was unaware at that time that this was to be my home for three years, and a very pleasant one it was.

My house tutor was Fred Smith, who was Head of English, but we never really saw much of him as he seemed quite happy for us to just get on with things. I think he only made an appearance on the students' side of the house about five or six times. The house was quite idiosyncratic in its layout though most of the rooms were very big and the staircase was impressive. There were no showers. There were, however, four enormous baths in the ablution area at the rear of the building and was also the back door entrance which most of us used. No one seemed to mind or bother.

Rowancroft House had one bugbear. It had a phone which actually became very problematical, especially as it was under the stairs and quite close to my room. When it rang, it was answered by the nearest student. All well and good if the student, for whom the call was destined, lived in the house but a nightmare if it was for one of the students in the court, especially if it was rainy and cold, and then, having discovered that the student you were looking for was out, you would have to traipse back to the phone and inform the caller and the usual response was to ask us to take a message, which would entail another walk back to one of the courts.

It changed from being a mere inconvenience to a total nuisance for there could be a dozen calls in any one evening. We complained with no result so we adopted a work to rule policy by answering all calls but only fetching the house residents to the phone, refusing to deal with 'court' calls. This brought down the wrath of Wally Rice, house tutor for Rowancroft Court, who seemed to hold me personally responsible for the boycott. I was very polite and held my cool but he did not. He spluttered and ranted and became very angry insisting that all calls should be answered and I just told him that I had not been given my place at college to be trained as a telephonist!

He blew up and stormed out amid cheers from the house residents which did persuade Fred Smith to come out of his flat for once. I explained what had happened and we all suggested that there could be three possible lines of action. He could take the phone out, or the 'court' could put someone on telephone duty each evening, or the college fitted phones in the court. The college authorities fitted phones in the court.

Rowancroft House lacked proper laundry facilities, and we did not even have a kettle, unlike our 'court' neighbours who had all facilities. We made a request for same to the Bursar's office and he did not even reply, so, with Fred Smith's permission, we organised a party in the house with a pop group and bar, which was a huge success, though Fred Smith did come out during the evening and suggest that we move the group and dancing downstairs for his ceiling was shaking. We made an obscene profit. We bought two washing machines and a big tumble drier, an electric iron and smart ironing board, a fridge and a huge toaster and kettle. Now we were the best equipped house in the college. The bursar decided that we should pay a contribution for the 'extra electricity' and we agreed, on the basis that it applied to all students in all the other houses as well. We heard no more.

Our food was provided by the dining hall, which was self-service, on the main campus so we walked down to breakfast every morning for a 'Full English' and there was also splendid lunches provided and a high tea. The food was good and I had no complaints. House tutors also ate in the hall at high table but they were served by the domestic staff. Grace was said before meals and always in Latin:

'Benedictus Benedicat, per Jesum Christum Dominun Nostrum.'

The kitchen staff had a black and white cat. His name? Benedicat, of course.

The dining hall also acted as a function space and was regularly used for 'bops' and dances. I usually attended chapel for morning prayers but there would normally be more staff than students, before attending lectures. This meant that there was a very small pool of students for the morning readings and I normally read at least once a week. After I had read three times in one week, I politely requested that the staff take their turn.

I was down to study Advanced English Literature and the course for primary teachers as I had thought that my levels of English Literature would not stand scrutiny in a secondary school. I was wrong but I am glad that I did the primary course as I now fully understand the mechanics and basics of teaching English and Math and also learned the tricks of interesting and lively lesson presentation methodology.

Up until a few years ago, preparatory heads knew little about teaching younger children as their normal promotion route was from a senior school housemaster to prep the school head which always seemed a little odd to me for they had no experience whatsoever of the prep school age group, or reading schemes or the teaching of math or the interests of such children. All that has changed now. I have inducted several heads into prep schools in the South West and believe you me, they are now totally switched on and have even completed the state schools' head teacher courses and that of the independent schools as well. They are highly motivated and very professional.

In addition to the above courses, we had to attend education lectures, RE and PE in the first year. The education lectures were held in the college theatre and all first students were expected to attend en masse. The lecturer was Sir Richard Ackland Bart MA Oxen, who was on the staff of the college by means of a peculiar arrangement. The ancestral home of the Acklands was Killerton which lies north of Exeter and is now completely in the hands of the National Trust because Sir Richard gave it all up and donated it to the nation in line with his socialist principles.

Sometime, somehow Jimmy Smeall became involved and during my time at St Luke's, the house was used ('?), rented ('?) or donated for use (?) by students who formed a completely separate community out of Exeter and were bussed in every day. I suppose for that area it was not unusual because Millfield School in Street in Somerset has dozens of satellite houses at many miles distance and their pupils are also bussed in daily. Sir Richard was apparently part of 'The Deal' and became a lecturer at St Luke's. His book We teach them wrong was on the required book list for all students and was included as part of our Educational Psychology Course.

I liked Sir Richard enormously. He was a kind and pleasant man of good character and we often discussed his politics and the very interesting aspects of his political past which was varied and lively. He was eccentric but that adhered me to him all the more especially as we held a joint interest in history. He was all these things but he was not a lecturer. He could not teach and his lectures were rambling, boring and lacking in pace and style. It is odd because he was a good orator and he could be very witty and funny.

Attendance at his lectures dwindled drastically and a three-line whip was put in place. We were ordered to attend, had to sign in for each lecture and could expect consequences if we did not. On one occasion, twenty students signed in over eighty students. There were no consequences…perhaps they were thinking of reintroducing corporal punishment again? This was my only criticism of Jimmy Smeall. He must have known what was going on and he should have acted rather than allowing the farce to continue.

Much later in my career, I employed a charming and vivacious junior school teacher who had a wonderful personality and adored the children but she could not keep order. My deputy head did not believe me because whenever she visited the classroom, the children immediately behaved. I had to tell her to go around the back of the classroom's outer wall and listen and then she was convinced. The noise was terrible.

Mr Norman Bull, MA Graduate of Trinity College, Oxford, was responsible for the delivery of the RE syllabus. He had been teaching at St Luke's since 1949. He was a large man with a shock of white hair and could have easily doubled for any prophet in the Old Testament. He had 'presence' and was a good lecturer. His battle was with the students who resented anything to do with religion which was rather strange for they had applied to attend a Church of England Teacher Training College. He was a great one for red herrings and much enjoyed going off the syllabus and would be much amused by my historical apocryphal interjections. He had a good sense of humour.

PE was a scream. The college, during my time, admitted many mature students who were mainly ex-servicemen such as lieutenant commanders, majors and wing commanders who were never really comfortable with modern educational methods and ideas and grumbled at us young lads, who had never done National Service, and who they considered to be too brash and energetic though never in an unpleasant way. It was interesting because some of the other mature students had done National Service and there was always that edgy feeling between the two groups.

Anyway, all students had to do PE and it was taken by one of the PE lecturers. It was the time of the new approach to PE involving making a shape or building up your own sequence of movement. Alas, it was beyond the physical stamina of many of the mature students and it was decided that they should be separated from us youngsters and attend another session. Their lecturer was a young and very attractive lady dressed in skin-tight Lycra and she gently eased the mature students around each session but she was very keen on educational dance and started to introduce it in her classes. The word soon got around and the biggest attendance for those sessions was us 'youngsters' up in the viewing gallery. On one notable session, alas the last, she had the mature students sprawled on the floor of the gym and then came the instruction.

'I am going to blow hard, and when I blow, I want you to flutter across the floor like autumn leaves. Ready? Blow.' The sight of the mature students rolling and bumping across the floor was too much. We burst out laughing but quickly quietened with her next instruction, which was to run lightly around the gym and as all the mature students lumbered into action, she shouted, 'Run lightly. That's it. On your balls.'

It was too much; we could not stop laughing. We were banned. The balcony door was locked but I reckon we did the mature students an enormous favour because their PE sessions stopped shortly after.

We were encouraged to do other sports and activities and rather foolishly, I decided to try rugby, and even wrote to my rugger mad godfather informing him of my decision. He was VERY pleased and wrote me an encouraging letter a little while later by which time I had to inform him that I had given up rugby, or rather, it had given up me and it had landed me in the police cells and the Royal Devon and Exeter Hospital.

I played wing in the training session which was fine when I had the ball and could sprint down the wing but what I did not like was those big, heavy guys trying to catch me to make me into mincemeat. I damaged my shoulder twice before I turned out for a practice match when one of those nasty forwards sat on my head and I was knocked out. In those days, you played on and I really did not seem to have any after-effects so my friends and I walked into Exeter that night and went to the cinema. We caught the last bus back.

In those days, all the last buses left the high street at the same time when an inspector stood out on the road and blasted his whistle three times. It was a splendid tradition and generations of students have cheered as the busses drove off together. We cheered as our bus left its stand and then I do not remember much else. Evidently, I flopped around on the seat and could not speak properly and rolled onto the floor.

The bus conductor was having none of that. He stopped the bus outside the Heavitree Police Station insisting that I was drunk. I was heaved off the bus, none too gently I might add, by two policemen and as I was unconscious, I could give no details so was thrown, and I mean thrown, into a cell. All this time my friends were insisting that I had not been drinking but the desk sergeant told them to clear off. There were no emergency numbers for students in those days and my friends did try to raise Fred Smith, my house tutor, but he was out on a date with one of his lady friends. Two tried to get back to me but could not gain admittance.

After two hours, the police doctor arrived. I was more alert and was able to tell him about being knocked out. He looked quite worried and rushed out of the cell leaving the door open. Five minutes later, I heard an ambulance siren and two ambulance men arrived, strapped me to a stretcher and I was rushed to the old Royal Devon and Exeter where I was immediately diagnosed with concussion and put in a hospital bed next to the nurses' station to be kept under observation for twenty-four hours.

I woke up (or came around?) quite late the next morning to find a police superintendent by my bedside. He was obviously concerned and well might he have been. The consequences of such actions nowadays would have been very severe. I was still muddled and I realised he wanted me to sign something. I was saved by the matron who was nearby. She gently took the superintendent by the arm, led him away and then proceeded to give him a real dressing down.

She returned to my bed and simply said, 'Don't sign anything. You understand?'

I was discharged late on the Sunday afternoon without any of the college authorities aware of what had happened and they did not know until the following morning when I reported to matron with the hospital medical report. That was that. I did receive a letter of apology from the police but nothing more. Anyway, thus ended my rugby career and the long cherished hopes and aspirations of my godfather. The college doctor gave me a thorough examination that morning. He was odd and quite strange in manner. Other students had warned me about him—I thought it was all made up but it was true.

As soon as you entered the college surgery, with whatever complaint you had, he made you strip naked with Matron in the room. I stripped and he proceeded to give me a thorough medical examination concentrating on the more intimate parts of my anatomy. I had suffered from concussion but he did not touch my head. Matron, Mrs D Butterfield, was pleasant enough but she was a fanatical bird watcher and it was best not to seek medical advice on the arrival of the avocets on the Exe Estuary. Her one cure and treatment was aspirin which was her panacea for all ills. I came close to very serious trouble in the following year because of her dilatory methods. I took my verruca problems to her but I think they were beyond her medical knowledge and eventually, I had them cauterised privately.

It was decided that it would be better if I tried rowing. It would build up my weak arm and it was not a contact sport. Now I know many of my friends will doubt the veracity of the next remark but I was told that my body frame was too light for the vigour of rugby. I duly presented myself to the captain of rowing at the college's boat house down at the Port Royal on the River Exe. The boathouse is now an extended dining room for the pub. I tried. Truly, I did try.

I could row after a fashion, and the technique was easy to master but I quickly realised that I could not see where I was going, only where I had been, and I detested being screamed and shouted at by some little puny guy sitting doing nothing at the back of the boat. I lasted for about a month, and then, one afternoon, I noticed two students lugging a canoe out of one of the lock-ups on the quay. I walked along the quay and made enquiries. It was the College Canoe Club and they needed members. I tried and I was hooked and I have now been canoeing for well over fifty years. I took to it like a canoe takes to water.

There was an anomaly at St Luke's. The college's collection of canoes was rather elderly hence the use of the term 'Collection.' They had several double canvas PBKs (Percy Blandford Kayaks) an NCK (National Chine Kayak) which was a very elderly plywood racing canoe and another old single of doubtful vintage. In these I learned to canoe on the River Exe and our 'waters' were from the quay upriver to the town weir and a short distance down river to the Port Royal Weir. It was good enough to begin with but there were no wet or dry suits in those days. We canoed in a warm top, always shorts with daps on our feet. Brrrrrr! It could get very cold in the winter and we had to wait to get back to college for a warm bath. Yes—in the 'walkthrough' bathroom.

Brian Woods was a PE lecturer at the college and was also the Olympic Coach for Canoeing but seemed to take no interest in the College Club. I politely tackled him about this one day and I think he was slightly embarrassed. We were then allowed to use the college workshop to build two Ottersport Single Plywood Canoes which was a step up from the wrecks we were using in the boathouse. We had no instructor and just muddled along. Pete Lee, the Chairman of the Exeter Canoe Club which was situated upriver near the town weir, was inordinately kind to us and gently nudged us along giving direct help, advice and instruction. It was he who roped us in for life saving sessions at the Exeter Pool and we also learned canoe rolling there. I gained the Bronze Medallion and the Award of Merit during those sessions.

Over a period of eighteen months, I managed to build up my canoe skill levels to a reasonably high standard and qualified as a British Canoe Union Instructor and indeed, at the present age of seventy five, I still hold a current award. I passed the actual test using a borrowed fibre glass kayak in the freezing cold waters of the River Dart in mid-December in a snowstorm at what was then the Outward Bound School at Holne—now the River Dart Country Park and I had to walk to and from the station as I had no transport then.

In company with the Exeter Canoe Club, we also made several descents of the Exe and once we had acquired fibre glass canoes, weir shooting became a favourite pastime. As soon as I qualified as an instructor we were able to offer canoeing sessions to what was then The Royal School for the Deaf, new college students and the Scouts and I also led several trips down the Wye for the Youth Hostel Association. The latter was paid. I have canoed on the Wye many times since then and often took my school groups on to the river. Their favourite part was always the Yat Rapids. Advanced college trips took in the Upper Dart and Teign but they became unviable due to the cost of repairing the canoes!

My favourite trip was always down the Exe to Topsham and out onto the estuary landing on the back of Dawlish Warren to avoid the tide race on the other shore at Exmouth. We would then paddle back up the Exeter Ship Canal to the quay. I once persuaded my mad Uncle Jack to pick me up one Sunday on his way through Exeter to Dawlish, including my canoe. The sea at Dawlish was quite rough and I enjoyed surfing for a little while and then my uncle insisted that he had a go. I told him it was too rough but he insisted and got into my canoe wearing…his suit. (I did warn you that he was mad) He capsized and blamed me. The previous year he had hired a rowing boat to go fishing offshore, lost both his oars and had to be rescued by a speedboat. His antics were the stuff of legend in Dawlish town.

What did we do for entertainment? Exeter is a pleasant city and I could easily live there and I know all its historical sites. I still visit it regularly in the course of my work and enjoy its environs. As students we visited the new Wimpy Bar of which we were devotees (the Exeter one could not serve bread—only toast), and the university and college held regular 'bops' in which drink was available, and we visited Bobby's the department store for afternoon tea and Muffins in the roof top restaurant. We drank, rather strangely, schooners of sherry, but I cannot remember being blotto or drunk. Perhaps we just did not have the money though I must say my grant was more than generous and I also found jobs in the holidays to earn extra cash.

I did have some odd jobs. I worked on a Market garden at Oldland Common as a 'Tomato Stripper'. This entailed stripping out tomato plants from 100ft greenhouses. It was very hot and sticky work and when I got back to the house in Church Road, I had to strip off outside the back door and wash my clothes, and me, under a hosepipe. Mrs Crewe, our next door neighbour, complained to my mother that she could see me naked at the back door while I was washing off which involved her having to lean out of a small upper bathroom window to see me thus. We also picked potatoes which was back-breaking. I also worked as a 'Faggot Stripper'. It is not what you think!

This was at Brain's Factory on Station Road leading up to Rodway Common. The faggots would be fast cooked in large ovens and when they were ready, I would stand by as each tray was automatically pushed vertically upright in front of me and then I would strip them off the tray into a basket with special gloves. When the basket was full, I would pass it over to the production line where two ladies would place two faggots each side of a foil tray, push it under the gravy squirter and another two ladies would place cardboard covers on top of the foil. That was the job, all day and every day.

There was light relief one morning when one of the ladies stated that she had lost a plaster on her finger. Every tray of that day's production was reopened until the plaster was found. After ten days of this, the foreman was going around asking for night shift workers to stack the faggots into the fridges at night. My hand went straight up and I started the night shift at 10 pm the next night.

There were only two of us in the factory. The chap who made the gravy and me. I had to dress in a polar suit and started pushing the stacked trolley into the fridges. After two hours, I took a break.

'How ya doin'?' asked the Gravy man.

'OK,' I replied. 'I've done twenty units so far.'

'You'll 'ave to slow up, son. Takes two men to do sixty units per night. Better go and 'ave a kip. Eh?'

I felt guilty but I did not want to give anyone a hard time so I used to sleep for half of the shift every night. My mother could never understand why, on my return from work, I only slept for four hours. On the penultimate night, one of the fridges broke down with fifty units inside so I shifted the lot into the next freezer and packed the day's production away as well but on my final night, the manager was waiting for me.

'Geoffrey. Thank you for saving the stock but who helped you last night?'

I assured him that I had moved it all and then he asked other searching questions. He got the picture. The company gave me a two hundred pound bonus which I declined initially because I felt that I had been cheating the company on 'time'. They insisted because I had saved thousands of pounds worth of stock.

Every December, I worked with the Royal Mail during the Christmas Rush. I was very lucky because I was put on parcel delivery with my older cousins with their furniture vans so I had a fairly easy time of it until we had to deliver to the Barton Hill Flats complex, where, of course, the lifts never worked. Carrying heavy parcels up twenty flights of stairs was not easy. During the second year I had a much more pleasant round out of the Kingswood Post Office around New Cheltenham Road and Syston Common.

Some weekends I would go back to Bristol. The procedure was always the same and better than any students card or bus pass. I would dress as smartly as I could, never casual, don my college scarf and catch the bus out to the then famous Exeter Bypass where I would proceed to stick out my thumb (do not try this nowadays). Within minutes I would be picked up and it rarely took me more than three hours to get back to Kingswood. It always worked like magic.

One day, I cheekily thumbed down a Rolls Royce and it stopped. The driver was a pottery owner from Staffordshire and to my amazement, the back seats of the car had been taken out to accommodate about fifty land drains which were his samples. I always caught the train back however on a Sunday evening. I thought I would give my parents a surprise one weekend and unannounced, I hitched up to Bristol and caught the bus out to Kingswood. When I walked down to the house, however, my key would not turn in the lock so I rang the bell. A strange lady answered the door and asked what I wanted. I told her that I was Geoffrey Allen and I lived there. She laughed.

'I am afraid your parents moved ten days ago' She smiled.

I had to phone my Uncle Jack from a phone box and he kindly came over to Kingswood to collect me and took me over to Valley Gardens to the bungalow that my parents had just purchased.

While checking up on my various facts for St Luke's, I made an unusual discovery. Before my time the campus was run as a holiday centre and to a very high standard including the accommodation and food. Staff and students were employed as required and they catered for hundreds of guests. The photographs of the period are quite clear and it was certainly not a cheap operation. I wonder how much it would have cost for the week?

I became secretary of the College Scout Club and we did meet occasionally but the role was mainly to link students who were Scouts to the various troops in the city requiring help though we did run a mammoth Senior Scout Marathon Night Hike to the South of Exeter with incidents and refreshment points and many non-Scout students joined in the fun. I was assigned to the 4th Exeter (Ladysmith) Troop that met in its Scout Hut, in the Clifton Hill area of Exeter (The hut is still there) not far from the college and I enjoyed the three years I worked with them. Strangely, I cannot remember camping with them.

I did assist on a youth hostelling trip to the Isle of Man one year and crossed the Mersey four times in one hour. I do not remember doing Scouting activities either on the trip. It was more of a holiday visit somehow. I did return to The Isle of Man in my business career and I judged the Miss Isle of Man Beauty Competition one year. That was a first.

Peter Spivey, the Scout leader, was very good at handling the troop, and though it was not a large unit the meetings were fun and friendly. We did make occasional visits to Dartmoor using the Okehampton Line, for Peter was a railway fanatic, and I was ever anxious after that to get up on the moor as much as I could. While Scouting in Exeter I also did my leader training at Chivenor and gained my Woodbadge—the Advanced Scout Leader Award at Gilwell in 1966 where I also worked for several summers mainly helping out in the stores helping to run the Woodbadge Courses.

Most courses at Gilwell were camping based and could spread over two sites. Everyone camped in patrols and they were always—wait for it—Owls, Peckers, Pigeons, Cuckoos, Curlews, Ravens, Wolves and Bulls. It was quite simply our job to organise and deliver all the equipment required for each session and then collect it at the end but there could be two or three courses running at any one time and some were more high powered than others.

The biggest challenge was locating all the various national flags of the course members who were in attendance which were flown on the main camping field. We held a large stock but it was very difficult to maintain and this was the time when independence was popular and new countries were emerging thick and fast.

The most memorable course was the Chief Scouts of the World Course and then the Training the Team Courses were quite intense. One summer, I also helped out on the Boys' Field which is the main campsite for the park. I was able to help out canoe instructing on Hitler's only gift to Scouting. A stray bomb from the Second World War caused an enormous crater in the Boys' Field which is now full of water and perfect for boating! While at Gilwell, I was chosen to take part in a Schweppes (Sh! You know who!) advert and was filmed with their MD at one of their new mobile bars.

While working at Gilwell, I had met Lady Olave Baden-Powell on several occasions as she still maintained an interest in her husband's work and on one day I had acted as her aide for several hours and we had several chats while she was walking around the site. She had a lively mind and I sensed her nervousness about the soon to be published Advanced Party Report with its recommendations for 'modernising' scouting. She was concerned about any possible neglect of BP's fundamental ideas and organisation

I was rather in the dark for none of us had seen a copy of the report but Lady Olave could not be put off by trite and bland platitudes. She had a very sharp mind and I could foresee problems ahead. I did speak in confidence to one of the camp chiefs about my concerns but there was nothing I could do. The report was published and she declared that she was not happy and threatened to move all BP's memorabilia to the Boy Scouts of America.

I was working in the store when one of the camp chiefs came in and asked to borrow me for several hours. We were busy and the request did not go down well but I was asked to put on a clean uniform and then we drove to Hampton Court in which Lady Olave had a grace and favour apartment. She was not angry but she was sad and we discussed the Advance Party's Recommendations in some detail. Finally, she asked me for my candid opinion.

I emphasised that BP was a man that was way ahead of his time and the Scout programme he had devised was innovative and totally appealing to boys and that was still the case today. There were millions of Scouts throughout the world and it was all due to his vision. His overall vision would not be changing because of the recommendations, it just had to be modified to fit into the modern world and if BP was still alive today he would be the first to adapt and keep up with progress because BP was progressive. I do not think that she was totally won over but she did agree to await developments and the mementos never went to America.

At college, we had to do a teaching practice at the end of our first year and then another towards the end of the three year course. I was at an Exeter Primary School for my first practice working with 10-11-year-olds which I much enjoyed and I was regularly left alone by the class teacher which was not supposed to happen. I was grateful for his confidence in me but he was slipping off for longer and longer periods. Not only that, he was very impressed by a large set of Maths work cards I had made on various topics and they disappeared. His report on me was important and it was glowing but I did report back to the college on his actions and he was not allowed to supervise students again.

My second teaching practice was at Bideford Grammar School—a teaching practice to which I was appointed. The A Level English teacher had a sudden serious illness on the run up to the actual exams and I was sent up post haste to try and plug the gap. I was lucky. The ten 6th Form pupils were studying the same course in which I had passed two years ago and it was for the same Southern Universities Joint Board that I had done.

We had four weeks to the exams so I set up a very controlled programme of revision and also saw every boy for a tutorial twice a week and I did much marking with full comments. This needed careful handling with the college as the grading of the Teaching Practice assumed that you were teaching a variety of lessons to show off your skills. I need not have worried because Jimmy Smeall had smoothed the path for me and I emerged with high credits. My 'A' Level grade predictions were 90% accurate. One student had a motor bike accident and did not do the exam.

You may perhaps have noted that I have not yet dealt with any aspects of my English course at St Luke's and I must hold you in suspense for a little while longer, I am afraid.

My great friend at college was Barry Cramp who was studying chemistry and was also a keen canoeist and he was a good friend because he had a motorbike. He was also involved in doing the make-up for the college plays. When I first knew him his home town was Taunton which he still blesses with the occasional visit, and then he married and lived in Chatham, after which I lost touch with him until several years ago when I tracked him down to Bunghay on the Norfolk/Suffolk border. He has happily remarried and both he and his wife are now ministers.

He taught chemistry for many years in the local schools and also specialises in Big Bang Demonstration Science lessons and fancy dress weddings. Well. Why not? Do your own thing, I say. When we meet, we talk and talk and then he tries to pinch as many of the silly songs I know. I believe the church to be much vitalised by them both. They are both very popular. They are both good cooks which is another good reason for visiting them as much as possible. The last time we met we got to talking about the various staff at St Luke's and I gave him some titbits about staff meetings.

Staff meetings! How do I know about the contents of staff meetings at St Luke's? They were all fully minuted and they are held in the Exeter Archives. They discussed very exciting things like staff parking and the fact that their studies were too small. They discussed the introduction of CCTV and the possibility of students sitting on the Academic Council. There was much discussion about a complaint from VSO and a Russian production of Hamlet at the Northcott. Please put the word 'Russian' on ice for a little while. There were warnings of student unrest. (I was there then and I do not remember being unrested)

There were heated discussions concerning Jack Straw's comments about religious-based colleges, 'which still have regimes more like that of a nineteenth-century seminary than a twentieth century institute for the education of teachers.' Of more interest, perhaps, again during my time, they discussed the punishment policy with reference to Gating. There were staff worries about the amalgamation with the university. There was another lively debate about the poor PE students struggling with their second main subject, ah! Bless!

And at the end of nearly every meeting comes the plaintive wail from Jimmy that there were too many meetings. The best item I gleaned from the archive referred to Sir Stanley Rouse, then a St Luke's student who had to climb out of a small window, thus breaking bounds, so that he could take his referee exam in the town. Another good friend was Peter Brewer who was also very keen on Scouting and helped in the College Scout Club. He organised a whole bunch of us to complete a Bessey Youth Training Course which also provided him with his future career as a youth leader. He and I were stopped by the police in Exeter City Centre late one night and they were keen for us to explain why we were carrying a park bench on our shoulders.

Peter was able to explain that he had won it in a raffle. That did not go down very well and we were advised to tell the truth. The problem was that we were telling the truth. Peter had won the bench in a raffle at a student dance that we had attended that evening and even had the winning ticket in his pocket.

During my second year at college, I still retained a room at Rowancroft and this one was a single. It had, however, two disadvantages. Firstly it was situated directly over the big bathroom below and secondly I had to walk through the TV lounge to access it but it was large and spacious and all mine. College life was now all routine to me and as all the preliminary courses had been completed and passed, I could enjoy myself much more and concentrate on my English Literature. It was not to be. I woke up one morning with stomach ache and decided, foolishly(?) to seek Matron's advice. By that simple act of making a college clinic visit, I nearly died. It was evidently a close run thing.

Chapter Nine
Reds Under the Bed

The English department at St Luke's was very well qualified in line with Jimmy Smeall's policy of recruiting the best that he could find. Fred Smith, my house tutor, was Head of Department and MA (Cantab.) and it was he who had conducted my interview prior to my acceptance. He also wrote what I would term, lightweight poetry, and he had a very pleasant light manner of teaching. He was apt to keep himself to himself, however, and I really did not get to know him that well. I only did one half termly session with him on the Romantic Poets.

The other staff in the department were of equal professional status and provided high class teaching with only one weak link. Edwin Crowle MA (Cantab.) and Reading was also a quiet and thorough tutor with whom I studied the War Poets. I met his wife at several college functions and was enamoured of her name, Marie-Theresa, which I knew to originate from Austria and was also the name for a chandelier!

I came across both of these tutors again quite recently. I was visiting the widow of Peter Spivey, the former Scout Leader of the 4th Exeter Scouts, for a pleasant lunch. Her house backs on to the Ide Church and graveyard in which Fred Smith is buried and there is also a commemorative plaque to his memory in the church. I was also lucky enough to contact Edwin Crowle, the only surviving member of the English Department, now aged 91 and still living quite close to the college. I spent a morning with him and his charming wife and we were able to reminisce about college life fifty years ago.

Donald Cross BA (Dal.) (Daloise College in Canada) and MA (Birmingham) seemed, to me, to be the weakest link in the department. I did not doubt his knowledge of English Literature but he was so boring. I had one half termly session with him on the Essays of Steele and Addison and his teaching was so uninteresting because he would read from prepared notes. I find it odd because he has since distinguished himself in the field of Mathematics by cracking some long-term Math problem associated with Fibonacci Multi Grades. I hope this makes sense to someone? Perhaps English Literature was not his first passion? He marked me down on the end of course essay, obligatory for all English courses, because my prose style was too hyperbolical. I was only trying to enliven an otherwise monotonous course.

We must now move on, however, to the two 'heavyweights' of the English Department. John Money, Jack, was a large shambling figure of a man, in his forties, and his clothes seem to hang loosely around him. He was a chain smoker and he smoked through every tutorial I ever attended with him so his study always smelled of cigarette smoke and so did we when we emerged out into the corridor. He had 'presence' which was backed by a wonderful baritone voice and he seemed to evoke a slight air of mystery.

It was rumoured that he spoke Russian and had a Russian wife and furthermore, spent most of his holidays in Moscow, to which I attached no special importance at that time. He was a tutor under whom I studied Shakespeare and the students particularly enjoyed reading the tragedies with him, especially King Lear and Macbeth and he was amazed at my knowledge of Macbeth, most of which I knew by heart. He was a natural actor and took over the part of King Lear with only two days' notice in one of the college productions and it was not until much later that I realised how he had managed to achieve this feat.

The Reverend Mr George K Booth was the other 'heavyweight' English tutor and I am sure he would be very amused at my use of the term because he was very lean and thin. Unlike the other English Tutors he did not have a study on the main campus because he was house tutor of Nancherrow just across College Road. He taught in the lounge of his flat which had big sash windows leading out into the back garden which was for the sole use of his Alsatian dog called 'Dog.'

'Dog' was a character and many a college myth developed around the animal. George took him for walks down by the canal most afternoons and let him loose on the towpath while he read poetry or a novel as he ambled along smoking his pipe. So absorbed was he in his reading that he forgot that the Exeter Canal has bends and walked straight into the water. 'Dog' just walked on.

One morning we were having a tutorial with George who was expounding about a wonderful wall chart that one of the more conscientious students had produced with a historical time-line linking arts, literature and music and was spread out all over the floor in its pristine splendour. Enter 'Dog' with a big leap from a very muddy garden. Dirty paw marks all over the splendid chart.

'Oh dear!' exclaimed George. 'What can we do?' and turned to me for help.

'Well, we shall just have to 'pause' for thought, George,' I responded.

That nearly brought the house down and I did not live it down for years either.

It is alleged that 'Dog' also assisted George in the marking of the PE students' English work. Fun was always poked at the PE students for their lack of academic prowess and George never looked forward to marking their essays. It is 'rumoured' that he read them to 'Dog' who would bark for approval and stay silent for poor work. One PE student had the temerity to question the low grading of his essay when he accosted George in the college precinct one morning.

'Nothing to do with me.' said George. 'You had better go and see Dog.'

George was related to the Liverpool Booths though there was a connection within the family to William Booth, the founder of the Salvation Army. He was born in 1912 and gained his BA in 1937 and his MA in 1940 at Pembroke College Oxford where he gained his 'Blue' on the hockey field. He attended Wells Theological College in 1937, was ordained as Deacon in 1939 and Priest in 1950. He had a terrible war.

He was curate in Portsea and Portchester in Portsmouth from 1939 until 1941 and this was the period when the area in which he was working suffered 67 air raids, 930 civilians were killed and there were well over a thousand casualties. Hundreds of houses were destroyed and 6,000 damaged. Thirty churches were bombed out. I cannot be sure but from my student conversations with George, it is possible that he suffered a nervous breakdown in this period for his next post was an assistant master at Epsom College, whose war time diaries are very interesting to read, before moving on to Felsted for one year and then Fettes College Edinburgh from 1947 until 1959 as House Master and English Teacher and where he produced many plays.

He then became curate of St Ninian's in Edinburgh for five years before moving as Chaplin to St Luke's where he also taught English. He retired in 1978 to live out his days in Rock, in Cornwall, where I maintained contact for many years especially when I had a bungalow in Padstow.

While at college, George would often spend his weekends at Rock. His mode of transport was a Series Three Land Rover—a very valuable vehicle these days, which he had purchased brand new. After a couple of trips down to Cornwall and back he was moaning, in our tutorials that it was too slow and it was taking him ages to get up and down from Cornwall. One of the mature students, an ex-army driver, agreed to go out with him for a test drive (he was a brave soul for George was a terrible driver), and quickly discovered the problem.

George was completely unaware that the Land Rover had three sets of gears and he had been driving in the lowest ratio gears up and down from Cornwall. I drove the Land Rover sometimes when I visited him in Rock and he used it once a week to go shopping at the top of the village. A mile up and a mile back. He had purchased it in 1965 and I was driving it then in 1985 onwards and it only had 10,000 miles on the clock.

George was a great tease and would often gently mock the college and staff. Donald Cross he called 'Little Cross' and he had nicknames for all the staff. With his wit and knowledge his tutorials could not fail to be interesting and everyone was keen to study with him just for the pure entertainment value. He would tease the PE students to their face and they loved him for it. They had a reputation for being, how shall I put it…non-academic? They strode around in their tracksuits looking athletic and full of vim and vigour.

When I was recovering from an appendix operation, I had to spend a couple of days in college until transport to Bristol could be arranged. The college authorities, in their wisdom, put me on the third floor (no lift) of the main campus block and it was quite a struggle to get down to meals. As I was slowly walking along towards the dining room one lunch time slightly hunched over, one of the PE staff came up behind me, slapped me between the shoulder blades and shouted, 'Straighten up, lad. You look like an old man walking like that!'

When I informed him that I was recovering from an operation, he had the good grace to look embarrassed but offered no apology.

George would hold 'Open House' on a Wednesday evening and though the session was initially well attended it dwindled to just me in the end, but Oh! the English that I learned was amazing. We launched into the Holinshed Chronicles, trotted on to Chaucer, through John Donne and the Metaphysicals, the Early Miracle Plays, Shakespeare galore, followed by Ben Johnson and Thomas Fletcher, then onto all the poets; Byron, Keats, Dryden, Pope, the eighteenth and nineteenth century novels, then back to Wordsworth, Shelley and Keats, Gothic novels, Victorian literature, not forgetting Dickens and then ending with First World War poetry, Gerard Manley Hopkins and T.S. Eliot.

I found it wonderful and all thanks to George for the teaching and it was fun. I still delight in English to this day and revel in its richness and quality. George always declared that he was writing a novel. From the year I first met him to the last time I visited him in 1995(?), the novel was allegedly still in progress. I searched the net but the book was nowhere to be found, but then Edwin Crowle confirmed that it had been published privately and described it as, 'scurrilous pornography.' Now I must get a copy somehow! George? The Porn Again Minister?

There can be no doubting that the English department at St Luke's was interesting, varied, professional and successful and I obviously enjoyed most of the courses.

I must just interject a rather amusing incident that occurred while I was at the start of my final year. The only good thing that came out of my appendix operation was that I started going out with one of the nurses who had cared for me. We had quite a good time together in the social life of Exeter and there was always a 'medical' party to attend on a Saturday night that added to the fun.

One evening I visited her in the Nursing Home and overstayed the curfew time of 10.30 pm 10.45 pm—panic. Matron was doing a night round. I had to quickly scramble out of a small lavatory window at the rear of the premises and cut my hand so badly in the process that I had to walk around the block and get it treated at casualty. Of course, everyone was in on the fact that I had been in the nursing home after hours and how I had cut my hand and they were all laughing about the mishap and teased me rotten.

In my final year, I signed up for a Children's Theatre Course, under the head of drama, David Kemp. He was a character larger than life and had red hair and a red beard. He was Falstaff to a 'T' and was an actor of some repute taking parts in many productions at the Northcott Theatre in Exeter and in the College Theatre as well. I recommend that all students do drama as it boosts confidence and the voice. Indeed, many university law students take drama as a second subject to help them with their legal deliveries and courtroom speeches.

We were aided by two mature students who were obviously 'actors' by their voice and speech—Patrick Verrier and Carole Tarrington both of whom I cannot locate now but they were the backbone of the play, Angel of the Prisons, by Brian Way with which we toured Devon schools during the whole of one term. Thus was generated my love of drama and I went on to produce many long and short plays in my teaching career.

My next statement is sure to get me into trouble because the drama course was successful in eradicating the last vestiges of my Bristol accent. Sorry Bristol, but it was proving a handicap, I am afraid. Many of my colleagues think that I now have a posh accent and it is definitely Middle English. I am well-spoken and it has been a major asset to me over the years. It has opened doors for me and I am able to access many people directly by phone on the basis that I am well-spoken especially in the educational world. I make no secret of the fact.

At the end of my final year, I faced an external examiner, and looking back now I realise that it had been arranged and perhaps even 'fixed' for I was interviewed by Ted Ragg from Exeter University. I suspected the 'hand' of George Booth and Jimmy Smeall. After checking my name and exchanging greetings, he pushed a copy of 'The Prologue' to The Canterbury Tales across the desk and asked me to read it aloud. He was somewhat surprised to have it pushed back towards him and I commenced aloud:

When that Aprile with its shoures soote
The droghte of March hath perced to the roote
And bathed every veyne in swish licour
Of which vertu engendered is the flour…

He was clearly taken aback. 'Do you know it all by heart?' he exclaimed.

I replied in the affirmative.

'And what do you make of this?' he asked, pushing a rather battered copy of Holinshed's Chronicles towards me. I explained it was Holinshed's Chronicles which Shakespeare had obviously consulted for the play Macbeth.

'Anything to add?' he enquired.

'Yes,' I replied, 'this copy belongs to the Rev Mr G K Booth.'

'Touché,' he said.

The game was up.

' By god,' he said, 'You should be at university. I can get you straight into Oxford now!'

I explained that George and I had already discussed this at some length. My parents had fully supported me through sixth form until I was eighteen and though my grant was generous and I worked in every holiday to help, they had also helped with my college education. Another three years was just not on.

'Well. If you ever change your mind…'

I think I have to thank Fettes College and George Booth for my first teaching appointment where he had taught for twelve years prior to starting at St Luke's. He had taught there with a colleague called Richard Stoker who was now the Headmaster of Wycliffe College Junior School and who needed an English teacher and more importantly, a Scout leader, to run the School Troop. At the interview there was no mention of English teaching, only the Scouting aspect of the job. The post offered free board and lodging in return for boarding duties and I was hired there and then. So concluded my college days and as they often say, all things being equal that would have been that—until I started researching this autobiography which may well have led me into muddy waters.

The more observant of you may well have noted one common factor concerning the staff of the English Department at St Luke's. For each tutor I carefully added their qualifications and universities so as to give my readers a hint. Apart from George Booth and Donald Cross all the others went up to Cambridge. Jimmy Smeall also studied at Cambridge. There could be no problem with this grouping. The 'Old Boy' network could simply be at work in its usual way and universities still maintain this system. I obtained my first job through the 'Old Boy' network.

Enter another factor. Jack Money could speak fluent Russian and 'may' have had a Russian wife. Having been educated at Radley College where he was a prefect and scholar he elected in English at Jesus Cambridge in 1942. He enlisted in The Buffs that year and served with them in France, Belgium, Holland and Germany eventually being promoted to Battalion Intelligence Officer, where he was 'mentioned in despatches.' He finished his war as a staff captain in the Army Educational Corps in East Africa. He came up again to Cambridge in 1946 and gained his degree in 1948 by proxy for his 'substantive war service.' It is 'possible' that his post as Battalion Intelligence Officer and his 'mention in despatches' may have involved him in clandestine operations in Russia and he 'may' have learned Russian during that period, or, just prior to that.

While at Cambridge, he was secretary of the Marlow Society and acted with them for two years. On 'going down' he joined the Shakespeare Memorial Company at Stratford and toured in Australia and Germany. This is why he was able to take over the part of Lear so competently and the source of his great voice. He lectured for the British Council at home and abroad and for The Cambridge Board of Extra Mural Studies. In 1959, he published a novel, The Impressario, which was well received by the critics. In 1966-7 he was seconded under the aegis of the British Council to Moscow University as Professor of Linguistics for which he had to brush up...fast. That was during my final year at St Luke's.

Where are we at? I was at college from 1964 to 1967. Before my arrival in 1963 Kennedy was assassinated and more importantly, Philby had defected. Philby was part of the Cambridge Five and was recruited by Anthony Blunt in the Thirties at Cambridge before the Second World War who had also recruited Burgess and Maclean and John Cairncross as the fifth member. They passed secrets to the Russians during the Second World War and into the early fifties.

James Smeall and Jack Money were at Cambridge before the war though Jack Money was there only briefly and Jack Money and Edwin Crowle were there after the war. Blunt was still recruiting all that time and it is possible(?) and/or feasible(?) that they were approached by Blunt, is it not?

We move on. It is a fact that Jack Money made regular visits to Russia during his time at St Luke's—there is no secret about that and it would appear that Jack Money was a crony of Philby and saw him regularly in Moscow after his defection. Philby seems to have had no restriction concerning visitors. Was John Money a drinking companion of Philby? The Gang of Five had always been regarded as 'drunks' and Philby hit the bottle hard having become disillusioned with Communism.

It is also fact that Jimmy Smeall visited Russia on many occasions lecturing firstly on eighteenth century poetry but mostly on English dockyards. I find that curious piece of information very odd indeed. This was the time of the Cold War. This was then time when the Russians were collecting as much information about England (and other countries of course) and they were doing so by any conceivable method they could. Inside the secret world of the Cold War the Russian mapmakers were producing highly accurate maps of their perceived enemies' territories. They were stunningly detailed and of an incredibly high standard. Many are for sale in cartographic auctions and they are beautifully produced. Surely Jimmy Smeall was giving them what they needed on a plate?

This was all under the aegis of the British Council. In a summit meeting between MacMillan and Khrushchev in 1959 the first Anglo-Soviet Cultural Agreement was signed and the British Council and Foreign Office were very keen on all types of exchanges to achieve Cold War aims. There was a massive input for the teaching of Russian which took place at Bodmin and Cambridge. The aim? One thousand five hundred Russian speakers.

One of the first priorities of the British Council was to promote an alternative cultural programme within Europe to balance the cultural propaganda of the Communist and Fascist countries. It was this programme in which St Luke's staff were involved and they even sponsored a Russian translation of Jimmy's book on satire. Researching further I discovered that for all visits the lecturers were briefed and debriefed by 'British

Council Staff' to glean as many details as they could about Russian life and activities. I am also assured by a reliable source that Jack Money would have been regarded as a very valuable asset to the Intelligence Services and he would have maintained regular contact with information gatherers in the UK especially as he was friendly with Kim Philby.

Strangely, however, though there are full records relating to British Council activities in the other countries in which they operated the Russian ones 'cannot be found.' How strange? The National Archives regretted that they could not help me further. Strange as well is the fact that there is no reference in any of the college's extensive archives at the university or the South West Heritage Centre which holds the records for Devon of any visits by college staff to Russia.

Chapter Ten
Poetry

The first three poems were written by me as a rather precocious teenager and I make but two comments on them. Firstly, and very unusually for that day and age, I was encouraged by my secondary school in my endeavours and secondly, they were entered in an adult poetry competition many years later, with most of my other verse, and the judges selected these first three as winners and ignored my adult ones.

Life

Sleepy mist coiling through trees
Snakelike as dragon without wrath.
Feeling trees, like slender spires erupt Stretch.
Pointed fingers reaching for the sky.
Stretch.
For the bleeding and watery sun
Casting beams on copper clouds.

Fireball sun with golden lust,
Searing heat turns life to dust.
Scorching. Squeezing. Shrivelling life.
Murdered by Phoebus's bloody knife.

Life turns to gloom. Shepherd's sky to bed.
Defeated sun like Phoenix dips its head.
Reluctant fiery rays are doused.
Instead pin lights sparkle and the grey-haired moon,
Foretells life is doomed.
Bringing—peace.

Defeat

Wind weary, battle worn.
Bows creaking, sails flapping.
Useless.
On deck, silence.
Defeated.
Rigging tangled. Masts jagged.
Carnage.
Blood with blood.
Hollow men shrinking.
Hideous faces.
Memories of agony
Twisted with fear.
Screaming.
Fighting fellow men
For life or death.

Prisoner

Four stone walls sweat beneath the naked light
Throwing harsh beams onto the door:
The door of death,
Scarred with scratches of human hands
In agony.
Hands that have clawed and clutched for hope.
But there is no hope.
Dawn will come and with it death.
Cold, clammy hands reach for cigarettes.
Waiting.
Waiting for what?
Death's herald.
O pray that it be quick.
Pray for hope.
But there is no hope.
Look!
The sun is rising,
Symbolising Death.
Your death.
Hollow foot treads.
Death's messenger is near
There is no hope.
An empty life.
Be brave.
Be strong.
Be resolute.
Death has come.

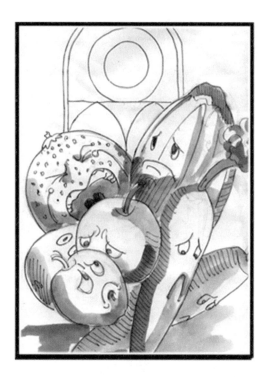

This poem was composed after a choir visit to a harvest festival service at a village church with no heating.

Sad Harvest

Polite coughing.
Reverent whispers.
Fidgeting.
Sad fruit and vegetables.
Congregation,
Shivers,
Huddles.
Vicar fusses around lectern,
Creeps to his draughty vestry.
Organist flaps at keyboard.
Organ starts with a protesting moan.
Coldly drags itself to life
With a groan of chilly pipes.
Reluctant pedals
Push wheezing music
Aimlessly amid the pews.
The withered fruit looks on.
The bright faced choir sweeps down the aisle
Followed by the lagging, shuffling vicar,
Files into creaking pews and rustling
Wait, watched by the coldly drooping veg.
'Dearly beloved brethren,' the vicar drones,
Intones in a wavering chant,
And a sniffle.
The attention wanders
To a stale looking loaf,
Half Hidden,
Huddling against a limp lettuce for warmth.
The choir trills into frozen song.
Hymn to Psalm.
Psalm to prayer.
Prayer to hymn.
Mind numbing sermon. '
The glorious harvest,
Wonderful fruits of the earth.'
Shivering fruit.
Shrivelled vegetables.
They know they will be eaten
But coldly smile
At the congregation turning blue
And a deep frozen sniffing vicar.
A chilly ecclesiastical indigestible meal.

A French Interlude

Beware all motorists Anglais
Of blaring horns and devilish ways
Of swerving cars and autobuses,
And, of course, the Frenchman's curses.
While travelling down a country lane
A car behind seemed quite in pain.
It swerved from one verge to the next.
Indeed the driver seemed quite vexed.
With cries of 'vache!', 'chien!', 'cochon!'
He waved his fist and blew his horn.
Leaned out the window in his anger.
Even called my car a 'Banger'.
I carefully drew in to the side
To let the Frenchman pass me ride.
With honking horn and clouds of dust
The Frenchman drove in to...a bus!

The idea is not original but I have added all the extras to amuse a generation of pupils.

Canet Plage

Bright parasols stand along the beach,
Children scamper through the waves,
While parents on painted towels lie
Slumbering beneath the Mediterranean sun.
Vendors pierce the chattering noise.
Le Figar-O, Le Figar-O.
Chou-chou, monsieur, Chou-chou.
Chocolateglace. Chocolateglace.

I survey the beach with half closed eyes.
The Club Sportique
Where bronzed young men display their feats
Attracting admiring glances from shy young girls.
Coloured lilos on the sand.
The antics of an amorous photographer.
Le Figar-O. Le Figar-O.
Chou-chou, monsieur, Chou-chou.
Chocolateglace. Chocolateglace.

The low hubbub of voices makes me turn away
To watch the sea.
Waves lick gently at the sand,
Crawl to shore, then slide away.
Swimmers laze in the sleepy swell
Disturbed only by the lumbering Pedalos.
Le Figar-O. Le Figar-O.
Chou-chou, monsieur, Chou-chou.
Chocolateglace.

The Red Arrows

Glinting demons rip the sky
Announced by a shattering whine.
Nine Red Arrows, glued in line.
Ecstatic thrill of screaming roars,
Flashing streaks reflecting silver.
Nine Red Arrows, thunder—quiver.
Looping diamond. Scatter. Break.
Swooping dives. Nerves a shake.
Fantailed smoke, all jets screechin
Climb, then plunge, eagles seeking.
Cascading plumes, plummeting earthwards.
Sudden twist, swerving dart.
Nine Red Arrows curve apart.
Corkscrew roll, bullet homewards.
Lazy silence, vapour shimmer
Nine Red Arrows wingtips glimmer.

This RAF display team never fails to thrill me whenever I see them. Their noise goes right through me and I love their displays.

I had booked a holiday at Canet Plage in the South of France but realised that one should never revisit old holiday haunts so I decided to move up the coast towards Sete. I found a delightful apartment next to a very hospitable French family who invited me to supper that night. I explained that I intended to visit Sete the next day to make serious inroads on the delicious seafood dishes that the restaurants served and they suggested that I walk up the beach to get there. They were obviously aware that I was unaware that my route would pass through the naturist village of Cap D'Agde!

Postcard from Cap D'Agde

Cap D'Agde?
Been there.
Done that.
Didn't buy a T shirt
Or a hat.
Very expensive.
Beachgoers friendly
Locals dour.
Supermarkets expensive,
Franc has power.
Dogs are a nuisance,
Don't feed stray cats.
Flies can pester
Get used to that.
Evening entertainment?
Beyond compare.
But not the restaurants,
Or their fare.
Nor the nightclubs,
Out of reach,
But in the mall
Exhibitionism reigns
Beyond the pall.
Dredged from boulevard
And kinky boutique
Emerge the erotic
Lurid, fantastique
A sensuous kaleidoscopic
Fashion parade.
Daring lingerie,
Exotic underwear,
Mostly diaphanous.
Nudity beware.
A gold mine of jewellery,
Chains abound,
Glittering displays,

Draped, festooned
In remarkable arrays
And piercings,
Unbelievable locations.
Ears, mouth and nose,
Genitals and breasts
Where else?
Who knows?

Cap D'Agde. Been there,
Didn't do that.
Preferred the beach,
Sunrises,
Sunsets,
Au natural.
Magnifique.

I realised, of course, that I had been 'had' but it was fascinating to visit the resort at night and stroll around looking at the site and sights! How was I dressed? Or not? I leave that to your imaginations. He who dares—wins!

I lived in Padstow, in Cornwall and its environs for many years and the best time was always the 1st May when the Padstonians paraded the Red and Blue Hobby horses around the town. I hope the tradition continues for ever.

Mayday

We arranged to meet
In Landwell St
On the first of May this year.
A friendly crowd
All talking aloud
A gathering of good cheer.
Then comes a shout.
'The 'oss is out!'
The crowd draws back in glee.
He prances out
Swirling about,
A splendid sight to see.
Then starts the dancing,
Skirt swaying, legs a prancing
To the accompaniment of the drums.
The accordions play
The 'oss makes its way
To the quay to start its fun.

The 'oss is sadly led.
With many a tear
It's hard to bear
The 'oss is put to bed.
A sad farewell
Words always fail
Until another year.
The best day yet?
Padstonians never forget
Their 'oss brings such good cheer.

Then around the town
Through the streets up and down
The 'oss parades in fashion.
Accordions playing.
Dancers swaying
A lively vibrant procession.
Right around the town
Until Stable bound.

Such a Bore

I have travelled throughout Europe on a Gentlemanly Tour
But Venice was badly flooded and Florence such a bore.
The Leaning Tower of Pisa I very much regret,
Was inclined to make me dizzy—an unpleasant side effect!

Now in Paris there is a tower, the Eiffel, I think it's called,
The view, they say, is splendid, though I was very bored.
But judging by the pigeons, for which the tower if famed,
Don't raise your head too often, the tower is correctly named.

On any tour of Europe Henley Regatta is a must.
Pimms by the gallon, and forget who is coming first.
A friend, however, insisted on yelling from the shore.
I pushed him in, it serves him right, for being such a bore.

I gambled at Monte Carlo at casinos by the score,
But luck obviously had left me and I found it such a bore.
It seems the wheel of fortune, Roulette, was to be my knell
For I not only lost my jacket but my DJ trousers as well.

I was asked to shoot in Scotland with the Laird of Anstruthoy.
His shoot of ten thousand acres is the county's pride and joy
At dawn, plus-foured and freezing, we set out across the moor
But all the grouse were hiding and I found it such a bore.

Supported by many Gluvines I tried skiing at St Moritz,
My control was somewhat lacking for I fell on every piste.
Even during Apres Ski which I found to be a bore
My arms and legs got tangled and I crashed on to the floor.
Every place in Europe made me most depressed,
Full of boring people determined to be best.

And now I am in Africa where lions nightly roar
But it is full of Europeans...and they are all such a bore!

I have spent many hours of my life underground in Devon, the Mendip Hills and the Brecon Beacons as well as in France and Spain. I have even caved in Africa. This poem was written as a challenge when some of my pupils said they would like to hear an amusing poem about caves.

Jim the Speleologist

Jim was a speleologist who loved to be underground
At home, at work, or in the pub, Jim never could be found.
Like the mole, the miner, or the rat, he was rarely above the earth,
He loved to burrow, delve and dig. His life was below the turf.

The bigger the hole, the better the hole, to Jim darkness was a delight,
The glare of the day, was, to him, dismay and he only saw his Mum at night, When
he would, occasionally, it is said, pop up for a buttie made of Spam,
Which he carefully ate in a special hole his dad had dug under the pram.

One auspicious night young Jim was engaged in a long and torturous crawl,
When, groping in the darkness ahead, he discovered a gigantic hole!
It was nearly as big as a double-decker bus and its width was incredibly wide,
It was long and black and high and dark and Jim could not see right inside.

'Perhaps it's the lair of some terrible beast?' trembled Jim with a shiver of shock.
Just then came a screech and a terrible roar and Jim's knees began to knock.
Two bright eyes loomed from the back of the hole, the monster came at Jim with
one bound.
'Twas then Jim realised it was the TUBE and the hole was the Underground.

I could not estimate the number of hours that I have spent canoeing and indeed, I still paddle now but I do remember my first white water outing on the upper River Exe in Devon.

White Water Canoeing

A moment of fear,
You catch your breath,
Push off from the bank,
Draw into mid-stream.
The bow slides around
And you swing,
Leap.
In one quick pull
You hit the surging foam
And powerful waves.
Rock!
Dodge it.
Steady the stern.
Shoot through a gap
Into an eddy.
Pause.
Slide into a haystack
Over the hump

Onto the sill of the weir.
Slowly, the stern tips
Scurrying you down the glass
With the stopper
Waiting for you.
A flurry of spray.
Paddle hard
Water pushing at your chest.
You tilt.
Support!
Only just in time.
Caught in the boiling turmoil,
Thrashing
Paddle straining,
You gradually pull out,
Scrape past a boulder
Sweep around a bend,
Safe.

I worked for many years in the tropical forests of Ghana and Liberia. I attended a job interview a little while ago and one of the interviewers was most visibly annoyed about my logging activities and my alleged destruction of the world's rainforest. In my defence, I would have to say that I was logging some forty years ago when the world was not expressing its concerns about the rainforests of the world. I did point out to the interviewer, however, that he was sat on a mahogany chair at a mahogany table and the mahogany came from West Africa!

Prime Timber

This forest is not noble though trees have stood for years,
Exotic beasts are absent, thus humans have no fears.
Lesser creatures that crawl and creep live in this tangled growth,
The snake, the spider and the ant, the insect and the moth.
The giants they are restless, unease in every leaf.
The forest sounds are muted—a deep, disturbing grief.
The giants they are restless for they know the noise to come
Is not the lonely hunter with trap, or knife, or gun.
Death is a clipboard and a pencil—a cross upon a map,
A blaze upon the tree trunk with bleeding, oozing sap.
From the moment of their marking the giants know their days are few,
Begins the listening and the waiting until their life is through.
A rumbling in the distance marks the beginning of the push.
The bulldozers are busy ripping access roads through bush.
Stuttering and grumbling they steadily advance,
Only in darkness are they silent in a somnambulistic trance.
Every mile they make a clearing and the giants need no telling
This is to be their graveyard for reception prior to selling.
As the bulldozers fade in loudness each tree receives a number,
The order of their cutting, for soon they will be lumber.
Then come the raucous fellers with the saws that scream and whine,

Severing the buttressed tree stump with the saw's teeth cutting line,
Until, with cracks of protest, the giant starts to moan,
Away scramble the helpers leaving the power saw man alone.

As the final cut is finished the feller makes a quick departure,
The tree begins to topple, slowly, at first, then faster.
With a rustling of shattering branches the giant swoops to the forest floor, Meets it
like a thunder clap…the giant is no more.

Its branches are shorn by the chattering saw that cuts, and slices and trims, Once a
proud tree, now a bare log, stripped clean of all its limbs.
The giant is checked and then must wait for its ride to the forest clearing Behind a
rumbling swamp skidder, hawser strained tight, slithering and twisting.
A final surge, a belch of black smoke, the giant is sucked from the forest,
A bewildering world of logs and men, paint, power saws, machines and lorries.
The tree is measured, crosscut and stencilled, into a line goes each section, Prime
logs, in rows, like an army asleep, awaiting the buyer's reception.
After purchase come the logging trucks, front end loaders fussing loudly,
A ballet with the volume high until the logs are chained securely.
With a steady roar the truck departs on its grinding journey to port, A jarring ride
on pot-holed roads; a journey without time to halt.
At the palm-fringed port, on a tropical shore, the logs wait with some foreboding
Until spun in the air they are placed in the hold of a freighter carefully loading.
For destinations perhaps yet unknown—Rotterdam? Marseille? Shoreham?
Tropical logs for temperate mills, Prime Timber for Northern Europe.

Every year I send out an amusing Christmas card and one was based on the alleged Bristol crocodile which had been seen by a bus driver and a lady jogger. Some joker even put up Beware Crocodile signs all the way down the Bristol Portway.

The Bristol Crocodile Rap

CONSTANT RUMOURS ARE RIFE THROUGHOUT THE CITY,
THAT A COCODILE'S AROUND BUT MORE'S THE PITY,
INSPITE OF SIX SIGHTINGS THE CROC HAS NEVER BEEN FOUND
THOUGH THE REPORTS STILL COME IN AND IT'S THERE I'LL BE BOUND.

FROM BEDMINSTER BRIDGE A BUS DRIVER HAD FIRST SIGHT,
AND THOUGH NOT ON DRUGS OR IN DRINK IT GAVE HIM A FRIGHT.
A LADY JOGGER THEN ESPIED IT ONE SUNNY DAY,
TOOK A QUICK PHOTO, AND RAN QUICKLY AWAY.

WE NOW KNOW THAT THE CROC IS ABOUT.
'COS FROM SALTFORD MARINA IN KEYNSHAM WE HEARD A SCREAM AND
A SHOUT.
THE CROC HAD TRIED TO EAT A RETIRED HEAD TEACHER
ALTHOUGH IN ITS DIET HEADS DON'T NORMALLY FEATURE.

THEY ARE ACADEMICALLY CHEWY, AND QUICKLY INFEST.
THE 'CROC BECAME ILL AND HAS BEEN ORDERED TO REST.
WITH RETIRED HEADMASTERS YOU'D BETTER NOT MESS KEEP YOUR
DISTANCE—THIS ADVICE IS BEST!

TO EAT A HEADMASTER IS CONSIDERED VERY NON-U!
YOU COULD BITE OFF MUCH MORE THAN YOU COULD CHEW!

Chapter Eleven
Bold and Loyal

If the first headmaster of Wycliffe College had been successful in his application for the Headship of Queen's College, Taunton, then it is likely that Wycliffe College might never have been founded. Having failed in his application for Queen's, Mr Sibly decided to go it alone and founded Wycliffe in 1882. I find his choice of site rather strange because Stonehouse, at that time, was quite industrial and had been heavily involved in the brick making industry since 1856 and indeed, maybe to tidy things up a little for my appointment at the school they demolished the huge chimney in 1968, just after my arrival.

I suppose one could suggest that I arrived with a Big Bang? Stonehouse was, however, mentioned in the Doomsday Book of 1086 as having a manor made of stone. Stonehouse. Unfortunately, it burnt to the ground in 1908. Stonehouse also had the distinction of supporting 33 pubs in 1838 and now there are three. All of which I had to inspect regularly, of course.

The school was named after the philosopher and theological reformer John Wycliffe (1320-84) who was responsible for translating the Bible into English and perhaps, as a nod to this, Wycliffe has never had a Latin motto. It has always been the English—Bold and Loyal. The Senior School Campus is spacious enough but it does seem to be squeezed in a 'V' shaped area between two main roads very close to the town of Stonehouse. This was of no real concern for me because I had been appointed to the staff of the junior school, a good half mile down the main road into Stroud and it had space aplenty including large games fields.

The main road split the Junior School in half and was connected by a pedestrian bridge which is still in place to this day. South of the bridge was Ryeford Hall, an imposing four-storey building, where most of the facilities were based and then a ring of classrooms surrounded a grassy quad. Below this ran the Stroudwater Canal, built in 1779, linking the Severn with the Thames, which was basically closed during my time at Wycliffe but great efforts are underway to resurrect it. North of the Bridge, around the periphery of the games fields, were the pre-prep department, and one junior and one senior boarding house behind which ran the main railway line to London. There was also an old cabin and a Nissan Hut, the latter to figure largely in my teaching career at the school.

Previous mention has already been made of how I was appointed to the post of master for the teaching of junior English and master in charge of scouting, namely that my favourite English tutor at St Luke's, Reverend Mister G K Booth, had taught at Fettes with Richard Stoker, the Junior School head and furthermore, both had houses at Rock where they socialised during the holidays. Strictly speaking Dick Stoker had a house at Trebetherick.

If you stand at 'Starvation Stile' just outside Padstow on the path to St George's Bay, you can see big houses marching across the horizon, totally despoiling the landscape—blots on the landscape. That is Trebetherick. I never received an invite to Dick Stoker's house and I do not think that his was one of those particular houses, but, one of the new houses further down into the village. I say this with some certainty because his new house was built near to others in a new development and it was only after it had been completed that he realised he had no right of way to his property and had to negotiate a high price for same.

Many local people claim that planning permission for these properties should never have been granted and I am aware that this allegation was included in an investigation of the activities of the North Cornwall Planning Committee sometime later. Most vociferous in their complaining were the Padstonians, across the estuary, who often poked fun at the activities of the people of Rock, or, 'Kensington-on Sea' as they nicknamed it. Mine was not the only appointment that year because Mr R D H Roberts was appointed as the new head of the Senior School. I mention this because his appointment coincided with the time that Harry Roberts the Robber was on the run and we wondered if the two were linked.

Dick Stoker was a tall, imposing man, ex RAF, who had flown Wellington Bombers during the war and had been awarded the DFC. He was kind and pleasant, though a little vague on occasions, and he was often to be seen galloping around the school with hammers, screwdrivers and pliers attempting minor repairs at which he was not too successful, I am afraid.

I did often wonder about the state of his Wellington and wondered whether he attempted the little repairs himself. My one abiding memory of him was that he would greet you in the morning of a summer term to recount the number of toffee papers he had picked up on the cricket pitch while walking over from his house. I have measured out my life in toffee papers. Then, prep school heads tended to do that sort of thing.

The Hole' by C.F.Simpson.

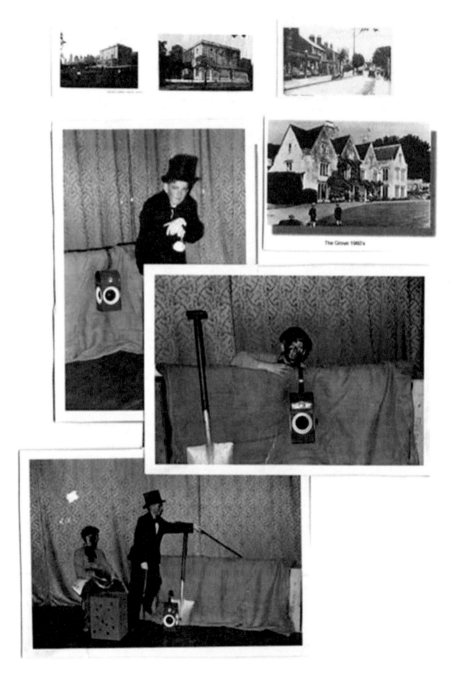

The Grove 1960's

Wycliffe College Junior School.

Many people have a preconceived picture of prep schools as a cross between Dotheboy's Hall and Greyfriar's of Billy Bunter fame with regular beatings, tuck shops and harsh regimes and I suppose some were of that model, even one not that far away from Stonehouse, but not Wycliffe College Junior School. The boys wore a fairly traditional uniform during the week with grey shirts and school tie, blue pullovers, grey corduroy shorts and long socks with black shoes. On Sundays they wore grey worsted suits. The games kit was practical and then all the boys had boiler suits for play and other activities; e.g., Art and Pioneers. It was sensible and practical. All staff conformed to the dress code of jacket and tie and some of the older members wore gowns but not mortar boards. The classrooms varied from elderly to very modern but all had modern desks…eventually.

If I recall correctly, the school was fairly equally divided between day boys and boarders. The day boys had a fairly long day though that was dependent on age, the seniors staying until after prep. Boarding was provided at Windrush for the junior boys aged 7-9 and then there were two senior boarding houses for boys aged 9-13—Sherwell and Ryeford—and they were as different as chalk and cheese. Sherwell's housemaster was Bertie Robinson who was a wonderful character and somewhat of an anti-establishment figure who held traditional views. He was continuously sniping at Dick Stoker and was not very keen on rules unless he had made them but he had a good sense of humour and provided much entertainment in the staff room. He was a great lover of mountains.

His regime at Sherwell was fairly lax and easy going and he was backed by Reg Helson as assistant house master who provided what discipline there was. The wonderful combination of 'Swilly' and 'Swatts' the two matrons, Miss Williams and Miss Watts completed the team. The boarding at Ryeford was more formal and I do not imply any form of severity or harshness.

David King, housemaster, adopted a much more disciplined and organised approach and everything ran smoothly and well. There were eight dormitories in Ryeford and most of them had ten beds. The boys always knew where they stood with Le Roi, 'The King,' and one flash of his glasses would quell any problems immediately. His matron was Miss Dinham who was a straight-laced slightly older lady who took good care of the boys, but, her one insistence was that all the boys should wear 'Y Front' pants for 'proper' support.

I started at Wycliffe exactly fifty years ago so my memory might well be at fault but I do not remember the boys being beaten. I am sure Dick Stoker had a cane but he used it seldom. He did use the slipper regularly when he was on dorm duty, however, as his discipline seemed lax. Indeed we always arranged to go out when he was on dorm duty because of the chaos. This was matched on the 'Staff's Day Off' which was an annual event. It was well meaning and the staff were very grateful for a day's holiday decamping to various restaurants of some repute. Dick Stoker would then attempt to run the school with the prefects. Result? Chaos. The admin, domestic and kitchen staff would look quite frazzled and exhausted by the end of the day.

There was a system of punishment and rewards. If you did anything well then you were given White Marks: if you misbehaved then you received Black Marks. On Fridays Bertie Robinson would take the miscreants with Black Marks over the bridge and they did 'rounds' (big lap right around the games field) according to the number of Black Marks they had received. The naughty boys were the fittest in the school and always came first in the cross country races.

I was given accommodation in Sherwell and walked the short distance to Ryeford Hall and back every day. I taught there but did my dormitory duties at Windrush—the Junior House. The boys and matrons were very pleasant but the housemaster was not. He really was not suitable to run a boys' boarding house at all. He was much too strict, severe and unfriendly and did not provide a good example to me. The boys were very young and did not need a strict disciplinarian who would have been more suited to looking after Borstal Boys. They needed light discipline and control and he created totally the wrong atmosphere. He was even quite rude to me on occasions, which I did not appreciate.

'Bed Puts' was quite a pleasant duty and the boys' behaviour was quite good though I did have one fright. I had agreed to stay later one evening as the housemaster and his wife were at a function and all the boys were sound asleep. The silence was shattered by an ear splitting scream from one of the dormitories and I switched on the lights as I rushed in. One boy was wailing and his bed was covered in blood. I thought he had been attacked somehow and carefully checked for any sign of entry, looking around carefully. Meantime, the junior matron had established that the boy had woken up suddenly and seen a white object at the bottom of his bed, which he had bitten hard. It was his big toe.

Though I am jumping along the time line somewhat, I took my concerns to Dick Stoker at the end of the first term who promised action about the problem. Nothing materialised so I informed him that I was not happy and was going to resign at the end of the summer term. Immediate action. I was given accommodation and senior dorm duties at Ryeford Hall in my second year.

The staff was amiable, friendly and helpful. Reg Helson, who taught geography, and Charles Ellis, science, helped me with the Scout troop and I could not have managed without their advice and guidance. Charles was unflappable and very supportive in my first year. Reg has remained a friend for life…whenever I have been in England.

Another colleague, Roger Kirby, became a good friend in spite of the fact that he loved soccer and I did not. He drove one of the new Ford Cortinas and was a very fast driver. We agreed one Exeat that we would 'swap' activities. He took me on a long hike along the Malverns…and back. I took him horse riding on Exmoor. I think he preferred the walking.

Prep schools do attract oddballs somehow. One such member was Willie Witts, Junior French and Geography. He would appear in the staffroom five minutes before a geography lesson and ask everyone what they knew about Meanders. Armed with our responses, he would then proceed to teach a brilliant geography lesson. He was a natural teacher. Another young member of staff arrived to take over some of my Junior English in the second year and then did a midnight flit and was last heard of running a sandwich factory in Reading.

Let it be understood that I was enjoying my first teaching post and liked the school immensely but for that junior housemaster. I was responsible for Junior English for the 7-11 year olds and having been primary school trained, I was very suited and qualified for the job as I understood the basics of reading and language teaching as well as the provision of properly planned lessons with appropriate material and books. John Shirley was my Head of Department, who was the archetypal English teacher, responsible for Common Entrance English, which was quite tough going for the less able boys. I asked him for details of the syllabus, current reading schemes and handwriting policy. He looked at me quizzically then said, 'Oh! Just teach them English, Geoffrey. You know? The usual stuff!'

That was it and I never received any teaching advice the whole time I worked in any of my other schools. Apart from formal inspections, no one ever checked my teaching or visited a lesson or checked my work. Things were slightly different for my staff when I started my own school. Val Hart, the teacher who took the pre-prep class up at The

Grove, came to my aid. She was very well organised and knew her stuff so I was able to build on her reading and handwriting schemes and the cooperation between us worked very well indeed. She was a wonderful teacher and the boys absolutely adored her. She had them wrapped right around her little finger. Misbehaviour with Miss Hart? Never. She was Joyce Grenfell personified.

The standard of English in the Junior School improved as it was now measurable in reading, comprehension and composition and I taught using the reward system—not only for achievement but also for effort. The boys did alternate weeks with comprehension and composition and we covered grammar and punctuation using the mistakes that they had made. This was backed up by using English Work Cards specifically designed to cover specific pitfalls and problems. Alternate lessons of poetry and drama were also very popular and I always read to the class at the end of two or three lessons a week.

I always tried to make lessons varied, interesting and fun. I enjoyed my teaching enormously and of course, I had all the enthusiasm of a new, young teacher. Most prep schools have half termly orders which normally takes the form of a small report card recording academic progress and effort and it was the job of the from teachers to organise and collate these. That was a fairy easy job but I did experience some difficulty at the end of my first term organising the reports.

You had to write your subject reports, then games and activity reports and then form master's reports. There were a lot of reports. I got into trouble because of my English. Evidently, my use of language and vocabulary was too advanced (I blame the dictionary that my godfather gave me) and Dick Stoker was worried that the parents would not understand terms such as persiflage, benefaction, facetious, hotchpotch, mordant and scintillate. At his request, I simplified my prose.

My relationship with parents was good and I attended the first of many parent meetings in which I always tried to be as constructive as possible about their offspring's academic prowess. The most difficult discussions were with the parents of the less able children when one had to be most diplomatic. During the whole of my teaching career I have always tried to involve parents fully in their child's school work and the policy has generally been very successful. There were regular Skittle Matches against the parents which were pleasant. Who used to win? The parents mostly, I think.

It was known that I was a horse rider and I did receive invitations to ride occasionally and I also tried hunting with the Badminton, by invitation, but I most enjoyed hacking then. Dick Stoker was the first of several headmasters to query my contact with parents as heads seemed to believe that only they should socialise with parents. This, of course was balderdash and he never mentioned it again when I asked if I should cancel my acceptance of an invitation to the Duke of Beaufort's Christmas Ball.

As with any school, whether it be a boarding or day school, or a mixture of both, one had to undertake duties and this entailed supervision of the whole site for a whole day with two other members of staff until the last day boy departed the school or the last boy had gone up to his dorm. The ringing of bells was normally performed by one of the prefects but they needed to be checked, then the boys had to be ushered into assembly and kept quiet for the start. You were on patrol during the morning, lunchtime and afternoon breaks and then…there was LOCK-UP.

What does it conjure up in your thoughts? At 7 pm after Prep, the boys had to decide which activities they would attend or what room they would be in for one hour and the school rule was that once decided, they could not move from that area. There were many activities on offer suitable for most of the boys at that time. It is a common feature of prep schools that they organise a myriad of activities, games and sport which I rate just as important as the academic progress.

A few of the boys thought it was great sport to break 'lock-up' and played cat and mouse with the duty staff. It did not bother me too much because I always carried a set of school keys when I was on duty and if I suspected that the boys were out and about, I would simply lock one of the corridor doors then walk in from the other end. Caught like rats in a trap.

One member of staff, however, a fit North country man called Neville Hopkins, took their bait and used to chase the boys all around the building at speed. It became quite a joke and we actually ran a book on how many boys he caught each duty night. Dormitory duties were quite pleasant as the matrons did much of the work and you just had to gently patrol the dormitories ensuring all was well. I used to allow reading time to extend past each dorm's lights out times if they had been good. Silly behaviour meant no reading time. In one staff meeting, the subject of pillow fighting was being discussed which rambled on and on until Dick Stoker turned to me and asked my views.

'It's great!' I exclaimed. 'I really enjoyed it when I was a boy! Are we perhaps forgetting that we were once boys?'

You see, this is a common trait with adults, particularly parents. They criticise the very things their children do knowing full well that they did the same thing when they were young. Children can get up to all sorts of things that are not acceptable and they must be admonished for same but there is no need for overreaction. Most children are naughty sometimes. I sometimes had parents in my study complaining about the attitude and behaviour of their children, fully expecting them to be paragons of virtue but I used to turn the tables by asking whether they would be willing for me to see their school reports. The fathers were normally the first ones out of my door.

Until a new policy was adopted a relatively short time ago, juvenile crime was said to be on the increase and then someone had enough common sense to study the type of crimes for which youngsters were being charged. An eight-year-old stealing a packet of Smarties, a girl breaking a window, begging on the streets, minor criminal damage. All petty crimes. The revised Youth Justice System now takes account of the seriousness of such crimes and deals with them using more appropriate methods and punishments so that they do not enter the criminal records system. Some schools are also retaining pupils who have been involved with drugs. The pupils are punished but are not excluded from education which could harm their prospects for the rest of their lives.

We also did weekend duties on Saturdays and Sundays. The Saturdays were fairly easy because there was Saturday morning school and games and matches in the afternoon. There would always be a film in the evening for the entertainment of the boarders. There was no TV.

Sundays were a little trickier. After breakfast, letter writing had to be organised and I always tried to treat this as light-heartedly as possible and gave direct help to the smaller boys. I would start off by offering a sweet to every boy who could think of a subject to write on the board to put in their letter and I would write phrases and sentences on the board to help them. The quicker boys would then help the slower ones when they had finished.

At 10.15 am the 'Crocodile' would set off. A long file of boys would walk over the bridge then up the drive past Windrush and The Grove then down to the Senior School Chapel where they had to endure a fairly boring service as the whole school was present and there was little effort to include anything for young boys. I am surprised that they all behaved so well. After the service, the 'Crocodile' would reform and we would return by the same route to Ryeford for lunch and afternoon activities.

Some of the senior boys were allowed out on bicycles during the afternoon and this led to 'The Battle of the Chopper' with Dick Stoker leading the opposition army and the Raleigh Cycle Company preparing their defences. One of the senior pupils owned a Chopper and came a cropper descending Frocester Hill, probably too fast, and though the boy was not seriously injured he did have quite severe cuts and bruises. Dick Stoker decided that it was the fault of the Chopper's small wheels and high handlebars and went to war with Raleigh. The skirmishes lasted for some time until both sides withdrew with no satisfactory conclusion. Choppers were, however, banned at Ryeford.

At the mention of Sunday lunch, I must say that it was always excellent and one of the highlights of teaching at the junior school was the food cooked by the caterer Louise Goymour. For breakfast and lunch, the staff always sat at the head of a table to supervise the boys' eating and table manners and tea was supervised by the duty staff. The staff had an evening meal to the standard of most restaurants and I had no complaints at all.

In a later conversation I had with David King, I recalled how the senior dormitories always used to try to be first down to the breakfast queue and he mentioned that he was never sure why. I had noticed the reason very early on in my time at the school. The boys queued up at the bottom of the stairs so that they could look up and see the young mini-skirted matrons coming down the long stairs, for Ryeford Hall was a four-storey building. My accommodation was at the very top under the roof. I was to become very fit going up and down the stairs. I used to sleep with my window open and one night I awoke to the sound of cries for help. I looked out from my lofty eyrie and peered into the November night. I could see nothing so dressed quickly, grabbed my torch and descended the stairs.

Outside, it was very cold and the cries were very weak but definitely coming from the canal. I walked down to the bridge and shone my torch on to the water. There was a car upside down in the middle of the canal with a very wet and bedraggled driver sitting on it. By this time several people had arrived and I asked if anyone had a ladder. One was produced and we managed to coax the driver along it to the safety of the towpath. The car was removed next morning by a large crane. I took some of my form down to watch and they wrote about it in their stories that day. Living English.

Then, one evening, the fire alarm sounded at about 10 pm so we quickly evacuated the building and did the rollcall, ensuring everyone was safe. The boys returned to bed just before 10.30 pm only to report that water was pouring down through the two senior dormitories on the top two floors! One of the resident staff had started to 'run' a bath before the alarm activated and had forgotten all about it.

I was employed to run the Junior School Scout Troop as part of my contract. I suppose really I was partially employed as a Scout leader. Wow! A professional Scout. The troop was quite large and I wish that I could say that this was solely due to the magnificent and splendid programme that I organised for them every Thursday afternoon. That might be stretching the truth slightly. The fact of the matter was, that, if you did not want to be a Scout, you helped the headmaster out around the grounds as a Pioneer digging holes and trundling wheelbarrows about the place. To be fair. Years before my arrival Dick Stoker and the boys dug a huge hole at the back of the school—the school swimming pool. One of his favourite expressions was, 'You can do anything with boys' a comment which would be considered a bit risqué now, of course.

During the colder months, we ran indoor troop meetings in the school assembly hall which was really too small for over sixty Scouts milling around with various activities and games but during the Summer Term and in September we tried to get outdoors as much as possible. I always kept spare 'wet' programmes just in case we could not get out. The Thursday afternoon outdoor sessions included fire-lighting up in the woods, building bivouacs, canoeing on the canal, pioneering with ropes and spars, cooking, a Good Turn, handcrafts, ironing and sewing, tracking, making plaster casts and erecting tents.

I built up the activities gradually as I became more competent and developed them much quicker after I had learned to drive and passed my test in Cheltenham. Up until then I had relied on a James Scooter to get me around but it was not the ideal vehicle for winter travel. We caved in the Wye Valley from King Arthur's Cave on the Doward and then through all the smaller caves along the sides of the valley and progressed to Port yr Ogof in Ystradfellte which provided a wide range of different type of caving, all in one cave, as long as you avoided the resurgence which was dangerous. We also caved sometimes up at Uley Rift Cave but it was not very suitable for school groups.

By this time, the troops, senior and junior school owned a seven-and-a-half-ton Ford Custom cab lorry which we could use to transport the Scouts and our equipment. The Scouts all piled in the back. 1967. It was all still allowed then. We used the vehicle for all camps and expeditions. My first Summer Camp was back at Brecon at Cwrt Gilbert Farm and I also had a green Ford van by then. We climbed the Beacons to Pen y Fan, and explored all the local waterfalls, swam in the river, completed many of the scout proficiency tests and explored the lovely town of Brecon. We also attended the Sunday Morning Service at the Cathedral much to the delight of the congregation. Alas, however, Reg came back from town from collecting supplies one morning having modified the near side of my van. He had come off worse in an argument with a dry stone wall in an attempt to avoid an oncoming vehicle.

The purchase of my van was the result of a gift of money from the bank. I banked with Lloyds in Stonehouse and one morning, Mr Stephenson, the Chief Clerk, left a message to call in asap. I walked up in a free period and was informed that a mysterious amount of £400 had been paid into my account. Did I know its origin? I did not. £400 was 40% of my annual salary. The manager insisted that it was mine so Bill Stephenson, on his instruction, wrote me a letter saying that the money was definitely mine which I invested in the green Ford van.

Six months later, I received a letter from the manager stating that there had been an error and he was taking £400 out of my account. I consulted Gadsden's Solicitors (Wycliffe parents) then directly opposite the bank, and they gave me a letter threatening to sue the bank as I clearly had a letter stating that the amount was mine. The bank dropped the case. Mr Stephenson led the way with telephone banking because I transacted much of my banking business over the phone until his retirement many years later.

Our second summer camp was at a farm near Exmouth where we were able to canoe on the Exe Estuary, try mackerel fishing at the mouth of the estuary, hike on Woodbury Common and visit the sites of Exeter. One whole day was spent watching television in the farmhouse—some programme about men landing on the moon!

From Wycliffe, I ran my first school ski trip to Reith near Brixlegg in the Tyrolean Alps which was just below Alpbach, the home of the British Army Ski Club, and we travelled by train on one of the famous Disco Expresses that travelled across the grimness of Northern France and through Switzerland to Innsbruck from where we transported to Reith by coach. In retrospect, I am unsure where I gathered the confidence to run a trip abroad but it was a great success and I ran another two trips to the same resort during my time at Wycliffe. All trips were backed by couriers in the resort and I learned my first German sentence:

'Voh ist der Courier fur Skiplan bitter?'

I used to take my patrol leaders up to Gilwell every year before Christmas because they ran an indoor Prep School Patrol Leaders Course. The courses were always great fun and much enjoyed by my patrol leaders. One year I happened to have the two grandsons of Arthur Negus on the course and had agreed to drop them off at his Cotswold house on Christmas Eve. There was a thick fall of snow and I did manage to get the boys to him but had to stay the night. My parents did not believe me when I phoned to say that I was staying with Arthur Negus.

One of the Wycliffe parents ran a hire boat company out of Braunston, which was central for many canals in central England, and his son Kenneth stayed in touch with me for many years. I was invited to help staff a Senior School Scout trip involving camping punts for the boys and a cabin cruiser for three staff, one of whom was the larger than life Geography teacher Frank Smith who later became International Commissioner for Scouts I believe and who still lives in Stonehouse.

The punts were about fourteen feet long but quite narrow with canvas walls and an outboard engine for power. I think there were four Scouts to each punt. They had basic facilities on board but I cannot remember the details. In good weather they were perfect. In bad weather and driving rain they were impossible to steer and the Scouts received a drenching. We were cruising down the Oxford Canal which has a high windswept summit around Wormleighton and that proved very tricky. It was an interesting trip but not to be repeated in punts I am afraid. For a couple of Easters after that, however, I did lead canal trips with one big cruiser and six Scouts which was much easier to control.

The Oxford Canal is some seventy miles long and is very picturesque. I must have travelled its length at least ten times but always in cabin cruisers not narrow boats. I even treated my parents to a week's trip which they much enjoyed for if you are cruising with adults there are enough pubs en-route for a lunchtime and evening mooring all very convenient. Our trips always attracted friendly attention because they were always under the 'Scouting Umbrella' as it were and we were shown much friendliness by other canal users.

On one famous occasion, I was phoned by the Braunston Yard a few days before an Easter holiday to ask if any of the Scouts and myself were free to rush a new cabin cruiser down to Oxford for a photo shoot. She was their new queen of the fleet, a luxury cabin cruiser, extremely well appointed and in pristine condition. I put out the word to the older scouts and six of my patrol leaders and their seconds took up the challenge. I had been told that she would be fully stocked, fuelled and ready to go, so, upon our arrival we were briefed on the essentials of the boat.

She was a beauty and far more luxurious than any other boat we had hired. I requested that extra fenders be quickly fitted and also took spares. She was a brand new boat and damage had to be avoided at all cost. We left Braunston at 10 am on a Saturday morning and we were in Oxford by Monday lunchtime. It was probably a record but it was hard work, especially as there were so many locks. Remembering Dick Stoker's advice that, 'You can do anything with boys, Geoffrey', we ran a simple system of two boys 'Locking' and 'Bridging' ahead, two crew on board, two boys 'Locking' and 'Bridging' behind, swapping around every two hours.

The canal had many bridges that had to be raised. The crew on board prepared lunch 'on the hoof' and a constant supply of chocolate, sweets and drinks throughout the day. The Scouts did well. We could not exceed the speed limit but we cruised at the maximum four knots and we cruised in poor light. We could not move at night but we were up before dawn to make as early a start as possible. In Oxford, the boys were extremely impressed with the models that were draped about the boat in swimming costumes and I suppose one or two of them still have photos of the event. We had to move the boat occasionally to include more shots of the city of Oxford but on the Tuesday morning, we started a much more steady return back along the canal to Braunston having had the stores and provisions replenished and the boys being given a generous amount of pocket money.

The Oxford Canal is rich in history and one evening we moored up at Cropredy where a Civil War Battle was fought and there is a wonderful pub called the Red Lion. We were cruising with one large boat involving two staff and six Scouts and it had been raining all day. I had popped in to ask the landlord if we could bring the boys in for a meal and a drink and he was more than happy to oblige and even made the boys a hot chocolate drink before their meal to warm them up.

After the meal, we were all gathered around a roaring fire and I was telling a ghost story when I noticed one of the Scouts sat at the bar talking to the landlord. He nodded that everything was alright and then told me afterwards that the boy reckoned that he really liked the pub and wouldn't mind being its landlord. Many years later, the boy became the landlord of the Red Lion in Cropredy.

Returning from Oxford late one evening, our two boats moored up in idyllic surroundings alongside the tow path in Lower Heyford. The weeping willows dipped into the clear water and the dappled late evening sun cast sunbeams on to the water lilies as we prepared our evening meal in total peace and solitude. We retired in the total silence of the night and slept like logs. We were awoken by a gigantically loud roaring noise that seemed to fill the channel of the canal with a deafening crescendo like the noise of a dozen avalanches and then a shock wave hit the boats pressing them down into the water.

We scrambled out to see what was happening thinking that we had been hit by a major catastrophe only to be greeted with another terrifying roar and as we looked up, we could see a huge aircraft passing low overhead. Twelve aircraft took off that night and then there was silence. We returned to our berths only to be awoken several hours later by all twelve aircraft passing overhead to land with the same horrendous cacophony of noise.

Lower Heyford is situated just below Upper Heyford. Upper Heyford was the home of one of the United States Air Force Bases and our much disturbed night's sleep had been caused by huge planes carrying fuel for a refuelling exercise over the North Sea. I discovered all this from a senior United States Air Force officer who had dropped in for some cigarettes at the local shop and post office where I was getting fresh milk and eggs. He was quite concerned about our disturbed night's sleep and immediately invited the boys up for a visit to the base.

Dressed in our Scout uniforms, we were picked up by a minibus, greeted by the Base Commander, profusely apologising for our disturbed night and shown around the base by a six foot tall Texan pilot. The Scouts were allowed to play with a twenty five ton fire engine and allowed to direct the powerful water jet two hundred feet along the tarmac and were then shown over several transport planes including one of the huge tanker aircraft and sit in the cockpits. Everything the amiable Texan showed us was smaller than what they had in Texas and the boys quickly picked up on this.

When we were in the Control Tower, amazingly one of the Scouts was allowed to 'talk in' a 'Goolie Bird' which is a strange small transport aircraft that seems to land like an awkward bird almost hopping along the runway. There were hares on the runway which could cause problems for landing aircraft and one of the Air Traffic Controllers pointed out a particularly big one on the far side. A couple of the Scouts responded.

'Big. That's not big! In Texas, we have hares that big,' and pointed high over their heads.

The controllers could not stop laughing. We were taken into the mess and the boys allowed to choose and eat what they liked and even my Scouts were surprised at the huge portions. Having been entertained, the boys thanked the Commanding Officer and thought the disturbed night was worth it for such a wonderful visit. Upon departure, as the Scouts were climbing back into the minibus, the Commanding Officer took me to one side.

'Excuse me, sir, but these guys. Are all English children so well behaved and polite?'

'Always,' was my reply.

One could not let the side down, could one? Especially in the interest of the development of Anglo-American relations?

Life at Wycliffe was never dull and it seemed one interesting round of activity. I was busy most of the time. I did much canoe instruction at the junior and senior schools and taught canoe rolling to the senior school boys. The more experienced Scouts' favourite trip was shooting the Symond's Yat Rapids on the Wye.

I was instrumental in the formation of the Gloucester Canoe Club and was its secretary for the first year. I also paddled from Ryeford Hall to Gloucester. It involved many portages on the Stroudwater Canal sections and a long, hard slog up the Sharpness Canal. No one believed me when I said that I had paddled up from Stonehouse. I was also expected to take games on some afternoons and I coached soccer though I could not have done it well as I am not a fan of the game even to this day. I was expected to referee inter-prep school matches and on one of those, I upset a visiting headmaster.

He was head of The Downs School at Wraxall and known for his volatility and bad behaviour on the touchline, about which I had not been warned. True to form, he started yelling and screaming on the touchline, shouting his abuse at his team, our team and me. I stopped the match, walked over to him and very politely told him to calm down and behave himself. Visiting parents looked astonished and when he objected, I simply told him that I would cancel the match if he continued his antics. He stopped and the match proceeded, ending in an honourable draw. He had obviously made a complaint about me to Dick Stoker who called me into his study and said that Mr Lazarus, headmaster of The Downs School, had complained about my rudeness to him.

I quietly explained to Dick Stoker what had happened to which he replied, 'Yes, Geoffrey but he is a headmaster and one cannot tell him off. You know what I mean?'

I simply responded saying that I was not refereeing to be shouted at and screamed at or abused. It was totally unacceptable. As a headmaster, he should be setting an example, not indulging in such behaviour. Dick mumbled something about a letter of apology and I merely stated that I would expect one in the post. Nothing more was said. Sometime later I had to deliver a rucksack to The Downs School that one of their Patrol Leaders had forgotten after the Patrol Leaders Course at Gilwell.

I offered to drop it in as I was in Bristol for Christmas anyway. Who should answer the door but Mr Lazarus. I explained my errand. He curtly demanded that I take it up to the dormitories upstairs. I am afraid I had to remind him about his manners and lack of courtesy, politely informing him that they obviously had not improved since our last meeting, dropped the rucksack at his feet and drove off.

I have to admit also that I did not seem to in the good books of the headmaster's wife for some reason. I never met her regularly but our meetings were always polite and formal but I was always wary in her company. She had a sharp tongue. The Scouts and I had agreed to decorate one of the wards in the Stroud Hospital for Christmas and I had thus to stay for one more night at Ryeford in my accommodation. She had organised a party downstairs for her son I think and was quite rude to me when I slipped in quietly through the back door to go upstairs. I was not in her way and I had raised no objections to the party.

She should have objected when I decided to get a dog. I can only say that it must have been in a moment of mental aberration and downright stupidity on my part and was totally impractical. It was a large shaggy-haired beast who answered to the name of Buster. I could not keep it in Ryeford so it slept in my van in its garage. It howled so I had to let it sleep in one of the huts. It howled. I took it home with me for my parents to look after while I went to the theatre.

They and the neighbours were highly amused at its antics as it dashed to the top of the garden, then madly careered down the slope, jumped the path straight into the kitchen, then did the whole thing again and again and again. They were not amused, however, upon their re-entry to the kitchen for Buster had eaten all the buffet supper that had been laid out. I think I had Buster for six days but had to return him to the rescue centre. I felt sorry for Buster as that was his fifth attempt at re-homing. Sorry, Buster.

The school organised a group of pupils to go on an educational cruise every year and I was asked to collect one such group from Liverpool with the Scout lorry. Neville Hopkins, he who had such fine sport chasing the boys at lock-up, lived in Wigan, and invited me to stay with him for the night before I drove down to Liverpool. When I arrived his mother was not well and he had booked me into the Grand Hotel in Wigan instead. Surprisingly their garage was large enough to accommodate the lorry which I parked and then we set off to visit Wigan Pier. It was hard to find and a little disappointing but we ate out at a splendid pub and I retired for the night 'fed up and well drunk'.

The next morning, I went down for breakfast. All the waitresses must have been over sixty and they were rounder than my Auntie Gladys but they were cheerful and good at their job, except they were under new management and a new regime was being imposed. A manager, whom I could not see, but clearly heard, was directing from behind the kitchen door. His main object was to stop the waitresses calling the guests 'chuck', insisting on Sir or Madame as a proper form of address. He was failing miserably though one or two waitresses did get as far as 'Chuck Sir', and I am sure I heard one say 'chuck madame' at another table. As I left the breakfast room, the manager appeared and apologised for the waitresses.

'That's alraht, chuck,' I said.

The Liverpool Docks were busy but I loaded my cargo, twenty boys and one member of staff and their suitcases and cautiously drove out through the dock gates. We were soon on our way back to Stonehouse, but, checking the vehicle at a services to allow the group to stretch their legs and make a loo call I noted that one of the tyres was low on air. We called out an emergency tyre fitter and he reinflated it and said it would be okay to take us the last thirty miles. We arrived safely and were only half an hour late. I left a note in the lorry and intended to phone the Group Scout Master in the morning but he phoned me first.

Guff Wright loved that lorry. He nursed and cared for it like it was his oldest son and I received a ticking off for driving the vehicle with a deflated tyre. I carefully explained what had happened and that the tyre was not deflated when I had driven in the evening before. Had he read the note? No. 'Please read the note,' I asked. He grumbled all that week and it only stopped when Frank Smith demolished a bus shelter with the vehicle the next weekend in The Forest of Dean while reversing. At least I was off the hook.

During half terms, Reg and I and two boys, to whom he acted as temporary guardian, loaded up his Morris Traveller with three canoes on the roof and drove down to Padstow, a place that I was now getting to know quite well. One of the boy's parents owned a holiday property in the town and we enjoyed our visits there. Padstonians are always very friendly and we always had a good time. Reg's Morris was not the newest of vehicles but it never let us down and it was always quite heavily loaded.

During my first year at Wycliffe, the school was building a big, brand new boarding house next to Windrush and the pupils were taken up for regular trips to see the progress. I worked quite a few English Lessons around the construction. It seemed well designed and its crowning glory was a copper roof. It was to be called Penwood and John Shirley was appointed as its housemaster with Reg as his assistant. It looked 'super-duper' to me and the boys moved in at the start of the school year. The character of the new house was much enhanced by the appointment of a new matron, Ruth Candy and her Basset hound dog, Lancelot, who was a bigger character than her.

Lancelot's sole aim in life was to amble over to the headmaster's house and have his 'way' with their Labrador bitch. Ruth reckoned he was very musical because he would climb into one of the baths and howl down the plughole. Ruth had come straight from the Royal Shakespeare Company at Stratford where she had been a very competent seamstress and her one claim to fame was that she had measured, made and fitted a leather jock strap to Paul Scofield. Dick Stoker had made the mistake of ordering new bedheads without consulting Ruth and she reckoned they were unsafe.

On the day of their delivery she phoned down to Dick Stoker, 'Your bloody bedheads have arrived.'

Dorothy Batley, who was Dick's secretary, was highly amused.

I was going to suggest that Dorothy was a bit of an enigma but I think that would be unfair just because she came from 'Uddersfield. She had a North Country accent but it was very refined and she had a good telephone voice and friendly manner. She was a good school secretary and was good company. Coming from a very large family I was quite surprised to learn that she had no relations whatsoever. She died quite recently and because of this fact, Wycliffe College arranged her funeral at the chapel.

Social life at Wycliffe was excellent. We would often go out as a group to the pubs of the surrounding areas, and visit garden parties and functions and theatres. We visited the Badminton Horse Trials every year and particularly enjoyed high quality restaurants. There were wine and Cheese and sherry parties and lunches and dinners. I also discovered Waterleigh Bottom where the parents of Ruth Candy lived and indeed, she now resides in Apple Cottage right down in the 'Bottom' where the last Cider Pub in England was situated.

I had a high old time and my salary seemed quite adequate for my expenses except that the school had an archaic system of paying salaries. You would receive an 'advance' at the beginning of term, a second 'advance' at half term and the full balance at the end of the term. Though I accepted it I did question its legality. It was instituted as a result of the income pattern that independent schools have but it meant that for most of the term one was working for free. It changed while I was employed there and we were paid monthly as is the norm.

This was the time that I part exchanged my van for a Red Austin A40. The local traders made a good living from Wycliffe and really looked after us very well indeed so there was no need for us to go out of the small town for most of our needs. When I mentioned to the local garage that I was thinking of changing my car the owner promptly took me around the back of the garage and showed me the A40 which had been his mother's but she no longer drove. He was looking for a van for the garage and we did a straight swop. It was a fair deal convenient for us both.

I am suffering from another 'brain wrack', in that I am fully aware that there was a sudden change in my existence and life at Wycliffe but I cannot recall how it came about. Suddenly in my second year, I found myself as Head of Art teaching art lessons to all the pupils in the school in the Nissan Hut in the woods near to the boarding house, Windrush. There is even a letter in my archive file, signed by the Chairman of the Board of Governors, thanking me for all my work in the summer holidays in converting and decorating the aforementioned Nissan Hut. How? I simply do not remember that or the reason. It may have been that they were planning a new Art Block on the Ryeford Campus and appointing a 'proper' Head of Art in the future and I was a 'stop gap'.

I can remember the lessons, for it was a splendid place to teach Art, away from any distractions, and I had swotted up on 'art teaching' so the lessons were varied, interesting and fun and it did not matter if we made a mess. The interior exploded with colour and we also did craft work and built paper costumes and masks. I know it was a success because it was one of the highlights of the 'Potential Parents' Tour as Dick Stoker ushered them into the 'Fun' Factory. I knew nothing then of drawing (apart from my money making Bristol pen and ink drawings that I sold) or colour palettes, proportion, tone and light. What were they?

The pupils just enjoyed working at their own standard, pace and level. Their art descended down the hill and burst into the school, for Ryeford was quite drab and needed an injection of colour. In my present work, I am involved with several art colleges and there is a growing school of thought that they are stifling the very creative essence and vitality of art as they are becoming buried under a mountain of legislation, regulation and frustration. The old, traditional art schools were full of creative enthusiasm and were quite Bohemian and freely producing superb artists. Guess which I would prefer?

I offered my meagre artistic talents to the staff producing plays and helped them with scenery and props and I was also allowed to produce a play, 'The Hole' by Simpson and that was the first of many. I use the term 'allowed' advisedly because one of the disadvantages of teaching in a prep school was the 'hierarchy.' Most staff, during my early teaching career, were 'lifers' and I suppose Freud put it most succinctly by describing the 'contented cow syndrome'.

Basically, they were quite happy to stay as they were. Most of the senior staff with whom I taught at Wycliffe stayed on until they retired. For the younger staff there was no real hope of promotion or advancement. They were waiting for 'dead men's shoes.' The older members of staff produced plays and I was very lucky to be offered a slot at all. I must add that one of my Wycliffe highlights was two senior members of staff appearing on stage during one production as 'Two Little Maids from School are We'. It was hilarious.

I made mention in the preface that this autobiography had led me to make all kinds of discoveries about my past and my friends and relations and the fact that these revelations led to more weighty considerations and matters of national and international importance. Such was the arrival of a young student at Wycliffe. I say young but he was only two years younger than me. His name was Robert Nairac and he made a spectacular entry into the front of the school in an open topped Austin Seven.

I think he had probably 'come down' from Lincoln College, Oxford where he had been studying Medieval and Military History and he had applied for a commission in the Grenadier Guards and was due to go to Sandhurst. Several members of staff had been treated by his father who was an eminent eye specialist in Gloucester and there was some link with Ampleforth College which Bob had attended. It was common for prep schools to take on students like Bob for short periods and they did cheer up the staff room and were much liked by the pupils. He was very popular and I believe he visited the school several times before he left for Sandhurst in 1971.

During my time at Wycliffe, the Irish Problem was brewing, gaining momentum and was obviously beginning to be a major headache for all parties concerned. Later news somewhat passed me by as I was abroad for over three years and we only received the details of the Northern Ireland situation through the World Service so it was not until I was back in the UK in 1977 that I heard about the death of Bob on the early morning news.

After passing out of Sandhurst, Bob, now serving in the Grenadier Guards, was posted to Belfast in 1973 tasked with searching for weapons and seeking out paramilitaries in the Protestant Shanklin Road area, and then, later, in the Catholic Ardoyne district. At the end of his tour he elected to stay on in Northern Ireland as Liaison Officer to the 1st Battalion Argyll and Sutherland Highlanders where he was nearly blown up by a car bomb in the Crumlin Road. In 1974 he volunteered for Military Intelligence, joining the 14th Intelligence Company whose sole task was surveillance.

It is recorded here that he often worked outside his remit, undercover, visiting pubs south of the border where he would sing rebel songs and he earned the nickname, Danny Boy. At this stage of his career concerns were raised about his work as he seemed to have no checks and no boss. Many considered him to be out of control. In 1975 he was transferred to Bessbrook Mill and his clandestine and secret undercover work continued. On 14 May 1977, Bob was by himself and undercover, drinking in the Three Steps Pub at Dromintee in County Armagh, under the alias of Danny McErlaine, a motor mechanic. He sang a song and left the pub. At 11.45 pm he was abducted from the carpark by the IRA and taken back across the border to Ravensdale Woods.

He was violently 'interrogated'—kicked, punched, pistol whipped and hit with a large post and he desperately tried to escape but he admitted nothing. He supplied no information or identified anyone. He was shot dead by an IRA assassin. His last words were, 'Bless me, father, for I have sinned.' His body has never been found and he is now one of The Disappeared. Liam Townsend was convicted of his murder and many individuals were also rounded up for their involvement in the years that followed.

He was awarded the George Cross posthumously in 1979. He may well have been a maverick but he was an exceptionally brave one. There is now a shaky peace agreement in Northern Ireland though there are political problems but I hope and pray that it will long continue and Catholics and Protestants learn to live with each other in full harmony. Several of my friends served in Northern Ireland and all commented on the strangeness of fighting in a civilian environment where there was extreme hatred on both sides. All of them have concerns, however, about legal reprisals, especially if they held high command.

I think I possibly do not have the full picture but I understood there to be an amnesty for both sides? It appears not, for retired British soldiers and officers are now being pursued while ex-IRA and paramilitary members have been pardoned. Is this justice? Is this right?

In my third year at Wycliffe, I realised that I would have to take a decision. I was happy at the school and though there were niggling minor details about such establishments, I could well have served out my time there, but there was no chance of advancement or promotion for many years and could I really have stayed for the next forty years? I think not. I was young, with no attachments or personal luggage to hold me back and my thoughts turned to Africa again. It was a gigantic leap into the unknown.

I applied, through the Crown Agents, for a teaching post and was selected immediately to take up a post in Lusaka, the capital of Zambia. Thus commenced my love affair with that continent which was to play such a large part in my future career, not all of it in teaching. My only regret was that I would not see my family and friends for three years, for that was the length of the posting and I did have plans to get back to the UK at least once during that time. The best laid plans of mice and men.

Chapter Twelve
Zambesi: The River of God

My summer of 1970 was busy. I had resigned from my teaching post at Wycliffe and in mid-July, I was required to attend a Familiarisation Course for my employment in Zambia, conducted at Farnham Castle in Surrey, under the auspices of the Crown Agents. The obvious aim was to ensure that we would know what to expect when we took up our posts abroad which was a good idea but the course embraced many countries and the mass of detail was overwhelming. I estimate that 80% of course time was a waste of time.

I became selective about my attendance and spent many a happy hour playing croquet on the front lawn with the administrator's wife who obviously loved the game and she was also kind enough to show me around the castle which had quite a history; ranging from the Normans through the Tudor period and the tenure of the powerful bishops of Winchester. The accommodation was good, the food excellent and the sessions I attended, of a high standard particularly Patrick Keatley's presentation on the Politics of Africa. He was Foreign Correspondent for The Guardian and really knew his stuff.

The atmosphere was friendly and informal and I had quite a relaxing time at the expense of the Crown Agents. It was possible to explore the whole castle and no one ever challenged me as I looked around. The Bishop's Camera was a lovely room in which many of the sessions were held and this had been the personal quarters of the bishops. The castle boasted two chapels. The larger of the two was designed by John Webb who built it to include the congregation as was the new Protestant thinking at the time. I was fascinated to discover the secret staircase that led up to the pulpit.

The Norman Chapel was much smaller and was my favourite. I liked the simplicity of Norman Architecture and its plainness of style. During the civil war the castle was initially in the hands of the Royalists until a troop of Parliamentary Dragoons arrived, blew open the main gate with a petard, and routed the defenders. The House of Commons then insisted that the castle defences be destroyed to render the castle into a total state of indefensibleness. (I have added this word to my vocabulary.)

The Great Stair was still in current use and the Shell Keep can still be seen under which lies the Hidden Keep which probably supported a four storey tower. We ate in the old stone hall—three substantial meals a day with the Minstrel's Gallery situated at one end. In 1962 an interesting discovery was made. The Tudor Wing was found to be half-timbered and it has now been restored to its original state. It is very impressive. The castle is unusual because some of its construction was in red brick including Waynepete's Tower. I was quite sorry to leave.

I had to get organised. I had to send off, by sea freight, my trunk, which did arrive safely and then I had to de-clutter my possessions and buy more summer clothing. I sold the Austin A40 but not before I had done a dutiful round of all my relations who obviously thought I was mad, going out to primitive conditions and mud huts.

My brother had finished his apprenticeship with Porter's and now worked as a sales rep for the South Western Electricity Board in their flagship building in the city centre. His job was to advise on the technical issues relating to the sale of electrical appliances and he seemed well settled. He still blew hot and cold but was now getting married, and I was asked to perform the duties of Best Man at Christ Church. He and Sue (his wife) moved out to Yate but I always experienced enormous difficulties finding their house as it was in the middle of a huge estate to the south of the shopping centre. It was like a maze.

My parents seemed well settled in their bungalow in Downend and my friends were sad to see me go including Jim Pickup who had been such an enormous influence on my life. My Uncle Jack reckoned that he was now going to have three years of peace before my return. Three years was quite a long time and represented a big commitment on my part.

At the end of August, my parents came to see me off at Heathrow. I was near to tears (I am very sentimental) and my mother was weeping but trying to put on a brave face so I did not wait too long before I went into the departure lounge of Terminal Three. It was my first experience of an intercontinental flight and the plane was crowded. I had managed to get an aisle seat as I like to get up and wander around during flights.

The plane was a somewhat elderly VC10 belonging to Zambia Airways—the national carrier—and the air conditioning had obviously not kicked in because the passengers on the inner seat were being soaked by condensation dripping from the luggage racks. I was sat next to another young man, Ralph, who was heading to the same school as me so we chatted as the flight proceeded and then I slept.

Mosi oa Tunya

Lusaka Boys' School

Livingstone - Real Road Dangers.

I awoke early. I was excited and started thinking about what lay ahead.

Zambia is shaped like a butterfly and is totally landlocked. It was bordered by Congo DR to the North then Tanzania, Malawi and Mozambique to the East, Rhodesia to the South, and Angola, Botswana and Namibia to the West. It has a tropical climate with three distinct seasons Cool Dry, Hot Dry and Warm Wet and the rains came in October and November and again in March and April. Being situated on a plateau at over 1,000 metres, however, the climate was very pleasant and there was none of the intense humidity that I was to experience later on the West African Coast.

Its vegetation is mostly savannah and grassland but there are forests in the more mountainous areas. To the south is the mighty Zambesi River from which Zambia derives its name and I was looking forward to visiting Victoria Falls as soon as possible. I also wanted to see the huge Kariba Lake and dam which was the second largest lake in the world when I was in Zambia.

I was also aware that Zambia had received its independence on 24 October 1964 and Kenneth Kaunda (KK) was its first president and still in office. Its main export was copper from the mines in the Copper Belt but its price was falling drastically. Zambia's people were as varied and numerous as the colours on a patchwork quilt as seventy-six tribes made up this exciting nation.

I think the direct flight took eight or nine hours and we landed at Lusaka in darkness where we were met by Mr Davidson (Dorcas), the Headmaster of the Lusaka Boys' School, who sold me a car. Strange as it may seem, I bought an unseen Wolseley 1600 from a member of staff, Hilary, who was leaving on our aircraft's return flight. Dorcas brokered the deal so I had 'wheels' ten minutes after landing. The drive in from the airport took about half an hour and we could not see much. The African night is very dark.

We arrived at the school site where we were to be living. Between the girls' school and the boys' school was a co-educational hostel where many children came in from the Bush for the term. They provided a multi-national sample of life in Zambia at that time. There was a good mixture of Expatriate and Zambian children. Their dormitories were upstairs well over a hundred yards apart while the staff accommodation was situated on the lower corridors, men on the main corridor and ladies in the lower wings. The hostel was slightly spartan but beautifully clean and tidy and my room was perfectly satisfactory.

In the morning, the children had not yet returned so the staff had a leisurely breakfast and I was able to gain more details from them. Graham Eadie was the Deputy Head and hailed from Scotland and Brian McCoy was a senior teacher from Ireland. Joan Balie was from South Africa and taught a junior class at the Boys' School. All the domestic staff were polite and cheerful and I much enjoyed their company and appreciated their work on my behalf.

I had yet to meet Miss Watson who was the Headmistress of the Girls' School and also in charge of the hostel. That pleasure was yet to come. I was a little surprised to learn that my dormitory duties were also to include the girls' dorm which I though was most odd and peculiar. We were only expected to do one night a week which was not onerous at all but we did have to do some weekend duties including the obligatory Sunday letter writing. Compared with Wycliffe, however, the duties were very light.

Graham drove Ralph and I down to the school. It was quite large and modern with all proper facilities and the single and double storeyed classroom blocks were ranged around a grassy quad with trees for shade. There was an assembly hall and large games fields and a beautiful swimming pool. The staff room was well appointed and everything was light and airy. I was introduced to some of the other staff-Paul Speed and Harold Love who were taking classes below mine. It was most pleasant. I liked it.

I was scheduled to take a Class Six who were eleven- and twelve-year-olds and the class size was 35. I was assured that it was an A Stream and they turned out to be a fantastic class. My classroom was on the second floor of the main block which were all 'intercom' linked. The headmaster (Dorcas) was suffering from cancer and could rarely make the stairs so had the intercom fitted in all the upstairs' classrooms so that he could contact us.

It was sometime later that I asked why Mr Davidson was called Dorcas and I was informed that he was named after the Dorcas Dressmakers' pins but there seemed to be no connection. I spent the next three days sorting out my syllabuses, lesson notes, books, exercise books, resources and classroom materials and putting the wow factor into the classroom. We did not have much space but we did make good use of the little we had. In the evenings we socialised. Dorcas and his wife invited all of us around for supper on the second evening and I quickly realised that the staff knew how to party.

There were at least two good parties every week somewhere and we also visited the big cinema in Lusaka or the drive-in cinema just out of town. There was also a film club that we frequented. Our favourite haunt was the Ridgeway Hotel though I must say that we stopped going there in the evenings at weekends because Mr Mugabe and his pals drank there and it was better to steer clear of their drunken behaviour. Many of the Liberation Movements were based in Accra at that time.

De Rigueur in Zambia was bush and safari suits for men and I was taken to a tailor in Lusaka to have some made up quickly. We taught in bush suits, that is to say, a light short-sleeved jacket and matching shorts and we wore long socks and sensible shoes. The safari suit was a similar jacket and long trousers. It was very sensible wear for teaching in such a climate because there was no air conditioning only ceiling fans which I grew to like. For socialising, we wore a smart shirt and slacks but this was to get me into trouble when I visited friends in Ndola.

Lusaka was a city of many facets and was quite large with very few high rise buildings when I was there. It is a different story now I understand. Its main street was Cairo Road which was extremely wide. Historically it was originally built so to allow the great wagons pulled by a span of many oxen to turn in the street. This was where we did our shopping though there were shops and supermarkets on the outskirts as well. Basically Lusaka is where the Great North Road and the Great East Road meet to which had been added Independence Avenue, Church Road and Haile Selassie Avenue, but, of course it had many side streets.

To these have been added Los Angeles Boulevarde which led out to the Leopards Hill Road and the Kabulonga Road. All the ministries were based in and around the city and there was a railway station, a bus terminal and post office. (I wrote home every week). There was also a splendid cathedral which was beautifully light and airy. I am under constant pressure to revisit friends in Zambia and I hope to go back soon.

On my second day in Zambia, I met Miss Watson who was in charge of the hostel. Now, we are always told not to go on first impressions but I immediately realised that I would have to be wary of her and though of diminutive stature, Miss Watson was very loud. Her 'normal' voice was about twenty decibels above the average range and she could shout louder than a Town Crier. She was an expert at hectoring like many old-fashioned headmistresses and she would shout at and browbeat the children. Her harangues could last forever and be heard all over the hostel. She was known behind her back as 'The Voice'.

I always maintained my civility to her (except once) but she was to prove very difficult to work with as she was most unprofessional in her approach. Evidently, she would berate her staff in front of their pupils and could be extremely rude to staff and parents. I never heard her talk normally to children or offer one word of praise. We maintained a wary truce while I was in the hostel.

This was my first experience of an African country and I had given no real thought to race relations and attitudes and it is only upon reflection that I realised that I was still one of a privileged expatriate community. The staff members of the Lusaka Boys' School were all expatriates, though the majority of the pupils were Zambian, and our social round was almost exclusively within the expatriate community. It is with regret that I now realise that I had no Zambian friends.

This was to be in total contrast to Ghana some seven years later where there was a true racial mix. I now have many Ghanaian friends with whom I am in regular contact. There were racial tensions in Zambia in the 1970s. As expatriates you could not criticise the government or individuals or insult the republic or its symbols. Indeed, in the next chapter, you will read how I inadvertently fell afoul of the authorities. This normally earned you a free ticket back to the UK on the VC10. Many of the white residents were experiencing difficulties in making the transition to full racial integration. Even I perhaps should have made more of an effort in my socialising.

I had a fantastic first year's teaching with a very bright class of boys with a dozen different nationalities and they really responded to my teaching and methods because I taught all subjects—Maths, English, Social Studies, Science, Art, RE, PE, Games and Swimming. It was quite hectic but fun. The class worked hard and played hard. I had discovered that the school held a huge stock of stage costumes so we produced a play every term—'General Custer's Last Stand', 'Jason and the Cyclops' and 'Zambia 2000'. That was in the first year and I produced another six in the next two years. They were performed in front of the whole school and seemed to go down well. I was under no real exam pressure as the boys would be taking their Secondary Selection Exam at the end of the following year and it was obvious to me that they would all sail through, which they did.

I have to mention, however, one problem that was bugging my teaching. It was the intercom. I fully appreciate that Dorcas could not visit the classroom and I did give him weekly reports on class progress so he knew what was happening but he was constantly disturbing my lessons and those of the other staff on the top corridor. I am afraid that we were guilty of removing the batteries and tampering with the wires to avoid the interruptions. Sooner or later, however, the school messenger would arrive at the door on the headmaster's orders and enquire about the breakdown in communications. It would be 'fixed' temporarily until another set of batteries 'ran down' again.

We had very little communication with parents especially as the Zambians regarded education very highly and had enormous respect for teachers, so, the odd complaint was mainly from expatriate parents, and British at that. I was called into see Dorcas one break who had a parent with him who alleged that I had insulted his daughter. I was very surprised. I did not teach his son and had never met his daughter to my knowledge and had not visited the Ridge Hotel for a couple of weeks where the incident was supposed to have taken place.

I explained this to Dorcas but the parent was adamant that I make an apology. I was totally bemused and politely rejected making an apology for something I had not done or knew nothing about. Dorcas accepted my word and I heard nothing more. Brian McCoy recognised the parent as a very drunken member of the golf club where he spent most of his free time and evidently, he started drinking very early in the morning.

Dorcas was my headmaster for only the first year after which his cancer laid him low and he died in the following year. I much enjoyed working with him and respected his authority and success. The last time I spoke to him he was not very well but I thanked him for all that he had done for me. He seemed quite touched but then slyly added, 'You'll be able to put the batteries back in the intercom now.'

He was replaced by Mr Jarvis whom we all nicknamed 'Jasper'. He was a good head to work under and much more switched on educationally than Dorcas but did not have the same presence or charisma as Dorcas. He did not hold parties either.

All staff had to contribute two activity sessions per week in the afternoons. School hours were from 7.30 am until 1.30 pm and activities were from 2 pm to 4 pm. Both your activities were timetabled for one afternoon per week. We had a pool, and a big one at that, so I took swimming. Many of the Zambian children were nervous of the water and we experienced great difficulty in coaxing them in and then teaching them to swim and I was quite happy to undertake this task while the other members of staff occupied the expatriate children who could all swim superbly.

I also trained the swimming team at Lusaka's Olympic Pool on Saturday morning. The manager of the pool was an expatriate called Skip Lyntott, whose real Christian name was Nelson. He was a pool manager, photographer and masseur and he also made money by providing shady loans for his pool staff and his massage business seemed to concentrate on young ladies. He was friendly enough and completed several photo shoots for me of good quality.

Many years later, I was visiting my solicitor in Bideford who had been Chair of Governors at Mount House School where I taught later in my career, and he asked me if I had ever come across a chap called Nelson Lyntott in Zambia. I told him I had and the circumstances under which we knew each other. The last I had heard from Skip was that he had fulfilled his lifelong ambition to own a big yacht in the Caribbean but it seemed that recently, he had come to live in Bideford where he was now up before the court on several charges of rape.

He had introduced hypnotherapy into his massage sessions and assaulted several ladies. He was found guilty and went down for many years. The Olympic Pool was also used for our school gala and the Lusaka Inter-School Gala which we always won. It is a small world.

The children at the hostel were very pleasant. Most of their parents worked out in the Bush on various projects and we had a large contingent of Yugoslavians whose fathers were working on the Kafue Dam in quite tough conditions. The heat there was very intense. We often struggled to find activities for them and I introduced a Pet's Corner but they were also quite happy to help wash cars because they could have fun with the hose pipe, an activity, of course, of which Miss Watson disagreed. No fun allowed. We also had use of the pool at weekends.

The dormitory situation was also tricky for Miss Watson's flat was directly below the girls' dorm and the slightest noise was enough for her to complain and start one of her tirades. Both the boys' and the girls' dormitories were large and had thirty children of primary school age in each one. The male staff obviously could not supervise the girls' dorm as easily as the boys' whereas the staff ladies could go into both. It was a situation about which I felt very uncomfortable.

As the term progressed, Miss Watson became louder and louder and one morning I thought she had lost it. The children were lining up for breakfast and her morning harangue turned into a screaming tirade. I went out to calm her down but she tried to involve me in her reprimands and I just walked away. I have never heard anyone talk to children like that in the whole of my teaching career.

For every celebration of Zambia's Independence Day on 24th October, the school was expected to send a contingent and the supervision of the pupils was conducted on a rota basis so I only had to attend one parade and I took my whole class. They made banners with the school's badge which we proudly displayed as we marched through the city's wide thoroughfares to the accompaniment of bands and musicians and all aspects of Zambian life were on display. The non-rota staff had set up water stations for us and it was generally most enjoyable and very lively. It was, of course, a National Holiday and after the official celebrations, everyone partied including the Lusaka Boys' School staff who had lost none of their expertise in this department.

My trusty Wolseley was giving me good service and was most reliable. I was purchasing it through a car loan scheme sponsored by the government and I must state that both the Zambian and the British governments treated me very well indeed for I had an ample Zambian salary in Kwacha and a salary paid into my UK account as well. It all seemed quite generous especially as one earned a gratuity upon completion of contract.

Many of the other members of staff had purchased new cars and the 'in' choice was the Peugeot 404. The problem, however, was that it was the first choice of the Accra car thieves and I did begin to wonder why the staff continued to buy them because one could almost guarantee that it would be stolen. They would generally be imported through South Africa because the staff were always keen to drive up through the 'Garden Route' back to Lusaka. My Wolseley was obviously not a target because it was the only one in Lusaka, it was second-hand and not very fast. I could leave the car open in the centre of Lusaka with the ignition keys in and it was sure to be there when I came back from my shopping.

We all had to take the Zambian driving test though I think we were allowed to drive on our English licence for a short period. I was advised to take two or three lessons with a Zambian instructor before taking the test which always went down well with the test centre. On the day of my test, I was greeted by a South African expatriate driving examiner who was younger than me. He asked about my driving experience and was particularly interested in the Ford custom cab lorry I had driven as his father wanted a similar truck for his farm. We drove about half a mile in the Wolseley and he told me to pull over outside a bar where he bought me a beer and we chatted for about half an hour before I drove back to the test centre.

'No problem,' he said, 'You've passed.'

I now had a Zambian driving licence.

Crime was quite rife in Lusaka and one had to be careful of pick-pockets, burglars and armed robbers. The burglars used long poles to fish out items from open louvered windows and the hostel staff lost several valuable items when they forgot to close their windows. Laundry could also be stolen from lines. Laundry also provided another danger—the Putsi Fly, though it also goes under the names of Mango, Tumbu and Skin Maggot.

It is a Blow Fly. If your laundry was not ironed properly, the Putsi would lay its eggs in the waist bands and collars and when they hatched, they would burrow into the skin causing a huge boil-like abscess from which the Putsi maggot would emerge. The first one I ever saw was on one of the boys in the hostel.

The matron had been off sick and Joan Balie was acting as her substitute one evening. She called me up to the surgery to look at what she thought was a large boil. We bathed it and obviously cut off its air supply because a big grub eased itself out of the boy's neck and left quite a big hole. I caught one on one of my trips to Ghana when I was teaching at Mount House. I knew what it was and visited the doctor telling him to put some oil on the spot. He checked this out and we waited for the grub to emerge.

At this point I was lying on my stomach on his couch with my trousers around my ankles and Rupert Gude, my doctor, decided that he must photograph the event. He returned with his camera and two other doctors and by the time the little grub appeared, there was a whole crowd of medical staff, most with cameras, recording the arrival of the Putsi. It must have frightened the poor thing to death. It was kept for medical posterity and my posterior has never received such fame and attention again. Actually, I tell a lie; it did, but some forty years later.

I was anxious to see as much of Zambia that I could and seized every opportunity. One of my first trips was to the Victoria Falls though I much prefer its real name of Mosi oa Tunya—'The Smoke that Thunders'. I travelled down with friends who knew the road which was long but easy to follow. Much of the road was tar strip, that is to say, there was only one strip of tar at the centre of the road and 'dirt' road at the sides. At the approach of an oncoming vehicle, you moved your two inside wheels on to the dirt and kept the two outer wheels on the tar, trusting that the approaching vehicle would do the same.

We drove directly south through Kafue and Mazabuka (that was to become my home town for a little while) which was often called the 'Sweetest Town' because it was the home of the Nakambala Sugar Estates, then through the small towns of Pemba and Choma then on to Livingstone where we were booked into the Livingstone Hotel directly beside the Falls. These were 'discovered' by the explorer/missionary David Livingstone who approached them from up river on the mighty Zambesi River. Completely ignoring the fact that they had been there for thousands of years and were worshipped by the local people for much of that time, he claimed them for Britain and called them the Victoria Falls.

I have always regarded this as a breath-taking act of sheer arrogance. The Falls are well over a mile wide and vary between 262ft and 304ft and it is estimated that between 20,000 to 700,000 cubic metres of water pass over them every minute depending upon the season. The spray from the Falls can be seen from thirty miles away. They are magnificent and truly one of the great wonders of the world. You can get up close by use of the Knife Edge Pedestrian Bridge and you can also climb down into the Boiling Pot and Devil's Cauldron to view them from below.

The height is impressive. Staying for a weekend, you are able to see the Falls at dawn and sunset and we just kept returning to them as many times as we could. I think I used up three slide films that weekend alone. Livingstone, I believe, used to be called 'Old Drift' and is a fine town with many genuine and attractive colonial buildings. It boasts a cultural village, a Game Park and a circular tourist route in the local area where I first came into contact with baboons. Our first meeting was civil enough but I was to become increasingly wary of them during my time in Zambia. They are quite unpleasant and can be very dangerous. Livingstone is attractive but earlier settlers found their lives very tough going as the graves on the banks of the Zambesi River testify. All seemed to have died of Black Water Fever.

On that first trip we faced one major wildlife problem…frogs. We were sleeping in air-conditioned rooms and there was the usual low hum from those but it could not drown out the frog chorus that was very loud indeed and much disturbed our sleep. The Hotel, in its wisdom, had built water gardens in its grounds and the frogs had moved into this very desirable upmarket accommodation and made it their own.

I visited the Falls several times. They were just too good to miss and on the second occasion, I drove down with Ralph, the companion I had met on the plane, and who also taught at the Boys' School and lived in the hostel. I think it fair to say that he and Miss Watson were 'at daggers drawn' from the moment they met and Ralph's forthright North Country accent and comments did little to sooth the warring parties' skirmishes. Ralph had joined me for a bit of recuperation because he had been involved in a horrendous car accident not long after his arrival. He had been invited to a party on the outskirts of Lusaka the venue of which was out on a dirt road.

Driving back, very late at night, in his Volkswagen Beetle, and probably having had a tad too much to drink, he failed to negotiate a bend, drove at speed into the Bush and the car hit a tree. In writing this account, I did, at this juncture, think long and hard about my use of vocabulary, because I maintain that Ralph was lucky. Normally, in accidents in Zambia, the victim was dragged from their car by well-meaning local residents and if that had happened to Ralph, he would have been paralysed or he could have died, for he had broken his neck.

Fortunately, a car that followed him out from the party and saw him leave the road was being driven by a doctor and his companion who was a nurse. They immediately took charge of the situation and very carefully extracted Ralph from the wreck and onto a board on the back of a pickup with the doctor and nurse staying with him until he got to hospital. Emergency ambulance? None. They did not exist in Zambia at that time.

Poor Ralph spent many weeks in his hospital bed with bolts attached to his head and in some sort of traction and we visited him regularly until he had recovered. He only returned to the hostel for a short time, however, and then moved out. On this trip we were able to visit the game park, which is quite small, but has a large variety of animals. We drove into the shade of some tall trees to drink our water and realised that we were surrounded by giraffes on one side of us and two leopards lolling around on the other. They seemed totally oblivious to our presence and we drove off very carefully. We also visited the cultural village but they were experiencing problems with a band of marauding baboons so our visit was very brief. I hate baboons.

There was a staff outing to the Kafue Gorge, at the invitation of one of the Yugoslavian parents, and we found the whole project very interesting but very hot. Temperatures in the gorge were in the high 90s and we had brought cold drinks and beer but also had to drink several toasts in Slivovitz (plum brandy), which really made us sweat. It was a relief to go underground into the tunnels which were cool and pleasant. We came back via a small resort on the Kafue River and were taken out on a pleasure boat which was very pleasant especially as we could see the crocodiles lazing on the banks and the hippos wallowing in the shallows.

Some weeks later, I visited the same resort at the invitation of another boat owning parent planning to have a picnic further up river. Within minutes, however, the boat hit an underwater obstacle and started to sink. I abandoned ship with all the alacrity of an Olympic swimmer and helped the boat owner and his wife out of the water at the bank. Looking around, the crocodiles were still lazing on the far bank and the hippos still wallowing in the shallows. Thank heavens.

Shortly after my arrival in Zambia, I was invited up to Ndola on the Copperbelt by the Williams family whose oldest son, David, was a boarder at Wycliffe. They had informed me that I was also invited to a 'do' at the golf club so I packed my best party clothes and set off Northwards, after, on serious advice, fixing a compass on the Wolseley's windscreen. In the UK this would have been regarded as naff but it was essential for driving in Zambia, particularly on dirt roads which could be straight, boring and extremely monotonous with no features or landmarks. One could easily skid and even turn right around and lose one's bearings. Some drivers had driven for twenty or thirty miles back down the way they had already travelled so I took no risks.

The drive went well and I was only about thirty miles from Ndola when I stopped to help a farmer whose small pickup had broken down. He could not get it going so I towed him three or four miles up the road and then a mile up a dirt track to his farm house. His wife insisted that I have some refreshment in their kitchen and thanked me profusely for helping her husband who had moved up from South Africa some five years ago and was making a real go of his new farm but she was struggling to home tutor their five boys. I had seen nothing of the boys until the mother bashed on an iron pipe outside of the back door.

Within a couple of minutes, her five boys arrived aged between about seven and thirteen, all brown as berries and they were all stark naked. The farmer explained that it was easier on the washing. I drove on with a small sack of corn on the cob as a present which the Williams family much appreciated. They lived in an old colonial house with a big garden in which their three sons were playing and they only wore swimming trunks as the weather was fine.

Unfortunately, the lack of dress for the boys did not apply to adults because when I came down the stairs dressed in my casual evening gear, I was greeted by Mr and Mrs Williams in full evening dress. Oh dear. I was to learn that life in the Copperbelt was much more formal. Mrs Williams begged and borrowed all the necessary formal wear from neighbours and I looked resplendent in my penguin outfit as we drove off to the Ndola Club. The whole atmosphere was one of formality and I did have the feeling, I am afraid, that I was probably seeing the dying days of the vestiges of the British Empire.

Everyone I met was very pleasant and polite but I had now become used to the informal parties of Lusaka. Nevertheless I had a pleasant few days in Ndola. We spent time relaxing informally at the Ndola Club and I was taken to see the Mupapa Slave Tree that was used by Swahili Slave Traders as a meeting place. One should note that Ndola is hundreds of miles from the East African Coast and the slaves collected in this region had to walk that route. It must have been a terrible journey. I found the sunken lake of Chirengwa quite boring I am afraid. I do appreciate that Ndola was not a tourist spot but it did look dreadfully tired when I was there. I was intending to go on to Kabwe and Mufulira but was warned that there was not much to see in either of those towns so I drove straight back to Accra.

One holiday, I drove to the Kafue Game Park with three other members of staff in the Wolseley. The park is relatively close to Lusaka and I drove out through Mumbwa to the Nalusanga Gate in one corner of the huge game reserve. Upon entry all vehicles have to be sprayed against the Tsetse Fly before one can proceed. Once in we headed for Kafue River Camp which was another twenty five miles on but the Wolseley overheated and we discovered the fan belt had frayed. One of the ladies kindly donated a pair of her tights and I quickly made a temporary repair but I had to get under the car to adjust it.

I suddenly heard all the car doors slam and the ladies told me to stay where I was. A group of inquisitive baboons had been attracted to the vehicle and tried to snatch at me while I was under the car. I told Graham Eadie to blow the horn and that scared them off and I emerged dirty and covered in dust, but safe. We limped into the camp which was quite well appointed with a pool, bar and excellent dining room, serving excellent food. I was ready for a cold beer, in fact, several cold beers. We sat watching the beautiful sunset sipping our drinks and wondered what animals we would see the next day. We saw plenty.

The Kafue Game Park did not seem to provide game in any quantity but we certainly saw lions, rhino, elephants, giraffes, hippos, all sorts of antelope including kudu, gnus and one lonely crocodile. One feels slightly nervous at first going out into the park in an open lorry but we were escorted by two armed guards who seemed to know their stuff and the next day, we took my Wolseley on a prescribed route along the Kafue River and saw many crocodiles and hippos in the water. We spent three days at the park and knew that we had seen a good few animals and that was the first of many trips that I made into game parks in Zambia and Ghana.

I am, perhaps, a little 'saturated' with game viewing and would certainly hesitate about being invited to visit another one. Strange to relate, I visited the same park and same lodge just over a year later. Friends of mine were at the camp and their vehicle was terminally ill. It would need to be towed out by arrangement with the rangers but I had agreed to go down and fetch them. I arrived late in the evening just before darkness fell and was greeted by my friends. The lodge was full but the warden had kindly arranged for me to stay in his bungalow which was exceedingly kind so we had dinner and I retired to bed early. The bungalow was quite well appointed and the warden had a bathtub so I soaked for a while, dried myself off and walked into the bedroom. There was a lioness on my bed.

I whipped back into the bathroom and called for the houseboy, 'THERE'S A LION IN THE BEDROOM!'

The houseboy walked into the bedroom and shooed the animal off the bed. 'That be Rosy—she be master's pet. She sleeps in here at night.'

That night I slept with the lion, if you understand my meaning, and I was not eaten once. Indeed, she was very gentle and loved being stroked and petted. Of course, my friends did not believe me at breakfast the next morning until the warden verified my account. He had reared her as a cub and she point blank refused to go wild again. Every time she was released, she came straight back. She was one big soppy lion. She was, however, beginning to be a problem in the lodge because a fully grown lion wandering around was not always good for business, especially if the clients had a weak heart.

I was quite keen to visit Malawi but it appears that my friends and acquaintances were not. They were set on visiting the big game parks and I was more interested in places, people and things and I particularly wanted to explore as much of the Great Rift Valley in which Malawi nestled, so I marched down to a travel agents and booked what my young cousins might call, 'a sort of, kind of, like, you know what I mean', two-week package holiday in Malawi which would enable me to see the various sights and also spend a whole week basking in the sun on the shores of Lake Malawi at the Club Makakola with cold beers, fine wine and fish suppers, and where, incidentally, the waiters delivered your drinks on water skis. I kid you not.

Having packed my suitcase with all necessary requirements, including long trousers (one was not permitted to walk the streets of Blantyre in shorts—nor could the ladies wear skirts above the knee) and visiting the barbers (only short hair allowed), I caught an Air Malawi Flight from Lusaka and landed at Chileka Airport some fifteen miles from Blantyre. This was some forty five years ago and the new airport at Lilongwe had yet to be built. A car was included in the package—a slightly elderly Morris 1100 and all formalities concluded, I drove into Blantyre.

My first impression of Malawi was how sleepy it was. There had been no crowds or hustle and bustle at Chileka. The few passengers on my flight had just sauntered through Immigration and Customs with the officials smiling and cheerful and the car hire agent was slow, polite and courteous. No fuss, no bother and no problem. The road into Blantyre was very quiet and I was easily able to locate the hotel that was of a good standard, book in, then, wearing my long trousers, walk into Blantyre.

There was not much to do or see. The people were very polite and friendly but did not hassle you in any way and I wandered around looking at mostly…Victorian-like architecture. Malawi had been a British colony and I had met members of the Colonial Service who had served in the country but theirs was not the greatest influence. It was the Scottish Presbyterian Missionaries that had left the biggest footprint in their buildings and culture, hence the no shorts/no miniskirts rule. I visited the museum and found it closed and the church with Livingstone connections was not all that interesting either. One must remember, however, that I was visiting the Blantyre of old. It is now a thriving capital with sprawling suburbs, industrial estates and big hotels and the inevitable traffic problems.

My second impression of Malawi was the friendliness of its people who were helpful, polite and cheerful. The country of Malawi is often referred to as 'The Heart of Africa' and all the people seemed to be in good heart and have big hearts. The service and food in the hotel were excellent and I had time for a swim and a couple of beers before I changed for the evening meal. There were only about six or seven guests so we decided to all join one another on one big table and had an interesting evening. We were international. British, German, South African, Zambian, Kenyan, Congolese and American and I am afraid that we did drink quite a lot of beer.

The next morning I drove out of Blantyre aiming to complete a circular route of the south of the country and headed for Limbe which was on my route to Mount Milanje. I was flagged down by a policeman who was very polite and courteous and wanted a lift to a village some ten miles down the road. I gladly obliged and we chatted as we drove down the dirt road. He was very interested in the fact that I was a teacher and he asked about my plans for the day. I told him and he said he would act as my guide if I wished.

I asked him about his destination but he seemed to have forgotten about that and guided me down to Mount Milanje Forest Reserve. The mountain is quite impressive. At over 10,000ft its bare rock stands in awesome majesty dwarfing all that surrounds it and it is most impressive. Brian McCoy, a teaching colleague at Lusaka Boys' School, once climbed the mountain and reckons he nearly collapsed on the ascent as a packet of cigarettes in his shirt pocket seemed like a leaden weight. I am sure it was nothing to do with his chain smoking.

My next destination was Nazombe to seek out rock paintings and I was glad that my policemen friend knew where they were because I would never have found them otherwise as we had to scramble through the Bush dive in under some low-hanging rocks…and there they were, perfectly clear after some one thousand years. Animals and hunters spread along the walls—a real blast from the past. I 'tingle' when I visit sites like this.

I am in awe of history and can clearly empathise with the artists and their lives.

We then drove on to a second set of cave paintings at Kapende before driving on to Chiradzulu and dropping my policemen off at Limbe again. I offered him a 'Dash,' but surprisingly he refused and would not take the money. I thanked him profusely and managed to buy him a couple of beers at a wayside stall. Naturally I had to join him out of politeness.

The next morning, I was off again to the Kapchira Falls, where, in 1859, Livingstone's Zambesi Expedition, was blocked and they turned back. They had been thwarted trying to navigate the Zambesi to get to the centre of the continent and now they were frustrated as well. Looking at the gigantic smooth boulders in the river over which the water roared and cascaded I could well imagine Livingstone's frustration but the power of the water was enormous.

According to my research, there is now a big H.E.P. Project there but I do hope that the beauty of the River Shire has not been destroyed. I was also able to locate Richard Thornton's Grave, the expedition's geologist, who had died of fever and also that of Herbert Rhodes, brother of Cecil Rhodes, who was 'Pioneering' in Malawi (Nyasaland then, of course) and was killed in a shooting accident.

I met a Malawian teacher quite by chance in Chikwawa and we discussed David Livingstone. He was quite surprised to hear that Livingstone's reputation was sometimes attacked in the UK. The people of Malawi held him in very high esteem. They praised him for his attempts to stop the Slave Trade and honoured him for bringing Christianity to Malawi. There are accounts of Livingstone attacking slave caravans on their way to the coast and releasing the slaves. Now I had to pack and head north for fifty miles to Zomba, the old capital of Malawi which had then been the residence of the President Hastings Banda almost right in the centre of the small town, for a small town it used to be. Malawi gained its independence in 1963 and barely eight years had passed since my arrival and now the new state capital is Lilongwe though Blantyre retains its reputation as the commercial centre.

Zomba was an attractive town and I wandered through its colourful market before preparing for my ascent that I knew was ahead of me for the town of Zomba nestles under the southern slopes of the Zomba Plateau. In geological term it is a Syenite Exclusion, that is to say, very hard rock. The plateau is about one hundred square miles and 7,000ft high and I was about to drive up the very steep and narrow road that winds up the 3,000ft bluffs that I could see clearly before me. It was quite daunting and I wondered if my little hire car had the stamina to do it.

I set off and climbed steadily and slowly because vehicles were descending and there were few passing places. Reversing was very difficult and many reversing cars had gone over the edge as there were no safety barriers. Eventually, I reached the top and drove into a completely un-tropical world. To all intents and purposes, I could have been in Scotland

There were beautiful pine trees and grassy banks with sparkling mountain streams and the weather was cool. I easily found the hotel because there was only one, checked in and set off in the car to explore. It really was like being back in an English forestry plantation and many tourists used the area for walking and hiking. I stopped at a lonely bridge under which was a deep pool which looked very inviting so I stripped off my clothes and plunged into the cool water. It was most refreshing and I swam around for a good half hour then turned to get out. Sitting not ten yards from my clothes was a baboon.

I hate baboons. It was not very big so I threw stones just in front of it and it retreated for about ten yards. I grabbed my clothes from the bank and holding them above my head, I swam to the other bank, dried myself off with my shirt and walked up to the car, during all of which the young baboon watched me carefully. I related my experience to the hotel manager and he stated that the baboons were causing many problems on the plateau and some action would have to be taken soon. In the meantime, they were advising their guests to wear trainers so that they could run away from the baboons.

I caught my own supper. There was a delightful stream at the bottom of the hotel gardens and guests could go down with simple hand lines and catch trout, to be cooked on the bar-b-que and eaten for supper. I caught four and scoffed the lot. I slept very well that night especially as there was no need for fans or air conditioners. In the morning I peered out of my hotel window to see real Scottish weather. A dank mist obscured all but the nearest trees and the forest looked quite forbidding in the half light of the dawn.

Generally, in the Tropics, or close to them, it becomes light at six in the morning and dark at six in the evening and it was just light when I decided to go for a swim in the hotel pool to refresh myself. I wish I could say that I did a brisk twenty lengths but I swam slowly up and down and then floated in a reverie for a while thinking on the day ahead for I was scheduled to head for the Club Makakola today for my week of rest and relaxation, but I meant to take a detour into Lilongwe first.

The trip down the escarpment of the plateau, was, if anything, worse than going up, and I stopped the car at least three times because the brakes were getting hot and it took a good hour to make the descent. I was soon bowling along the main road, however, heading up to Dedza and the Chongoni Cave Paintings. Upon arrival, however, the guides were on strike and protesting about wages and conditions so I reluctantly drove on to Lilongwe. The Old Town is very interesting and traditional buildings abound. Many of the towns and cities that I visited in Zambia and Malawi had not then learned to make the most of their attractions and places of interest and Lilongwe then was a good example of this. I wandered around for some time before driving across to Salima and on to Monkey Bay then down the western shoreline of Lake Malawi to Club Makakola.

This was a lovely resort. Set right on the beach of the lake it was like a tropical paradise. Accommodation was individual Rondavels that were airy, light and air-conditioned, the main bar and restaurant was tastefully decorated and the food was superb. The service was excellent and indeed, I only left the complex once to visit Monkey Bay and its interesting shoreline architecture but it took some effort to do that. I would swim in the early morning, take a very leisurely breakfast, then laze and swim all through the day. I had a good supply of books and read at least four of them, but I did not let the reading interfere with the excellent lunches and my afternoon siestas on the beach in the shade.

The only break in the routine was at three in the afternoon when the ski boats would start up and it was traditional for the waiters to perform a display for you. You could swim out to the raft about fifty yards out, after the show, and have a beer delivered by a water skiing waiter. The manager was most kind and realising that I was British, he insisted that I took afternoon tea on the veranda with cucumber sandwiches. It was a very lazy week. All good things must come to an end, however, and on the last day I drove back to the airport in the morning finding time to stop off at the famous St Michael's and All Angels Church at the airport road junction.

This was a red brick edifice, quite ornate for a Scottish Presbyterian Church I thought and it had been built in a most unusual manner. A missionary clergyman had arrived with absolutely no building experience whatsoever. He cleared the site and put down a foundation on local advice and then ordered a pile of bricks. He cemented one brick to the foundation, then another and carefully started to build up walls brick by brick. There was no plan and no finished concept. If he did not like what he had done then he would knock that section down and rebuild. S-L-O-W-L-Y, the church began to take shape and you would never guess that it had been built on such an ad hoc basis.

There is a small sequel to this trip. At a staff party one of the ladies, who was Scottish, was kind enough to ask me about my trip and I described the Zomba Plateau to her. She was doubtful about there being a part of Scotland in Malawi but she and her husband took a short trip to see for themselves and were totally smitten by the place making an annual pilgrimage there every year.

Readers may be sceptical but I did have much free time in Zambia. I had been used to working seventy- or eighty-hour weeks at Wycliffe and now I suppose that I did no more than forty hours at the very most. What to do? I was still interested in horse riding so I cast around for a stable and made contact with Barbara and Ricky Cripps, who had a small riding establishment not far from the agricultural showgrounds on the other side of the Great East Road. It would not have conformed to the standards of an English Riding School but there seemed to be no other establishment so I commenced riding there. I think much local legend had evolved around Barbara and Ricky because they were too outspoken to be popular but it was hinted that Ricky was a relative of Sir Stafford Cripps, who was despatched out to the colonies because he was too wild at home, and both of them had been part of the failed Groundnut Scheme in East Africa before they drove over to what was then Northern Rhodesia to live. I can find no verification of this anywhere, but suffice to say, they were 'interesting' characters.

They lived in a very untidy, and dare I say, unhygienic bungalow, on the same plot as the riding school and seemed to employ no domestic staff which merely meant that no domestic work was ever done in the house which was full of dogs and cats. Barbara could cook, however, and I sampled many a fine roast with them. She had about twenty horses and I rode two or three of them quite regularly and I did pay for their hire but it was very cheap. Barbara did have a temper, however, and took umbrage at the slightest things so I cannot say our friendship was very stable and she threw several tantrums in my direction. It made her even more annoyed because I would not react or respond to her shouting. It was always Ricky that poured the oil on troubled waters as he was the peacemaker and I got on with him very well indeed.

Unfortunately, I was bitten by one of her monkeys. We could not establish which one so I had to undergo the full course of anti-Rabies jabs, fourteen in all and in a ring around my stomach. It was some party piece for a little while. Gradually, however, my rides became fewer and fewer and I was asked if I could try to open a riding school in stables adjoining the agricultural showgrounds with a miscellany of horses and ponies belonging to an acquaintance. We smartened up the stables, paddocks and general area and launched the school. It was quite successful and we took on a young expatriate lady to run it when I was working.

Many expatriate ladies become bored out of their minds because their domestic staff do all the housework and gardens so are always keen to find something to do. I have forgotten the financial arrangements surrounding the school and it was certainly not making a profit but it did provide the owner with funds for feed and other bills. He was quite a bombastic chap anyway and he could be quite unpleasant. I saw the writing on the wall and realised that I was not really enjoying the riding school. It was just not my scene. The owner was very angry when I told him but I was taking none of his bombast or unpleasant behaviour and that hardened my resolve and speeded up my departure.

It was during my time at the riding school that I met Denise who partook in show jumping on the show grounds and she had borrowed Geronimo to enter one jumping event which he and she won. We became quite friendly and we were often to be seen out together and I even met her parents. Her father was convinced that I was trying to seduce his daughter (which I was) and kept on asking me about my intentions. Unfortunately, at the moment of seduction, in one of my leave houses, friends called in uninvited and Denise's chastity was maintained. She and her parents emigrated to New Zealand sometime later but I did not follow.

Still as keen on horses and riding as ever, it was Polo that saved the day. Friends at the showground persuaded me to try, and a good friend, Desmond Caines, kindly let me have a go on his pony, Sox, one afternoon, during slow chukkas. I was hooked. I was, initially, completely flummoxed by the rules and set plays and it takes some time to get used to changing ends every time a goal is scored but I gradually learned.

I joined the Accra Polo and Hunt Club as a playing member and never looked back. The club was quite close to the school so I could be there in a couple of minutes and polo started to fill my free time but I never let it interfere with my school work though I did referee a staff versus boys football match one afternoon on one of my polo ponies.

I was not destined to set the polo world afire but it was the game of kings and the king of games and I loved it. It was to become a continuous thread woven throughout the various stages of my life and I still maintain contact with many polo friends and families. Many people outside the game consider it to be for the rich and aristocratic, a game played by snobs and the nouveau rich but I have never found it to be so. The polo players with whom I have been associated are friendly and very sociable and I have only ever received a good welcome in any club which I have visited or joined apart from Soto Grande in Spain which is full of unsociable and arrogant players, but, perhaps they have seen the error of their ways now?

My initial challenge was to find polo ponies because they did not come on to the market very often. I was lucky. I became friendly with one of the club members who had recently come up from Southern Rhodesia, Nick Conner, who had one grey for sale. Knowing my fondness for greys he gave me first option and I rode it and snapped it up immediately. I purchased some second-hand tack including a saddle, borrowed two sticks and I was ready.

Someone had given me a Pith helmet to wear while playing and in Lusaka, we played in black boots. Everywhere else in the world one had to play in brown boots so one did not dirty the other chaps' white britches. My grey pony had not yet been named but it acquired one during the first chukka I played him. By some fluke, I had probably been in the wrong position, the ball came my way and I scooped it up and lofted it along the wing in front of the clubhouse. I charged up the wing, hit the ball again and it went to the centre. There were loud cheers and laughter and everyone shouted 'GERONIMO' and that is what we called the horse. I bought another grey a short time later and he was already called Prince so we stuck with that. I occasionally had use of other horses but the two greys were my gun platforms and they gave me very good polo for I was still a novice.

Nick had sold me my horses below the going rate and I could not refuse when he asked if I could accompany him to Salisbury to collect two horses that he had left at the Polo Club there though there was a problem. I would have to obtain a permit—a permission to leave Zambia and enter Rhodesia because this was the time of UDI. Rhodesia's Unilateral Declaration of Independence. I duly applied and had to attend an interview. I was amazed to find out how much the interviewer know about my father's service record, facts that even I did not know until I was researching for this autobiography and he knew all about my family. I had certainly not disclosed any of this information in my application for teaching with the Crown Agents.

He then amazed me by showing me a photocopy of a postcard that one of my friends had sent me before Christmas wishing me a 'Black Christmas'. I explained that I had already written to that particular friend admonishing him for his stupidity. He then asked me about my polo—he seemed to know all about that, then asked why I wanted to go to Rhodesia. I explained that my friend was anxious to settle in Zambia and leave the Smith Regime with its problems but wanted me to go down to Salisbury and help bring back his horses, adding that I did not really want to go but I was returning a favour. He issued the permit.

Nick had a large Ford Falcon and we set out early one Friday morning (it was a school holiday) towing a three sectioned horse box, heading for the border crossing at Chirundu where we completed all formalities quickly and then sped into the unknown as far as I was concerned. We passed through Makuti, Karoi, Chinoyi and on to Salisbury. We left the trailer at the Polo Club and drove over to one of Nick's uncles who was a cabinet minister in Ian Smith's government but we did not talk politics. Nick took me out for a meal and we decided to go to the cinema afterwards. It was segregated and that simple fact hit me for a six. My feeling of discomfort was quite marked and I could just not come to terms with this act of segregation at all. I did not sleep much that night, especially on the realisation that this was happening throughout Rhodesia and South Africa. I really felt very ashamed that I had been a party to such an act.

Politics was discussed at breakfast and Nick's uncle, a thoroughly nice man it would seem, talked Nick through the latest negotiations and developments. I did not ask any questions or join in the conversation. We lunched at the club and my day was made worse by being introduced to Ian Smith who was exercising his dog on the field. We attended a party that night and there was no mention of politics or UDI but then there were no black guests there at all. Driving back from Salisbury to Lusaka, Nick sensed that I was uneasy and asked the reason. I discussed the issue with him and he pointed out that he was of the same mind and that is why he was moving up to Zambia. Fair enough.

I was kindly invited down to Kariba one weekend with friends from the Polo Club. One could not enter the water because of Bilharzia and crocodiles. It was really just one huge boozing weekend. We went out on a short fishing trip but the rest of the time was spent eating and drinking. Now I enjoy a party but was not a big drinker and the whole weekend was a bit of a trial…and I caught Bilharzia. It could only have been from the baths and showers. One injection eradicated the bug but I never went anywhere near Lake Kariba again.

By now I was in my second year at the Boys' School and was teaching the top class, all of whom were very bright indeed. It was a big class. I enjoyed teaching them and they enjoyed my teaching so it was a good year for them and they all passed their exams. I continued with the Drama, Swimming and Swim Coaching and was quite happy to help out whenever needed. I was, however, teaching next door to Brian McCoy who was teaching a less able senior class. He would sometimes roar at his class but all the boys liked him. He played a constant 'cat and mouse' game with Dorcas. Brian would slip one of his class out through a side window to run over to a nearby store to get his cigarettes and Dorcas always tried to catch the boy. Brian was promoted to be a headmaster at a school on the Copperbelt.

During this year, I started teaching adult classes at a night school out on the Leopard

Hill Road. The adults were keen to learn and I tried to help them as much as I could but it was obvious that several of the class were experiencing difficulty with English and would not be able to take the exam. I offered extra help but it was a dismal set of exam results—five out of the twenty passed. Success rate? 25%. The night school manager was delighted. No one had passed from that class before. Some were on their fifth attempt. I was amazed to see, one evening, when there was a swarm of flying ants, the class were plucking them out of the air and eating them.

I also got to know some of the parents of the boys in my class. One boy, Kirti Solanki, was top of the class and very intelligent. He suddenly turned up one morning with a wedding invitation for his sister's wedding at the Lusaka Temple which I accepted and attended, seemingly as a guest of honour. Placed on the front row I was offered sweets, drinks and delicate pastries until I was quite full, all this as the ceremony was progressing. I thought the whole thing had finished until I was ushered into a big room where a banquet was laid out and had to eat more. I could not face food for several days.

I was experiencing a slight problem with the Wolseley with the suspension and I was surprised to see Mr Vlachos, a Greek transport company owner, approach me before school one morning. I had been helping his son and one or two of the other foreign expatriate children with extra English lessons. He insisted that I give him the keys of the Wolseley and took it away for the whole morning returning it with the problem fixed. He was offended when I offered to pay because I had not charged for the extra English lessons. He became the guardian of the Wolseley after that and I only ever paid for spares, not labour.

My friends always say that I look at life through rose-tinted glasses but I was perfectly content and enjoyed my work, socialising and polo. I was, however, not content with the dormitory situation. The hostel was probably one of the best in Zambia but I felt that we were not treating the children correctly. One has to be careful that one does not transpose and impose more modern ideas and methods into the past but those children should have been much better treated especially as many of them were so young. Any attempts to put fun ideas and adopt a less austere regime were thwarted by The Voice.

The staff were not unkind but The Voice dominated everyone and everything. I was accosted one evening down at the Polo Club by Alf Francis, the treasurer and stalwart of the club, who simply told me that I was going to look after his house while he was away on a month's leave. I asked him how he was sure that I could do it.

'Simple,' he said. 'You're the only person that the dog will allow in!'

This came about when I had to deliver some photocopying to his house one evening and I just walked in through the garden and knocked on the door. A large Rhodesian Ridgeback came running up to me and when Alf opened the door, he was rolling over on his back and I was tickling his tummy.

'How the bloody hell did you do that?' asked Alf.

'Do what?' I replied.

'Geoffrey. You are the first person that dog has let in through the gate…ever!'

Thus commenced a new aspect of my life—looking after Leave Houses. I think I handled my departure from the hostel diplomatically by having a quiet word with Jasper, the head, and explaining my concerns about the hostel but emphasising that I did not wish to make a fuss. The Voice refused to talk to me but I did not mind that. I stayed until half term then took over Alf's house immediately after. It was quite a large house so I was now able to entertain friends and hold small parties and dinners so I was a very popular chap at the school. The staff had been very kind to me and I was now able to thank them properly and I was also able to help out by supplying some food for the Polo Club at functions and events. It was starting to be a very good life indeed.

It was at this time that I was invited to an equestrian cross-country event on the outskirts of Lusaka. It was off season so several of the polo crowd were attending and it was well supported. Forget any images you might have of Badminton and Burleigh. This was a twenty low fence course cut through the bush and it was fun. Geronimo was quite a good jumper and we came a respectable ninth. After the event, we were all helping with the clearing up when someone accidently hit me with a fairly light pole on the back of my head. I felt alright, finished the clearing, had several beers…and then…I remembered nothing until I woke up in Alf's house on the bed the next morning, still dressed in my boots and riding gear.

Concussed (again), I had gotten into my car and driven back on the main road, through Lusaka and back to the house. How? It was even more remarkable because the cross country had been out on the Kafue Road coming back from which was 'The Murder Mile'. This was the most dangerous stretch of road in Zambia and was the scene of horrendous accidents and fatalities. I had obviously navigated it somehow.

Driving in Zambia was bad. There was a very high death toll on the country's roads and one always drove carefully. On pay day at the end of the month most of us kept off the roads because of the combination of Chibuku (local beer drank out of plastic buckets) and Appeteshi—a local spirit.

I learned a great deal about horses and my mentor was Alf. I helped him in the stables quite a lot and I could always be around in the afternoon to arrange bedding and feed and keep an eye on the grooms. I think we played practice chukkas on Wednesdays and Saturdays but there were also matches at the weekends either at Accra or in Mazabuka. The Accra Club is still on its original site and in those days it led straight out on to the Bush and the area of Kalingalinga so we could ride out for recreation.

The Club had a bar and I must say I did enjoy many a Castle Beer from its fridges after which I would go back to a good evening meal provided by whoever's cook I was 'House Sitting'. I also now had to go shopping. The supermarkets and stores did not offer a great variety but I generally coped. My one abiding memory of Zambian supermarkets was the constant shuffling noise made by the customers' flip-flops as they walked around the aisles. Most of the houses that I looked after had a cook, houseboy and garden boy. One had a maid but I arranged for her to take her leave at the same time as her employers to avoid any potential problems.

No one really sat me down and explained the game and tactics of polo, nor was I coached in the game or offered direct advice. One learned 'on the hoof' as it were and I am afraid that new players were often yelled at if they did something wrong. Most rules were based on 'The Line of the Ball'—which you could not cross but it took ages for me to grasp all the concepts of the game. Penalties were numbered according to their severity, a Number Five for minor infringements and a Number One for a serious offence which resulted in a goal being given to the opposing team.

The most amusing penalty was a 5B which was awarded when someone fell off. 'They had dismounted without the umpire's permission!' The rules were really concerned with safety because the game was fast and furious and could be extremely confusing. You had to keep your wits about you all the time. Not long after I had started playing I galloped down the field and scored a goal about which I was very pleased, except I had scored at the wrong end. It happens to all new players. One thing is for sure. I was now quite certain that I had found a team game that I really enjoyed and I made the most of it.

The club was full of wonderful characters. There was Alf, of course, who worked for Coopers and Lybrand then, who had been in Zambia for years. I think he attended Kabulonga Boys' School and the Gilbert Rennie School in the early 50s where he played rugby and was good at athletics. He was also one of the founders of the Baobab School in Lusaka. Even when I was in Zambia he had accountancy contacts with the big safari companies and he still disappears to increasingly luxurious safari camps and lodges. We still remain in contact and we always meet up when he is in the UK. He is still in the same house which I looked after in 1972.

Kurt Schellenbeck was a German player and his occupation in Zambia was as head of the Mercedes Group selling cars and trucks. He had a big chestnut polo pony called Bushfire on which he used to charge up and down the field shouting for everyone to get out of the way. He used to shout, 'TEACHER' at me and I would yell back, 'CAR SALESMAN', which he found very amusing. There was a father and son duo, the Vechys, whose company I much enjoyed and for some reason, they went back to Zimbabwe not long after I had left Zambia.

Desmond Caines was another stalwart of the club and was a good friend and I think he was involved with an electrical company. His wife and son were often at the club because Noreen liked her beer. The Miller family was much involved with the club and Peter was its chairman for many years. When I first knew them they were farming out at Lilayi, south of Lusaka and growing wheat, maize and soya beans. The farm had been purchased in 1866, I think, by Peter's great grandfather(?) and the wonderful old farmhouse sited there was built in 1924 which became a national monument in 1996. Peter and Annette, his wife, had five sons and I know that many of them still play polo but not in Accra.

In 1989, the family developed the Lilayi Lodge and Game Farm and also formed the Lilayi Polo Club which is another venue for matches in Zambia. Jurgen and Ross Gutnicht were also members and Jurgen worked as a mechanic for Mercedes. He was a beginner player like me and we often commiserated with each other over our various failings while downing a beer. Theirs was another house that I looked after. Klaus and Lena Winter and their young boys were also members and I was great friends with them. Lena used Geronimo and Prince to teach the boys riding. They were both bomb proof so there was never a problem. They bought both horses when I finally left Zambia and I am still in touch with them and I met Klaus in Accra some time ago. The boys are all now over 6ft and are good polo players.

The club had social members many of whom were hardened drinkers but I rarely stayed in the club after 7 pm so avoided the maudlin drunks. One social member was not such a one, however and like me was always away by seven. Harry was a slight mystery. He never actually told anyone what he did. It was rumoured that he was an advisor at State House but that was never confirmed, however, I do believe it was he who was instrumental in sorting out a tricky situation in which I had inadvertently become involved. Incidentally I did also look after his house as well.

There was one very sad incident involving the Kenyan Team which I had accompanied to Victoria Falls at the start of which one of my friends had played a trick on take-off. We were escorted to the Falls at Livingstone by an expatriate who was the chief accountant for a tobacco company. He took us to the hotel for lunch then drove us around the game park before his wife served us a delicious tea after which he drove us to the airport for our return to Lusaka. Upon arrival, Alf took me to one side and informed me that after he had dropped us off, he had stopped on the airport road and shot himself. Evidently, he had been embezzling the tobacco company's funds. It was very sad.

Chapter Thirteen
President Balls

Here is an important announcement. To honour his Excellency's birthday, it has been decreed that the title of President shall be reserved exclusively for his Excellency, the Head of State; and all clubs, businesses and institutions, must, with immediate effect, amend their laws, regulations and constitutions accordingly. No other persons, goods or services may use the title President. This decree issued by the Minister of State. Long live the Republic. The radio announcement concluded. This announcement was followed by the rather pleasant strains of the National Anthem.

Another law. Another rule. Another edict. The government and state house were forever formulating new ones and though they had to be obeyed, they rarely caused any problems, if one conformed. Failure to do so invariably led to trouble and any direct criticism of the President or the Republic guaranteed a free ticket home on the VC10 owned by the national airline. On the whole, however, we were living in a beautiful country with a wonderful climate.

As a member of the polo club, we were able to play for ten months of the year in glorious sunshine. We worked only five hours a day, during the morning, so much of my life was spent at the club and on the field living all the thrills and excitement of the game. Critics might, perhaps, suggest that a little too much time was spent in the club's bar drinking Castle Beer but it was absolutely vital to maintain ones liquid intake because of the heat. It was a very good life.

Sitting in the filthy 'cage' at the Central Police Station with a varied assortment of thieves, villains and prostitutes, I reflected on all these pleasures. I had been arraigned before the desk sergeant on duty who had insisted that I hand over all my valuables and every item in my pockets including my handkerchief. To my embarrassment, I was made to remove my shirt and shoes and socks before I was thrust into the 'cage', near the desk and open to the public, to the obvious amusement of all its inmates.

There was a loud chorus of, 'Sorry Boss', as I sat dejectedly on the bare floor clad only in my shorts. The 'cage' was very public and I was soon the main attraction for an inquisitive group of spectators and passing policemen. I presented an interesting sight—a half-naked European who, by now, was sweating profusely as it was sweltering in the late afternoon heat. I carefully pondered my predicament. The facts were clear.

That morning the secretary of the polo club had sought me out immediately upon my arrival at the stables and asked if I could go to the airport to collect a consignment of one gross of polo balls. We had been very short of balls now for over a month and I knew that the shipment was well overdue, besides, I could call in on a friend on the way to the airport for a snack lunch and a beer. Taking the folded weigh-bill, I walked over to my car. This was a somewhat elderly Wolseley 1600, which, surprisingly, I had purchased within half an hour of landing in the country some months ago.

My employer, who had met me at the airport, introduced me to my predecessor who was still trying to sell it before her imminent departure. We clinched the deal there and then. To be frank, I had never really liked the car but it did give me better service than most. It had only broken down once when the fan belt broke in the middle of a game park and we had to make do with a pair of one of the girl's tights. While under the car I heard all the doors slam and everyone told me to stay under the car and not move. A pack of inquisitive baboons came up to give us their attention and saw me under the car. They tried to grab me so I was wriggling and writhing to escape their attentions. My friends yelled and shouted and eventually drove them off. I was a little shaken, dishevelled and very dirty but the repair held until we reached the camp.

But I detract. To my knowledge it was the only Wolseley in the country and whether it was because of this, or the car thieves doubted its mechanical reliability, it was never stolen. All my friends had their cars stolen on a fairly regular basis but I could leave the keys in my ignition in the main street and the car would always be there on my return.

I drove carefully out of the club and on to the main road. The airport was twelve miles outside the city but I had plenty of time. It was necessary to exercise extreme caution on the roads anyway as the general standard of driving was appalling. As most vehicles were grossly overloaded with goods and passengers, when an accident did happen, the injury and death rate was very high. At the end of the month the combination of beer, gin and the local drink, Chibuku, led to a further increase in the accident rate which was one of the highest in the world.

I quite enjoyed the drive out, however, and quickly reached my friend's house which was about one mile from the airport. He was the cargo manager at the airport and his company had the difficult task of running the national airline. Irish, he had a wonderful sense of humour and was a great practical joker. He was a qualified pilot and often flew the old DC3s on internal flights. On one occasion we were taking a visiting Kenyan team on a day trip to Victoria Falls and the DC3 was on the route that day with my friend as pilot. We boarded the aircraft, made ourselves comfortable, double-checked the seatbelts and…waited.

Our attention was drawn to an obvious drunk at the back of the plane who was cradling a bottle of gin and loudly demanding when we were due for take-off. He took big swigs from the bottle and became even more vociferous and cantankerous. He was demanding to see the pilot. By this time, I had 'twigged' what was about to happen and had extreme difficulty in suppressing my laughter. The unsuspecting Kenyan team and the few other passengers began to look very uncomfortable as the stewardess (obviously in on the ruse), tried to reason politely with the drunk—my Irish friend. Suddenly, he staggered to his feet, brushed the stewardess to one side and wobbled very unsteadily up the aisle.

He leered at the Kenyan team, belched loudly, hiccupped and then exclaimed, 'Noh pilot! Gonna fly de plane myshelf!'

He wobbled up the aisle, threw open the cockpit door and lurched to the pilot's seat and controls. The engine roared into life, we taxied down the runway and took off. The faces of the Kenyan team and other passengers were a sight to behold as they sat rigid in their seats. I had to explain the joke and everyone relaxed and a chorus of jeers and boos met the pilot's first cabin announcement.

My friend greeted me enthusiastically and ushered me onto his cool veranda which overlooked a big bend on the airport road. I was wary, suspecting a practical joke, and slid carefully into one of the cane chairs. He shouted for his houseboy who, for some strange reason, was called Motorcar, and requested two beers. When the beers arrived and I had quenched my thirst, he started to explain the reason for his excitement. We were disturbed, however, by the faint sounds of sirens approaching at high speed.

'Here they come!' he shouted and leapt to the edge of the veranda.

I joined him just in time to see two columns of police motorcycle outriders, accompanying a black stretch limo, at very high speed. The motorbikes were very powerful white BMWs, a gift from the German government. I recalled reading about them in the local papers. A huge international conference was about to take place in the capital and though the motorbikes had been a gift, there was criticism of the massive expenditure on limousines, extravagant villas and the specially built venue. I thanked my lucky stars that I had not been on the road as these cavalcades gave you little time to move over. Many motorists often ended up in the huge storm drains on either side of the road.

'Wait for it!' shouted my friend as he jumped up and down in anticipation.

The cavalcade swept into the bend at well over 60 miles per hour and it was then that I saw the two leading police riders begin to wobble. The outer one recovered but the unfortunate rider on the inside could not hold his line. He braked hard, swerved and crashed straight into the storm drain. His companion, directly behind, also swerved to take evasive action, hit a big pothole, then catapulted over the handlebars of his bike landing in a tangled heap on the grass verge while his beautiful motorbike slid along the tarmac, metal screaming in protest, for a good forty yards. Chaos ensued as the rest of the cavalcade tried to stop.

Everyone eventually came to a very haphazard and sudden halt. There was much shouting, loud recriminations and wild gesticulations as the whole group argued and squabbled though not one of them gave any thought to their injured colleagues. Suddenly a crash-helmeted head popped above the parapet of the storm drain, bobbing up and down, loudly berating everyone on the road. It was extremely comical and I also noticed that the other rider, who had landed on the grass verge, was also standing up, apparently unscathed. They must have been pretty tough because these guys never wore protective leathers—their uniforms consisted of a tunic and shorts. Both uninjured riders were bundled into the back of the limo and the two motorbikes placed carefully at the side of the road. The cavalcade reformed, the sirens restarted and they all drove off at speed.

'Brilliant!' laughed my friend. 'Happens every day. A pick-up truck will come and collect the bikes. Did you count the bikes?'

'No, but I think there were twelve,' I responded.

'Right—and they started with twenty-four!'

I felt very sorry for the outriders. They were obviously not used to such powerful motorbikes and were not being given much time to practice but I bid my farewells and drove to the airport. I found a parking space in the cargo handling area and strode into reception. I knew the staff well and indulged in much banter while the documentation was being processed efficiently.

The supervisor accompanied me to the warehouse while another member of staff ran off to find a customs officer to clear the consignment. We quickly located the crate and two strong handlers moved it to the huge doors. The customs officer was there, waiting for us, dressed in an immaculate, neatly pressed white uniform. He was courteous and friendly, bade us all a good afternoon then proceeded to check the documents. Suddenly, his smiling face froze and he looked at me severely.

'Anything wrong?' I asked innocently.

'Yes sah. Please read the description of the goods,' and he pointed at the documents. I glanced down at the part of the weigh bill which he had indicated.

'ONE GROSS OF PRESIDENT BALLS.'

It was then that I remembered the radio announcement and I started to get nervous. I tried to explain that the goods were ordered before the announcement thus we could not be held responsible. The consignment had been ordered four months ago and the decree had only been issued that morning!

The cargo handling staff and supervisor also argued on my behalf but the customs officer was adamant and marched me off to the controller's office. He declared it a matter for the police. Two policemen promptly arrived and arrested me.

Now, in many African countries, the solution would have been relatively easy. A 'gift' would have changed hands and I would have been released. Not here, however. Such an offer would have led to even more trouble. I was frog-marched out of the office, but then came the usual problem of transport. The police had very few vehicles, no money for a taxi and no one was willing to give them a lift. I offered my car. This was turned down so I sat in the hot sun awaiting a solution. The two policemen started to argue, the outcome of which was that they agreed that I could use my car.

They insisted that I remove my shoes. I glanced at the expanse of the gravel carpark and inwardly winced. I informed them that it was illegal to drive without shoes, which was strange but true. There was another conference. I was allowed to keep my shoes. A strange procession then walked to my car. I took the lead, the two policemen followed, two handlers carrying the crate with the polo balls came next and then an assortment of official handlers protesting at my arrest.

At this point my friend arrived and also protested at my arrest and treatment but it was to no avail. It proved quite a struggle to manhandle the crate into the boot of the Wolseley and I waited patiently while this was in progress, pondering upon my fate.

When it was finally achieved, one of the policemen tried to handcuff me to one of his wrists but had not worked out how we could get into the car or how I could change gear! Another conference. The handcuff idea was abandoned.

The drive was completed in silence and I was very nervous as any remark might have been misinterpreted. It seemed a very long time before I drew up outside the Central Police Station and my arrival caused quite a stir. It was not every day that a European was arrested.

I was formally arraigned and charged.

Perspiration was pouring off me and my shorts were sticking to my thighs. I desperately wanted a pee and I managed to attract the desk officer's attention. I explained my predicament and he went back to his desk. He was probably looking for the key I thought. Instead, he came back with a battered tin and shoved it through the bars of the cage. I had to pee in front of everyone and it was very humiliating. I sank into despair.

The desk officer had hinted that my case was very serious and could probably not be heard by a local magistrate, involving a trial of some sort. Were Europeans given special treatment? There were persistent rumours that the two Germans in the main prison at the moment (they had misjudged a currency deal) were running the prison soccer team and were actually being treated very well but I knew very little about soccer. Perhaps I might be deported on the old DC10? My mind was full of vivid ideas, thoughts and consequences.

The cage started to receive new inmates; a veritable assortment of pickpockets, shoplifters and car thieves and it was becoming very crowded though it was probably safer for them in the cage as 'Instant Justice' was normally the rule on the streets. If you were caught 'red-handed' a mob normally administered a bad beating. Guilty or innocent it did not matter. Everyone was very considerate towards me was but it was unbearably hot and crowded.

Unbeknown to me, however, wheels were beginning to turn on my behalf. My friend at the airport drove down to the polo club and informed them of the situation. Several members had wondered about my absence and the grooms had missed my afternoon check of the stables as well. The secretary was very upset at the news and a hurried council of war was convened. Some of the members seemed to be more worried about the potential loss of the polo balls than my problem but they did discuss the best course of action, I understand. The meeting was interrupted by the arrival of George.

George was a much liked and respected non-playing member of the club. He was pleasant, amiable and polite and courteous but no one really knew what he did for a living. It was rumoured that he was an advisor to the CID but it appears that he was probably much higher placed than that dealing with state security. He listened carefully then drove off in his car telling everyone not to worry.

Meanwhile, I had been spotted by a mounted policeman who had come to visit his telephonist girlfriend at the Central Police Station. Realising that he could do nothing, he disappeared to find his inspector. I had done many favours for the mounted police and was attempting to train them for polo. Their tactics required much more honing but their enthusiasm was second to none though our practice sessions looked more like the Charge of the Light Brigade. There had been several bad falls but just like the police motorcyclist, they always sprang back into the saddle to play on. We played on a dirty and dusty practice pitch and there was probably more dust on us than on the field at the end of each session.

I understand that the chief inspector of mounted police was in his 'office'. This was a bar at one of the local hotels. When informed of my problem, he jumped up and to his eternal credit, without even finishing his beer, drove straight to see me. Nodding to the desk officer, he walked straight to the cage and called my name. He attracted my attention by shouting as the cage was becoming even more crowded and noisy due to the arrival of four drunks. The cage was now very cramped as we were all squeezed and packed in tightly.

I explained briefly what had happened and he reassured me that he would do everything to release me immediately. He marched back to the central desk then harangued the officer insisting that he be shown the chargesheet. Then ensued a heated exchange and tempers flared. The shouting match was very loud. It caught the attention of the inmates of the cage and after a brief silence, they all started to jeer and/or cheer the two officers as they ranted and raved. As the noise rose to a deafening crescendo, the main doors were flung open and in strode the chief of police escorted by his security.

SILENCE.

The chief inspector of mounted police and the desk officer sprang to attention standing absolutely rigid. Now I would be in for it, I thought. This could only make matters worse but to my total amazement, the chief of police strode up to the desk, quickly glanced at my chargesheet…and then ripped it into little pieces. The pieces of paper fluttered to the floor and I waited with bated breath.

'Release him immediately!' the chief of police barked.

The keys were quickly located and the desk officer released me from the cage amid applause from my fellow inmates. I felt a trifle incongruous as I approached the chief of police clad only in my shorts and thanked him sincerely. He was charming.

'Rather,' he replied, 'we should apologise to you for our overzealous policemen. I will ensure that all officers concerned will be disciplined in the morning,' glaring at the desk officer who visibly blanched.

He turned to the mounted police inspector and shook hands, requesting him to ensure that I would have all my property returned. Turning to me, he apologised once again, shook my hand and strode out through the doors. It dawned on me then that this had been no chance visit. He had come specifically to release me and I was jolly grateful.

The location of my property proved a Herculean task. It took a good fifteen minutes to locate the tattered brown envelope containing my wallet and watch and amazingly, all my money was intact, even the small coins. My shirt, socks and shoes had vanished and I suspected that they were well on their way to a market stall somewhere in one of the townships, so we decided to go. Two sturdy policemen carried the crate of polo balls to my car and I followed the chief inspector out. It must have been an odd sight. A chief inspector of Mounted Police in full uniform, including well cut breeches and black leather riding boots and me only in a pair of grubby shorts. The inspector promised to follow me to the club.

'Good idea, 'I remarked. 'I have no wish to be caught driving in bare feet!'

'At least you can ride bare back!' he retorted, looking at my naked torso.

We drove off in our separate cars and it was only then that I realised that my Wolseley had been parked outside the Central Police Station for well over five hours, open, with the keys in the ignition. No one had stolen it…again!

Our arrival at the polo club provoked a noisy welcome and my attire, or lack of it, prompted a whole series of ribald remarks and comments. Several ladies were present, thank heavens, otherwise vulgarity might have raised its head. One of the grooms very kindly cycled back to my bungalow to collect a set of clean clothes and some footwear while I was plied with cold beer. The secretary could not stop apologising and kept on topping up my glass. He was quite upset and distressed that he had given me so much trouble.

'Pity about the polo balls though,' he mused. 'I suppose we can claim on the insurance and order another batch?'

'No need,' I replied. 'Thanks to the chief inspector here and the chief of police, the balls are in the boot of my car!'

There was a very loud cheer and everyone rushed out to my car. The balls were carefully carried in and locked in the storeroom and their safe arrival was an excuse for an impromptu party.

'But…but…I don't understand,' stuttered the secretary. 'How come the chief of police came to release you in person? How did he get involved?'

'I really have no idea,' I laughed.

Just then, I caught sight of George perching on a stool up at the bar. He gave me a slow wink and put his finger to his lip. Now I know.

'Let's all have a ball!' I yelled, and leapt on to the dance floor, starting to gyrate with the youngest and prettiest lady in the clubhouse.

'I think we've had enough of balls for one day!' shouted the captain. 'And you're in danger of being barred from the club as you don't have a shirt on so you are improperly dressed!'

Chapter Fourteen
Safari

I was now in the third and final year of my contract in Zambia with the British Crown Agents. My teaching was flourishing, my polo was enjoyable and I was still busy caring for leave houses, most of which were all approximately less than a mile from the Polo Club. On my second Christmas I had looked after the bungalow of Lieutenant Colonel Ernest Achey Loftus who had distinguished himself in various actions during the First World War and commanded a territorial unit during the Second. Between the wars, he had been a headmaster and he was now teaching at Matero Boys' School in Accra; History, I think.

It was a school bungalow that I had to look after. He was very pleased that a teacher would be looking after his house and staff and I was invited to formal afternoon tea with him to finalise all arrangements. He had one cook/houseboy and a garden boy who worked mornings only and whom I rarely saw. The cook/houseboy was very elderly and also very slow. He moved at a snail's pace everywhere and was seemingly incapable of any sort of fast pace or speed but I was in no hurry and he was a good cook. He was also called…Motorcar.

The Colonel was 73 when I met him and amazingly, he only died 30 years later, in Zimbabwe, at the grand age of 103. He was quite a character. You can look him up in the Guinness Book of Records because he was the world's most durable diarist which he kept for ninety-one years.

I hope that I did not come as a shock to Motorcar and I did try to keep to a timetable so as not to upset him but if I was dealing with an injured horse or a stable emergency, I was sometimes late but he would always wait politely and I made sure that he was suitably recompensed at all times. I paid him a dash over and above his weekly pay that the Colonel had left in a series of dated envelopes. He was well rewarded. I approached him about the possibility of cooking a 'British Christmas Lunch' and his eyes lit up.

He assured me that he could and I worked out a shopping list for me which he carefully checked when it landed on his kitchen table. I had invited friends for the lunch and they duly arrived. They were quite surprised by the meal and even more pleased when I produced two bottles of Ouzo afterwards which they proceeded to despatch a mite too quickly and fell asleep for the afternoon. I told Motorcar to pack up the remains of the Christmas lunch and take them home to his family and also gave him a generous Christmas box as a Thank You. He had done well.

The next house that I looked after was beautifully decorated and quite modern and I was there for two months. My friends were much impressed by its facilities and it even had a small pool. I did become quite wary of the staff, however, who tried to take advantage whenever possible but my friends had told me to contact the company for which he worked if there was any problem, so I did. The Personnel Manager arrived and gave them all letters of warning and that did the trick, except, one week later, the cook disappeared. The houseboy stated that the cook was sick and I sent him down with a note to tell the cook to contact the company so that he could go to the clinic.

The houseboy came back and simply stated that the cook was dying. I drove down to his house. He did not look well at all so we bundled him into the car and took him down to the hospital. The doctor examined him and simply stated that there was nothing wrong with him but that he would probably be dead in a few days. The cook believed that someone had put a spell on him and was dying of self-induced apathy. If he believed that he was going to die, then he would die. I notified the company immediately and handed the whole thing over to them but the cook died. While at this house, I formed an idea for a small business.

Most expatriates had large gardens and many of them had avocado trees in them that produced more avocados than they could possibly eat. I had taken my class out for a visit to the airport only two weeks before and the cargo manager had clearly stated that the cargo holds returned empty to the UK and thus cargo rates from Lusaka to London were very cheap. I checked on the rates…they were cheap.

I thus organised all the expatriates I knew to drop off their green avocados at a nominated house where one bored expatriate lady sorted and packed them and took them out to the airport to be sent to a wholesaler in London. The scheme worked well and we all made a profit because there were so few expenses. I handed the whole enterprise over when I moved to Mazabuka but it ceased when there was a huge increase in cargo rates sometime later.

Coming right up to date, my next house belonged to Jorgen and Ross Gutnicht. The surroundings were not so salubrious but perfectly adequate and I was tasked with training two young Doberman dogs while they were away. They were old enough to be let out at night and roamed the compound which the night watchman did not like because it was disturbing his sleep.

THE BIG FIVE

It appears to me that there was a tradition in all the African countries in which I resided or visited that your night watchman slept and you would often hear the loud tooting of horns at gates in the owners' vain attempts to wake them up when they came home late. The Gutnicht's night watchman slept across the front gate so anyone could have gone over the fence at the back without him noticing.

I heard a disturbance one night and found the dogs in the garage with the doors closed. I let them out. The night watchman did not see me in the shadows, threw stones at them and hit one with a stick. I just opened the gate and indicated for him to go out and then locked them again. Next morning, I phoned Mercedes and they sent a replacement. When the Gutnichts flew in, Mercedes had arranged the airport pickup and I was at a 'do' at the Polo Club. I was not late back but I walked into the house and found Ross and Jorgen trapped in their lounge. The Dobermans had let them in but they would not let them out.

'Didn't I train them well?' I asked.

Dobermans had two traits. One was as above. They might let someone on the premises but they would then keep them there and the second trait is that they are generally silent until they are facing an intruder. They are silent attackers.

Now I had a problem. There was a gap in my leave house programme. I was homeless. I knew this was bound to happen sometime and now I was faced with the event. Ross and Jorgen kindly let me stay with them for a short time but neither I nor they realised that we were about to be descended upon. My mother had vaguely mentioned that cousins were in Egypt attempting a 'Cairo to Cape Trip' in an old camper van and I thought no more of it until they turned up at the Lusaka Boys' School and their vehicle was ailing. I managed to get them on to the showground for the night but that was a temporary measure.

Ross and Jorgen came to the rescue. They let my cousins and their husbands bring the lorry up to the house and then Jorgen helped them locate the part that they needed in the back streets of Lusaka. He even helped them fix it but it took two days. It was very nice of my cousins to call in but I did feel guilty about us all landing on Ross and Jorgen. I took them out for a slap up meal at the Mukumba Room at the top of the Hotel Intercontinental as a thank you. It was a lovely meal and Ross said that I could stay as long as I liked. But I was off on a Christmas Safari and I was even going to see snow in Africa.

Some weeks before, Graham Eadie and Helen Mantell, two good friends, had invited me to join them on a trip to East Africa. I am not sure if they needed me for ballast or sharing the costs but I leapt at the chance and we had been busy preparing everything for our departure. I think we had to obtain some visas and permits. The Christmas holiday in Ghana was really the summer break because we were off for six whole weeks and we calculated that the trip would probably take five at least.

It was no small undertaking and we were going in a quite small Peugeot 304 which was carefully prepared for the journey with spares and water carriers and first aid and an entrenching tool. It was Helen's vehicle and I think she was very brave to use it for the safari though she did have one or two wobblies. I do not think my Wolseley was capable of such a journey and we could not have used Graham's 404 for fear of it being stolen but even the 304 came close to that.

The day after term ended and the school closed for Christmas we left Lusaka and headed up the Great North Road through Landless Corner and Kabwe to Kpiri Mposhi. I knew the road for I had driven the same route to Ndola but now we branched eastwards to Mkushi and Serenje facing a 600 mile drive to Dar es Salaam. We were on the infamous and notorious Hell Run on a road used by hundreds of trucks every day importing goods from Dar es Salaam into Zambia seeing that the southern routes were blocked by the Rhodesian crisis. In the 1970s, it was Zambia's lifeline.

It is over forty years since we made the trip but I think we stayed at Mpika for the night before heading for the Tanzanian Border the next day. We passed many signs for the Luangwa Game Parks as we made slow progress with awful road conditions to Chinsali. The road was littered with broken down cars and trucks and there was a steady stream of traffic heading towards us. Our next target was Isoka and there we became stuck. It was getting dark anyway and it would be a terrible risk and very dangerous to make any attempt to drive at night but there was also no fuel. The filling station had not been resupplied.

The petrol tanker had evidently rolled off the road at Nakonde. Three or four cars were stranded with us so we held a council of war. It was agreed that the biggest car, a Toyota Estate, should drive back to Chinsali, using the last remaining petrol from all our tanks, about a seventy-mile round trip, with five forty-five gallon oil drums which we would fill there, and then drive back to Isoka with the fuel. I offered to go back with the owner of the Toyota with quite a fund of money.

Leopard – Secret Valley Lodge   Mount Kilimanjaro

Elephants – The Jamil Mosque – Nairobi   Humphrey from Pemba.

It took about six hours and we had to open the windows of the Toyota to disperse the fumes. I had to prevent the driver from lighting up inside the vehicle so we had to stop every so often for him to have a cigarette. Everyone was relieved to see us on our return and we quickly siphoned the fuel into the cars' tanks. I insisted on having a shower and a change of clothes because I was covered in dust and as it was too late to cross the border that night we found a small hotel in Nakonde. We crossed the Zambia/Tanzanian Border early in the morning having passed the stricken tanker still resting on its side. Welcome to Tanzania. Jambo.

The road on the Tanzanian side was no better and constant and major road works impeded our progress. Surprisingly, however, there was much game crossing and grazing on the verges with no apparent fear of the noisy traffic at all. We were held up by them crossing the road and had to wait some time for one small herd of elephants to move on. Car horns did not scare them at all. Slightly pressed for time, we could not stop at Mbozi where the 8th largest meteorite in the world had landed. There were big enough holes on the road anyway. All the towns through which we drove displayed German colonial architecture which was looking very sad and the towns were unkempt. Mbeya was situated between mountain ranges including Mount Rungwe to the east. We pushed on to Makambako with the Usanga Flats to our left and the Mpanga Game Reserve to our right with its peak of Kirenganye at 7,000ft.

At this point the road was alongside the TANZAM or TAZARA railway that was being constructed at that time and it took from 1968 until 1976 to complete. Originally, it had been a dream of Cecil Rhodes but now the Chinese had been brought in to complete the project as a matter of urgent necessity as a critical link to the sea. It was also known as the Freedom Railway or the Great Uhuru Railway. At over 1,500 miles long, it was to have 22 tunnels and 300 bridges. It was using 15,000 Chinese and 45,000 African workmen and was on schedule.

Meanwhile, the Hell Run had already claimed seventy lives as a result of overturned vehicles, drivers falling asleep, exploding tankers, plunging into ravines, stuck in mud, skidding and inexperience. There was a severe shortage of drivers and nearly anyone could get a driving job. We had a similar situation in the UK. When 'Coal was King' in South Wales there was a huge demand for ships, captains and seamen in the South Wales ports and all sorts of unsafe boats and crews were in demand to move the coal. As these crews steamed out into the Bristol Channel they turned right and the first obstacle they could hit was Lundy Island…and they did! Lundy's seabed is littered with the wreck of colliers.

The Hell Run was a 1,200 mile drive and this was the road on which we were travelling. I just wondered if any of the drivers had died because of accidents involving game? Livingstone was notorious for serious accidents to drivers as the hippos would come up and sleep on the warm tarmac of the roads at night. The road was grey and so were the hippos. Serious accidents would often occur when cars ploughed into them. Indeed, we came across this advice for tourists in Tanzania:

Don't leave the vehicle—you might be lunch.
Don't shoot unless it is with a camera.
Don't feed the animals.
Don't offend the locals.
Don't smoke in public. Don't litter.
Don't take the experience for granted.

The interesting thing was that we never saw a Chinese person the whole time we were on the Hell Run. Poor chaps, it would appear that apart from work, they had to keep to their camps which were rather like prisoner of war camps, perhaps? The Zambians and Tanzanians quickly latched on to the fact that money was to be made out of the Chinese who were ever anxious not to offend, so when a family member died they would take the body to the railway line and place it in front of a train, claiming that the engine had run over their dearly beloved. They were paid a thousand dollars in compensation with no questions asked.

We made our way through Makambako to Mafinga and then on to Iringa which was a pleasant town high on a plateau. Though the streets were full of low, squat houses the Jacaranda trees lining them relieved the monotony somewhat and it did have wonderful views over the lowlands to the west. We were tired and thirsty. We had tried to stop off at two lodges on the way.

They were superbly designed and fitted to very high specifications but both were empty because there were no tourists and manned by caretakers so we had to buy drinks from the roadside and some refreshing oranges. On the modern maps of Tanzania there are lodges and attractions clearly labelled but on our trip there was little to see except for the wonderful scenery and landscapes which was far more interesting than Zambia.

Tanzania gained its independence in 1961 and its first president was Julius Nyerere and indeed, was still the president at the time of our trip. He was a socialist and applied this to the social and economic development of the country. He was much respected on the international stage and achieved massive improvements with a reduction in child mortality rates, an increase in life expectancy and literacy and a 25% increase in primary school places.

Under the famous Arusha Declaration, he instituted the policy of Ujamaa (familyhood in Swahili) which involved the collectivisation of village farms and land which was not only hugely unpopular but led to a big decrease in agricultural production on which the people depended. The country was then hit by huge oil price rises and a worldwide downturn in trade. This was the Tanzania through which we drove. The country was nearly on its knees hence the general overall depression through which we passed.

The next day we were hoping to get to Dar es Salaam. The route was very scenic as we threaded our way through the Rehebo and Udzungwa Mountains but the road was still not good and we all drove with care. There were very steep descents down the twisting and winding roads and there was heavy traffic. At every descending twist and turn would appear, clearly written on the road in big letters—'POLY-POLY.' Swahili for Slow-Slow. Basically we drove in one hour shifts which was a good system and we tried to stretch our legs every couple of hours anyway. We came out of the mountains in Mkumi and then followed the edge of the Mkata Plains to our left and we then had the Uluguru Mountains to our right. Its highest mountain is Kimhandu at nearly 10,000ft and it towered over the town of Morogoro. From there it was a very easy run in to Dar es Salaam. Our first objective had been reached.

I do wonder how I afforded such a trip. My salary was sufficient but not overly generous and my polo was an added expense which was a slight drain on my resources. We had budgeted for four types of expense; namely fuel, food and drink, entry fees (mainly for game parks) and hotels, and our expenses would be pooled anyway. We deliberately chose the cheaper accommodation and hotels in most cases and I think this helped and I do wish now that like Colonel Loftus, I had kept a diary.

In line with our pockets, we found a reasonably cheap hotel in the old part of Dar es Salaam and sat on the veranda with a cold beer. The Hell Run had been hell. It had been a driving nightmare and we vowed that we would not return by that route. It had shaken our nerves and bodies badly and Helen had constantly worried about her car even though we had driven carefully. Graham and I were aware of her concerns and we were both defensive drivers anyway but we agreed that a garage should check it over next morning. It nearly did not make it. We agreed that a celebration was in order to celebrate our Hell Run and decided to have a slap-up meal in the rooftop restaurant of the Hotel Africa near the beach and opposite the harbour.

We had finished the meal and were at the coffee stage when I glanced out the window and saw three men poking around the car. We dashed down and as we emerged from the hotel's revolving doors, the thieves ran off. They had clearly tried to force a window but there was no other damage. Helen was now VERY worried about her car, so ever the gentleman, I agreed to take it over to Ocean Drive, park by the beach and sleep in it.

I was disturbed once by police but I explained why I was there and they let me be. I returned to our hotel for breakfast and the owner was most upset that I had not slept in my bed and only charged me for the breakfast which I thought was most magnanimous of him. Furthermore, he recommended a garage to us to check over the car which also kept it locked on its premises for the following night so I could have a good night's sleep.

I liked Dar es Salaam. On the far side of the Kurasini Creek was all the hustle and bustle of a modern harbour catapulted into frenetic activity by Zambia's urgent need for imports at an ever increasing rate. The harbour was jam packed with cargo ships and the unloading continued throughout the night 24/7. It never stopped. On the opposite bank was another world. The old part was tranquil, calm and even peaceful. This was the Dar es Salaam founded in 1862 by the Sultan Seyyid Majiid of Zanzibar (I like pronouncing that name) and its name means, 'House of Peace', which I found very apt.

It developed as part port and part fishing village and became the largest city in Tanzania. It was its capital until the title was awarded to Dodoma sometime after our visit. Its people are a mixture of African, Indian and Arab but everyone was happy and helpful. We had arrived when the bougainvillea trees that line the streets were in bloom and the town was at its best. We visited the Lutheran Church and St Joseph's Cathedral and wandered the streets and the markets particularly fascinated by the huge variety of sea shells for sale. We strolled along the beach to view the fish market and then drove around to Ocean Drive so that I could swim in the Indian Ocean.

No one had warned me about the coral and Graham spent the early part of the evening extracting it from my feet! I stuck to swimming pools after that. Dar es Salaam, like its harbour and docks, has been catapulted into the modern world and is now a sprawling, fast developing city like many others on the continent. I am very glad that I had seen the real Dar es Salaam with its old-fashioned atmosphere and air of peace and tranquillity.

Our route was now directly north, following the coast and our first target was the Kunduchi Ruins of a fifteenth century mosque surrounded by Arabic graves of a much later date. Chinese pottery had also been located on this site bearing witness to the ancient Oriental Trade. We found the ruins slightly disappointing as they were much overgrown and the site was difficult to even walk through. History was being slowly eroded and neglected.

Many years later I met the Danish Ambassador, by chance, as I was taking a small party of school children around an old Danish slave fort at Old Ningo east of Accra in Ghana. We were both disappointed. The site had been much neglected, with vandalism and theft rife and even worse, it was being used as a toilet. We did not stay long but the Ambassador had been invited there in the hope that the Danish government would fund its restoration. In the circumstances the request was rejected because of the disrespect of the local people. I am very pleased to say that the Kunduchi Ruins have had a make-over and now well worth a visit as you can see them in all their glory.

We now drove up to Bagamoya which is forty five miles from Dar es Salaam. As soon as I got out of the car I felt an immediate atmosphere of doom and gloom, almost sinister in nature, and the feeling has often been repeated whenever I visit locations with a connection to the Slave Trade. Bagamoya was once the major port for East Africa and was the terminal for Arab caravans coming out of the interior with ivory and slaves for which it was a holding port before they were shipped to Zanzibar.

I had seen the starting points for these slave caravans in Zambia and Malawi and now I was at one of their ongoing destinations. The posts to which the slaves were chained were still visible on my visit and there had obviously been no cover for them at all. They were treated like animals in a totally inhumane way. This vile trade continued in Africa long after abolition in the UK and even under the colonial German rule of the nineteenth century.

It was from Bagamoya that Burton, Speke and Stanley headed off into the interior and it was to Bagamoya that Livingstone's body was brought to rest at the Catholic Mission before returning to its burial at Westminster Abbey. The port has a strange mixture of German and Arabic architecture and we were able to visit the German Colonial Headquarters and the Catholic Mission. As we left the site I pondered on the fact that the early German officers and colonists attended the Lutheran Church every Sunday as devout Christians presumably but let the tail end of the Slave Trade continue with all its inhumanity.

We were now on the trail of 'White Gold' (Sisal) but had to cut inland as the 'main' road avoided the Saabani National Park and ascribed a big loop through Muheza to Tanga where we had planned to stay the night. We had visions of staying at the Planter's Hotel but it was too expensive so we moved downmarket a little. The Planter's Hotel in

Market Street was a very ornate edifice full of arches, columns and plinths and in the past it was alive with Greek Sisal plantation owners gambling and drinking their nights away and as often as not, their plantations as well because many an estate was lost at gambling.

Tanga was the centre of the Sisal Industry and in 1961, Tanzania was the world's largest producer of Sisal which employed approximately one million people in the industry. It was called the 'White Gold'. Many people of Greek descent were employed in the area. Why? Many Greek engineers had come out to build railways in the past and when they had finished the work, they turned to Sisal production though many Germans were also employed as a hangover from the German East Africa Company. There is still a Greek Orthodox Church in the centre of the town.

Tanga is Tanzania's most northerly port and was also a major player in the Slave and Ivory Trade but any historical traces of it seem to have been eradicated. Its status has changed from town to city now and you know which I would prefer. We had located a cheaper hotel whose individual rooms led out directly on to the veranda. We had eaten a reasonable meal, towards the end of which there was a tropical deluge. African rainstorms announce their arrival with a rustling in the trees, in our case, Palms, the sound of thunder and lightning and then the deluge from the heavens and they can really be quite violent.

We were sitting on the veranda watching this particular storm, drinking our beers, when an apparition loomed out of the rain wearing a black plastic cape and an old bush hat, pushing a bicycle. He must have gotten word that we were the only tourists in town that night because he wanted to sell us carvings, which, quite frankly, we did not want. The car was quite heavily loaded so we had agreed to cut down on all such items. He was most persistent and I gave him five African shillings to go away. He insisted that I take one carving and he gave me 'Humphrey'.

'Humphrey' stands before me now on the table. He is an ebony carved stooped old man, wearing a long loincloth, carrying a load on his head. It is a good carving, slightly caricatured, but it has a certain aura about it with which I empathise. I have since learned that he is a valuable Makonde wood carving, of some value, and now I feel very guilty so I always keep him with me as a reminder of this trip and for me to be kinder to my fellow human beings perhaps…

We were to visit the Amboni Caves the next day which was on our route to the Kenyan border. These are a huge complex of naturally formed caves formed some one hundred and fifty million years ago during the Jurassic Period and it is estimated that they were under water until twenty four thousand years ago. The locals believe that supernatural 'Mizimu' live in the caves and they were still being used for fertility rites. Unfortunately there were none on that day when we arrived so we had to make do with a guided tour into one section of the cave. The whole complex is vast spreading over many square miles. Having been used to the Cheddar Caves it was quite pleasant to walk through relatively unspoilt caverns without the paraphernalia of subdued lighting and sound effects.

We stopped at Manza to look at a local market and buy a soft drink and we noticed our first Maasai Warriors who were also buying food and Cokes. We thought no more of it until we reached the border at Lunga. We were stopped on the Horohoro Tanzanian side by a police sergeant who stated that one of the Maasai Warriors had insisted that Graham had taken a photo of him in the market at Manza. You were not permitted to take such photographs and we all knew this and Graham insisted that he had not. We were turned back, however, and had to retrace our way to Manza Police Station.

We did not realise until we arrived that two more tourist parties had also been forced back in case it was them. In the end the Maasai Warrior was called in to face the three tourists and he could not identify the culprit nor could he identify the camera all three of which were laid out on the desk. Case adjourned. We drove back to the border and entered Kenya. In a couple of hours we were on the Likoi Ferry crossing through the busy Kalindi Harbour into Mombasa and booked into a hotel right beside 'The Tusks'. I am not sure if it was appropriate or not but the huge double pair of aluminium tusks straddling Kalindi Road had been built in honour of Princess Margaret who had visited Mombasa in 1956 so they were not that old.

The history of Mombasa can be traced back to the ninth century and it was a centre for spices, gold and ivory and was also a slave port. It was originally mainly Muslim and had much Arab influence and thrived on the monsoon pattern of trade. The many boats would arrive from India and the East when the easterly monsoon winds blew and then return on the westerly winds later. Outward goods included millet, sesamin and coconuts. The boats used were the traditional dhows and we were pleased to see that several were beached at Tudor Creek and still carrying on the trade.

It was a 1,000-year-old trade at least. Mombasa was visited by Vasco da Gama in 1498 but he met with a somewhat chilly reception and did not dawdle. In the late fifteenth century, 1589 to be exact, the Portuguese arrived in an attempt to control the vast Indian Ocean Sea Trade and then followed a seesaw series of bloody and riotous occupations right through until 1695. The Portuguese built the huge Fort Jesus which dominates the seashore and towers over much of the old town. It was well guarded and protected and its big guns commanded Mombasa's approaches.

We visited the fort and its museum and once again, learned about its role in the slave trade in the second half of the nineteenth century. It maintained huge 'Holding' dungeons for the captured slaves arriving from the interior and they were shipped out through a doorway to the sea on its Northern Wall. Thousands of slaves passed through that gate and one can only imagine their fear and terror as they were herded into the boats. Unfortunately, I was to undergo this experience many times again visiting the Slave Forts that ran along the Ghanaian shore and always felt that same sense of foreboding and dread. Mombasa eventually fell under British Rule and the British East Africa Company and the rest is history.

We confined most of our attention to the Old Town with its narrow streets squeezing under the towering ramparts of Fort Jesus and it was then I noticed that many of the fort's guns were 'trained' on the town for the Portuguese feared attacks from the land side as well. We visited a very old mosque that was also used as a Musalla where the young Arab boys learned the Koran by rote (why not the girls?) and we were taken right around it and even climbed one of the Minarets. I was a little disappointed that they had modernised the Call to Prayer by fitting a tape recorder, loudspeaker and amplifier. I wonder if one of their young boys had ever dared to change the tape?

Like many African towns and cities it had a lively, bustling and colourful market in which we tried pomegranate fruit juice…delicious. I think it was when we were in Mombasa that I fully realised that there were different charges for residents and non-residents. The government wanted to encourage its people to visit all the various sites and game parks and they paid a very low fee while non-residents paid larger sums but as they did not seem to be that high, it did not really bother me. Perhaps we should adopt the same system in the UK?

Making our way out of Mombasa the next morning, we headed over the concrete

Ngali Bridge on our way to Gedi. I understand that a new super Ngali Bridge is in the process of being built to alleviate the horrendous rush hour traffic jams but we had no such problems and we happily drove up the coast relatively traffic free. Most of the coast had not been developed at all on our trip but now there seems to be a hotel or resort every hundred yards perhaps leading to over development? North of Malinidi the resorts are suffering from a lack of tourists because of shootings and kidnappings spilling across the border from Somalia.

Gedi is a mystery and almost an anachronism but certainly enigmatic. It is a beautifully constructed stone built town, though now in ruins, with excellent architecture, streets, running water and even flushing toilets. The ruins are almost hidden away by a lush forest near the seashore and I was impressed by its air of tranquillity and peacefulness. It was probably built in the thirteenth century but no one seems sure and I think the identity of its builders is yet to be proved. African? Phoenician? Arabian? Indian? 'Now who would have lived in a city like this?'

Having spent two days in Mombasa it was time to move on for our next destination was to be Nairobi about 280 miles north-west of Mombasa. This would make our total distance travelled to date at approximately 2,000 miles but we believed that the road was tarmac and quite good so aimed to do it in one day. We did not adopt a tight schedule for the trip or even a detailed itinerary. The route had been roughly planned but we found it better to take things as they came and did no pre-booking. This meant that we could change our plans according to circumstances.

The first part of this leg was over the coastal plains of Maunga, Dika and Ndara then we passed the Sagaal and Taita Hills to the west of the road before arriving at the Manyani Gate to the Tsavo National Park which now spread for miles to the east and the west. We stopped at Kibwezi for a late lunch then drove on through Emali and Sultan Mamu. Emali was known as 'the town that never sleeps' because traffic passed through it all night and Sultan Mahmud was evidently in charge of Zanzibar in 1902 but why a town 250 miles inland bore his name we could not discover. By late afternoon, we passed through Machakos and then to Nairobi which we reached just before darkness fell.

Now I am not sure if the road followed the railway or the railway followed the road but they appeared to be glued one to the other for the whole of the journey. Construction of the Uganda Railway commenced in 1896 with a workforce of 25,000 British India labourers as Kenya was too sparsely populated to provide such numbers. 2,498 died and 6,724 elected to stay in East Africa. It was backed by the British Government, just, as it was of strategic importance in maintaining British dominance in the region and was part and parcel of the Scramble for Africa at that time. Its critics labelled it the Lunatic Railway and Henry Labouchere MP (Sounds a good British name. Must have been a Norman.) wrote the following:

What is the use of it? None can conjecture.
What it will carry there is none can define.
And, in spite of George Curzon's superior lecture.
It is clearly naught but a Lunatic Line.

It was finished in 1901 and ran up to and including the time of our safari though it was in a shambolic state. The railways of Kenya are being rehabilitated as I write and I understand that there are now fast trains operating on new lines—in both senses of the word.

Blood was shed in its construction. In the Kedonony Massacre, 500 workers were killed by angry Maasai Warriors when two Maasai girls had been raped. The second you can read all about in John Patterson's book The Man Eaters of Tsavo, when two man-eating lions terrified the workers and actually brought construction to a halt. I was reading the book on the trip.

There have always been tensions between the Maasai and the Kikuyu Tribes in Kenya. The Maasai are nomadic and their way of life has always clashed with the Kikuyu who are farmers and the two ways of life are not compatible bearing in mind that the Maasai are fairly warlike anyway. We saw no real signs of confrontation but the Mau Emergency had only ceased in 1966—eight years before our trip. Present-day politics still tend to run on tribal lines and the old friction is always there.

We stayed at The Garden Hotel on Kenyatta Avenue not far from the cathedral which was pleasant, quiet and relaxing with a nice pool so we chilled out on our first evening with a swim, sundowners and a pleasant dinner. Our drive up to Nairobi had been uneventful but I had been impressed by the vastness of the plains. Mainly to the east there had been mile after mile of savannah grassland and beyond that there was a huge swathe of the country almost devoid of people and habitation.

Studying a map of Kenya reveals that most of the Eastern part is devoid of roads, houses or people. The vastness is unbelievable. Nairobi is not an ancient town though it does have some colonial buildings and the original railway station was still there. It is more a city of landmarks and for that reason is quite pleasant to wander around but it was also modern with up to date services and modern shopping centres. My favourite Nairobi haunt was the Thorn Tree Café in the Stanley Hotel. It is said that you will always bump into someone you know at the Thorn Tree and I found this to be true but I did have an advantage.

Kenyan Polo teams often visit Zambia for matches and so I knew many of the players and their wives and saw several of them including a very good friend to be, Francis Erskine who ran Sanctuary Farm in Naivasha. He had been in Kenya for most of his life and was quite a daring young man. In his youth he drove an MG sports car and accepted a bet to drive it around the circular lobby of the Norfolk Hotel and out again. He did—but the next morning he was called up in front of the magistrates, who of course, all knew him so could not conduct his case, so poor Francis was escorted out to one of the townships where there was a magistrate who did not know him.

The policemen read out the charge. The clerk asked whether Francis was guilty or not guilty and Francis answered truthfully, 'Guilty.' The magistrate then leaned towards Francis and enquired the sum involved in the bet.

'Ten shillings, sir,' was the reply.

'Francis Erskine—I find you guilty as charged and do hereby fix the fine for the offence at ten shillings. Case dismissed.'

I did pop into the Norfolk Hotel to see the lobby and it was big enough for a car but Helen would not let me try with the Peugeot 304.

We visited the Botanical Gardens and had a splendid afternoon in the Nairobi Game Park even visiting the animal orphanage. We also found time to visit the magnificent Jamil Mosque in the centre of town guarded by wonderful black elephant statues.

That evening we went to the cinema and as soon as we had parked, we were mobbed by small boys who wanted to look after it for us. Graham chose two. It was always better to employ such boys on the basis that if you did not, your car might be deliberately damaged. We conformed to the tradition which reminded me of a similar incident in the

UK. A driver had just finished parking his Range Rover in a side street at the back of Leicester Square in London when he was greeted by a small boy.

'Luk avter yer car for ya, Mister?'

'No thanks, sonny. Just look in the back,' said the driver and pointed to a big black dog. 'He looks after my car.'

The young lad looked the driver in the eye.

'This 'ere dawg. Gud at putting owt fires, is he?'

We travelled out to the Ngong Hills to do a little walking and much admired the views before we visited Karen Blixen's House and were able to go around it. Her book Out of Africa is now quite famous and it was another volume with which I was travelling. The film, however, was not filmed at the house which has now been presented to the people of Kenya as a permanent museum by the Danish government. Alas, however, Nairobi's housing development is nearing its doorstep.

We were now to embark on the Tourist Route, for, every day, minibus loads of tourists leave Nairobi and do a circular tour of the region north of Nairobi visiting as much as they can in the ten hours of daylight which they can use. In a way it was quite amusing to watch as we would be lazily viewing flamingos by a lake and two or three minibuses would speed up, disgorge their passengers to take photographs, load up again and career off to the next point of interest. It seemed very frenetic to us for ours was a slower pace and we took three days to complete the tour.

As a geographer, I was keen to see the Great Rift Valley part of which I had already travelled in Malawi and I now recalled that this enormous structure was 6,000 miles long commencing at the Beqaa Valley in Lebanon right down to Mozambique. The Western Rift is the home of Lake Victoria and the Eastern Rift that we were about to travel contained many lakes, mountains and volcanoes. Going over the edge of the Rift Valley was quite spectacular as one could clearly see its vast cross section and the road down gave good views down to the valley floor.

Our first stop was Longonot and the car had to make quite a steep ascent before we arrived at the viewing point at some 9,000ft to view the caldera crater which was some 3,000ft below and spread over an extensive area. The view was spectacular and shimmered in the hot sun. Being truly British we made a cup of tea and sat admiring the scenery as we dunked our biscuits. Next came the Njorowa Gorge or Hell's Gate where we walked down a narrow ravine to visit hot springs and we saw a duiker. We were on dirt tracks for a while but it was dry so they were all in good condition and very firm.

Lake Naivasha was picturesque and we saw some marabou storks bur we did not see any hippos or fish eagles so we did not stay long. Naivasha was only a small market town but it was the home of the 'Happy Valley Set' and the infamous Lord Delamere who bought 100,000 acres of land in a rather dubious deal with the Government and became involved in a couple of murder cases.

As we drove up to Nakuru through Gilgil for the night, we had splendid views of the Aberdare Mountains to the east and as the sun began to set and the shadows lengthened, I suddenly remembered that the Mau had been very active in this area, a fact that I am to deal with when I relate some interesting discoveries about my friends, of which I had been blissfully unaware.

Lake Nakuru is the home for well over a million flamingos and they form a huge pink band around the lake shore as you approach and then they materialise in to thousands and thousands of individual birds all slowly strutting along the shallows. It is quite a sight and their stately movements a pleasure to watch, spoilt only by minibuses arriving and blowing their horns to make the flamingos take off for purposes of photography. The lake has Acacia forests as its edges and seems to attract large game including leopards of which we saw none.

Leaving the Rift Valley, we drove up through a very scenic approach of valleys, farmland and forests to Nyahururu and Thomson's Falls. The site was well looked after and the falls pretty but not spectacular. We had come from Zambia, remember, and Thomson's Falls were very small (74ft) compared to ours. Size matters.

We were beginning now to catch glimpses of Mount Kenya as we travelled through the Aberdare Range on to Nyeri where we headed directly north up to Nanyuki. We had promised ourselves a visit to the Mount Kenya Safari Club and what better view of the mountain could you gain from sitting on the club's veranda sipping drinks and tucking into a delicious lunch?

The Safari Club was the height of luxury and was the haunt of millionaires, celebrities and royalty which meant that we could not possibly afford to stay there. The main building was an old property called Mawingo from which extensions had been added and all the rooms had balconies overlooking Mount Kenya. It had manicured lawns, on which peacocks and cranes strutted, semi-tame sacred ibises and marabou storks stalked and kudu grazed. Everything was very grand and formal and it is said that even the insects dressed for dinner.

It provided very gracious living and Mount Kenya with its snow-capped peak was its backdrop (snow on the equator). William Holden and his friends often stayed there and he always used Room 12B which is now much more expensive than any other room. In 1967, Mr Khashoggi had given the club to his son as an eighteenth birthday present and all I received was a new sports coat for college.

Mount Kenya stands at about 1,700ft. It is the highest mountain in Kenya and the second highest in Africa. My mountaineering friends assure me that it is a good mountain to climb because it has a variety of different climbs. They need not have worried for we had discovered a different route.

We were going up Mount Kenya that night. Not climbing you understand but in a four-wheel drive ex-army truck that took a party of us to The Secret Valley Lodge. This was similar to Treetops but on a much more basic scale and standard. From the veranda one looked out over to three platforms that had all been baited with fresh meat which was to entice leopards.

We waited some time and did spend half the night leopard watching as they devoured the meat and one even fell asleep, and so did I. I did wonder about the ethics of this. We are often told not to feed wild animals and wondered if the leopards would lose their hunting skills and survival instinct. It did seem quite artificial and perhaps misguided? I am often uncomfortable about the treatment of animals in this way and for that reason have never been a great fan of zoos.

We had crossed and recrossed the equator the previous day and now we headed for Meru that many people, including generations of school children, have probably seen on TV, for, just south of Meru, is The Man in the Car Park, who demonstrates the Coriolis Effect. In the north end of the carpark, i.e., North of the Equator, he pours water into a bowl and lo and behold: it drains in a counter clockwise direction. He then goes to the south end of the carpark, i.e., South of the Equator, and repeats the same process and the water drains out clockwise. I believed it. My friends believed it. All the other people there believed it. I hate having my illusions shattered but it has been scientifically disproved. To gain that effect, the man in the carpark tilts the bowl by sleight of hand

and the whole thing is a trick. How will I stand to have my illusions shattered like this?

We drove back to Nairobi and I was pleased that I could see the Rift Valley again, if only briefly, and we returned to the Garden Hotel. It was a magnificent trip and we had added another 300 miles to the two thousand already achieved. We were now on 2,300 miles and had yet to get back to Zambia. We were going back avoiding the Hell Run but the roads ahead were unknown and we would have to be prepared for trouble of any sort. It was slightly daunting.

We had a straight run out of Nairobi the next morning through Athi River and then across the Athi and Kaputier Plains, due south to Kajiado and into the Maparasha Hills, emerging to see the snow clad summit of Kilimanjaro directly ahead. There, in front of us, was the classic view of Kilimanjaro with elephants in the foreground moving across the foot of the great mountain. Stunning. We were at the Kenyan/Tanzanian border crossing at Namanga well before lunch and the formalities were fairly speedy as they were keen to encourage tourists.

We kept to the main road, passed through a rather sleepy and quiet Arusha and arrived at our accommodation in Moshi which I think was under the auspices of the YMCA but it was perfectly clean and tidy and well presented. Nowadays this area is very busy with hundreds of tourists, busy roads and an International Airport nearby. It buzzes with activity but while we were there it did not buzz and had no activity. We explored the southern flank of Kilimanjaro, sauntered around the town of Moshi that seemed fast asleep in the hot sun and went early to bed as we were heading for the Ngorongoro Crater in the morning.

Our first objective was Lake Manyara National Park where we saw the tree climbing lions slumped fast asleep along the branches. Skirting the northern end of the lake we headed straight for the crater and to our accommodation that was slightly set back from the crater rim. It was a beautiful lodge and specialised in serving all forms of game meat and all the different types of animals seemed to be on the menu. I think I aimed low and had a kudu steak which I much enjoyed and the beer was very refreshing. We had to be up before dawn next morning, had a quick breakfast then five of us packed into a Land Rover which carefully made its way along the circular track that ran right around the crater until we came to the track that led to the crater floor.

The Land Rover was open topped and we gulped as it slid over the rim of the crater and inched its way down a death-defying steep track with hairpin bends and very rough surface. We breathed a huge sigh of relief as we drove out onto the crater floor. The track was the only way down which also meant that it was the only way up so a time system had been devised with simple African logic. Down in the morning/up in the afternoon. The crater was huge and our guide simply asked us what we would like to see.

There was a big variety of animals and we saw them all. I think it is the only game park that I have ever visited with such a diverse variety. The Big Five. Elephants, Buffalo, Lion, Leopard and Rhino. There were all types of the Antelope family plus Gnus, Giraffes, Hyenas, Vultures, Monkeys and Baboons. The list was never ending. We saw a kill in action and were also taken to see a lioness with her cubs. To one side of the crater floor was a Maasai Camp and I wondered how they managed to live alongside the wild animals, especially as they were herding cattle. At lunch time we ascended the torturous track again and spent the rest of the day relaxing by the pool, eating and drinking, for tomorrow we were heading into unknown territory. Were we about to jump out of the frying pan into the fire using a road about which we had little knowledge?

Skirting Lake Manyara we hit the main road at Makuyuni, turned right and drove south. Road conditions were good as we drove between the Tarangiri National Park and the Malbadow Escarpment as far as Babati which was a very small town jammed in right under Mount Kwaraha. Here we faced a choice. By adding fifty miles to our journey we could have continued on the good road as far as Singida but then there was a long loop around on a second class road to Dodoma or we could just head directly south to Dodoma on a second class road.

We took local advice and headed directly south. We were now on a dirt road which was in good condition. Traffic was light and there were none of the hazards of the Hell Run. The weather was dry and we made good time to Kondoa. We were used to driving in such conditions in Zambia and had no concerns until south of Kondoa it started raining. Road conditions now turned hazardous and dust turned to mud so we had to drive slowly to avoid skidding while maintaining momentum to obtain grip on the slippery surface.

The heavy rain also made visibility a problem and we crawled the sixty miles to Dodoma. Four miles outside of the town, there was a deluge and we had to pull over but darkness was approaching and we had to push on. I remember this clearly because I was driving. I was forced to a halt two miles further on where floods of water were now sweeping across a short causeway. It looked extremely hazardous especially as the light had faded and darkness was upon us. I studied the situation carefully and noted that one man had waded across and seemed OK. It was Helen's call. It was her car.

She agreed to go on and I carefully edged into the water. We all held our breaths as I edged through the flood and we were mighty relieved to get to the other side. Phew. Still in driving rain, we drove into Dodoma and quickly found a hotel. It had taken us six hours to drive 55 miles and we did wonder what the morrow might bring.

The hotel manager was insistent that the route we had hoped to take was impassable for anything other than four-wheel drives so we had no choice but to drive the sixty odd miles down to Iringa and re-join the Hell Run. That alone took five hours and it took us another two days to get back to Lusaka. Altogether, we had travelled well over 4,000 miles and returned with no damage to the car or ourselves. It was some achievement and now I am able to boast that I have been on safari.

This is what happens to Crocs who eat cattle.

Sammy.

Chapter Fifteen
Mazabuka

One colleague, many years ago, in my first teaching job at Wycliffe, berated me for not coming to terms with my environment, with the obvious inference that I should have stayed at Wycliffe, slowly declining into a Mr Chips-like role in my dotage. I do recognise, however, that, even nowadays, I am ever eager to expand my horizons, to move on and always take whatever opportunities that life offers. I like to grab life with both hands and live it to the full and I believe that I am the perfect example of growing old disgracefully.

I offer a dangerous example to my students and should never be regarded as a perfect role model. I do not think that I could possibly have settled into the normal routines of life and perhaps it is fortunate that I have never had the responsibility of raising a family because that would have made my life choices impossible. I do have, however, the most extensive portfolio of relations and friends and many of my former pupils remain in contact. I am never alone though I do enjoy solitude but can also be very gregarious. To me it has been a perfect life balance which I continue to enjoy.

My final year at the Lusaka Boys' School was as enjoyable as ever, my social life was interesting, my leave house 'minding' as busy as ever and my polo ever expanding. I must add that I have never been extremely talented on the field and mostly held a handicap of one or two. The highest handicap is ten and very few players hold such an accolade. I was always a Low Goal Player and would have needed more money and ponies to even struggle up to a three I think. I was playing in games and matches, helping to manage the polo stables, with Alf ever in the background in support, and I was always prepared to help out whenever the need arose. The Castle Beer Brewery may have expressed concerns about my contract coming to an end and I was beginning to think about that myself.

I had played many matches at Mazabuka. The club members there were mainly farmers who owned vast farms, kept huge herds and grew large amounts of maize. They lived hard and played hard and were generally more enthusiastic and devoted to the game than us expatriates in Lusaka but they were always extremely friendly, hospitable and generous. Many of their sons and daughters still own the farms to this day though most have added other income streams.

During one Mazabuka versus Lusaka match weekend, I was approached by Bevis Coventry who owned a farm on what we called the Back Road near the sugar estates. He asked if I might be interested in a job as cattle manager at the Anchor Ranch. I was slightly taken aback but assured him that I knew nothing about cattle. He reckoned that my knowledge of horses would suffice. I asked for time to think it over which I did. I had nearly ten months left on my work permit, I was enjoying the lifestyle in Zambia tremendously and such an opportunity would probably not arise again.

By arrangement, I drove down to the Anchor Ranch to spend a weekend with Bevis and see the ranch. There would also be an opportunity to play some polo as well which would be appreciated. From the gate on the road at the bottom of the ranch I drove up the mile long track to the farmhouse which was most attractive and fairly extensive but it had not been when Bevis first arrived. They had lived in what was now the kitchen at the back, and Bevis, with a couple of labourers, had fenced the whole of the 355,000 acres himself. This was not just around the perimeter but the whole area had to be divided into 'camps' for the cattle. It had been a Herculean task.

Furthermore, Bevis had also constructed a whole series of stone cattle crawls near the farmhouse that had also been a mighty task. We rode down to the main road, for the farm extended down to the Kafue River on the other side, and one camp was given over to large scale maize production which was not included in my possible job description. We caught some women stealing the maize and Bevis chased them off. He knew who they were and said he would deal with them later.

Much later in Ghana, I was to discover that a major handicap to agricultural development was theft of crops and many Ghanaians had given up farming because of it. Bevis had anywhere between 3,000 and 4,500 head of cattle depending on calving which were mostly indigenous but maintained a superb herd of Brahmans, many of which were fine bulls. They were gentle but huge and you could mingle with them though one had to be wary with them at all times. My job would be mainly checking each camp's water supply daily, both north and south of the main road, and helping with the Spray Race and de-ticking for this was a weekly chore to include every calf, cow and bull. Then there was also the de-horning and castration of the calves.

Bevis assured me that I would be able to cope so I agreed to join him, accepted his salary offer and free food and accommodation, for Myra, his wife, was a superb cook, and we spent the rest of the weekend playing polo. There was one problem and that would be in the near future—the renewal of my work permit. Without that, I could not be employed.

My biggest problem was to inform my parents of the decision and I was never really sure of their reaction but they did know that I would be back in the UK in the not too distant future because I had an air ticket which I could use so they would see me for a short time at least. The staff at the school were a mite surprised and everyone at the polo club said that I had only taken the job to get free feed for my ponies.

During those last few months at the school, I had also played host to Francis Erskine and Ted Nightingale, two of the Kenya players, and Francis had stayed another couple of days when he was on business in Accra so I got to know them both quite well. They were very pleasant and mild in manner, extremely generous and kind and friendly. Now all I knew is that they were both farmers and played in the Naivasha Club for years and Ted Nightingale had been the player who had introduced Simon Kaseyo to polo, indeed, Simon had held Ted's ponies when he was but a very young boy. I was aware, because Alf had informed me, that Francis Erskine's father, Sir George Erskine, had been Commander in Chief of the British Army Forces in Kenya during the time of Mau Mau and his army career was very distinguished indeed. Neither of them gave me any inkling of their past lives but is so fascinating that it cannot be omitted from this book.

The Mau Mau were a secret African society originating in the Kikuyu Tribe committed to expelling the European settlers from Kenya by means of brutal killings, massacres and atrocities and their uprising lasted from 1952 until 1966. They were brutal killers and much feared. They slashed and burned and also dismembered their victims. It is no consolation but only a comparative few white people were killed, 32 is the official number, but eleven thousand loyal Kikuyu were slaughtered in their cause because they fought with the Europeans against Mau Mau. 69 of the security forces were killed. 1,090 Mau Mau were hung.

Francis was obviously in the middle of all this as his farm was at Naivasha, part of the White Highlands where the uprising was at its worst. He was a captain in the

Kenya Regiment and as early as 1953 he was awarded the MC for bravery and then his story takes a very unusual twist. He and Bill Woodley introduced the concept of Pseudo Gangs. Loyal Kikuyus turned themselves into Mau Mau members by dressing and acting like them and the disguise had to be 100%. They used human hair, lived as close to animals as possible and even used animal excrement to spread on themselves so that they were Mau Mau in every sense of the word. They were totally indistinguishable from the Mau Mau members.

They then infiltrated the Mau Mau bands and achieved much success in gradually capturing them and whittling down their numbers. They would lead the regular soldiers to each group who would ask for their surrender or they would gun them down and hack off a hand for identification purposes. My story does not end there, however, because I have learned that young white Kenyans also ran with these Pseudo Gangs and one such white Kenyan was Francis Erskine and he did this for several years. By using this method the Mau Mau were eventually cornered and surrendered after the capture of their leader. Francis was one of the white farmers persuaded to stay by Jomo Kenyatta after Independence. He was highly respected in Kenya.

Ted Nightingale was the gentlest of men and I really had no idea of his history and achievements in life until I blundered across a reference to him on my Kenya research. He was educated at Rugby School and he was the son of a Devon vicar. He went up to Emmanuel College, Cambridge where he bumped into a former graduate who was serving in the Sudan. Rather like me, I suppose, he was taken with the idea of going abroad and applied and was accepted for the Colonial Service.

In 1926, he sailed for Port Sudan armed with six white lawn shirts with detachable collars from Hawkes in Saville Row which he never wore. Under his conditions of service he had to learn Arabic and Law during his two year probationary period and he was duly examined and passed. He served in Darfur, Gezira, Nyala and Khartoum as well as the Bahr-el-Ghazal District in Southern Sudan among the Dinka tribesmen. Trekking on camel for ten hours a day he acted as magistrate, assessed taxes on crops, settled grazing disputes between disgruntled tribes and also found time to fill in the blank spots on the totally inadequate maps with which he had been issued.

During the Second World War, he was ordered to stay at post but his life was not all travel and drudgery. The cattle-loving Baggazara Arabs of South Darfur often called on him to shoot the Lions and Leopards that attacked their herds. He once bagged five lions before breakfast. (They could have done with Ted on the Tsavo section of the Ugandan Railway.)

He also had two cheetahs as pets and they were called Swan and Edgar. He retired from the Colonial Service in 1954 after serving as Governor of Equatoria Province in Southern Sudan. Determined not to leave Africa he and his wife, Billie, the loveliest of ladies, moved to Kenya to be close to her parents and bought a farm on the Kinangop Plateau at the height of the Mau Mau Uprising. Nine years later they were forced to leave it and bought another farm at Naivasha and started the biggest turkey farm in Africa which produces 30,000 birds per year.

In 1994, Ted and Billie celebrated their golden wedding at which Billie surprised Ted and all their children and grandchildren by donning her wedding dress and dancing until the small hours. Ted died in 1996 and Simon Kaseyo was at his graveside in England. He picked up a clump of earth and stood over the grave. Trickling the earth onto the coffin, he simply said in Kiswahili:

'Kweli, hii ne kwaheri ya kweli.' (Truly this is the last goodbye.)

Sadly, Simon was to follow his mentor and friend only a short time later when he was killed in the Abidjan air crash, the details of which I have already described. It was a very tragic incident.

There were various parties and farewells in Lusaka before I was able to say my goodbyes, jump into the Wolseley and drive down across the Kafue Bridge to the Anchor Ranch. I had a travelling companion, a dog called Sammy, who was a pedigree Red Setter and who had been used to high living being fed only on chicken fillets and other tasty snacks. The Speed family, who were returning to the UK, had begged me to take the dog with the idea that it would have a perfect life on a farm. The Speeds obviously had no idea that Bevis bred Alsatian, Doberman and Bull Terrier crosses as cattle dogs and the leader of the pack was Butch, who was rough, tough and quite nasty. Nor did they realise I am afraid that all the dogs lived on a simple diet and there was no hope of chicken fillet.

Sammy had a wretched three days and was bullied and refused to eat but he wolfed down what the other dogs ate on the fourth day and the other dogs, including Butch, accepted him and were relatively friendly after that. He followed me around the ranch every day.

I have to say that I was somewhat surprised when Bevis informed me that he and Myra were going away for a two-week holiday the next day. I was left with no real instructions, completely on my own and no advice. It was daunting. Bevis obviously thought that throwing me in at the deep end was the best approach. I knew that checking the water supply was absolutely essential so was somewhat surprised that there was none when I tried to have a morning shower in my little rondavel the next morning. This was an immediate emergency so I phoned a friend—the young farmer next door—and explained my plight.

He came straight up to the farm and spent some hours fixing the main pump which supplied all the water for the house and cattle crawls and Spray Race and I was able to drive down with one of the farm hands to check the waterholes and troughs to the south. On my return we had water so I left the foremen at the crawls dealing with the three herds that were scheduled to be checked, sprayed and de-ticked and did the water check to the north. On my return, I checked the stables and horses, found all was well and got to the crawls just as they were finishing the last herd.

The foreman assured me all was well and I asked him to check the 'camps' for water tomorrow while I got used to the crawls. He said that would be OK and I headed back for a shower and supper but I was met by another groom who was anxious about one of the horses which he thought had colic. Now I had checked all the horses not half an hour ago and they had all seemed alright to me but the groom led me out to a small paddock at the rear where there was a fine polo pony which was showing signs of Colic. It was evidently Bevis's favourite polo pony and he had not even told me it was there.

The horse was not trying to lay down to roll so I immediately fetched a couple of warm beers which we forced the horse to drink to relax it and then searched in the office for any horse medicine, or Mooty as the Zambians call it. I located some Acetol Promezine and gave the horse one and a half cc, telling the groom to stay with it until I came back in one hour. I needed a shower, several cold beers and supper and in that order. Much revived I went out to see the horse again and it looked OK but I did check it several times during the night. It was down and sleeping so I left it and it was fine in the morning and scoffed all its food. No problem.

Bevis had set up the crawls very well and seemed to have devised a workable system. We had about twenty herds on the farm each one with about 175 head. Three of these herds would be driven in by their Lozi cattle herders every day and would go through the Spray Race and at the same time, would be checked and then have the de-ticking grease on their backsides and in their ears. The Lozi herders were very good at telling us if any of the cows had a problem anyway so I was helped by them. The Lozi were tall and were well known for their knowledge of cattle. They knew each beast well and were good at handling them.

There was one important rule. If a cow was missing then it had to be found and if it was dead then either Bevis or I had to inspect the carcass and we normally conducted a post mortem to check the cause of death. That was not a particularly pleasant task but I gradually got used to it. It was my job to put down any suffering cattle and this was achieved with an old Colt 45 which I learned to use cleanly and efficiently. One day I was going down to the far end of the farm when I noticed a pack of wild dogs heading towards one of our herds. I stopped the pickup, loaded the gun, and fired at the dogs hoping to scare them off. I actually killed two and the rest ran off. That increased my street cred with the farm hands.

I enjoyed the day in the crawls. The cattle were well looked after and the day proceeded at a leisurely pace. The cook brought out a tray all set out with bone china cups, teapot, milk and sugar which I drank in the shade of an overhanging roof. Bevis had tea at least twice a day like this and cake in the afternoon. Under the shade of that roof, he was able to touch and check every one of his cows and he obviously enjoyed them and the ranch. He was a true cattleman. He was hoping now to have more free time and his hobby seemed to be welding. He loved tinkering in a small shed which he used for the equipment and was really quite good at it making troughs and feeders mostly for the cattle but his wife also got him on household repairs as well.

By the third day, I was becoming more confident and learned much about the work and methods. Bevis's sons popped in to see if all was well and I felt that I was now able to accept many of the invitations to visit other farms for braais and parties. By the end of the first week, I was enjoying myself. I worked hard on the farm from early morning until mid-afternoon then practised my 'stick and balling' on the rough field below the crawls and then I would eat in or go out to a neighbouring farm to eat, for all the farmers were very hospitable. I played polo at the weekend and was now a member of the Mazabuka Club. I found their enthusiasm infectious but I could not keep up with the better players or their drinking rate in the bar afterwards.

On the following week I had just returned from checking the waterholes below the main road when Lionel Coventry phoned me to warn of a fire right at the bottom end of the farm. Bush fires are dangerous and any reduction in our grazing posed a serious problem. The foreman and I drove down to the fire and commenced to beat up its backside. We had no water carts and beating was the only method which we did all morning. I sent a driver back to the farm to get drinking water for the men and he came back with a message that Lionel Coventry had stationed himself on the main road and was organising a controlled backburn. It did the trick but we had to leave several men to ensure that the fire did not restart with any sparks.

It was no small fire. It burned out well over three thousand acres. I had also to brush up on my Chilapalapa. Zambia was a country of some 67 tribes and it was thus necessary to adopt the universal language that they used in the mines in South Africa to communicate with all workers. My Chilapalapa was very limited as I had only used it with the grooms at the polo club but it was important now to for me to learn a larger vocabulary quickly.

By the time Bevis returned, all was well and I had written him brief reports on a daily basis so that he knew about his horse, the broken down pump and the fire but I was pleased that he could now take responsibility for the ranch and I could relax a little but I was in trouble. Myra asked if the painter could repaint the office while they were away and I agreed to check on his work. I reckoned that he had done a pretty god job and the office was now fresher, cleaner, brighter and full of light. Unfortunately, in his enthusiasm, he also painted the old wooden office chair. It was a family heirloom and Myra had not wanted it to be touched. We managed to strip the paint off and it looked quite old again. Myra handled the financial affairs of the farm and was very good at it and she was very firm but fair with her staff and the farm workers.

Bevis was known as Bwana Mukuba (Big) and I was known as Bwana Picinney (Small) and it was impossible to break that habit. We had a large workforce on the farm of over 100 workers mainly Lozi for the cattle and Tonga for the horses and other farm jobs. It was not hard to earn their respect as they knew I was good with horses and I always handled the cattle well. I showed respect for them and they had respect for me but it was often necessary to be firm rather than friendly.

We had our amusing moments. An Australian sales drive was initiated in Zambia for the sale of Ford Falcon pickups. I like them but they are long and low and completely unsuitable for the bush or even dirt roads but the salesmen insisted that we trial one for a couple of days. I took it to the other end of the farm and it became stuck in a Donga (a dried river bed) and when the salesman came back, we showed him where it was and he had to arrange a heavy tow truck to extract it.

I was sitting down to breakfast one morning and Bevis was tucking into a big steak, his usual fare before he went out to the crawls. He suddenly stopped eating and looked at me.

'This steak is not so good. Which cow did you kill for this?' I told him that I had killed one of the indigenous cattle. 'Geoffrey. Next time please kill the best animal on the farm. I always eat the best please.'

The Mulungushi Non-Aligned Conference was scheduled to take place and we were requested to supply twelve head of cattle for it. Bevis was not happy to see his good beef go so he arranged to buy twelve head from Harry Nkumbula's farm that bordered ours to the South-east. Harry Nkumbula was an opposition politician of some fame who had previously worked with Kenneth Kaunda and then split to form his own party. His stronghold was to the south of Zambia. Our neighbour could be quite troublesome but I did find him very interesting.

Like all African politicians, he had spent some time in jail and regaled me with stories of the high security facilities. Somehow he had managed to herd the twelve cows up to our crawls. They were all pretty wild and obviously not used to being handled. Our Tonga boys struggled to round them up and load them into the lorry. It took them over an hour but it provided good entertainment.

Bevis's son Keith married while I was on the farm. His stag do took place at the polo club and was nearly washed out due to a tropical storm. I had nearly driven into downed power lines on my way to the club. It was a wild night in all senses of the word and Castle Beer sales shot up that evening. I left just after midnight and I was due a lie-in next morning but the cook woke me quite early. Problem? Keith had been stripped and left in the bush, which I understand was a tradition in Mazabuka, and had not yet returned.

I dressed and we gathered some of the men to search the ranch and along the main road while one small boy was sent to tell all the Lozi herders to look out for him. Keith had actually been sensible I am told and spent the night with one group of Lozis on an adjoining farm. He walked back the next morning in time for the ceremony but I knew Bevis had been worried.

I had two accidents off the ranch. The first was when I dismounted without the umpire's permission; i.e., I fell off. I rolled but chipped my vertebrae, so the X-ray revealed, which was very painful for a couple of weeks and I had to be propped up with cushions whenever I was driving around the ranch. The second incident started off as funny but nearly led to serious consequences.

In the off season, many of the polo players tent-pegged. This was a sport which was very popular in the days of the Raj in India which British Cavalry adopted. A small wooden peg was knocked into the ground towards which you galloped and then attempted to lift the peg out of the ground either with a sword or a lance with one sweep of your arm. The sword was the preference in Mazabuka. I had practised this on the farm with varying degrees of success but on one weekend I was paired with Jock McCay, the local vet, who was supposed to be experienced.

We galloped down towards the two pegs set apart and I struck and carried mine but as I did so I felt a blow below my left knee. There was a small hole in my breeches, and blood! Jock had stabbed me and not the peg. It was only a small wound so we applied first aid and I thought nothing of it. Two days later, Bevis's daughter, who was a doctor, rushed me into the hospital in Lusaka with severe blood poisoning. Initially, I was put into a ward and then forgotten and it was not until the day after that I was rediscovered, treated and I was soon back on my feet.

I can still recall now the names of the Mazabuka polo players at that time. There was my boss, Bevis Coventry, of course, then Charlie Stubbs, Cedric Oates, Ewen Pinkney, Gordon Kirby and Paul Taylor. All these players held good handicaps and were fine players but the amazing thing is their names still appear on the club membership because all their offspring play. Furthermore, most of the original ranches and farms remain in those families and I spent no little time in tracing them through to the modern day and some of the players even have their own polo fields. All Bevis's offspring, male and female, are still active in the game and there is also strong equitation activity in the area.

Paul Taylor, during my time, was the first player, I think, to introduce bigger thoroughbreds into the game and remarkably, his mother had a farm on the outskirts of Portishead near Bristol to which I was invited for afternoon tea. Gordon Kirby was a 'Go Ahead' farmer and he was the first one to acquire a plane for his ranch. His driving was pretty dangerous so we all worried for our safety if he flew overhead. I was able to trump him some years later when I had a plane…and pilot. It was safer that way. He had a large dambo on his farm which is a large lake and we duly started water-skiing on it. I had water-skied before and I could drive speedboats so he roped me in on Sunday afternoons after polo. No one seemed to have any concerns about crocodiles and Roger insisted that the lake was safe.

Several months later, however, we were enjoying a braai on the bank and one of the ladies threw some scraps of meat into the water. Chomp. Chomp. Two crocodiles ate the scraps. No more water-skiing. I have since learned that crocodiles can travel large distances on land between water sources. Our Lozi cattlemen always kept a careful eye on the cows when they were drinking from the Kafue River as a crocodile could easily grab the snout of a drinking animal and drag it into the water. Other game could also cause occasional problems such as hyenas and the occasional leopard as they regarded our stock as 'Food on the Hoof'.

It is necessary here to mention another player called George Bender. I do so hesitantly because, before my arrival, he had a major disagreement with the Accra Polo Club which was serious enough for him to leave the club. And that was in 1968. There was much acrimony at the time and his name was mentioned only in whispers in the clubhouse. I met him only rarely in Lusaka and he and his wife were nothing but pleasant to me and treated me with much civility and kindness. Paul Taylor then persuaded him to play at Mazabuka. George accepted the offer and used to drive down from the outskirts of Accra to Mazabuke most weekends for seventeen years.

I met him many times at the Mazabuka Club and I gradually patched together his life story. He was born on a farm in Sesotho in 1927 and only learned English and Afrikaans later in his school life which he often regretted and he spent much of his boyhood with Lozi friends and loved to ride horses. He started work on a farm and was literally pushed around a lot by the owner. One day George retaliated and pushed the owner back. He got another job on another farm and worked very hard indeed but when he asked for time off for his 21st birthday party, the farmer insisted that he do his week's work before he could go.

George worked four days and four nights nonstop, without sleep, and got to his party. He had the ability to go without sleep throughout his life. When he was 25, he joined an agricultural company as their rep in the Transvaal and moved to Lusaka for the same company in 1954 having previously married his wife Eileen. He drove well over 70,000 miles a year on gravel roads and did well so in 1964 he founded Kapuka Tomatoes and was the first farmer in Zambia to produce these and Brussels sprouts and broccoli that he grew on three farms which he managed personally.

He was also highly praised for his charity work which was never ending and to which he devoted much time. His story was not unusual and it set me to thinking about colonialization and the bad press that many settlers received for these were remarkable men. The settlers I met in Ghana had not seized land—they had purchased it. They worked incredibly hard in poor conditions and built up their farms and businesses. They employed many people and everyone has benefited. Different countries had different policies and I know that in Kenya and Rhodesia land was offered to retiring service personnel so the situation was not the same but it is interesting, that, when Mugabe started his 'Land Grab,' Zambia and Nigeria offered the deposed farmers land to farm in their countries because of their skills and expertise with the proviso that they would pass on their skills but not their farms. Many aspects of colonialization were not good but the good should be retained.

During the whole of this period, I had been struggling to renew my work permit. I had no agricultural background or experience and in the background to all this there was a political problem. Bevis was the Chairman of the Zambian Cattle Association which had questioned the forecast production figures published by the Government. Avoiding any publicity they had simply told the Minister of Agriculture that the Cattle Association disagreed with the statistics and the argument rattled on for some time. I was aware of the disagreement but thought no more of it at the time.

It was obvious, however, that a renewal was not on the cards and Bevis and I had to face that fact. In a frank discussion we had it was agreed that I should go, to enable Bevis to get someone to take my place as quickly as possible. The experience had been invaluable to me but I am not sure how valuable I had been to the Anchor Ranch. I had only been able to learn the job by watching and doing and learning from my mistakes. Direct help and advice would have been advisable but none had been forthcoming.

The farmers in Mazabuka are mostly descendants of farmers and they, of course, had lived on the farm anyway so they learned during their younger days. Bevis took a gamble on me and for that I am grateful and I am sorry it did not work out. For my part, if I had been able to get a work permit, I would have stayed and who knows where that might have led? There was no acrimony involved with my departure for I had to go anyway and Bevis did ask if I could help Charlie Stubbs out for a week or two. Charlie's wife was seriously ill and he needed someone to look after the farm while he took her up and down to the hospital in Lusaka. Not sure if I would be much use I agreed and moved the next day.

By arrangement, I took Sammy with me because Charlie's wife was going to take him on. She loved him and Charlie was very happy about that. Now Charlie ran a dairy herd which was beyond my experience so I had to take him aside on the first day and carefully explain that I required explicit instructions while I was with him. I did not know the geography of the farm, his stock or his workers. He seemed to accept this but on the third day he asked me to build the concrete foundations for a bore hole. He did give me rapid instructions but my alignment was a few inches off and he had to adjust it himself the next day.

I was about early in the mornings to check the 'milking' and was very amused to find that all the milking hands were bigger and fatter than the other farm workers; otherwise, I helped with the normal care of the cattle and checking the water. Charlie's farm was much smaller than the Anchor Ranch and the checking only took a couple of hours. His farm did have one huge advantage over the Anchor Ranch. It had a swimming pool so I was able to relax. I had only been able to skinny dip before in our concrete reservoirs down beyond the main road. The last night I was there we had a braai and party and I joked with Charlie whether it was a thank you or a celebration of my leaving!

Back in Lusaka again for one week, I stayed with Harry who seemed to be working all the hours of the day and night on…? One never knew. One night, however, we were having dinner and the subject of my work permit arose. Harry asked why I had not turned to him for help. Quite frankly, I had not thought about it but then he did indicate that my former employer and the Cattle Association were not in favour with the current government and that had certainly hindered my application. He asked me to think about it and contact me from the UK if I wanted to return.

I thanked him but I would still like to know what he did in Zambia. In a flurry of farewells and goodbyes and final arrangements, the week passed quickly. I popped into the school and was invited to a party that evening; the staff had lost none of their partying spirit. I spent some time at the Accra Polo Club and bought some very expensive jewellery for my mother and grandmother and presents for the rest of the family. I arranged with the Winters to deliver Geronimo and Prince, and Keith Coventry collected my Wolseley on the last day which he had purchased the previous week. I was ready to go. Alf drove me to the airport and I had an uneventful overnight flight back to the UK, sleeping for most of the time but also thinking about my future, bearing in mind that I could return to Zambia if I wished or…?

St. Bede's School.

Chapter Sixteen
St Bede's School

I returned to the UK just before the Christmas of 1974 and I had forgotten about the cold. I had been in a pleasant warm tropical climate for three and a half years and now I was back to the joys of an English winter and I had no warm clothes. I caught a train back to Bristol from Heathrow and took a taxi to Downend but not Valley Gardens. My parents had moved to a large semi-detached house at 67 Downend Road with the Mormon Church as neighbours and a large Masonic Lodge to the rear. They had informed me by letter so I did know about the move but I have never fully understood the reasoning behind it.

Foord Bros, the electrical contracting company, had moved out of their premises at 18 Warwick Road to Two Mile Hill which left my grandmother rather out on a limb as she was still living in 14 Warwick Road two doors down, and I am not sure if they knew then but the whole of that area was eventually demolished and redeveloped. Most of my grandmother's children had moved to Downend as well and that could have been a factor but for some reason my parents had sold Valley Gardens and moved into a larger house so that my grandmother could move in with them. There must have been some financial agreement because my parents could never have afforded such a move.

I was also aware that there were niggling problems that slowly developed. Nana Foord would turn the heating full on, not fully appreciating the cost, and my father always stated that Nan's contribution to the household budget did not cover her expenses. My parents were also socially handicapped. My mother had taken the brave decision to learn to drive quite late in life, for which I much admired her. She passed her test and she bought a small yellow Ford Anglia which she liked but only she drove.

My father held a full driving licence but never drove. He had obtained it before the Second World War and you were just given one. It did mean, however, that he could take my mother out to practise her driving all legally and above board. Every time they went out in the car Nana Foord expected to be included. Mum would drive her everywhere and she and my father found this very limiting and restricting. My parents much enjoyed walking but Nana Foord could not walk far.

My parents gradually became a little resentful of this inhibition all the time. Other family members did, however, take Nana out for outings and invite her for weekends and look after her while my parents were on holiday, but it was my parents who shouldered the majority of the burden. More friction arose when my father retired from a Bristol college where he had finished his working life as a porter, and where he had much enjoyed the work and made many new friends.

Nana Foord and my father home alone all day did not work well. A huge demarcation dispute broke out over the dusting. My father did all the dusting while my mother was at work but Nana Foord insisted on helping out by insisting that she did the living room. A tenuous truce was negotiated by my mother and Nana Foord dusted the living room every two weeks. When my parents retired, their decision to move to Cornwall was partly due to their need to shed the responsibilities of Nana's care but my grandmother was not amused.

Everyone seemed glad to see me and I was allocated the small bedroom at the front of the house so that I could enjoy the Christmas festivities with the family. I was financially stable and had no worries concerning that and for the first time in my life, I had enough capital to buy a new car. I looked around for a few days and settled on a white Triumph GT6 which I collected two days before Christmas and was, initially, very pleased with it.

I visited my brother and his wife, Sue and their new baby son (I was an uncle) in their new house at Yate and experienced even more difficulty in locating it after an absence of over three years. I dutifully did my round of the relatives and was treated as the long lost relative back from deepest, darkest Africa and of course, my Uncle Jack asked when I was going back…soon?

I enjoyed that Christmas because I had missed home life and being out of contact. There were no instant telephone calls those days and any call I had made to my parents entailed going down to the post office in Lusaka and getting the operators to connect you. Mobile phones had yet to appear. There were times in Zambia when you had to face problems alone and that had certainly developed my character and made me very independent but it was pleasant to talk to my parents again. What an audience I had!

I could relate all my stories and experiences to all of my relatives. I even enjoyed attending the Boxing Day party at Auntie Gladys's on Fishponds Road and there I could catch up with the Allen side of the family. I must have bored everyone to tears.

Come the New Year, I needed to make a decision about my future. The invitation to return to Zambia was a possibility or I could return to teaching in England again or…? I did look at other jobs and occupations. Several teaching posts were on offer from local prep schools but I would still have been regarded as a junior master with no short term prospects and to be frank, I was better than that. It would have been foolish to rush in and accept any teaching job and most of them at that time of the year were temporary posts anyway so I decided to hang fire and start applying for the September batch of posts that would be advertised in the coming months. If that had not worked out, I would consider Zambia again.

That decision made, I had a few months in which to occupy myself and a job found me. I walked into the Fishponds post office one morning to buy some stamps for my mother and met the sorting officer manager with whom I had worked a couple of times on the Christmas rush. He was pleased to see me and was fascinated that I had been out in Zambia for a few years and asked what I was doing now. I told him that I was biding my time for a few months before applying for teaching jobs in the UK.

'Why not come and work with us for a few months?' he asked.

The following Monday morning, I presented myself at the Fishponds sorting office at 6.30 am and became a postman for several months. I was given a buddy for a couple of days and I found it ironic that I was given help and training to be a postman but never for teaching. In essence, the job was simple.

First thing, at 6.30 am, all the postmen would sort a van load of parcels in a small warehouse at the back. At 7 am, we would sort our bundle of letters that had been previously sorted overnight and left in front of our 'walk' pigeon holes. A 'walk' was your route of delivery. The pigeonholes were about fifty in number and it was so organised that the start of the 'walk' was at the top left hand corner and finished in the bottom right hand corner with all the roads and big delivery receivers clearly labelled. We would sort out the letters according to their roads and then quickly sort the letters into order in each pigeonhole, bundle them up with elastic bands and then load our post bags in reverse. The last letters first and the first letters last. For two days, I delivered using a post office bike as did my buddy.

I reckon the sorting officer manager had been kind to me for I had a very pleasant round in Frenchay and it provided a true cross-section of English social life. I started with a prefab estate. These were little concrete and asbestos one storey bungalows that had been temporarily and hastily erected to overcome the housing shortage due to the Blitz. Thirty years on from the war they were all still in use, the great problem being that their occupiers liked them and were loath to move out. From there I moved on to a council estate that was really very pleasant and not at all run down in any way before delivering to a private estate.

From there I delivered to the big modern houses down one side of Frenchay Common before diving down Pearces Hill to the River Frome at the bottom. It was a very pleasant 'walk' and my clients were all helpful, kind and friendly…except one. I quickly learned to walk the 'walk' and I was also required to deliver to a very small first class only 'walk' off Lodge Causeway later in the morning. On my third day, I zoomed around my 'walk' with the GT6, much to the surprise of the local residents, and strolled back into the sorting office at 9.30 am.

The sorting office manager looked aghast and quickly took me to one side explaining that my 'walk' was allocated a three and a half hour schedule time so I should come not come back until 11.30 am. For the rest of my time with the post office, I completed the round in double quick time, then returned to the house and had a leisurely full English breakfast before returning to the sorting office at 11.30 am. The afternoon 'walk' was so short that I always walked it and was generally finished by 1.30 pm. Due to the GT6, I became quite well known on my deliveries and was often given small gifts and titbits and invited in for coffee.

I often mused on the post office in Lusaka which was a vast building and where one experienced huge queues. Generally, the post office there was largely run by members of the Bemba Tribe so they often gave priority to their own people so could be in the queue for some time. Of course, there was no house delivery of mail. All incoming mail was delivered to your post box, for which you were allocated a key, and you had to collect. Mine was collected by the school messenger who faithfully rode off on his bike every morning to collect the post. Zambia did experiment with postmen and house deliveries but it failed because the postmen threw all undelivered mail away at the end of their rounds. The post was sacrosanct, however, in most cases, and it was unheard of for anyone to interfere with your letters in any way.

I had one unpleasant incident. My very last delivery was to a small cottage down by the River Frome where there was a small terrier dog that was snappy and went for your heels. I did manage to deliver once or twice but the animal persisted and the manager wrote a letter of warning to the occupant and the dog was kept indoors. Unfortunately, as I pushed the post into the letterbox, the dog jumped up and bit my fingers. It drew blood and I had to pop up to the doctor's surgery to have it treated and have a tetanus shot. The occupant was contacted by phone and told that she would have to collect her post from the post office and pay compensation to me. I made no deliveries to the cottage again.

Twice while I was working for the GPO, several of us were sent down to the Central Sorting Office in Bristol. We hated it, I am afraid, because you had to work flat out there all the time. It was a very busy and hectic work place and one was constantly under pressure. We always appreciated getting back to Fishponds with its quieter pace of life. Postal staff work much harder nowadays than I ever did. They walk further and deliver far more mail and have a very strict routine of deliveries. They seem to be under pressure the whole time because of the volume of mail and modern working practices in spite of super automatic sorting and advanced technology.

That was the last manual job that I experienced and looking back over the many part time jobs that I had done I realise that they had provided me with extremely valuable work experience and a fuller appreciation of different vocations and occupations. To this day I am appreciative of everyone's work in the community and never belittle any employee's efforts and contributions in their place of work.

While I was happily delivering letters, other postmen were delivering my letters of application for teaching posts in the South and West of England. I had decided to aim for a Head of English post except they were few and far between and I was beginning to get concerned. It was obvious that the 'Old Brigade' were still at their posts. Preparatory schools at that time did have a propensity for eccentricity and quirkiness and I was somewhat surprised to receive a phone call from a school secretary in Eastbourne to which I had submitted an application at least two months since. She was anxious to check my swimming and Scouting qualifications. There was no mention of English. An interview was arranged for the following week.

St Bede's School was a tall, imposing conglomeration of a building situated on Duke's Drive at the Beachy Head end of the promenade, between the main road and the sea and all of the many rooms facing south had a wonderful view of the sea in all its magnificent states. I parked the GT6 just across the road and walked over a small drawbridge-like structure to the front door where I was greeted by the school secretary, Cecelia Wise.

I had begun to realise, even this early in my career, that the school secretary was the main linchpin in any preparatory school, and Cecelia was no exception. She had an exceptional 'telephone voice' which was an asset when dealing with parents, knew the school inside out and could charm the hind legs off a donkey. She ushered me into the headmaster's study and I met Peter Pyemont, not only the Headmaster but also the owner of the school until it was turned into a charitable trust.

He was fairly young; indeed, he was a headmaster aged 25, had a charming personality and was very friendly. He did not conform to the classic model of a prep school head but I liked him all the more for that. The interview did conform to the classic prep school format, however; namely, we only briefly mentioned the teaching of English though he was impressed by my reference from George Booth, my college tutor. We focused on swimming.

The school had a good-sized pool on site but had failed to win the Annual Eastbourne Prep Schools' Gala or even be highly placed. He was clearly very ambitious to do so and grilled me on my experience and knowledge. He wished to know what it would take to win. I responded by asking what it had taken to be so successful with the school's rugby, the cups for which had been proudly displayed in the trophy cabinet in the front hall.

'Hard work, much training and coaching,' was the answer.

'That would be my method as well regards your swimming. The pupils would need to be in the pool for half an hour every morning before breakfast improving their swimming standards, technique and endurance.'

He was keen that Scouting form an integral part of school life and obviously had read the Scouting section of my CV which had impressed him no end and there was no doubt of my ability in that area. We did a full tour of the school. It was like a rabbit warren inside. It was at least five floors over which the offices, classrooms, specialist rooms, chapel, dining room, dormitories and staff accommodation were all spread in a glorious conglomeration of split levels, narrow stairs, peculiar angles, hidden storage and interlinking corridors. In the basement area was the rather magnificent pool.

Outside was a new pre-prep department and a small games field that reached out to the sea, the end of which was fenced. There was another games field about three hundred yards down the prom and a huge games area along the path to Beachy Head called The Hollow. I met many of the children and they seemed quite lively and switched on, all of whom looked you in the face when you were talking to them, which was a good sign.

Like many prep schools, there was no entry exam and children of all ability were accepted and taught to the best of their individual ability. I had lunch with Peter and his wife, Elspeth, who was slightly domineering in a pleasant way. She had a good sense of humour but I had the impression that she was not totally devoted to the role of being a typical headmaster's wife and could be very independent in her thinking. We were to clash occasionally, but neither of us bore a grudge and respected each other's views. She and Peter lived in a house on the corner just above the school and the path to Beachy Head. I am afraid that Elspeth used to constantly worry about anyone looking slightly unusual on that path in fear that they were heading to the cliff to commit suicide.

I knew I had gotten the job just based on the swimming alone and Peter confirmed that at the end of the interview. It was getting rather late and they enquired about my journey back but I assured them that I was staying with polo friends at Cowdray for the night and had arranged to play chukkas next day.

There had not been one question about my English skills or about reading schemes, handwriting policies, views on the teaching of grammar, essay writing, literature or the Common Entrance Examination. I was not sure what the school policies were on these but I quickly found out. None.

I had a good weekend at Cowdray Park and played a couple of chukkas on borrowed ponies which was most enjoyable. England, however, is normally the odd one out in the international polo scene. When visiting teams or individuals play abroad, it is traditional that they be mounted by the host club.

When Kenya visited Zambia, we would provide a pool of ponies for them to ride and the Zambians would have the same arrangement when they played in Kenya. I observed, however, that British teams, though in receipt of this arrangement when visiting abroad, did not reciprocate when the foreign team visited England and on several occasions visiting foreign teams had to hire ponies to play matches. I had polo playing friends in England who were quite happy to give me a couple of chukkas but I had to pay for the hire of match ponies in England.

I returned to Bristol with the guarantee of a teaching post and immediately started to swot up on the task ahead. As Head of English my job was not only to teach my classes but also to organise all aspects of the department and I needed a pretty good idea of the ideas and methods which I needed to introduce for there was no groundwork in place at all at that time. I was not a slave to method but I was determined to ensure that there was continuity throughout the department.

The 'just teach them some English, you know?' philosophy was not acceptable and I scoured the university education departments for the latest schemes and books to blend them in with my prior knowledge of such material. All schemes and plans had to be practical and easily managed and I wanted the teaching to be interesting and fun whenever possible though the Common Entrance Exam and Scholarship Entry Papers were a more serious proposition for the 13+ age group.

There was also the added complication of the state 11+ exam for which some of the pupils entered, then the 11+, 12+ and 13+ entries for the girls for independent schools. It was quite a mixed bag of targets all of which required advanced planning. In some areas I had to go back to basics for there was no library and class libraries were very unsatisfactory indeed. That was to prove a major challenge. Nor were there many selective course workbooks in existence.

Peter Pyemont contacted me to request that he would like to teach his two English classes in their final year if I did not mind and I readily agreed as this would give me a year's breathing space to get organised lower down the school. In agreeing to this I also sent him my initial outline proposals with an estimated departmental budget which was the first time he had ever received one with the further request that I arrive a week early to organise myself and the department. I also needed to upgrade my personal library which had been somewhat neglected during my sojourn in Zambia.

My GT6 was not the largest of cars so I had to resort to sending my trusty trunk in advance to await my arrival in Eastbourne. Thus arrived one Head of Department in sunny Eastbourne at the end of August 1974 ready for the task ahead to which I was looking forward especially to the teaching as I am in my element in the classroom.

I was quite surprised to be given a small two roomed flat on the south side of the school both with superb sea views though I was to learn that strong winds and gales rattled my windows quite often. It was small and comfortable and perfectly adequate for my needs and had more than enough bookshelves for my scant collection of books but my personal library rapidly expanded during the course of the year. Peter and I had ongoing informal discussions most days over coffee, lunch and afternoon tea so he was able to make immediate decisions about many of my queries and concerns.

I was welcomed with open arms by the pre-prep department as they seemed slightly out on a limb with little or no coordination with the staff and older classes. To me, it was just common sense. They were switched on with literacy and reading schemes so all I had to do was follow on from there, check that the first forms were building on that foundation and extend it all up into the main prep school. There were discrepancies.

The pre-prep was teaching reading by phonetics while the first form was using the Whole Word method while the teacher taking the nine- and ten-year-olds did not teach reading! I also had concerns about the standard of English of our students from abroad as they had not received any direct help in class. They had been given extra help for only about half an hour a week from a retired teacher but this was far too little. They needed a much better programme especially as they were expected to cope with the full range of subjects. This was a matter of some urgency. The school was forthcoming with funds.

Generally, new staff have that 'Golden Slot' in their first year when they can ask for money and I did make the most of my requests. Peter Pyemont was very fair in his allocation of funds, however, and had the advantage of being able to make immediate decisions on all matters financial. St Bede's funding, salaries and bonuses were quite generous which I appreciated and vowed that I would operate the same policy if and when I was running a school, and I did. As a comparison, when I worked at Wycliffe College Junior School, I received £1,000 per annum in salary which was a good starting wage at that time but it was at St Bede's that I passed through the same £1,000 mark per month and the gap was only six years.

Further into that first week I was beginning to meet the staff as they trickled in from their holiday, the most important of which was George Cole, the chef. He was a chef of some physical stature as becomes a chef and he must certainly have enjoyed his own food. He made his kitchen look quite small. He served excellent food to the pupils and staff especially the staff suppers which were of gourmet standard. George was Coxswain of the Eastbourne Lifeboat and when the maroons were fired he deserted his kitchen, jumped into his old car and sped off down the seafront and it was often Peter and Elspeth

Pyemont who took over the cooking.

As a prep school head you are expected to turn your hand to everything and cooking was definitely on Peter's CV. He also maintained the prep school headmaster's long standing tradition of carrying out minor repairs and was often to be seen wandering around the premises with a collection of tools, doing odd jobs dressed in his baggy tracksuit bottoms. The school building was rather elderly and needed constant care and attention. Peter was also a great expert in repairing dormitory beds and the plasters on his fingers normally bore witness to his struggles and tussles with them.

The two staff that I had first met, namely Cecelia Wise, the school secretary and George Cole, the chef, were loyal, trustworthy and both good at their respective jobs but I was to learn that there was one day when they could not be trusted to do anything. Both had been in the Royal Navy and they both celebrated Trafalgar Day in chef's pantry by opening a bottle of rum and emptying it during the morning drinking toast after toast.

Elspeth had an unusual arrangement with a local agency about the employment of au pairs that were employed to look after her children but also worked in the kitchens in the main building. It is alleged, and has become one of the legendary myths of the school, that they once indulged in after school activities in the school hall with an unnamed music teacher on the gym mats.

Frank Bagnall was the Senior Master who was always pleasant and helpful towards me but constantly moaned about the methods, or lack of method, concerning the school. He was a good Maths teacher and most of the children enjoyed the subject. His one particular moan was about The Hollow and the girls' games wear. The Hollow, if readers will recall, was the huge games field situated in an even bigger natural amphitheatre half a mile from the school along the cliff path to Beachy Head.

The girls wore a one-piece white flannel short-legged jumpsuit which zipped up at the front. This meant that if they needed a pee in the bushes at The Hollow, they practically had to take the whole thing off. It was tricky and not really acceptable at all but the headmaster's wife apparently would not alter the girls' kit.

As soon as I met John Cousins, I knew that we would get on well. He taught Science in a somewhat elderly science lab which had that usual science 'stinks' smell. He could become slightly exasperated but in a very pleasant way and always showed immense kindness and consideration. I had him earmarked from the first day I met him for a very special position in the school's future plans. I just knew that he was perfect for such a job.

There were very few lady members of staff in the middle school. Yvonne Kavanagh taught French and was in charge of the Girls' Boarding which was eventually housed directly across the road from the school in a very pleasant building. She eventually became involved with my activities by no design on my part, which was unusual.

John Smith was an ex-copper and taught PE. He often regaled us with outrageous stories about the London Metropolitan Police some of which were quite hair-raising. He had a blunt, bluff manner which did not necessarily go down well at staff meetings but did add interest to the proceedings and he was always very keen to join in the adventure activities of the Scout troop. He was appointed as a JP while he was a teacher at St Bede's which gave me the added advantage of not being able to appear before two JPs in Eastbourne as the chair of governors was also on the bench. I did wonder sometime later how Peter Pyemont had managed when he had backed one of the school minibuses into a restaurant's front window in Rye and was found to be over the limit. I would love to know the outcome of that case, for I believe Peter had influence.

I had never started a Scout troop before and I thought this might be problematical so

I had contacted several of my Scouting associates with whom I had worked at Gilwell Park and they were a great help in assisting in the formation of the 1st Eastbourne Sea Scout Troop based at St Bede's School. We were allowed to start anyway as the local District Commissioner allowed us to catch up with all the paperwork after we had started. I do not think the process would be so easy nowadays but I was a qualified and experienced Scout leader and that helped enormously as I had all of the skills.

Our biggest problem was a place to meet. There was a school hall but this was in constant use in the afternoons. On my walks down to the beach I had noticed a quite large and long flint stone building which seemed to house nothing but junk. It needed decorating but was structurally sound and would be highly suited as a Scout headquarters. There was even a lean to shed for our equipment and an outdoor concrete area for boats. Perfect. I put the idea to Peter and he agreed straight away. Within two weeks the junk had been thrown out, the roof checked and the interior re-painted. We had our Scout headquarters and very smart it was too.

I deliberately did not do a high-powered recruitment pitch to the school because I had visions of being swamped. I was aiming for an initial troop size of 36 eleven- and twelve-year-olds which would eventually form six patrols. I did not see much point in taking final year students for one year only. The boys who wanted to give it a try fell roughly into that age structure but I continued to be careful as I thought it best not to put them into uniform until the second term in case any dropped out. None did.

We struck lucky with uniforms because one of the parents supplied the school uniforms and he was also the local Scout uniform supplier. He visited the troop on the last meeting of the Christmas Term, measured all the boys, and their uniforms were ready by the first troop meeting of the Easter Term. That particular meeting we had invited Peter and Elsbeth Pyemont to attend the Flag Break and opening prayers followed by a mass investiture and they were very impressed with the smart and magnificent turn out. The boys wore navy blue cord long trousers, sea scout jerseys, the troop scarf and lanyards and the sailor's hat. Even I was impressed. We had agreed that a Cub Pack should also be started in the second term and John Cousins would be leader. He worked with me during that first term to gain some of the necessary skills and I acted as his assistant when he took over the Cubs as Akela.

We were soon to face another problem, however, the same one that Baden-Powell had when he started the Scout Movement. Girls. They wanted to join as well and quite rightly so, except they could not join the Scouts then. A Girl Guide Unit would have to be formed and Yvonne Kavanagh agreed to be the leader but I am afraid its start-up took much longer than that of the Scouts due to formalities, procedures and rules and regulations, so they did not get underway until the September of the following year.

The term started and I was obviously very busy with my teaching, duties, Scouts and social life. I was starting my day at 6.30 am with senior swimming and then took the juniors at 7 am followed by a quick breakfast, tutorials and then teaching until 3.30 pm which was games time during which I took swimming, one afternoon of football and Scouts, then supervised tea and prep after which we were expected to organise activities or be on dorm duty.

I had not realised that I was expected to supervise the girls' dorms again though we only checked them after lights out. Peter did explain that this was only temporary until the girls moved to their new house in the Easter Term. I also had weekend duties on a rota basis. Now life in prep schools can be very busy for all staff and it was normal to work seventy-hour weeks but I quickly realised that the pace was even faster at St Bede's. Its pace was near to hectic and I reckon that I worked eighty or ninety hour weeks.

The only saving grace was the fact that you did get nearly the whole of one day off per week and one did get long holidays but I would have wished for a slightly slower pace. My day off was a Wednesday which coincided with chukkas at Cowdray Park during the summer months. I must say that the resident staff worked far harder than the staff who lived off site but then we did receive free board and lodging.

I produced major productions at St Bede's which also took up much time. Against all advice I took on The Royal Hunt of the Sun by Peter Shaffer. I had read the play in the summer and had immediately become fascinated by the medieval Inca world in which it was set; a world where gold was as common as wood and the growing relationship between Francisco Pizarro and Atahuallpa which provided a central theme and a pivotal focus throughout the two long acts and I liked the Inca music. I was lucky, because in the auditions, I found the two perfect characters for the parts of Pizarro and Atahuallpa.

It will be appreciated that the two boys were only thirteen years old at that time but they produced outstanding performances with a wisdom and understanding far beyond their years which seemed to lift all the other actors on to a higher plain as well. All pupils aged eleven and over took part…a cast of thousands, you might say, but there were about seventy in all and we used up a London supply of body paint for the three performances though the matrons were not too happy about the dirty bed sheets! We were slightly hampered by a traditional stage but we brought the action down into the auditorium and the Mime of the Great Ascent was conducted along the wall bars.

In my long association with this age group, I have constantly been astounded, astonished and amazed at their ability and talent in all areas whether it be drama, adventure activities, music, dance, singing or sport. What is even more incredible is their ability to switch their talent on and off so a choirboy who has just sung a beautiful solo in one of our great cathedrals will the next minute be seen playing football outside or another whose acting entails him or her to be on stage on and off during a production will be playing with an electronic game in the wings in between his appearances.

I recently managed to see, at long last, the London production of Billy Elliot, last year and the talent and versatility of the particular boy playing Billy that night was unbelievable. I enjoy teaching all age groups but I have to admit that the ten to thirteens are my favourite. They are full of life and fun to be with. Anyway, The Royal Hunt of the Sun was a success and highly praised. Bolstered by this, I produced the Golden Masque of Agamemnon by John Wiles, which was his attempt to bring all the stories of Agamemnon together in a lively stage production, and I accepted this as my next challenge, again, using all the senior pupils for speech, song and movement.

During rehearsals, it became clear that it lacked pace somehow and I could not quite achieve the atmosphere for which I was looking so I took the daring chance of adding James Bond theme music, and it really lifted the whole production. Thank you, James Bond.

My next production attempt was a flop! I had hit on the insane idea of writing a play using the text of Richard Adams' Shardick and letting one of my English forms write the script. As a class project it was quite good; as a school production, it was not and I should really have had the sense to have seen that bearing in mind that I had set the bar very high on the previous two productions. We learn by our mistakes. It had been hit hard by a particularly virulent flu outbreak in the school anyway and many of the actors looked washed out. Most of the staff were barely standing as it was, let alone the children.

It was all 'water under the bridge' when I produced my final play as I had enormous fun in producing Androcles and the Lion by Bernard Shaw. The 'star' of the show was the lion costume that I had hired from a London costumer. It was magnificent and the pupils, staff and parents thought it was fantastic as it had facial features which could be manipulated. The poor actor in the costume did not exist as far as the audience was concerned—the lion stole the show. It did get me into a spot of bother with the law however.

First, I have to mention that I no longer owned the GT6. It was obviously a 'Friday afternoon-Monday morning' vehicle and was giving me all sorts of problems. After a sustained series of complaints to the Triumph dealer in Eastbourne, I made a direct complaint to Michael Edwards which he attempted to ignore but my threat to go public resulted in another car offer, namely a red Triumph Stag, which I test drove and loved and took it in a direct swap. I drove it for about a year and still liked it very much but it developed a starting problem which the dealer put down to my bad driving. I did not see the connection and the problem persisted. I was lucky enough to bump into its previous owner at a parent's house one evening purely by chance and he asked how I found it. I told him about the starting problem.

'That's why I got rid of it,' he said.

The St Bede's Chair of Governors was a solicitor and wrote a formal letter to the Triumph dealer threatening them with legal action for selling me a vehicle with a known history of poor starting. I was offered a direct swap for a brand new Triumph Spitfire which gave me excellent motoring until I sold it a year later. Here I must admit to being an older boy racer I am afraid for many of my friends owned sports cars and though we never raced against each other, we had a set course time trial. We would go out on the Brighton Road, turn left at the Birling Gap junction then race back over the Beachy Head cliff road and return to the pub. Please do not try this at home. I do regret being so daft now, of course.

The police were aware of this activity but only had Pandas and could never catch us. The local police sergeant was a Welshman and he always reckoned that he was going to catch us one day, and I was caught, but not by him, which must have broken his heart I think. A good lady friend of mine, that I knew from Wycliffe College Junior School, where she had been a matron, had applied for the job of caterer at St Bede's. She came down for an interview, which was successful and I took her out for a meal before she took the last train back to London and The Royal School for Soldiers' Daughters near Hampstead Heath.

Alas—the last train had been cancelled so I had to drive at 'some' speed in the Triumph Stag over to Brighton to catch the last train there. At Brighton she jumped out of the car and dashed for her train. My attention, however, was drawn to two police Panda cars flashing their 'Blues' and sirens wailing, which…drew to a halt in front of me. I wondered what on earth was happening.

'In a bit of a hurry, sir, are we?'

I explained what had happened and added a sincere apology adding that the lady had to be back at work the next morning to serve breakfast to the soldier's daughters. I gave him my details. Evidently, the Eastbourne police had started chasing me (Tom, the police sergeant?), given up and had radioed ahead to Shoreham to warn them of my speeding but their cars could not catch me.

'Oh dear! That's ruined Tom's evening now,' I commented.

'You do realise that I could book you for serious speeding, sir.'

I informed him that I was fully aware of that fact and he must do as he thought fit. I had broken the law. He let me off with a warning and I drove carefully back to Eastbourne and lo and behold, there was Tom sitting in his Panda opposite the school. He wound down his window.

'I hope they threw the bloody book at you!' he shouted.

'Lady in distress,' I said. 'I was let off with a warning.'

But back to the lion's costume. To return where I left off. The chairman was holding a barbecue on the Sunday lunch time after the play and one of my friends thought it would be a good idea if I turned up in the lion costume in my Triumph Spitfire. OK. We did drive around the centre of town for a little while until we were stopped by the police. It was Tom again.

'I bet you a thousand pounds I know who's under that costume,' he said.

I took off the head and greeted him, enquiring jokingly that under which Eastbourne by-law he was proposing to summons me now. Tony, my friend and driver, who was very witty, suggested that he 'paws' for thought before booking a lion on the 'Mane' Road while I put the head back on and fluttered my eyelashes at him, giving a loud roar at the same time.

'Bugger off!' he said. 'And get out of the town centre NOW!'

I would hate you to think that I was in constant trouble with the police but one of my lessons was disturbed by Peter Pyemont one morning who informed me that Tom, three policemen and a detective inspector, no less, were in his study and wanted to see me. Cecelia and Elspeth were agog to know why I was a 'wanted man' and thought that I had upset the local constabulary again in a big way.

The Detective Inspector was rather formal and asked my name and then for any proof of identity. I informed that I had not any on me because I was teaching and he seemed satisfied. I was then asked about my movements the evening before. That was easy. I had been putting the boarders to bed. I had been on dorm duty. Did I have any witnesses to that? I confirmed that sixty boarders, two matrons and the headmaster would probably be able to witness the fact. Had I gone out at all in the night?

I explained that a dormitory duty entailed being 'On Call' for the whole of the night in case of any emergency and I could not leave the premises under any circumstances whatsoever. Did I own a boat? Yes, I did. I had purchased a small cabin cruiser, Indaba, with a 60 horsepower engine, as a bolt-hole to get away from the school when I could and she was berthed at Shoreham. Had anyone permission to use it. Again, affirmative. The yard owner had permission to use it for fishing trips on the understanding that he charged me a reduced rate for the berth and maintained my engine. It was a Gentleman's Agreement.

Now for the crux of the matter. Indaba had been intercepted by French Customs during the night on the French side of the channel, eluded them by use of her fast engine but was caught coming back into Shoreham in the early hours of the morning by British Customs. There were two men on board and they had been smuggling Normandy Butter. Evidently, one of the men arrested worked at the yard where she was moored and had 'borrowed' her for the night.

The owner was terribly embarrassed and gave me free mooring for six months and made sure he hid the spare keys somewhere safe. It took me three weeks to get Indaba back. George Cole teased me mercilessly about the incident and thought it very funny to offer me Normandy Butter with my toast in the morning. I often went out with Indaba. Shoreham had a tricky entrance. It was OK exiting the harbour but there was a big sandbank just before the entrance and one could end up surfing in at certain stages of the tide. I made several trips to France and though I could navigate I often drove up to Newhaven and followed Senlac, the cross channel ferry across to be absolutely safe but I would only make the trip if the weather was set fair.

Captain Debuisne was the skipper of the Senlac and I know this because I taught his son Freddie and he was also one of my patrol leaders. I acted as a somewhat loose guardian to Freddie and it was often my job to drive him over to Newhaven to catch the Senlac for the 11 pm sailing if he was going home for a weekend Exeat, or half term or school holidays. He used to love eating in the Taurus Steak Bar in Eastbourne. He was as thin as a rake but could eat like a horse though he disliked French food.

One evening, Tony, the owner, was offering a big mixed grill and if you ate it all you could have another one. Freddie scoffed one down and got half way through the next then asked for a big ice-cream for a sweet. Tony let him have the whole meal free! Captain Debuisne helped me out when I was looking for a French campsite for the Scouts. I drove over in the GT6 and Freddie accompanied me down to Brittany to visit an uncle's cider farm who said he would be delighted to see us in the summer and all was arranged.

At the end of my first year, all was progressing well and I was now very clear about my progress and plans. I taught in a very large room looking out over Duke's Drive and much of it was surplus to requirements so I hit on the idea of converting half of it into a school library. Some of that summer was thus spent repainting the whole room in light colours, erecting book shelves, buying books and Peter ordered me some new teaching tables and chairs as well. The pupils were delighted and the library was generally crowded and well used.

The governors were also pleased and gave me a £500 bonus to show their appreciation. We also introduced a new English scheme called Clard. Comprehension, Language and Reading Development. I am not an innovator but I had researched this scheme very comprehensively and rated it quite highly. Basically it dealt with all aspects of English including the foundations of comprehension and creative and sustained writing in a series of easily attainable levels. Every pupil could work at his/her own pace and level and it catered for English as a second language and quite remarkably, it was of our scholarship standard at its higher levels. I rated it highly, especially as it provided easily definable levels of attainment.

Teaching had become more serious as I was now responsible for all the Common Entrance and Scholarship English Candidates. The Common Entrance was fairly straight forward and the exam consisted of a comprehension paper and a second essay paper though some of the Girls' School Entry papers differed in format. Generally, if a candidate could cope with comprehension questions and write any form of essay then they would be alright in the exam though grammar, punctuation and spelling were rated and marked. The Scholarship Papers were much more formal and generally, harder and much more intellectual in content. They were difficult and varied from school to school so it was vital that each 'scholar' practice on past papers. This led to rather stilted teaching but it was on a very much individual basis which was a great help.

St Bede's held no formal entrance exam so one had to work at a variety of levels particularly helping the less able and foreign pupils. The teaching was difficult from that respect and I learned all about Individual Learning Programmes long before they were invented. Towards the end of my second year I was nervous about possible results and was anxious for the pupils. Pupils had a somewhat easier time with Common Entrance. Generally 50% was a reasonable pass mark but this criteria slid up and down according to the reputation of individual schools. Some would accept 35% upwards, others would insist that the pass mark was over 60%.

After the Common Entrance exam results were published, there would much haggling and trading by Preparatory School Heads and other factors would be taken into account such as Music and Sport. There was also the individual to consider. 35% would have been a very satisfactory mark for a pupil who was learning English as a second language for instance. If a pupil had taken a Scholarship Exam and been awarded nothing then they would automatically be offered a place at Common Entrance Level. There was much wheeling and dealing and many factors to take into account.

It was something I was to get used to later. My results were satisfactory but two or three pupils had scored slightly below my expectations and one had mysteriously crashed. I learned that he had been receiving coaching at home during the evenings and weekends which had undone most of my work. The interference of outside tutors was a difficulty that I would constantly face later in my academic career. They can cause enormous problems.

But what of our swimming training? What happened at my first Inter-Prep School Gala? Did all the early morning training sessions bear fruit? It was quite embarrassing really. We swept the board and took first and second in every event including the diving and we did exactly the same in the next two galas as well, and unfortunately, we did not win by small margins either. We were yards ahead in all races. Our girl swimmers were particularly good and two of them eventually managed to get swimming scholarships to Millfield in the next year.

Peter, of course, was delighted. He was over the moon but I did not get a bonus—I received a fairly substantial pay rise instead. I was pleased for the pupils who had turned up every morning. It had been hard work for them and they enjoyed their success. I treated them all to strawberries and cream. In modern days this raises the important problem of specialisation with our children for if a child is to succeed to a high standard then they have to start and train early whether it be for sport or music or chess.

We had one lively member of staff, Piers Chater Robinson, a name of which you might have heard, who was very keen on composing and show music. He persuaded Peter Pyemont to let him produce a musical, Mr Chips (I think) using the pupils and staff. This was all very well but what Piers had failed to understand was that the staff were already flat out with a thousand other things on their minds and rehearsals did not go well. It was 'pulled' almost immediately and I was never sure if it was ever produced or not but I do know that Piers is now well established in the theatre industry in London and had produced many musicals. He had obvious talent and deserved to get on.

The pressure on staff was ever increasing and many of them would get edgy and bad tempered at the end of term in particular. One major cause of this was lesson interruptions which was raised in many a staff meeting. The one thing that I could never understand with prep schools is that pupils were taken out of lessons (which were paid for by parents through fees) for music lessons, for which the parents paid a termly amount again. It was not logical but there were other anachronisms.

Extra academic lessons were also run in lesson time and sports coaching often took priority over teaching. Pupils were taken out of lessons to go shopping in Eastbourne when it would have been better to go down in games time. The girls were often taken out all morning to have bras fitted and missed three or four lessons. I merely requested that the situation be reviewed to see if we could gain more lesson time because it had truly reached a ludicrous situation. Elspeth got on her high horse and burst through my study door justifying the fitting of bras. It was not the right reaction but I burst out laughing and she stormed off. It was so funny and had to be included in this book.

The Scouts, Cubs, Guides and Brownies were now in full swing and my Scouts were progressing well with all their tests and proficiency badges and the Patrol System was running splendidly. Many of the staff had noticed that the children were kinder and nicer to each other because of the influence of Scouting and Guiding. Even during the holidays I was busy running camps and expeditions and these were varied and far reaching. Most of my patrol leaders had attended the camp in France during the long summer half term and they all coped well in spite of much rain.

I was able to pass on leadership skills which they would need at our big summer camp in the Peak District. Sometimes they were ahead of me. Two of them every morning walked down to the farmhouse to collect our milk and I began to notice that they seemed quite excitable on their return and even more unusually, they were always eager to volunteer for the milk round. Freddie's uncle had been giving them a glass of rough cider each when they collected the milk.

Though we were in France, I managed to teach them about Patrol camping. They learned quickly as I was able to treat them to a couple of French meals out at restaurants which they all enjoyed…except Freddie. We packed up in the rain and drove all the way up to Dieppe in the rain and when we arrived at Newhaven—it was raining. We drove into the Custom's Shed and the Custom's Officer looked us over and saw how wet we were.

'Been camping in all this rain then?' he asked.

The soggy boys answered in the affirmative.

'Drive on. If you are smuggling anything, you blooming well deserve it!'

I am still not sure of the logistics of the troop camp we held at Edale in Derbyshire. We took thirty-six Scouts, masses of equipment and I seemed to have plenty of staff. We were camping in an open field under the western flank of Kinderscout in a big natural hollow. I had found the site when I had taken two of the patrol leaders up to Edale during the term and we were involved in a very strange incident. We had found and selected the campsite very soon after our arrival and we had set up two small hike tents as we were staying overnight. I drove around to the village and we had a meal in a pub.

I noticed a man in a raincoat was watching us and as we left, he followed us out. I had the GT6 still and sped out of the car park and he followed us at speed. He appeared to be following us. I turned right and he was still behind us so I accelerated right past the campsite, turned off my lights and reversed well back through an open gate. He roared by and I drove back to the campsite without lights and the boys quickly opened the gate. Still without lights, I drove across the field and down into the dip. The boys went to bed and I watched as the car drove slowly up and down the lane twice then it went back on to the 'B' road and drove off in the direction of Liverpool. I could clearly see its lights as it approached a high pass and then it disappeared.

I did stay up for most of the night, however, just in case. We had much to do at that first troop camp and I wanted it to be fun. Unfortunately, my Troop Leader slipped on a bank while collecting wood, and broke his left wrist. I had to drive him into casualty in Sheffield. We phoned his parents and I asked if they wanted him to come home which I would be able to arrange the next day. His father, who was the youngest Lieutenant Colonel in the British Army, spoke to his son who wanted to stay in camp, to which I agreed.

On the first full day, we had flag break and then concentrated on setting up and improving the patrol sites and helping the scouts with camp craft and cooking. Each patrol had one adult who basically stayed with the patrol for the whole of that day, only advising and nudging. We were then able to start our activity programme, which, on the first day, involved patrol hikes up and over Kinderscout and down into Edale where they were shuttle-bussed back to camp. The patrol had to plan the hike themselves and they and their equipment were checked before they left. Every patrol was 'ghosted' by an adult who walked well behind each patrol and only moved in if something was going astray.

It was their first experience of hill-walking and they liked the lunar landscape type terrain on Kinderscout. The next two days they had a choice of four activities except that they had to do all four. They pony-trekked, caved in the famous Jackpot System, visited the Blue John Mine and did a river ascent that ended up in a gorge. We had one wobbly.

On the second day, the last party was late coming up and out of the Jackpot, went over their emergency time and I had no choice but to call the police who called out the Cave Rescue Team. I drove up to the cave and met the party coming out safe and well. The cave leader's watch had stopped so I left the Cave Rescue coordinator to deal with him. It was quite a scare. The next day we were having a Back Woods Day but without the Patrol Leaders. The seconds were taking over the patrols while the Patrol Leaders were going off on their First Class Hikes. The Patrol Leaders set off in two groups of three and we ensured that they were properly equipped and knew their route.

Progress would have been easy to check if we had mobile phones. We had no direct means of contact so in an emergency they were told to contact the police who would find us. Peter Pyemont and the Chair of Governors visited us that day and were staying one night. They were mightily impressed with the camp and the bivouacs that the Scouts were building to sleep in that night and then we set off to locate the Patrol Leaders. I knew the area where they expected to camp but there was no sign of them there so we continued up the lane for a short distance and were suddenly surrounded by four men with guns. They were obviously special forces or SAS.

When we explained that we were looking for our scouts, they informed us the Patrol Leaders had also entered the security cordon and had been directed to a campsite about a mile away. It was an upmarket site and the Patrol Leaders were looking very pleased with themselves living in the lap of camping luxury. I checked with the campsite owner who said he was pleased they were on the site and they could stay free of charge as he had once been a scout. He even gave them bacon butties in the morning.

They all passed their First Class Hike. The bivouacs had also survived the night with their occupants. On the last day we walked around to Edale Village as some of the Scouts wanted to buy souvenirs and then we had to start packing up the camp ready for our departure in the morning. The weather was set fair so we fulfilled every boy's dream by letting them sleep out under the stars.

We also made regular trips to Plas y Brenin, The National Mountaineering Centre in North Wales every Christmas where we hill-walked, climbed, skied on the dry slope and met many of the up and coming climbers of the day. Parents often joined us and they enjoyed the five days more than the Scouts. On our second trip, however, two of the boys contacted chickenpox on the final day and we had to make a somewhat hurried departure as the Director was worried about his staff. I drove down in the minibus with the Scouts, overnight, three days before Christmas Day.

All went well until we came across a police check, south of Croydon, at about 2 am, where they were directing vehicles into a lay-by for breathalyser tests. I tried explaining our circumstances to the young officer but he was brusque to the point of rudeness and insisted we join the queue. Eventually, a more senior officer approached the minibus and I explained that we had Chickenpox cases on board and he backed off and said we could drive on—but we couldn't! We were hemmed in among the drunken drivers of Croydon and it took a good fifteen minutes to extract us from the check. One of the Fathers who was with us muttered, as we eventually drove away, 'I hope they all catch chickenpox!'

I took a small party of six Scouts canoeing in Devon and we camped just north of Exeter on the banks of the River Exe. Surprisingly it was beautiful weather and we had two wonderful days on the river, and another on the Exe Estuary out of Topsham. On the final day, I awoke the scouts very early. None of them had been up so early before. I had cooked bacon butties and we took the minibus and loaded trailer down to Dawlish Warren.

At 5 am on a beautiful sunny morning Dawlish Warren was like the Mediterranean. The sea was absolutely flat calm, blue and crystal clear and there was not a soul about. We quickly unloaded the canoes and stripped down to swimming trunks, donned our buoyancy aids and launched off the beach. Everything was perfect and we canoed gently around to Teignmouth. Paddling back, the conditions were so safe that I just let the Scouts play in the water which was cool and inviting. The whole experience was idyllic and the Scouts thought it was fantastic.

We rounded the red cliffs approaching the Warren and to our amazement, the beach was packed. There was no beach, only people. We had a very difficult time landing and a struggle to get the canoes to the minibus and trailer but thankfully some of the holiday-makers on the beach thought it was great fun and helped us. We had a fish and chip lunch then headed back to the site. We had canoed on a perfect day in perfect conditions. How many times does that happen in England?

Two of the boys on that trip were from Zambia. I had not met them while I was in Zambia but their father worked for a mining company on the Copperbelt where they had done some sailing. I got to know them through the school and again, became a temporary guardian for them, during half terms and exeats just because of the Zambian connection. One exeat I took them down to Bristol to stay in my parents' static caravan at Portishead overlooking the Bristol Channel.

We took a school Mirror Dinghy and three canoes with us for I had arranged for the brothers to sail at Portishead Sailing Club in some Junior Races and I could take them canoeing on the River Avon. I explained to the Youth Officer that the boys had only sailed on a small lake in Zambia and would probably need much help, advice and assistance and he invited me go out in one of the rescue boats to keep an eye on them if I wished. The brothers won both races and in quite difficult conditions. The Portishead Club members were amazed and so was I.

We also ran regular canal cruises for the children and parents both on the Oxford Canal and the River Thames and we managed to get some of the girls on board about which they were thrilled. Of the two trips the Oxford Canal was the safest because the Thames is wider, busier and has larger locks. Sometimes, however, the parents could be more difficult than the children. I maintained that the children generally did as they were told whereas adults…!

The parents at St Bede's were most supportive of the Scouts and Guides. On the official opening of the tennis courts down behind the flint stone building that we used for meetings were two new courts with a high brick wall at the sea end for obvious reasons at the back of which I had built a climbing wall with the aid of Mr Driver, a local builder of some repute. I had first met him when he had lowered the level of the swimming pool by about nine inches and then got in and fitted a hand rail for the beginners with an electric drill.

I was asked to produce a display for the official opening which I did with ill grace, I am afraid, for it was yet another burden on my daily round. At the official dinner that evening, I danced with Billie Jean King and then met Mr Brandon Bravo who was a local businessman and who had children at the school. Let this be a lesson to you. He had been so impressed with the display that he offered an open chequebook for the coming year to buy anything that we wanted. It was a magnificent gesture and I could never thank him enough. Another parent was keen to offer us a Martello Tower further along the coast, one of the many that had been built to repel Napoleon, and we all struggled one weekend to remove the concrete blocking the door with huge power jacks. We did succeed but it was declared structurally unsafe and we could not use it as a boathouse.

We had one sad incident involving the Scout troop. I had come across an injured serviceman who resided in a home for the Service Disabled in Eastbourne and we became good friends. He could drive with a specially adapted vehicle and we often met up with friends in restaurants and pubs. He was an ex RAF pilot and had the distinction of being shot down before he had even taken off; a story that he often told with much good humour. I knew he had suffered from depression in the past and had made several suicide attempts but he and his doctors had put this in the past now.

He became the Treasurer for the Scout Troop and was very efficient and popular. He attended a small party that I had organised in the Grand Hotel at Bristol one weekend and he seemed to be on top of the world until I accidently received a copy of the troop's bank statement. It had been nearly emptied at the rate of thirty pounds a day to fuel his alcoholism about which I had not been informed. I faced him with the facts and he admitted that he had embezzled the funds.

I made sure that his medical staff at the home knew what we had discovered and specifically asked that a watch be kept on him. His family quickly reimbursed the money and I thought that was the end of the matter but unfortunately, he committed suicide some months later.

I have to mention that I was involved in an alleged scandal at the school. One of the mothers was going through a divorce and I had socialised with her and other parents as well as joining in a theatre visit to London which she attended. I was confidentially warned by a friend that rumours were circulating that I was the cause of the divorce and the gossip machine was in full swing. I traced everything back to a vindictive member of staff and put the facts right and put him right in no uncertain terms and he also received a letter from my solicitor threatening legal action. I escorted many ladies while I was at St Bede's and there was never any impropriety.

I ran two ski trips while I was at St Bede's. Peter Pyemont and his family were all excellent skiers but did not attend my trips. Our first trip was to Sauze d'Oulx which was a big Italian resort but quite well organised and offered two lessons a day. I believed myself to be fairly experienced in the organisation and running of a ski trip and indeed, all went well. The children had a good holiday and there were no problems. I returned to the hotel twice after that but on a personal basis. The hotel owner was getting married and wanted his fiancée to learn English. He asked if I could go out for a couple of weeks to give lessons to his wife to be and then I travelled out in the summer holiday for the wedding which I am sure had been taken directly out of a scene from the film, The Godfather.

It was an evening wedding so one had to wear a DJ. I was frisked at the door and was escorted around the room to be introduced to the many guests by a young relative who was obviously to be my 'minder' for the evening. I was not unhappy to receive such kind attention and I had a splendid evening at the ceremony and a scrumptious meal afterwards. It was Italian hospitality at its finest and I was grateful and I met many interesting people. For my second trip I had again booked for Sauze d'Oulx but the ski company was in trouble so I pulled out before there were any financial consequences and on a parent's advice, booked to go to Macugnaga in Italy.

I think the parent must have gotten her resorts muddled because a good resort it was not. The hotel was basic and so was its food, the ski instructors offhand and not very conscientious and the lifts looked third or fourth hand. It was a struggle to hold the trip together especially as snow conditions were deteriorating and some of the children were getting minor injuries. By the end of the penultimate day, I pulled the plug on all skiing and we took the children tobogganing instead. One of the girls crashed and though it was not broken, she had damaged a tendon, or so the doctor thought, and he put the leg in plaster. Of the five ski trips that I had run, that was my first casualty.

I also had to deal with a very strange case on that trip. One of the boys was from Sierra Leone and one of the instructors said he was behaving quite strangely in ski class. I suggested that it was probably the fact that he may not have seen snow before and the environment was completely strange and alien to him. I kept a careful eye on him and actually 'ghosted' his ski class the next morning. The instructor was right. The poor boy would suddenly freeze and go into a trance-like state. Initially, I suspected Petit Mal, a minor form of epilepsy, but if you tapped him or gave him a gentle shake, he came out of it. The children said that he had done it quite often and often shook him to bring him around.

That night two of his roommates sought me out saying that their door had been locked, they could not get into their room and the hotel owner alleged he did not have a spare key. He should have done. I had to force the door and found the boy in a deep trance, shaking and talking to himself(?) Someone(?)

I gently shook him to no avail and had to give him a sharp smack on his cheek. He came out of his trance-like state immediately. I strongly suspected that I knew what it was and I should have realised earlier. As Westerners, we do not understand the cult of 'magic' associated with the spirits of ancestors and other primitive rites and rituals, indeed, we are normally highly sceptical of such practices.

While in Zambia, I had come across two such cases, the one in which a houseboy had died through auto suggestion or self-induced apathy and the other was a pupil whose behaviour was not good and the parents had consulted their local medicine man who had turned the boy almost into a walking zombie. It does not matter whether we believe this to be true but many African people do. Child 'possession' is a problem on the African continent and witchcraft is still much in use to this day. We were scheduled to leave next morning and had an awful job to get out through the neck of the valley due to overnight avalanches. I had to physically carry the girl who had her leg in plaster and we had enormous problems getting her onto the plane.

I made a full report to Peter Pyemont on my return about the whole trip, specifically mentioning about the problems and suggesting he might be wise to speak to the parents of the Sierra Leone pupil. I had the distinct feeling that he thought I had not handled the trip well and there were two outcomes. The first was that he suggested that he and his wife take the next ski trip to which I happily agreed. Three children came back in plaster and one of the adults slipped and broke his leg getting out of the bath…the headmaster.

I cannot say I was pleased but I could not help but feel an inward sense of smugness perhaps.

The second outcome was the arrival of the boys' parents who were unhappy that I had slapped him for misbehaviour. I asked whether they would like to discuss this in private or in front of the headmaster and gave them a knowing look. They had another young man with them who was obviously from the Sierra Leone Embassy and asked what I would recommend. I suggested a private interview and they agreed.

I apologised immediately for the slap that I had to give their son but I went on to describe his trance-like behaviour and proceeded to list my understanding of its causes. It was obvious that the parents had resorted to some sort of traditional advice but it came as something of a shock to the man from the embassy, who assured me that he would sort everything out and there would be no need to report anything to social services about the whole affair which I had hinted at. The boy returned to school and appeared to be absolutely normal and had no further problems.

It was very pleasant to be on a short direct line to London and that added an extra perk to the job in Eastbourne as I could easily pop up to see friends or see a show and we could also take the pupils up for day trips which I often did. I also took the Sea Scouts up for a trip to visit HMS Belfast, staying at BP House in Kensington so we could visit the museums as well. My social life was very busy to the extent that I maintained two DJs for one was often at the cleaners.

I am afraid that I did sometimes turn up for early morning swimming in my DJ which amused the pupils no end. My friends, the police, from constabularies ranging from Cowdray to Eastbourne, regularly pulled me over thinking that a driver in a DJ, driving an open-topped sports car at three o'clock in the morning must be a cert for a breath test but I was always stone-cold sober.

I was playing polo irregularly at Cowdray. I could not keep on using my friend's horses so I hired polo ponies from Peter Grace who lived in Ambersham House with his lovely wife, Elizabeth, and four daughters, Janey, Pippa, Victoria and Katie, who were very young when I first knew them, in fact I even changed the nappies on the youngest, Katie, when baby-sitting one holiday. It is alleged that Jilly Cooper based part of her book, 'Polo', on the four girls and their antics because they led very exciting lives when they were older and played as a ladies team under the name 'Amazing Graces'.

I started by playing chukkas on the Ambersham Grounds under the auspices of Cowdray Park Polo Club and I first met the Earl of Cowdray in rather unusual circumstances. I was making my way down to the Ambersham Grounds on my first visit and I was driving the Triumph Stag slowly and carefully in case I met any horses. Suddenly an old beaten up Land Rover hurtled around the bend coming straight for me and I swerved into a gateway but it was a near miss. The driver was an elderly gentleman so I enquired if he was alright, which he was, and then I did give him a gentle warning about driving so fast because of the polo ponies trotting down for chukkas. He nodded and drove off slowly.

When I reached the polo grounds, I met up with Peter Grace who insisted that he introduce me to the Earl. It was the driver of the Land Rover. When Lady Cowdray heard about our meeting, she gave John Cowdray a good telling off but I do not think he mended his ways. I got to know both of them quite well and liked their company. John Cowdray would often work in the garden of Cowdray House in the scruffiest of clothes and liked being incognito and watching the visitors because everyone though he was a gardener. John only had one arm which did not seem a great hindrance to him and he used to umpire chukkas on his faithful grey horse which knew his ways and was a willing and gentle mount and I much admired and spoiled it because it was grey.

I was truly staggered one day when Peter Grace came up and told me that I was to umpire and I could use the Earl's Grey. (Get the pun?) I thanked John profusely and often umpired on the horse the whole time I was at Cowdray. Peter quickly realised two things about me. Firstly, I could play polo and was not one of his novice clients and secondly, and more importantly, having returned from abroad I could drop my handicap to a Minus One, so I was worth one goal before I even rode onto the field.

Thus, I sometimes played for his Rangitikei Team with him and Tony Devich who was playing off a Four Handicap at that time. I played up at Number One but I followed Tony's instruction explicitly. If he shouted 'Man', I rode off the nearest player; if he shouted 'Ball', I hit the ball; and if he shouted 'Wing', I moved out to take a pass. Sometimes I was told to stick with one opponent and constantly ride them off to foul up their game and I really enjoyed that. I liked playing with him and Peter who always played a solid Number Four in defence. Tony would also give you a briefing after the game and I learned much from him and Peter.

It could be quite amusing because most matches were covered in the Horse and Hound which was one of THE magazines that Elspeth read on a regular basis and she was amazed to find my name and sometimes my photograph, if we had won a match or tournament. I would also pop up at Hunt and Polo Balls and the occasional society wedding, then the Guards Club and the Cavalry Club on Piccadilly. Whenever she was conducting potential parents around the school, I would be introduced as 'Geoffrey Allen. Head of English and he plays Polo.'

I was always included on the tour of the school. She was very impressed with a photo she had cut out of some magazine of a team we had put together for a charity match—Ronnie Ferguson, Terry Hanlon (the voice of polo), Jimmy Edwards and me. When I was buying horses at Cowdray some ten years later, I met all three drinking tea at the refreshment tent.

'Not seen you for a little while, Geoffrey. Been away?' asked Terry Hanlon.

'Only for ten years!' I retorted.

Jimmy Edwards was a big man and could not play that well but he enjoyed the game and that was fair enough. He came off in one game I played against him and we could clearly hear the thump as he hit the ground. The poor man was totally winded and could not play on. He drove a big Jaguar with a big boot that always contained a case of gin, several crates of tonic water and ice and lemon. He was very popular.

Ronnie Ferguson was Prince Charles's Polo Manager and the Prince would play chukkas when his duties allowed. He was always social and polite but this was the time when he was being—Hunted? Harassed? Chased? by one, Davina Sheffield, who considered the Prince to be her future husband. Unfortunately she had a red sports car so whenever I appeared driving on to the grounds the Prince would hide behind a horse box until he was assured that it was me and not the huntress. Prince Charles was delighted when I swapped the Stag for the Spitfire.

Many years later I happened to be in Tavistock when Prince Charles was opening a newspaper office and as he emerged onto the street, he saw me and gave me a hard look.

'I know you!' he called. 'Where do I know you from?'

I crossed the road to shake his hand.

'Cowdray Park, Sir. Ambersham Grounds. Red sports car.'

'Ah!' he said. 'I remember now. I hope you are well?'

We shook hands again and he walked off. I could not mention Davina Sheffield because he was with a lady…Camilla.

I was introduced to the Roberts family and a small romance blossomed. Squadron Leader Roberts, D.F.C. and Bar (flew Lancaster Bombers in 514 Squadron during the War) and his wife Phyllis lived not far from the Ambersham Grounds at Grafham and had two daughters and a son. The son was Charles who drove dangerously everywhere in a dark blue Porsche nicknamed 'The Inkpot'; one daughter was Lavinia, who was a good polo player, and then there was Beverley who was the youngest.

I would stay for the occasional weekend when I was playing and Bev and I spent some time together. She was the quiet member of the family probably because she could not be heard over her brother's noise. We attended the Polo and Hunt Balls and she liked the fact that I could dance. She was not, however, impressed with my tennis. Her father only had one lung and one had to hit the ball to him at tennis but my shots could not be produced with such accuracy and I was often in trouble. Her parents were both forceful characters. They owned two big Labrador gundogs and they were definitely not pets. One day all hell was let loose at the front gate. The dogs were attacking an elderly couple who were trying to enter.

'What the hell do you want?' Phyllis yelled.

'We were just passing and wondered if you knew that your chimney was on fire.'

Squadron Leader Roberts sprang into action. Unrolling the garden hose, he pulled it in through the kitchen but it only just reached the door of the lounge. The fire was extinguished by pouring a crate of tonic waters into the grate and the fire service came and checked the chimney.

Squadron Leader Roberts and I often chatted together. He and his brothers were milkmen. They had started Jobs' Dairy in North London and it had grown into a big supplier and Charles was hoping to take over the business but I think they sold the company in the end. He had also had a distinguished career in the RAF. I, incidentally, always called him 'sir' and I did so on the phone one day within my father's hearing.

My father objected to this but the next day he took a call from Squadron Leader Roberts and addressed him as…Sir. Phyllis seemed to like me too. She did not shout at me as much as the other members of the family but I was not sure if I wanted to make any commitment so Bev and I had a long talk and agreed to hold off for one year with no conditions. She fell in love with an accountant from Birmingham and I attended their wedding at Westminster Abbey.

As Phyllis walked down the aisle, she whispered to me, 'Bloody fool. It should have been you!'

The reception was at the Ritz and I had to drive Patrick Churchwood's Rolls Royce back to Sussex because he was so drunk.

While at St Bede's, I was befriended by a Lebanese family who often sought my advice on all matters British. They had a daughter and son and I was invited to dinner one evening in their large house In Eastbourne. It was the first time I had eaten Lebanese food and as I was eventually to discover, I was quite unused to the massive amounts on the table. They were surprised to learn that I played polo in the UK and had also played in Zambia and they asked me about my Zambian experiences.

The Khourys, for that was their name, lived mostly in Accra, Ghana, where, Alain assured me, they played polo, not half a mile from their house. It was played by many Lebanese, Ghanaians, Expatriates and about twenty other nationalities. Alain assured me that it was a big club. It was a pleasant evening with more thoughts of polo in my head.

Storm clouds were beginning to develop, however, and I felt the wind of unrest. The overall pace and pressure at the school had not been alleviated, lessons were still disturbed and cancelled and there was no effort to establish a professional ethos in the school. Everything was fragmented and central control was lacking. There was no overall view of what was happening and what the school was aiming for in educational terms. It was simply unprofessional. I had serious doubts about a general lowering of standards and concerns for the selection of Scholarship Candidates in particular.

I had a long discussion with Peter who immediately went on to the attack and became quite annoyed about my comments. I refused to continue unless he considered this as a discussion and not a fight. Peter had always controlled everything and probably knew for what he had been aiming but he had never shared his vision and was unwilling to listen to other views. Furthermore, he resented any criticism of his school. He had completely ignored the problems which the school faced because he had been so busy himself.

He always looked down and never up or around. He had no idea what was going on in lessons. He had never seen any of his teachers teach. He was a wonderful charismatic leader, a brilliant teacher and loved the school dearly but he was blind to everything else. At this stage my concern was for my Scholarship and Common Entrance Candidates, who, were time-tabled for five periods of English but were averaging only half of these. He did promise to rectify this but nothing happened so I had to resort to a tutorial system which put more time pressure on me again.

It was obvious to me that things were not right. The school had been owned by the Pyemont family for many years and was started, it is alleged, because his father had a big win in a casino in Monte Carlo and the only way he could get it back to the UK free of Tax was for educational purposes. It is probably not true but it is a good story, almost as good as polo players starting a school. Family members seem to have drifted in and out of the establishment and left no real footprint and this even happened when I was there.

There was also the strange arrival of a housemaster from a well-known girls' school bringing a whiff of scandal about skinny-dipping with senior girls. Probably, again, it was not true but he was never formally introduced to the staff and no one seemed to know why he was there. My feelings of discomfort grew as the term progressed and I was not entirely happy.

I am an organised person and everything was a hectic muddle. Out of the blue came an invitation to visit the Khourys, not in Eastbourne…in Ghana…and I eagerly accepted the invitation for the Easter Holiday for I had no ski trip to worry about, had I?

I was a little surprised to see the Check in Queue for Ghana Airways at Heathrow Terminal Three. I had arrived early and so had everyone else. I had one suitcase and most other passengers had three or four and copious amounts of boxes, crates, prams, car parts and four or five pieces of hand luggage as well. I was led to understand that this was all normal but the queue did stretch right back through the terminal building and moved very slowly as each passenger argued and cajoled to exceed their baggage allowance and then struggled off carrying their load of hand baggage.

I managed to check in and made my way through to the departure lounge for a coffee where the Ghanaians were easily identifiable by their bright colour clothing and mountains of hand baggage! My attention, however, was drawn to a group of Jamaican ladies seated at one table who were dressed beautifully in what I would call traditional British Ladies' Outfits with very smart frocks, cardigans, matching shoes and handbags and dazzling hats. Though they were of an age, they could easily have just popped in from Ladies' Day at Royal Ascot and they were sipping their tea in a very refined manner with the little finger cocked.

It was a wonderful tableau and they were all going to Jamaica—Montego Bay. My flight was called and there was a general rush in which I did not join as I had my seat so there was no problem. Actually, there was because someone was sitting in mine. The steward found me another seat but it was in the smoking section and I did not want that so I was upgraded to Business Class. I would thus like to thank the nice Ghanaian lady who pinched my original seat.

The flight took about six and a half hours and for much of it, of course, we flew over the vastness of the Sahara Desert which I could clearly see below, and we landed at 8.30 pm in Accra's Kotoko Airport. As soon as one steps out of the aircraft, you are 'hit' by the very hot and moist air of the West Coast. I had been used to the dry heat of the high plateau in Zambia and Accra's heat was a different matter. Immigration formalities took some time so it took an hour for me to get out of the airport and join the melee and scrum outside but Alain was waiting for me and we drove the very short way to his house that was in Airport Residence barely half a mile from the airport.

In Africa, nights are very dark and so I did not gain any real impression of my surroundings on arrival but the Khourys lived in a fairly large, though plain house, which was air conditioned and cool. Alain, Hilda (his wife) and Adelita and Albert (daughter and son), were all there and I sat down for another Lebanese meal which was very difficult as I had eaten more than enough on the plane. My room was comfortable and cool so I fell asleep wondering what the next day would bring.

Alain had an assistant, a young Frenchman, called Bernard, who he asked to drive me around Accra in the morning so after a light breakfast and my ritual coffee we drove off to see the city. On the outskirts of Accra, then, there was much open land and few buildings and the drive down Independence Avenue was really quite pleasant. The houses and buildings, however, were beginning to become more dense, until, eventually we were driving in narrow crowded streets jammed with traffic and people. The whole scene was noisy, crowded, bustling and lively and all the traders and street vendors were shouting, laughing and busy.

I was quite impressed by some of the original Colonial architecture of the old town but it was quite neglected rather like their counterparts in East Africa. Bernard was quite impressed by the large colourful balloons in Makola Market until closer inspection revealed that they were condoms. We drove around to Independence Square and the Ministries and back up through Oxford Street in Osu which was incredibly busy and was a riot of colour and noise. Bernard then drove me out to the university bookshop with which I was not impressed.

I managed to buy a couple of books about Accra which were slightly dog-eared and faded and also bought two books for the children that were in a slightly better state. The university then was almost in a rural setting and had some fine views of the Aburi Hills.

We returned for lunch and I was introduced to a Munir Captan, who was a polo player, and invited me down to the club later that afternoon to 'stick and ball'. The children were riding so I happily accepted and after a small siesta, we all jumped into the car and drove down to the Accra Polo Club. There introductions came thick and fast as most of the club's players were down checking and riding their horses, or drinking beer. I met Munir's older brother, Nour, and several other Captans (it was a big family) and Nour kindly let me ride one of his horses on which I very gently stick and balled for half an hour and came off the field covered in sweat but at least the club members knew that I could ride a horse and hit a ball.

I was thirsty and needed a beer. The Captans generously included me in an invitation that they had received for a party over at Tesano and kindly collected me. It was at the large house of Armeen Kassardjan with a huge garden and it was already in full swing when we arrived. I was swamped with about another fifty introductions and it looked like being a good party. You will recall that I had received good training in Zambia and my social life in Eastbourne, Cowdray and London had not been that quiet. I loved to party. There was lots of fun and high jinks and during the evening I tied Nour to a tree and sprayed him with champagne. I got back to Alain's house in the early hours and had a very good night's sleep.

When I came down to breakfast in the morning, however, I noticed that Alain was looking at me rather oddly. Munir had already telephoned that morning about some deal involving tyres and had mentioned their, and my, antics of the night before and I was invited over to the Captan House for dinner that evening. Alain could not stop laughing and Hilda was even more amused. The family took me up to Ada which was at the mouth of the Volta about sixty miles from Accra where we could water-ski, swim, eat and drink cold beer. Alain was somewhat surprised to realise that I could drive a speed boat and water-ski. It was somewhat cooler in and on the water and it was a very enjoyable day but I had a dinner to attend and a superb meal it was too.

Captan House was on Sir Arkuh Korsah Road in Airport Residence and Alain drove me around. They had a 'tiger' at their gate! The watchman was called Tiger for some obscure reason and was with Nour for many years and was always hopelessly inefficient. If he did not hear the car horn one had to shout, 'TIGER!' very loudly and he would then generally come very slowly. Captan House was big and was spread over three floors. Nour lived mostly on the ground floor while Mounir and Nour's parents (when in residence) had bedrooms on the top floor and there was always a variety of dogs outside and an African Grey Parrot, which amazingly is still alive to this day.

The downstairs lounge was huge and had a big dining room table. Ten people were sat down to dinner that evening and I was under the impression that it was a special meal. It was not. The family dined like that every night and sometimes there could be fourteen people. The food was delicious and plentiful for this was the Lebanese way and I had a lovely evening. Nour showed me his paintings and they were all on a Ghanaian theme which he had picked up at reasonable prices but were now worth a great deal of money.

He had received many an offer for them but preferred to keep them for his enjoyment. In one corner of the room was a pair of enormous elephant tusks that his father had purchased many years ago and which, I presume, were extremely valuable but were now stuck in Accra because of the Ban on Ivory. After dinner Nour showed me the middle floor of the house that was really one big reception room, beautifully and tastefully decorated in a classical style with wonderful tapestries and chandeliers and decorated furniture. It was quite stunning.

Alain was keen to show me his work in Ghana and what it entailed. Many Lebanese business men were traders and Alain followed that tradition. At the time of my visit he was importing rice into Ghana which entailed the laborious process of applying for work permits through the Ministry of Trade. He also owned a sawmill in Sunyani and a cinema in the same town. He also imported steel rods and acted as a freelance agent for Krupps of Germany. He admitted that the sawmill was a burden at that time but hoped to receive a loan for its revival.

I also spent a day with the Captans. Nour's father had been in Ghana long before Independence and had established stores in Accra and then moved into the cinema business in Accra owning seven or eight cinemas in the town. The Ghanaians liked the cinema and trade was good. He was a friendly advisor to Nkrumah, the first elected president of Independent Ghana, and was given much land that spread down from the airport right down to the seashore so the family owned many properties in Accra.

Nour seemed to handle the cinemas and properties from his office in Opera Square and Munhir had his own electrical appliance store called Radio Shark because he had traded in shark fins in the past for the Japanese market. Getting to their office in the early morning rush hour traffic was quite a challenge but there was always a coffee upon arrival. I believe that this was the time that they were also planning to open a salt project with refinery to process salt. This was an ancient trade and much respected in Ghana. In the days of the old kingdoms Ghanaian gold was exchanged for its same weight of salt from the Tuareg Tribesmen in the north.

I liked Ghana immensely. The people were very friendly and open and there was no hint of a racial divide. There was no 'Expatriate Community' in existence and I met, not only many Ghanaians socially, but also many other nationalities from the African continent and further afield. It seemed a truly multi-racial society. The polo club had over twenty different foreign nationals and it was real pleasure to mix in such an environment and talk to everyone and anyone with no bias, favour or concerns. I met ministers, civil servants, international businessmen, teachers, lawyers, tradesmen, shopkeepers, labourers, drivers and domestic staff and conversed with them all.

I particularly liked talking to the drivers as they drove me around for they knew much about Accra and its environs and were very keen to help and advise. They were ever patient and kind and fiercely protected you in all manner of situations and they knew much more about their employers than their employers realised. I was often out in a big American sedan, a blue Ford LTD, with Alain's driver, Martin and got to know him quite well. He was a large and very patient man. He had to be because I was often out late at night for which he was richly rewarded.

There was thus a constant round of parties, lunches, dinners, visits to restaurants, stick and balling, chukkas, polo, bush riding, swimming, sun-bathing and water-skiing though I did make visits to the museum and the Aburi Botanical Gardens on a couple of days. It was a whirl of socialising and I really enjoyed myself. I did indeed fall in love with Ghana. The heat could be quite oppressive but the cold beer alleviated that problem. During my last weekend we played a friendly Ghana versus England Beer Match at the polo club and the winners won a crate of beer. England lost six goals to three and I had to buy my own beer.

I have to say that I had accepted Alain's invitation willingly and thought that it was just that; a kindness for which I had been extremely grateful. His family had been very hospitable and I enjoyed being with them very much indeed. It had been a great trip. On my last day Hilda had taken the children into town to do some shopping before they and I flew back the next day and Alain and I were talking, over coffee, in his side office. He asked what I thought of Ghana and I was able to make many of the comments which I have written above in strong support of the wonderful country. It was also obvious that I had made many new friends and had really enjoyed the trip.

'Would you consider coming back and working with me?' he asked. 'I really need someone to help with the business here and in the UK as well.'

I politely requested that he give me a brief outline of what it would entail and an outline of terms and conditions. All seemed very fair and reasonable and I accepted there and then. This may have seemed very rash and hasty. In truth, however, I did not wish to continue at St Bede's. I am well able to work long hours and withstand pressure even in difficult conditions but I am, essentially, a very organised person and experience great difficulty in working in an environment that is not. There was no real sign of any improvement in this respect at St Bede's so you may say that I jumped at another opportunity. There, again, I knew I was taking a chance and stepping into the unknown. I was adventurous and willing to take the gamble. I could also get back to full time polo. All was agreed. I would join Alain in mid-August in Accra.

On my return to St Bede's at the start of the summer term, I formally handed in my resignation and the staff were slightly taken aback that I was heading out to Ghana to work with one of the parents. Several were amazed. I was anxious, however, to ensure that I worked conscientiously and properly for the whole of that term and my absolute priority was my examination and scholarship pupils to whom I devoted much extra time, for timetable disruptions were still rife even on the lead in to exams. Peter took over the early morning swimming and that enabled me to hold more individual tutorials and give extra help.

I continued to run the scouts and adventure activities at weekends while fulfilling all my duties and obligations and I even ran the Scout summer camp at Brecon and 'ghosted' a North Wales expedition that John Smith ran before my departure and I continued with my polo and social life. Individual exam results were good in Scholarship and Common Entrance especially as many of the candidates had started from a low academic base and I was pleased. I could hand over with a clear conscience.

Peter gave me a good send-off and a formal farewell with all the Scouts, Guides, Cubs and Brownies assembled and I still have the presentation salver in my motor home. I left St Bede's with no ill-feeling. I had enjoyed my time there and it was a very valuable experience as part of my teaching development; an experience from which I had learned much and was to put to good use in the future. I made many friends and had liked my time with the pupils and staff.

A few days before I left, I was in Eastbourne and bumped into Tom, the police sergeant, who had heard of my departure. I ventured to suggest that life would be easier for him after I had gone.

'Oh yes, Geoffrey, but the fun was in the chase and who will I chase now?' He winked, we shook hands and he walked off with a slight spring in his step.

There is a 'rider' to my time at St Bede's. It later faced an Independent School Inspection which rated the school with serious failings and it was given time to improve before there was a Follow-Up Inspection, which it also failed, and it took a third to bring the school up to the required standards of the Independent Schools' Inspectorate. Many of the problems that they failed on I had identified in my last year and they were major then.

Trigger   The Door of no Return

Condemned Cell –
Elmina castle.

Kakum

Chapter Seventeen
All Grist to the Mill

The Hawker banked steeply then levelled out for its approach to the airfield at Sunyani. It was a beautiful morning. I was fortunate to have a window seat and as I peered through the scratched glass carefully. I could clearly see the sawmill that I had come to manage even though I had no experience of logging at all. It covered a large area but much of the roof was missing and there was a miscellaneous collection of rusting machinery and broken equipment scattered all around the perimeter fence. There was no sign of life. I could not see any logs. I could not see any workers and I could see no sawn timber. I wondered what I had let myself in for.

The Hawker made a smooth landing and taxied over to the main building which was really a large shed to one side of the runway. The airport's ground staff were wearing big smiles and were very welcoming and friendly. They seemed to know all about me.

'Mr Allen. Welcome to Sunyani. Welcome to the Brong Ahafo.'

The manager escorted me into the lounge which comprised of three old chairs and a long wooden bench. We chatted pleasantly while the handlers unloaded the baggage. I had travelled with only a small suitcase but this was far from the norm. The poor handlers struggled with huge suitcases, big cartons, large crates and a varied assortment of animals all of which had to be manhandled on and off old trolleys and pushed into the building. He helped me find my bag and I turned to the exit where I hoped I could find a taxi to take me to the mill.

'It's alright, Mr Allen,' explained the manager, 'the taxi has gone to Kumasi so we will give you a lift with our transport. We will not be long. Please wait.'

The taxi—I mused. This must be a very small town if it only had one taxi. I gazed out across the runway. At ground level, the mill was hidden by a screen of trees but the 70' jib of the crane stood out against the blue sky. This crane was to be the bane of my life but I was not to know that until later. Just then, our transport announced its arrival with a constant blowing of its horn and a cloud of dust as it hurtled towards us. It was a trotro, or mammy wagon as it is sometimes called.

Imagine a Bedford truck with bench seats across the back section and a wooden canopy. They were often well decorated and always had a slogan or biblical verse written over the cab. This one had 'Sea Never Dry'. As we were some four hundred miles from the sea, I wondered why the driver had chosen such a slogan.

I clambered up into the back of the trotro with the other staff refusing all offers to travel in the front cab. By tradition this seat was always the most expensive but it was also the most dangerous seat in the event of an accident. It was nicknamed the Suicide Seat by expatriates. The two stewardesses were remarkably smart and pretty and kept us amused with their jokes and banter. The trotro lumbered along rather like a charging hippopotamus hitting every pothole it could possibly find which sent jarring jolts straight through the hard wooden benches and up one's spine.

No one had warned me about my accommodation. I had been told, rather vaguely, that there was a manager's house, the condition or state of which I was unaware and held out no great hopes after seeing the poor condition of the mill. To my astonishment the mammy wagon drew up outside a palatial bungalow with a gravel driveway and huge wrought iron gates. It was freshly painted, smart, clean and...huge.

'This is it!' shouted the manager above the noisy engine of the trotro.

I thanked everyone profusely, had my small suitcase handed down to me and the Trotro roared away. By this time, the gates had been opened by the houseboy dressed in an immaculate white top and shorts. He beamed.

'I'm Peter sah. Welcome sah. Welcome to Sunyani.'

He led me on to the front veranda which ran the whole length of the bungalow. It was nearly the size of a tennis court. Peter, still beaming, pushed open the front door and ushered me inside. The lounge was even bigger than the veranda and contained wonderful, old mahogany furniture that had been made in the mill. Everything was spotlessly clean. Peter noticed that I had been taken aback by the size.

'For cinema sah. For film, you see?'

The cinema. I had forgotten the cinema. This was supposed to be another of my responsibilities. My employer also ran a small cinema in Sunyani. Whenever he was in residence, he would show the current feature to special guests in this lounge. After supper, when the first reel had finished at the cinema in town, a messenger would jump on his ancient bicycle and deliver it to be shown in the bungalow. He would then return for the second reel and perhaps a third! His record (The Life of Christ) involved the transportation of six reels. And this was an annual event every Easter! This meant the poor man had travelled twenty-four miles on a potted dirt road in the dark. The messenger was still employed by the company and he did look very old and weary.

I toured the rest of the bungalow. There were six bedrooms, all en-suite and including a huge master bedroom, a gigantic kitchen and a spacious office overlooking the mill. The garden was half an acre and surrounded by an eight foot high wall. Peter pointed out a pet deer in the garden which my employer's wife had hand-reared. Rather unimaginatively it was called Bambi.

Just then we were interrupted by a shout from the other side of the garden gate leading to the mill. Peter opened it and in trooped the sawmill manager, Joseph; the mill electrician, George; the sawmill clerk, Matthew, and the carpenter, Kwasi. After introductions had been made I ventured to ask why the mill was not working.

'No electricity,' replied Joseph.

'No logs,' remarked Kwasi.

'No bush operations,' sighed Matthew.

'No transport,' said George.

'But, apart from that, there are no other problems,' added Joseph seriously.

Evidently, the cinema was also closed due to the electricity outages and a brush with a new Fire Safety Officer. 'May I show you the mill?' requested Joseph.

'I would love to see the mill, Joseph, but as there is no food in the house and there seems to be no rush, Peter and I should, perhaps, go to the town and market and see what we can buy? The only problem seems to be transport as your Sunyani taxi appears to have travelled to Kumasi.'

'By car sah,' beamed Peter.

'The Pontiac is ready for you, sah,' said George. 'I spent all last week working on it.'

'The Pontiac?' I queried.

In a solemn procession, we trooped around the end of the house to the garage. Peter pulled the doors open and there stood a huge, bronze-coloured Pontiac Coupe. It had seen better days but was sparked with a slight delay and a muffled roar so Peter and I quickly jumped in, assuring Joseph we would return soon for the tour of the mill. Though untuned, the Pontiac had an extremely powerful engine and we sped into town, pulling up outside the National Trading Company's store. We both walked in.

The NTC in those days was a state-run company and was intended to supply basic and essential commodities at affordable prices for the general population, so imagine my surprise to find that they only had four sets of commodities on offer, but, what commodities! By the door there were cheap wooden toys which were of no interest to me. The other three commodities were amazing.

At one end there were dozens and dozens of bottles of tomato ketchup. At the other end, from floor to ceiling, were stacks of strawberry jam, while all the way down the far wall were bottles of champagne. All were of a very good quality. Sunyani was proving to be very interesting. It should be noted that all commodities were in very short supply in the country and this could possibly be regarded as a veritable gold mine. I was approached by the manager.

'Mr Allen. You are welcome.' Even he seemed to know about my arrival. 'Do you wish to buy anything?'

Yes, I did, I thought. I was not concerned with anything. I wanted everything! It was going to need careful handling and negotiation. Inspiration was directly at hand.

'May I treat you to a bottle of your champagne?' I asked, grabbing a bottle from the shelf. 'So that we can toast my arrival in Sunyani?'

'Yes, of course,' he replied. 'Please come into my office.'

Three bottles of champagne later, I had concluded the deal. The manager had willingly agreed to my purchase of all the stock at a knockdown price and would transport everything to the bungalow in the morning. I remember that one bottle of champagne actually cost me less than a small bottle of beer in the UK. I kept one case each of the ketchup and strawberry jam but then sold the rest to my friends who ran shops in Accra. As for the champagne, for over two years, every night I was in Sunyani, I drank one bottle of chilled champagne with freshly squeezed orange juice…Buck's Fizz… delicious! Every weekend I would fly back to Accra with two cases of champagne for partying over the weekend. What a tough life!

Peter and I concluded the more mundane shopping and drove the rumbling Pontiac back to the bungalow. By now it was quite late so I sent Peter to tell Joseph that I would tour the mill in the morning then asked Peter if there was any water for a bath. We were plagued by water shortages in Accra so held no real hope for the same.

To my surprise, Peter asked if I would like my bath hot or cold. He proceeded to run my bath, providing clean towels and then checked the bed and mosquito net while I wallowed in the bath with a glass of champagne. A good dinner awaited me so all was well with the world. I had good food and drink, wonderful accommodation and a very efficient houseboy.

'Peter!' I called.

'Yes sah,' responded Peter, popping his head around the bathroom door.

'I think I am really going to enjoy Sunyani.'

Peter woke me at first light in the morning with a cup of tea. I sat on the front veranda watching the dawn lighten before deciding to go for my early morning jog. Donning a pair of light shorts and my trainers (it was already too hot for much else), I jogged around the perimeter fence of the mill's big compound. Several well-trodden paths ended at our fence, an obvious sign that our spares were being sold to other parties. I also encountered a chestnut horse. He was about 14.2 hands and sturdily built. He eyed me carefully as I stroked his nose and checked him over. He was slightly neglected but I could find nothing wrong with him. He was roaming completely untethered and seemed very content with his lot.

I was hot and sweaty upon my return so I showered under the overflow of the large outdoor water storage tank at the side of the bungalow. I was under the impression that I could not be seen from the road so this became my morning habit. As soon as I had finished my run, I would strip and shower in the garden. Peter collected my running gear for washing every day and left a towel. I dried off then had another cup of tea before dressing for work. My ritual had not gone unobserved, however, for on a later flight to Sunyani, I had been gently teasing one of the stewardesses who was anxious for Kwasi, our carpenter, to make her a cheap bed as she was about to be married.

'Is your man handsome? Is he strong and well built?' I taunted.

'Oh yes,' she retorted, 'almost as good as the one I see sometimes taking a shower in the garden whenever I pass in the early morning!'

Touché.

Joseph, Matthew, George and Kwasi arrived promptly at 7 am for my conducted tour and I left Peter to supervise the arrival of all the goods from the NTC store. The buildings were all poorly maintained and the machines had all the appearance of a recent Luddite attack. Huge mounds of sawdust were dangerously piled everywhere. Sawdust presents a great risk to sawmillers. If a fire started it could smoulder and burn for many days and could even travel underground, so, when you had thought the fire was finished, it could reappear a hundred yards away. It would be necessary to dispose of it quickly.

Joseph had been sensible in maintaining the workforce on a casual basis so they were paid only when the mill was in operation. This seems a little unfair but the staff actually liked the system as they could tend their farms in the downtime. Indeed, we extended this system by allowing them to have garden plots within the perimeter. In electrical outages, they would till their vegetables and then be at hand when the erratic electricity supply was turned on.

The huge crane, one of the biggest in the country, did work but the clutch slipped constantly which was rather alarming as it swung logs into the air. We maintained a running battle with the Civil Aviation Authority who did not like it beside the airport. They insisted that we should place a light at the top of the 70' jib to warn low flying aircraft at night.

As there were no night flights at Sunyani, we argued a strong case but a compromise was reached. We agreed, that, in the event of our watchman hearing the approach of a low flying aircraft at night, he would climb to the top of the jib and swing his lantern to warn the pilot(s). It must be mentioned that our night watchman was the very old messenger who used to transport the films on the road at night and he was, by now, very ancient indeed.

The sawmill offices were clean, tidy and bare, so we moved on to the lines of rusting vehicles and machinery. They were no good but we agreed to sell their parts to generate a small immediate income. I would have wished them a better fate. We moved on to a set of three large garage-like structures the doors of which were locked. Joseph produced the keys but the first lock was very stiff and would not budge.

The key turned easily but the hasp would not open. George, ever the practical one, picked up a rock and struck the padlock. It flew open. It took some time to tug the doors open against an accumulation of weeds and debris and at first, it was quite difficult to see inside. After the bright sunlight of the main yard, the interior was very dim. Eventually, I made out a strange shape.

'What the blazes is that?' I exclaimed.

'A snowplough!' they all chorused in unison.

'A SNOWPLOUGH?' I queried. 'Why do we have a snowplough 100 miles from the Equator. I have heard of climate change but do we now expect snow in the Tropics?'

'Gift from the Russian government, sah,' answered Joseph. 'They thought we could extract logs with it, sah.'

'Have we tried it? Does it work?'

'No sah. It came with no engine!'

I had experienced the vagaries and complications of foreign aid before but this seemed an exceptional example of idiotic thinking.

I was now anxious to move on to the next locked garage as I was fascinated to establish what might be there but, out of the corner of my eye, I could see a column of men, dressed in dirty white smocks and shorts, marching through the main gate. Matthew explained that they were prisoners who came to collect firewood on a daily basis. What was remarkable was their formation. At the front of the column walked five prisoners in a single file followed by a smartly dressed prison officer. Obviously, there was nothing remarkable about this, except, behind the prison officer walked another prisoner, who was obviously a trustee, carefully carrying the prison officer's rifle on his head!

The prisoners turned out to be a great asset in the end as we used them to clear the mounds of sawdust and paid them in firewood. The prison governor was quick to seize an opportunity and eventually, the local prison set up a poultry farm and took all our sawdust. We were happy and they were happy and I received a regular supply of eggs for the sawmill workers and myself!

We turned our attention to the next locked door. It took all four of us to heave and prise the heavy doors open and after my experience with the snowplough, my imagination was beginning to run riot. Perhaps there was a Centurion Tank inside, or a dune buggy, or a double-decker London bus! It was none of these. Parked inside was a brand new Peugeot 604. It was covered in dust but the bodywork was unblemished and the interior had that wonderful leather smell.

'Boss just told us to leave it there, sah,' commented Joseph.

It appears that it belonged to my employer who had the luxury of two Peugeot 604s and this one was the spare. It was moved down to Accra eventually and sold.

As we walked back to the office, we met the old watchman leading the chestnut horse I had discovered earlier. 'Can it be ridden?' I asked.

'The watchman often does,' responded Matthew, 'and there is a saddle and bridle in the office.'

I rode Trigger most evenings. He was not what you might call a fast steed but he was steady and reliable and he would, occasionally, shamble into a slow canter. I much enjoyed the evening rides. Indeed, during acute petrol shortages, or when the Pontiac failed, I often rode him into town to do the shopping much to the amusement of the local inhabitants.

'Right.' I turned to Joseph. 'Let's go and see the bush operation. How do the workers get there?'

'Tractor and trailer sah,' answered Joseph.

'OK. So if we go in the Pontiac now, everyone will be working hard?' I asked.

Joseph looked at his watch doubtfully. I noticed his hesitation and the two-way look that passed between the four of them.

Later, in my logging career, I was responsible for huge forest operations including 30 trucks, ten Caterpillar machines of various hues and descriptions, twelve Komatsu Bulldozers and eight front end loaders as well as a huge fleet of miscellaneous vehicles from tankers to pickups—all of which worked reasonably well apart from the occasional breakdown. We operated in over one million acres of forestry concessions and ran four bush operations to feed two sawmills and one moulding plant as well as exporting thousands of cubic metres of round logs to Europe.

I often think back to my first day in the bush, near the small town of Berekum, as what followed, was pure pantomime. The workers had only just arrived at the site and were now busy physically unhooking the trailer so that the tractor could pull an ancient yellow Land Rover into life. Once sparked the Land Rover and tractor, combined in tandem to spark an elderly Timberjack, which in turn, pulled our one logging lorry into life. Then, and only then, was everybody ready for logging.

There was no huge concession here and logging was by agreement with the local cocoa farmers who spent hours haggling over the price of individual trees. We were lucky if we managed to extract one tree per day—about thirty cubic metres whereas my later company could extract 1,000 cubic metres a day! Thirty cubic metres would supply our antiquated sawmill for one day. The whole process was painfully slow. Breakdowns were a regular feature, spares were very difficult to source and the logs had to be physically loaded on to our lorry using the Parbuckle Method assisted by the tractor and Timberjack. Even then we were not home and dry.

The Mercedes Lorry was on its last legs, or tyres you might say. The tyres were all home-use and purchased second hand. I also realised that the trailer had no braking system and the driver was actually very skilful in handling the truck under such circumstances.

On this particular day, the truck had been loaded the night before, but, as its lights were not working, it could not move until the morning. We followed it keeping well clear of the huge cloud of dust it was spewing out behind.

At the mill, the workers suddenly appeared as if by magic and rumbling and protesting, the crane juddered into life and unloaded the three sections of the tree trunk. One was directly placed on to the carriage of the circular saw which made short shrift of the logs cutting them into broad boards that were then fed through the Edgers to obtain the measured timber required by the customer who was standing by with a pickup.

After three hours, the whole process was finished, the machines fell silent, the workers disappeared and all was quiet again. It was no wonder that the business was not a success. It had extreme difficulty in meeting its payroll obligations, the machines, vehicles and equipment should have been in a museum and the buildings were ramshackle and poorly maintained. After all the fun and surprises of the previous day I was coming down to earth with a bump. How could I improve things? Now I was wondering what I might find at the cinema that evening. I decided that the sun was, at least, within reach of the yard-arm, and returned to a relaxing bath and a Buck's Fizz. I lay wallowing in the bath and gradually, a plan began to form in my mind.

By 7 pm, I was ready to visit the cinema which had been allowed to reopen for a short period of time as I had promised to sort out the outstanding problems. We had been given a month's grace by the local fire officer as a goodwill gesture, but I knew he was not popular in the town because of the enforced closure of the town's only cinema. It was operating in an old beer hall with only one entrance.

The projector was in the same state as all our logging equipment, that is to say, ancient and decrepit, but, as mentioned, this was the only form of entertainment in Sunyani and thus had much potential. There was a kung fu film showing that night which was very popular. The people of Sunyani adored the martial arts films and anything with a religious theme.

While the film was running, I spoke to the owner of the beer hall. I explained the fire officer's need for another exit which we would be willing to finance. Strangely he refused. He was taking a good rent from us, more than enough for a reasonable income for himself and his large family so I failed to understand his reluctance until Matthew explained, that, when the film finished the cinema reopened as an illegal gambling and drinking club. One door could be easily watched—two would be a problem. It was his loss, for, as part of the master plan I was slowly evolving, we moved to another, cleaner, more central premises that automatically conformed to fire regulations.

The owner was very enthusiastic and helpful and delighted with the arrangement. Our film supplier, in Kumasi, was urged to send us only popular films with the martial arts and religious themes and the cinema never looked back though I do often wonder whether Jesus would have approved of being on an equal billing with Bruce Lee.

On my return to the bungalow that night, I had a visitor—the Chief Superintendent of Police—and I wondered if I had fallen foul of the law, but, evidently it was purely a courtesy call. I did wonder if he had caught wind of my champagne deal and he certainly enjoyed a whole bottle for himself and accepted another one as a gift. He became a good friend and made regular 'courtesy calls'. He had the distinction, at a later date, of being attacked by my cat, a half Siamese called Bulldozer. The superintendent had settled into one of the easy arm chairs in the lounge and we were discussing the incursion of an Ivorian logging company into one of our areas as we were very close to the Ivorian border. He had slipped off his shoes and was waggling his toes as we talked.

This was too much for Bulldozer and she leapt at his toes with her claws fully extended. With a howl of pain, the superintendent sprang out of his chair and hobbled around the room in pain. Peter appeared at the door, somewhat alarmed at the noise, so I requested the first aid kit and some clean water. We spent the next half hour applying first aid to his toes and he started to see the funny side of the incident and of course, another bottle of champagne helped the smoothing process!

He proved a valuable asset when the inroads of the Ivorian loggers became too blatant to ignore. He organised a night time operation to seize all their vehicles and equipment. The Ivorian company protested to no avail and though it reached a diplomatic level the company lost their machines for six weeks and had to pay a large fine and compensation to us. Our sawmill did very well out of the whole affair.

One night, or, perhaps I should say, one very early morning, the phone rang. Sleepily, I padded into the office to answer the call. It was the superintendent.

'Surprise! Are you awake?' he enquired which I thought was a rather silly question at the time. 'Prepare to meet your wives,' he chuckled. 'I am bringing them to you now. We will be there in half an hour.'

The call had come at 2 am. I wrapped a kente cloth around me and waited outside on the veranda. By 3 am I had convinced myself that the superintendent had been drinking and went back to bed.

I was gently shaken awake by Peter.

'Sah? Sah? Superintendent outside with ladies, sah. Please come.'

I jumped out of bed and started to make my way along the corridor.

'Excuse me sah,' Peter smiled and held out my kente cloth. 'Ladies might be embarrassed, sah!'

As I slept with no pyjamas during the hot tropical nights, the ladies, whoever they were, might well be shocked. The surprise might well be on them. I walked outside to greet my visitors being careful not to tread on any soldier ants as their bites were like red hot burning needles. Patiently waiting on the veranda were the police superintendent and a police sergeant and two other figures who were covered from head to toe in dust. They had both been crying and one of the apparitions ran up to me and hugged me like some lost cousin. Long lost cousin. That was probably the answer, I thought. I had been visited by a cousin before when I had lived in Zambia and they had been on a trans-African trip. I peered at the apparitions, carefully racking my brains, wondering who they were. As my family is very extensive, I was truly struggling.

'Who are you?' I tentatively enquired.

'I'm Barbara and this is Sarah!' shouted the figure who had given me the dusty hug. 'Thank God we found you.'

All was clear now. Both worked for the British High Commission and I had met them at the polo club. They had talked about doing a tour of the north around Tamali. They had been advised that there was some unrest in the north but they were keen to go so I had given them my contact details in case of any problems. They had been involved in an accident 50 miles north of Sunyani and their car was evidently a write off. They were very lucky not to have been injured.

The superintendent and sergeant sat in the lounge drinking champagne while the girls showered and cleaned themselves up. The dust had infiltrated their suitcases so I instructed Peter to give them some of my shorts and t-shirts. By 5 am we were all eating breakfast and washing it down with copious quantities of champagne. At 6 am the superintendent and sergeant left to resume their duties and Barbara and Sarah crashed into bed.

They slept all day and had only just woken when I came in from the mill that evening. They were amazed to find that all their clothes had been washed and ironed and their luggage and bags thoroughly cleaned. While they had been sleeping I had arranged for their wrecked car to be taken back to Accra and had also booked them on the early flight to Accra for the next morning.

They were worried that they had finished my champagne supply so I showed them two of the other guest rooms full of dozens of cases so they had no more qualms about drinking more and I think it fair to say that we almost poured them on to the flight the next morning. The superintendent was a little sad that I had not kept one of them as a wife!

By this time I had sorted out many of the mill's problems. We had stopped bush operations and were now buying logs at the gate transported by other bush operators for which we paid cash in the local currency. We concentrated on maintaining one production line cannibalising all the spare machines to keep it in full and proper operational order. We then sold sawn timber to buyers from Ivory Coast and the north and were paid in foreign currency. We sold nearly all the scrap vehicles but the snowplough remained. I wondered what happened to it eventually!

The staff and workers were now much happier. They were receiving regular pay and there was lots of work. Everything was looking a little more prosperous and we even repaired the mill roof. Success can bring sadness, however, as I was now to be moved on to another project. I felt very sad on my last day as I gazed out on the mill from the office window though I was looking forward to the big champagne farewell party that night for all my staff, workers and friends in Sunyani.

Just then my thoughts were interrupted by a taxi driving through the front gate. Apart from the driver there were four passengers three of whom were occupying the back seat. My attention was drawn immediately to the passenger in the middle at the back. He looked very drunk and was lolling about everywhere. His two companions kept on propping him up trying to support him.

'Matthew!' I called. 'Please go out and don't let that man in the middle of the backseat out of the taxi. He looks very drunk.'

Matthew rushed out. We had experienced drunks in the mill before. With all our machinery, it was very dangerous if they started to wander around. In a few minutes Matthew came back laughing.

'Please sah—he's not drunk. He's dead sah. They have come to buy timber for the coffin.'

'Perhaps he is going to choose his own wood!' I replied.

Sunyani was still full of surprises.

'We regret to announce the delay of Flight 318 to Kumasi and Sunyani. This is due to adverse weather conditions.'

Kotoka Airport Accra. A usual and familiar situation for me on a Tuesday morning awaiting the 7.30 am flight to Sunyani, where, no doubt, there would be the usual crop of problems waiting at the sawmill. Our vehicles and machinery could hardly be described as modern, nor for that matter, perhaps, could our staff, but, somehow we kept in spasmodic production and had actually started to make a profit in the last few months.

From the departure lounge, I could see the Hawker waiting patiently on the hot tarmac with the mechanics, technicians and security personnel lounging around it in whatever shade they could find. Inside the lounge my travelling companions a fair cross-section of Ghanaian life including cheerful market women, stern faced Hausas heading for the north, smart businessman and the inevitable lawyers involved in the interminable land disputes that seem so much part and parcel of West African life. Other companions, in the form of sheep, goats and chickens had already been loaded on to the aircraft leading me to recall one incident I had experienced with the Sunyani flight.

The Hawker normally was the scheduled aircraft for the flight, but, very occasionally an F28 was used if the Hawker was being maintained or serviced. Sunyani was famed for its geese. They were big, fat birds and highly suitable as Christmas presents for my friends, so, one year I had instructed the mill carpenter, Kwasi, to make a suitable crate for six geese, having first checked that: a) The Hawker was on the flight and b) that we had the correct dimensions for the cargo hatch. The live geese were carefully placed in the crate and we drove to Sunyani Airfield.

To my utter consternation, the plane landing was the F28. There had been a fault with the Hawker which was grounded. The ground staff valiantly tried to load the crate but it was too big. I was concerned because I was now delaying the plane, but, never one to give in without a fight I quickly offered to buy all the passengers a drink and then despatched my clerk back to the mill to bring the carpenter, Kwasi, back with his tools. He returned quite quickly and he sawed off 4" from the top of the crate and nailed the lid back on. So we all boarded the plane and were only ten minutes late for take-off. The passengers enjoyed their free drink and my friends enjoyed the geese.

COO!

My usual pattern of flying was to travel to Sunyani on a Tuesday morning and return on a Friday evening. This left me the whole weekend to enjoy with friends in Accra and the whole of Monday to deal with many other aspects of our businesses when my boss was away, and also hunt for spares. It was my habit to check in early on the Tuesday and then spend half an hour or so with our Accra clerk, Ignatius, setting out his programme for the week. He was extremely tall and very polite and efficient.

With the inevitable delay, we were able to devote much time to the week's programme and became completely oblivious to everyone and everything around us. The usual sounds of the airport faded into the background. There had been no coup attempts for over a year now and the new government seemed very keen to assist business and the timber industry in particular. Foreign buyers and investors were beginning to make overtures and the mill had already completed one export order. Things were looking good.

Suddenly, a hail of bullets spewed over our heads and the next few minutes became very confused as I pushed Ignatius to the floor besides the chairs whilst registering the screams and yells around us and along the concourse. Everyone was running for cover.

'Must be another coup,' I shouted at Ignatius as he lay on the floor.

He made no sound or movement and I really thought he had been shot so I punched him on the shoulder. He rolled over and looked at me. This had taken a few seconds and bullets were still flying around. I forced Ignatius to crawl along the line of chairs to a corridor which we sprinted along, crashed through a door marked NO EXIT and found ourselves at the edge of the car park. Our car, a very elderly blue American Ford LTD, was parked on the incline so we frantically jumped inside and I drove like a maniac back to the house which was very close in Airport Residence.

We were well used to coups by now and had adopted a fairly standard procedure in the several we had experienced already. The car was locked in the garage with the keys hidden somewhere safe, the front gates were padlocked, all the staff were warned, the dogs kept in the office, the drinks hidden and the documents put away in a secret safe.

Having completed all this, I was shaking like a leaf. The house was reasonably secure so I sat down with Ignatius to take stock. The silence seemed very ominous. We could hear no signs of sirens, shooting or mortar bombs. It was very eerie—none of the sounds of a military coup. By this time we had been joined by the cook, the garden boy, the laundry girl and a driver. They were all dependent on me for their safety.

We all jumped nervously when the phone rang. I nearly jumped out of my skin. No problem, I thought, it is probably the British High Commission phoning their nationals warning us to stay indoors. I picked up the receiver.

'Geoffrey Allen,' I said quite curtly to hide the shaking in my voice.

'Mr Allen,' purred a pleasant voice on the line. 'How are you this morning?'

'Fine,' I replied very cautiously. 'Gladys?'

Gladys was my favourite air hostess and was a regular crew member on the flights to Sunyani. She was always polite, helpful and cheerful but I was having some difficulty relating her present cheerfulness to the current situation.

'Gladys,' I whispered. 'where are you?'

'At the airport, of course,' she replied. 'Your flight is ready to depart so please come now.'

'But…but…but…Gladys,' I stammered, 'we have just come from the airport. There was shooting. Bullets were flying everywhere. We just managed to escape through a back door. We were jolly lucky. Are you alright? Are you safe?'

'Oh, that,' she laughed in reply. 'A foolish soldier dropped his gun with the safety catch off so when it hit the ground, the bullets went everywhere. There are big holes in the roof but it was lucky that no one was hurt.' She chuckled. 'See you in a couple of minutes!'

There was one amusing incident that happened at the sawmill that I have not covered.

When I first arrived, I did have some problems with the senior staff. Joseph, the manager, was rather hesitant about taking decisions; George, the electrician, was a strong union man who was quite belligerent at times; and Kwasi, the carpenter, could be downright cantankerous on occasion. There was no clocking in or clocking out system at the mill when I arrived and some of the hours claimed were wildly inaccurate leading to arguments about pay. I had to insist on a signing in and signing out book but George was very unhappy about this and persuaded Joseph that the workers should go on strike.

They marched into my office one morning and declared that they were going to hold wildcat strikes. I was very patient but finally requested that if they were going to strike, could they please inform me when they were actually on strike. When they asked me why I had made the request, I just told them that I could not tell the difference at present! This broke the ice and they laughed. There was no strike action.

Joseph was not a strong manager and was posted as a clerk to our Accra office where he did well but I was left with no manager. I knew who I wanted so I called George in one morning to discuss an electrical problem. He was grumpy and bad tempered so I led him on and told him he was fired. He became very annoyed and said he was going to the union rep and I told him that he could not. When he asked why I told him that senior staff members did not have a union. He looked puzzled and then I told him he was appointed manager. He was delighted.

Immediately I took over any company, I always cooperated with the unions by doubling labour law and always consulted with their reps. Even with our large work force in Liberia, some 1,000 workers, we never had labour problems.

Scenes from Ghana.

Chapter Eighteen
Ghana

I had regularly played polo in Ghana for three years, mostly at weekends. I had acquired two mounts—Mazda and Kalabuli—upon arrival in Accra. Mazda was a small bay pony and Kalabuli was a black and white piebald horse with an ugly head that Nour hated. He was called Kalabuli because the name was associated with trickery and crime and he did look a rogue! They gave me reasonable polo but these were the days when thoroughbreds had yet to be introduced from abroad in Accra. Nour bred his own horses and very fine they were too.

He had a famous race horse called Sandring which had won the Gold Cup on the track and that was later brought into the game. Polo at the club was very fast and furious and there were often great arguments and very occasionally fisticuffs off the field. Shall we say that the players devoted much passion to the game? Every player appeared to know the rules better than any other players and much better than the umpires. Few could play to the whistle and play could be quite ill-disciplined but the better players did excel in club and international matches and there were quite a few of these.

The biggest 'Needle' match was always Ghana versus Nigeria for the Nigerians were very well mounted and loved to display their expensive mounts. It was traditional for them to display all their sticks and tack before a game but the club also played against the UK, Canadian, Indian, Kenyan, Irish and South African sides. Prior to my arrival a Zambian side had also played at Accra and evidently experienced slight problems with the multi-national nature of the club.

Several ladies played as well and it was traditional that the first chukka of the New Year played at 8 am on 1st January, at the Captain's Breakfast, was ladies versus the gentlemen. There were never more than four ladies and they played against at least twenty gentlemen, but, it was traditional that the ladies should always win. I never recall the gentleman even scoring one goal. That may have had something to do with the fact that they had been seeing in the New Year at an assortment of nightclubs.

There was always much fun and high-jinx in chukkas and club matches. In one match, a very good friend, Jose Beckley, who was an excellent polo player, suddenly found the 'Captan Team', his opponents, deliberately passing the ball to him and it freaked him out and he lost the game. In another match I took a water pistol out on to the field and squirted it at Nour every time he went to hit the ball. The Captans retaliated the next week. At the 'throw-in' to start the game, they all had water pistols and gave me a thorough soaking. I was drenched but it was very refreshing in all that heat!

We had one very large American friend who rode a very large grey horse called Colombo. Like Jimmy Edwards, he fell off and we could hear the thump as he hit the ground. Richard, for that was his name, had the habit of snacking and fridge-raiding in the middle of the night about which his wife, Peggy, constantly complained, so I came across the ideal attachment for his fridge when I was in London. The next time Richard sneaked down to his fridge and opened the door—he had a shock!

'CLOSE THE DOOR, FATTY. YOU'RE NOT EATING AGAIN!' screamed a very loud voice. Peggy and the children were amused. Richard was not!

I made many friends with whom I am still in contact for everyone was very hospitable and I maintained a hectic social life, especially at weekends. My friends reckoned that

I was like a cowboy coming off the ranch when I came down from Sunyani on a Friday flight but everyone looked forward to my arrival as I had a never ending supply of Champagne. We would visit La Reve or Le Cave nightclubs and party all night. One famous New Year, I had donned a kilt because I was dining with Scottish friends then turned up at La Reve and did a Highland Fling on one of the tables.

What does an Englishman wearing a Scottish outfit wear under his kilt? Amid cheers and catcalls, I...S-L-O-W-L-Y lifted my kilt higher and higher to reveal...a pair of Union Jack underpants! I was booed and had beer thrown at me. That was the night that Nour was feeling really wicked and put a dressed lobster into the Spanish Ambassador's wife's handbag without her noticing until she retired to the ladies. There was one almighty scream and we all cheered again.

Driving back from the club we started off in convoy because we wanted to do more serious damage to the champagne at Captan House. Jose Beckley was driving his Mini in front of me and he suddenly disappeared—the Mini just vanished. I stopped and everyone else stopped. Where was he? He had taken his eyes off the road and driven straight down into one of the enormous storm drains at the side of the road. He and the Mini were stuck.

We got him out but had to leave the car until the next morning. The convoy drove on to Independence Avenue where the roundabouts were of peculiar construction. They had roads going straight through them for the sole use of the President and visiting dignitaries but it was another tradition that we also drove straight through as well during the early hours of the morning.

The Captan family was large and initially, I only met the Ghanaian branch. It was even bigger in Lebanon. Nour would very often introduce me to another member who was a distant relative four times removed and the relative's mother was also married to a second uncle who...well. There was a lot of them. I met Feisil, Muhammed and Ahmed. Munir I already knew and then he married Anita Kubchandani who lived in Glamour House across the road from the Captans and they had two children, Kareem and Lara.

Her parents were Kichen and Rosy and they had two sons, Jimmy and Ricky and they all had children. Sammy Captan was often in the house and was a chef in New York. He was quite a character and had worked in Liberia and walked back to Ghana! He had walked around the long way by exiting Liberia through Grand Jeddah in the north-east then walking along the northern border of Cote d'Ivoire and down through the north of Ghana to Kumasi then on to Accra. It was quite an amazing feat and a tough walk in the 50s, I think. Then relatives would often be visiting from Lebanon and Nour's brother, Eid, would also briefly pay a visit.

He was a great lover of ballet and opera and a massive intellectual on many subjects. He was very fastidious, however, and considered Ghana to be very unhygienic and dirty. He would spend hundreds of pounds on roses for opera singers in London and nearly swoon if he met ballet personalities anywhere. He was not entirely sure about me. I had no real interest in opera but I could hold my own in the sphere of English Literature and he was quite surprised at my knowledge.

I quite liked talking to Nour's parents as well. His father was quiet and extremely courteous and his mother was quite a character whom I was to see several times in London later. Visitors were a regular occurrence and most of the polo players, business acquaintances and other friends would be constantly popping in and out. It was a lively and wonderful house and there was barely a dull moment especially when Nour had foreign polo players staying with him and then it could become a madhouse!

Nour had many Canadian friends. He had played with a Ghanaian team there several times and the Canadians loved to visit Accra because of the superb hospitality and social life. They became regulars. Ross Farge was a particular friend and he and Nour were great buddies. Strange to relate I did not meet him until I was visiting Ghana many years later though I had bought two of his polo ponies in the UK which he had left with a friend. Ross and Muhammed were incorrigible. They were both avid beer drinkers and they were famous for their drinking antics. When the bar closed at the polo club they would get their grooms to saddle up two horses, ride up to the airport, get a policeman to hold their mounts and continue drinking in the airport bar until they were thrown out of there.

Food supplies in Ghana were not of the best though one could live perfectly adequately.

We often drove up to Lome to buy essentials and luxury foods and it sometimes was required of me that I do the run. Alain was very friendly with Colonel Sam at the border so once I was introduced I did not worry about visas and re-entry permits. One night I came back through the border and Colonel Sam was not there. I was refused re-entry by a lady border guard and she would not accept a dash to let me in.

'Sister,' I said. 'You are refusing me entry into Ghana where there is no toilet paper, no potatoes, no rice, no wine, no sugar, no electricity and no water and you want me to go back into Lome where there are wonderful French restaurants and wine and cold beer in splendid hotels? Oh! Sister. Thank you. Thank you. Thank you!' and I gave her a big kiss.

'You are a foolish man!' she responded. 'Drive on!' and she opened the barrier to let me pass.

I did the same trip some weeks later and Colonel Sam was in attendance. He ushered me into his office and asked if I wanted a beer which was odd. I normally had to buy the drinks. Just then, however, the barriers went up on both sides of the border and six cocoa lorries roared through with no checks. Ghana was in the position then of being the largest producer of cocoa in the world but the second largest exporter because there was so much cocoa being smuggled out. There was regular smuggling going on all the time I was in Sunyani into Cote d'Ivoire.

In truth, there were many border posts in Ghana but it was quite easy to walk over the border in the bush and I used the same method in Liberia quite often. The Lome Road extended through Benin to Nigeria and Nour had used it in the past but there were many roadblocks with the police, customs, para-military and army, all of which demanded a dash to continue.

On the last barrier, near Lagos, a policeman was all smiles and welcomed them to Nigeria and hoped that the polo match would go well and there was no demand for cash. Just as Nour's driver was about to drive off the policeman tapped on the roof and asked, 'Nothing for being nice?'

I did play one polo match at Ibadan in Nigeria but we flew to Lagos. Landing late at night we were met by a group of security guards who escorted us along the Airport Road, where there had been many robberies, to the house of Vassilis where we were staying and guards stayed with us for most of the trip. We were playing at Ibadan and the drivers were told not to stop for anyone going up and going back.

In spite of all this, it was a pleasant weekend's polo. Flying back into Accra on the Monday morning, however, it seemed that we were coming back into a sleepy village after the hustle and bustle of Lagos. The change in atmosphere was quite dramatic. But drama was in the air!

I was up at the mill when The Bank of Ghana devalued the local currency, the cedi. It stood at £1 to one cedi then was devalued to 1.2 cedis to the pound. We were all horrified but we should not have worried because it moved on to 1,500 cedis to the pound some years later. In spite of this the mill had been doing well, and I had been building up funds steadily so that we could cover the payroll and fuel and electricity bills on time. But Alain kept on drawing on the money for Accra expenses and leaving me with insufficient funds, so I opened a workers' account and always ensured we had enough in Sunyani before any went to Accra.

Some mention must be made of the political climate, and to simplify the situation, it is perhaps easier to say that it was unstable, prior to my arrival and later while I was there. In 1960 Kwame Nkrumah was the Head of State leading Ghana into independence. In 1966 he was deposed by Major General Joseph Arthur Ankrah, Commander of the Armed Forces who was responsible for saving the life of Patrice Lumumba in the Congo in 1961.

He, in turn, was deposed in 1969 by General Akwaasi Afrifa, who was deposed by Akufo who did hand over to Busia, but, in turn was deposed by Colonel Ignatius Acheampong. He was in power when I first visited Ghana and it is said that he was the best of the leaders and did much for the people. I know Alain had met him several times but I had not. From this point on, however, I did meet and knew every Ghanaian president until the present incumbent Nan Akufo Addo in varying circumstances and in different locations.

On 5 May 1979, Flight Lieutenant John Jerry Rawlings came to power in his first coup. I was woken by gun fire and Nour phoned me to let me know what was happening as I was alone in the house apart from the houseboy/cook and two English Collies which I kept inside. We stayed in touch by phone, until eventually, a group of four soldiers came in through the front gates and amazingly a boy Scout. All were drunk, high on wacky baccy and carrying guns even the boy Scout.

Our front entrance was a large patio glass sliding door and one of the soldiers shouted to me to open it while another one shouted, 'HANDS UP!'

I looked at them and shouted back, 'YOU WANT ME TO OPEN THE DOOR OR PUT MY HANDS UP? I CANNOT DO BOTH!'

It was a ludicrous predicament. I opened the door. They were very drunk and almost totally insensible. The dogs were barking in the office and I told the soldiers to hang on while I locked them in a cupboard. That done they muttered something unintelligible about having to search all houses but they were out to loot. I merely stated that I was a British visitor and obviously not Lebanese. All they got was some cheap wine and Albert's air rifle and they left. I phoned Nour but he was in serious trouble.

The soldiers were picking on the Lebanese and they were in his house at that moment. A soldier answered the phone. I told him that I was from the British High Commission and was checking to see if the Captan family was alright! By this time the drunken soldiers had taken Nour and Munir up on to the front terrace and were threatening to shoot them. It was a very unpleasant situation indeed. The soldiers were paid off, which was probably what they were after in the first place, but Nour and Munir were taken in for questioning at Burma Camp (Army headquarters) and released later.

I risked driving around that evening and was stopped once by soldiers but I waved my British passport, said I was from the High Commission again, and drove on. It did occur to me that night, as I listened to the World Service then tried to get to sleep, that Alain had not tried to phone me which I thought was a little strange. You could not get direct international lines out of Ghana in those days. We all had our own contacts in the telephone exchange whom we paid an extra salary to put through our calls.

Next morning I phoned ours to see if she could get a call through to Alain, which she managed within the hour. Alain seemed strangely out of touch with events and I gave him a brief resume of the coup. He seemed a little irritated that I had let the soldiers into the house and I did ask him what he would have done with four guns aimed at him by drunken soldiers. He was worried, however, by the fact that there was a very real harassment of the Lebanese community and many of them had been taken to Burma Camp. Their goods and cars were being appropriated and impounded. Many were heading for the Lome border.

While I was talking to Alain, there was firing in the background and we could hear the sound of loud explosions as the mortar regiment from Ho was pounding part of the city but Alain still did not seem to understand the enormity of the situation. I asked him to phone me the next morning and perhaps things would be a little clearer. The next morning a party of soldiers came in through the gate and were looking for Alain. They were not drunk and they had been specifically ordered to take Alain down to Burma Camp for questioning. I carefully explained that he was in the UK and obviously, would not be back for some time.

I invited them to search the house and showed them around. There was nothing to hide and they left. That afternoon an army captain arrived with another soldier and requested to search the office which was a waste of time because we kept all important documents elsewhere but then they turned their attention to me and I explained who I was and what I was doing, namely running a sawmill in Sunyani. This was backed up by Benson the cook and George the houseboy.

There was bad news on the third day, Rawlings had executed three former Heads of State and five high-ranking military officers by firing squad on Labadi Beach which he later much regretted but regarded it as an act of necessity during the coup. I was concerned now, however, because there had been isolated incidents where employees of the Lebanese were now being sought out and I did not want to be one of them. Nour told me to get out as quickly as I could. I locked one car in its garage and told the driver to hide the Ford LTD, packed and was on a British Caledonian flight out that night.

As I was boarding the flight, a soldier seemed to recognise me and started to hold me back but I just ran up the steps and into the aircraft. The flight crew would not allow him on board. Well, he didn't have a boarding pass, did he? I was very grateful for their actions. The next morning I arrived at Alain's house at Eastbourne but, again, he seemed slightly critical that I had flown out.

Within two days, however, there was a warrant for his arrest, his Ghanaian staff had been rounded up and held for two days and all his cars located and impounded. The sawmill had also been sealed off. I must now admit to all that I had a wonderful summer because the Accra Polo Club came to London. Everyone was fleeing the coup. Some had been held under arrest and fled by any means possible after their release, others had simply travelled out with their families and some of them had been smuggled out. Nour got out on a United Nations flight via New York and I was very pleased to see him.

I estimate that there some twenty members of the club in London that summer and we made whoopee. I was still working with Alain with his British clients mainly in London and I met up with politicians, businessmen and investors. Most of our business was conducted at the London Hilton where I spent my mornings before going to join my friends, though I did visit Germany and France several times as well as a trip to Switzerland to help collect Alain's new Mercedes, to which he had fitted a gold Mercedes badge on the bonnet. It was stolen the first night we took delivery of the car. We met a business associate who resided in the Hotel Bristol in Geneva and who had whisky and peppers every day for his breakfast (and who did me an enormous favour later in my career) and we visited the offices of Krupp Stahl Export in Dusseldorf, I think, who gave me an unusual assignment which I later completed with full marks and a hefty bonus.

I think I did more clubbing and enjoyed more night-life that summer than I have ever completed since. The Accra Polo Club hit London in a big way and I am ashamed to say that some fine establishments in Central London thought our celebrations for escaping the coup were a little rowdy. We were asked to leave several such locations and we were even evicted from a Greek restaurant and a Chinese take-away.

I experienced several unusual business trips during this period. Alain had somehow become interested in the WDA (Welsh Development Agency) who were offering many grants and incentives to start up industry in Wales to encourage investors and I had to make three visits to the Isle of Man to discuss same with a financial advisor through whom Alain, I think, had mortgaged the Eastbourne house. I had visited the island for a Scouting trip before, so knew my way around and was the recipient of good hospitality.

On my third visit, I was asked to be a judge for the Miss Isle of Man Beauty Competition in Douglas and our first choice was fifth in the Miss World Competition. It was a tough job but somebody had to do it! Twice I had to do an extended road trip to Paris then on to Dusseldorf before returning to England via Belgium.

We were keeping a close eye on events in Ghana all the while. My friend Jose Beckley was one of the first to return and I phoned him nearly every day. He had friends in the army and was making low level enquiries for me. He stated that he could help and I should return so I flew back into Ghana in mid-August. Jose had made contact through an army captain and I was escorted down to Burma Camp to see him. Surprisingly our cars were parked near his office.

I was not entirely sure of the situation and merely explained that Alain was anxious to return to Ghana but needed to know about the warrant for his arrest and the situation about his property and sawmill. All papers relating to Alain were on his desk including the documents for the cars. The captain assured me that the whole problem could be cleared up very quickly so I went outside to discuss things with Jose.

'How much does he want?' I asked.

'Five thousand dollars!' Jose replied. 'How much do you want to offer?'

'The pickup?' I asked.

The Peugeot pickup was good and it was worth that amount. Jose went back inside and came out almost immediately. 'Deal done and completed,' he laughed and we drove away with the cars and documents.

The documents I burned immediately and both cars were checked over and in quite good condition. I phoned Alain the next morning and he was astounded that all had been completed but I had to tell him that we had lost the pickup. For once, he did not seem to mind.

Accra gradually returned to normal and there were a few problems with soldiers but nothing to be concerned about except the curfew at 7 pm which made a big hole in my social life for a while until it was stretched to 11 pm. I was travelling up and down to the mill as usual and then working on Mondays in Accra. Alain was spending more time in England and I was dealing with his clients in Ghana. Import permits were now even more difficult to obtain. There is a very old steam engine wreck outside what was to be a Museum for Science and Engineering in the centre of Accra.

One of the clients asked what it was and I informed him that was the last import licence that had been issued! We had also spent some time helping ships into Lagos, where the unloading situation was dire. Ships that could not afford the premium docking fees could be off-shore waiting for two or three months. I flew out to two or three of them and arranged for the ship to be unloaded for the value of ten percent of the cargo and we would cover the docking fee.

This was working well until one Greek ship unloaded and ran. I shadowed the ship along the coast and from a light aircraft I photographed the crew manning the lifeboats with all their possessions, including record players and food then scuttling the ship. Alain approached the insurers and we received a very handsome profit from the whole affair.

I have mentioned that I was given an assignment by Krupps in Germany while I was in Dusseldorf. They had formulated a Feasibility Study for the Rejuvenation of the Steel Works in Tema which consisted of two large volumes and they wanted it presented to the President, or failing that, the appropriate minister. I could not do this until Jerry Rawlings handed over the reins to Hilla Linman not long after the Coup late in 1979.

I had a contact, and an appointment was arranged with him at The Castle, which was then the Presidential Office. The President had a British lady secretary that we all nicknamed Moneypenny and she duly ushered me into his office. President Linman was a gentlemen, as I have found all his successors to be, and bade me welcome. I wished him a good morning and was about to launch into my spiel when he asked if I would like some tea. I accepted and he moved to the phone.

'Tell me, Geoffrey, is it true that you all call my secretary, Moneypenny?' he asked.

'Yes. I am afraid it is, Your Excellency.'

'Ah! Moneypenny, could you please bring in a tray of tea and some ginger biscuits? Thank you so much.'

He was interested in the operation of the sawmill and had much concern for the poor state of the timber industry in Ghana. The tea was duly brought in and he said it would be alright if I wanted to dunk my biscuits and he did the same. He was interested in my opinions on Ghana and when we had finished our tea he asked my business. Rather pompously I said that I had the pleasure to present to him a Feasibility Study for the Rejuvenation of the Steel Industry in Tema from Krupps in Germany and put the volumes before him. He stood up and walked over to his enormous bookcase and slid the doors back. Floor to ceiling the shelves were packed with Feasibility Studies.

'Might I politely suggest that it be added to your collection, Your Excellency?'

He thanked me and I was ushered out saying a big thank you to Moneypenny as I passed. I was thus able to assure Krupps that I had delivered the Feasibility Study directly into the President's hands and absolutely guarantee that they were in his office right in front of him for which I was paid a fee of $2,000. Thank you!

Nour was a very good friend to me but it was only when he was absolutely convinced of my integrity that he started to advise and warn me that all was not what it seemed with Alain Khoury. It was hinted that he had tricked several members of the Lebanese Community on business deals and very slowly, I was beginning to discover flaws and anachronisms with some of his deals.

When I had been concentrating on the mill in Sunyani, Alain had been going about his business deals alone and I was not too happy with some of his associates. It seems that many had mentioned nothing to me but I could see that he was having financial problems and it was all beginning to snowball. I became aware that he was behind on school fees, owed an enormous amount to American Express and had borrowed large amounts against deals that had not materialised in London and Germany when I was with him in the summer and that he was keen to get back to Ghana. He also owed his gardener and his builder.

On the evening of my departure back to Ghana, after the coup, my builder friend stated that Alain had showed him some small gold ingots from Ghana. I was now very concerned. Back in Accra Alain had become involved in the sale of twenty ex NATO Army Trucks that were now en-route to the West Coast, on board a ship bound for Nigeria, but the deal fell through and they were off-loaded in Liberia sometime in late September and there they stuck. No one really wanted them and they were difficult to adapt for logging anyway.

I shuttled back and forth to Monrovia visiting prospective clients but it was going to be a hard sell though I did make several contacts while I was there which stood me in good stead later. I returned home for Christmas and Alain was back as well but we now seemed to be dealing with a new group of businessmen in central London involving a Dane called Paul Kring and an American Ben Barone. Suddenly Alain was talking about making million dollar loans to Sierra Leone and Liberia. It did not add up. Alain told me to concentrate on the Accra side of the business, that is to say, steel rods, rice and car tyres so I spent a lot of time at ministries and I did manage to get two permits for the importation of rice.

In early April, Alain asked me to visit Liberia to have another try at selling the trucks. I stayed with the accountant of Contimba, one of the companies I had made contact with before, whose house happened to be nestling right up under the wall of the Executive Mansion. I spent several days touting the trunks around the town but to no avail. On 11th April, the accountant and I turned in early. That was the night of Sammy Doe's coup and we were part of the action.

We were awoken just before midnight with the sound of explosions and automatic gunfire. It was very loud, very close and very frightening. I suggested that we move downstairs and we spent the next two hours under the dining room table. There came a hammering on the front door which I cautiously opened and soldiers burst into the house with a machine gun which they mounted on the veranda and started firing into the grounds of the executive mansion. Fire was returned from the mansion and all the upper windows were shot out.

One of the soldiers did come down to check if we were OK, however, and we stayed under the table. By first light there was only sporadic gunfire and to our amazement, an armoured car pulled into the front drive of the house. The soldiers told us to quickly pack some things, I had my suitcase anyway, and we jumped into the armoured car and it drove us up to the general manager's house in Sinkor. They had been worried about us and were glad that we were safe and unharmed. I did make time to thank the soldiers for their help and I realised, later, that they were under strict orders not to harm expatriates, visitors, foreign nationals or the Lebanese and I knew the reason why.

During my visits to Monrovia, I had been introduced to Doctor Kassas who was part of the Lebanese Community and who had a surgery in the city centre. He had raised Sammy Doe from boyhood, cared for him, educated him and helped him in every way and Sammy Doe was very grateful hence the order to respect all non-Liberian Nationals. Doctor Kassas had advised me carefully during my various trips and I respected his opinion and suggestions. He was a very valuable and highly regarded contact...and a good doctor!

The general manager's house was very small so I moved out to Paynesville, to the house of Victor Hannig and his family. I welcomed the move because Victor was a brilliant host and also had a swimming pool. We had met some months before when Alain and I had visited several logging companies in Liberia with a view to selling them trucks and according to Alain, to look at investment potential in the logging trade.

A couple of days after the coup, I accompanied Victor to the Contimba Office in Paynesville. The first thing we saw was the huge safe in the middle of the road. It stood six feet tall and was incredibly heavy. The soldiers had, somehow, managed to get it into the street and then had tried to shoot the door open. One soldier had been wounded by a ricochet we were informed. The really amusing fact was that Contimba had never used the safe and were always intending to get rid of it. The office was relatively untouched.

During that day, the radio stations kept announcing that an invasion force had landed in the harbour area and we did wonder what would happen next until Victor suddenly realised that it was not an invasion force, it was our trucks! I did not want Contimba involved so I got a taxi into Monrovia to see Dr Kassas. We were stopped at several road blocks but as soon as I mentioned I was going to see the doctor the soldiers immediately let me through. The doctor's office was very crowded so I slipped him a note with the word URGENT on it. He cleared his office and I explained about the 'Invasion Force' mix up.

He was not sure what to do initially but when I explained that the trucks were insured against all contingencies including war and riot, he caught my drift and smiled. The trucks were seized by the army and the insurance paid out. I had finally got rid of the trucks. There was nothing I could do in Liberia so I flew out on the first available flight. British Caledonian must be thanked again for flying me out of yet another coup. There were still problems on the Airport Road out to Robertsfield but I bypassed all that by chartering a light aircraft from Paynesville to Robertsfield and had no problems.

I hoped that I would not have to call on British Caledonian again to fly me out of a trouble zone. My nerves were still somewhat of a jangle, however, because when a car backfired on the Downend Road outside my parent's house I dived to the pavement! At least I could look my father and uncles in the face and state, that, I too had been under fire!

Two weeks later, I headed out to Ghana again to meet a glum Nour. There had been an outbreak of horse sickness in the stables and many horses had died including several of Nour's. It was all very sad. I had lost Mazda, but what really annoyed Nour was the fact, that, every day he visited the stables, there would be Kalibuli, alive and well, looking out at Nour from his box. Kalibuli was a survivor. I eventually sold him to a riding establishment at Tarkwa and I know he lived on for many years as a ceremonial horse for one of the local chiefs. I think he had the last laugh on all of us.

On Alain's instructions, I made ready to go over to Liberia again but was again surprised when I was asked to arrange a ticket and visa for our clerk, Martin, to visit the UK, before I left. As my laptop sometimes flashes at me, 'something is not right'; and I questioned Martin carefully. I had treated him well in the last few years. I asked him directly if he had been asked to carry gold. He vehemently denied it but I could not think of any other reason for his UK trip and warned him of the consequences to him and his family.

I realised that I could be getting out of my depth with all this and the next week was to prove me right but I boarded a Pan Am flight to Monrovia the following evening and Contimba had requested that I bring over some shipping documents from one of their shipping agents which I had in my briefcase. I was seated right at the front of the aircraft and the overhead lockers had been filled by the excessive hand baggage of other travellers.

The air hostess insisted that my briefcase would have to go into the hold which I declined explaining that there were valuable shipping documents and the chief steward also insisted that the case must go into the hold. Now, I played polo with the Accra Manager of Pan Am and I always joked with him that his cabin crew must have been trained by the Gestapo because they were invariably abrupt, rude and terse. I was warned that if I did not comply I would have to leave the aircraft to which I retorted that it would give me great pleasure to do so giving me the opportunity to fly with an airline with polite and pleasant cabin crews. They reopened the doors and my friend came in. I let the cabin crew explain and I did not say a word. There was a discussion in the galley and the chief steward escorted me forward to first class for the half-hour flight to Robertsfield.

I was anxious not to involve Contimba in any of Alain's business at present so I stayed in the hotel at Mamba Point for several days, situated only a little distance from Salvatore's Restaurant which served excellent food, especially the surf and turf. Alain contacted me to meet him at Robertsfield in two days' time also requesting that I make an appointment through Doctor Kassas to meet the President Sammy Doe. I discussed this with Dr Kassas, very carefully voicing my concerns about the visit. He suggested that we go ahead but to proceed with caution and he agreed to meet with Alain, before we met Sammy Doe, to warn him against any ill-advised offers or actions.

Two days later, I was out at Robertsfield with a hired car. I checked the arrivals board. No scheduled flights were due for two or three hours. A little while later a private Lear jet landed and out stepped Alain, Paul Kring and Ben Barone. They had chartered the Lear jet out of Stanstead and had flown to Sierra Leone the day before to meet the President, Siaka Stevens, who was infamous for his corruption and exploitation and offered multi-dollar loans to his Government. They were a mite surprised that we were due to see Dr Kassas first until I explained that Sammy always went by the doctor's advice.

We had a pleasant half hour with him then moved on to the Executive Mansion and into a lift leading to the fifth floor and the President's office. Dr Kassas was with us and Alain explained that they were offering multi-million pound loans to Liberia on soft terms and subject to legal documentation, they could be arranged in a matter of weeks. Now, we were all aware that Sammy Doe needed such funds, as Liberia was much impoverished, and development was a priority and after the meeting, Alain thought all had gone well.

He was very happy as I drove them back to Robertsfield and they flew off in the Lear jet back to England. I returned, by his request, immediately to Doctor Kassas, who asked me outright, should Sammy Doe accept the loans, also pointing out that any future aspirations I had in Liberia would depend on my answer. I gave an emphatic NO. I was worried about the source of the money, its terms and conditions and the characters involved in the offer.

'That's good,' said Dr Kassas, 'because I have already told Sammy not to accept the money!'

He had not trusted the trio at all. He then asked me about my plans and I merely stated that I would return to the UK where I was safe and under British Law, resign, then consider my options from there on. I stayed on for a day or two and joined Victor and his family on the beach for one of them, then flew straight back to Heathrow where I found two Special Branch officers waiting for me.

They escorted me to an interview room and asked me about my activities during the last week, then about Alain's activities and that of his family. They informed me that our clerk, Martin, had been arrested the previous week for gold smuggling, at the airport and was now on remand at Wandsworth Prison. They also informed me that Interpol had a careful watch on Paul Kring who they suspected of laundering Mafia money but they did not mention Ben Barone which I thought was a little odd.

Colonel Ben Barone was in the USAAF and had a very distinguished career in the war and after and he had held various senior appointments including an involvement with military intelligence. Somehow he did not fit into the picture and I just wondered if he was not still in active military service intelligence? They had a full list of Alain's deals, double dealings and debts that astonished even me. He had obtained all his cars by fraud, and the house in Eastbourne, which he had mortgaged to the hilt anyway. He had deceived suppliers into advancing him money throughout Europe, obtained loans under false pretences and was a gold smuggler.

He owed over £23,000 to the Hilton and he had not paid the £72,000 for the charter of the Lear jet. He was a rogue and con man! He had always been kind, pleasant and affable to me. He had generally treated me well and I had always been paid my salary and expenses. Thank God, however, that I had not been directly involved in his nefarious activities. My hands were clean. I assured the Special Branch officers that I would give them all the cooperation they required in the course of their enquiries but they seemed to know everything there was to know about Alain and he never came back into England again.

I sent him a very simple letter of resignation and he responded with a lengthy telegram that could well have been a contender for the Guinness Book of Records, threatening me with legal action to which I responded that I would welcome such an action when he was next in England. I phoned all the contacts with which I had been dealing for Alain and warned them about any further contacts with him and explaining my resignation. The world was my oyster still, as I could return to Ghana or Liberia or Zambia or stay in the UK.

I had an urgent phone call from Contimba, based in Lemgo, near Hanover. Could I pop over for a discussion about Liberia? I accepted and flew over the next day. I understand that Victor had suggested that I was just the man to spearhead their investment into a sawmill in Liberia, which was very kind of him. Would I be interested? They were very honest because they knew of my association with Dr Kassas and Sammy Doe and considered that a valuable asset in their new venture.

Rudi Bunter and I agreed to the very generous payment package, far higher than anything I had received prior to this employment, and I would work five months on and one month off, which suited me fine. I was thus appointed general manager of the Libco Timber Company.

Chapter Nineteen
Liberia

In the Cause of the People the Struggle Continues…

The very first time I landed in Monrovia, I hired a taxi to take me into the town and we loaded my luggage. The taxi driver turned around to me and said, 'E ma'—gota sto' n ge' gah—ya er?'

'Pardon?' I queried and he repeated the comment. I repeated that I could not understand.

'Ga! Ga!' he shouted. 'Gota ge' ga. Yu don sek engli?'

I twigged. He wanted to stop and get gas (petrol) for the taxi. And he was also accusing me of not being able to speak English. Thus was my introduction to the Liberian patois. It really was quite easy. All you had to do was leave off the end of each word.

It is absolutely vital that one should be aware of the founding of Liberia, especially as its essential character was moulded by its manner of inception but let it be clearly understood that the Krahn peoples had inhabited the region prior to the creation of the state long before it was called Liberia. In 1821, at the commencement of the Abolition of Slavery, the country was founded by freed slaves from America, and from that time throughout that century, particularly in the 1840s and 1850s thousands were arriving until, by 1870, it was estimated that there were 30,000 freed slaves settling in and around the main city—Monrovia.

These were largely Americo-Liberians (Conga) who were obviously more sophisticated and worldly wise than the indigenous Krahn peoples who were quickly marginalised even though they formed 90% of the population, and the Conga quickly set up the True Whig Party in 1877 which ruled Liberia for just over one hundred years as a 'One Part State' with a somewhat distorted form of democracy. Government was under the control of the big families such as the Roberts, Coopers, Bensons, Talbots, Warners and Gardiners. Very few of the Krahn ever held governmental office and this had to a long simmering resentment. Sammy Doe's coup was the result of this deep seated resentment, for the Krahn wanted their rights and fair share of the wealth of Liberia. During the coup, many of the big families fled.

The name of Liberia is also associated with the registration of foreign ships under its Flag of Convenience. This was a lucrative form of income for Liberia and it is second only to Panama for its popularity. Basically it was cheaper, easier and more convenient to register with Liberia, the ship owners were faced with fewer regulations and their sailors could be lower paid and work longer hours. Ships could register even if they had never visited Liberia. It was very profitable.

When I returned to Liberia, the 17 main coup participants had formed the PRC—the Peoples' Redemption Council under the slogan—'In the Cause of the People—the Struggle Continues,' and its chairman was Sammy Doe. Its members, strongly believing in their cause, very often acted unilaterally with no joint decision making and very often, handed out arbitrary punishments and beatings. There was still a night curfew from 11 pm until 6 am and the town was full of road blocks and checkpoints manned by soldiers. They were, however, operating the strict policy of non-interference with the Lebanese or foreign nationals. They were polite, courteous and relatively efficient.

All the soldiers were quite nervous because they were wary of an invasion of foreign troops either from other African countries or America. This was mainly rumour. Their main cause of worry was a counter-coup from dissenters in their ranks. There was also friction between the army and police which often led to disturbances. Spasmodic firing would occasionally break out because some troops had the jitters and one would often see Jeeps and army vehicles dashing around the town.

I did have a near miss. All over Africa and in the Near and the Far East in any war zone the most common armed vehicle was a pickup with a machine gun mounted on the back cargo floor. I was popping down for a haircut at my Lebanese barber's but as I turned the corner from Randall Street to visit his shop, one such pickup approached at speed with the 'gunner' clutching the two hand grips of the machine gun and with his feet apart and braced on the floor.

The pickup spun around the corner at speed and the gunner lost his footing and was spun around at the same time gripping the firing button and a volley of bullets shot over our heads smashing signs and windows in a hail of bullets. The pickup stopped and a very heated argument ensued between the gunner and driver, which at any other time I would have considered very amusing but I was shaking like a leaf and had to steady myself in a doorway.

The bullets had passed barely two feet above my head. I told my barber that 'It was a close shave!' Some ten years later I was in Accra, in Nour's office and his barber walked in to give him a haircut. It was my barber from Monrovia and we recognised each other immediately. The Civil War had been too close a shave for him as well.

I was based at the Contimba Office in Sinkor but was forced to play a cat-and-mouse game with the outgoing General Manager who was leaving in one month and was quite upset that I had been appointed in his place but the nature of the job had changed and would now involve working in difficult bush conditions and forceful negotiating. Contimba had agreed in principle to take over Libco Timber Ltd which was a large mill with forestry concessions, South of Zwedru, in Grand Gedah County, in the far North-east of Liberia and this was to be achieved as soon as possible.

I sat down with Victor on my first night, for I was staying at his house (with the swimming pool) and we set out a rough timetable. The absolute priority was reaching an agreement with the creditors because there were debts of over one million dollars spread over about twenty companies and banks and there was no way we would be willing to take that amount on. Head Office suggested payments in full settlement at between 30% and 35% to be paid in instalments over one year but I wanted to go in at 20%, and Victor and I then agreed to pick off the order of approach.

The more creditors who agreed initially, the better would be our chances with the more stubborn ones. At the same time, we would visit the Forestry Development Authority (FDA) to sanction our takeover. The latter visit was much easier than I thought. The Director of the FDA had travelled so we saw his deputy who was more than reasonable. Libco only owed current forestry taxes which had not yet been invoiced and I was told to pick up a letter of permission to proceed that evening. Here I must admit to the use of a little chicanery which may, or may not, have influenced the deputy's co-operation. He had asked where I was to be living and I honestly replied that I would need a house but had not started looking yet. He owned four flats at Paynesville and the ground floor flat next to his was vacant so we agreed to meet him there to have a drink, look at the flat and receive the letter.

The general manager insisted that he join our negotiations and our first day was a disaster. He was quite rude and abrupt with the creditors, talking them down and insisting that they accept the deal. Victor and I visited Emmanuel Emeh, the Deputy Director of FDA, on our way back to his house, were given the required letter, shown a very nice flat, haggled over the price of its rent, agreed terms and finished off with a cold beer. A good evening but a bad day.

I phoned Rudi Bunter, my boss in Lemgo, Germany, that night and told him that we could not proceed with negotiations if the general manager was present. He would blow the whole project. Victor took over the phone and talked for a full fifteen minutes and as it was in German, I did not understand much.

On instructions from head office the general manager was to take over Victor's work for a couple of weeks while he and I conducted negotiations, planned the take over and completed all the legal work. That suited me fine. At time of writing Britain's Brexit negotiations are proceeding, or not, in the full spotlight of the national and international press, under the scrutiny of Parliament, in the face of public criticisms and the negotiations are analysed and dissected in the smallest details. It is plain daft. No one can be expected to negotiate in such conditions.

Victor and I became a double act. I would go in hard and play immovable and stubborn and Victor would be conciliatory and allegedly on the client's side. Different creditors required different methods. I would sometimes walk out of an office and Victor would stay and be sympathetic stating that he could probably get me to raise my offer a little. Victor walked into one office alone when we were dealing with an ultra-difficult client who wanted to know where I was. Victor informed him that I was on my way to Robertsfield to catch a flight to the UK as I was fed up. The ruse worked. The client settled.

It was very hard work, extremely tiring and I was normally exhausted at the end of each day. We obtained an agreement of 25% payable over one year with every creditor… except one bank which stuck. I visited Doctor Kassas and we discussed my problem. He knew that I knew that where the sawmill was situated was near Sammy Doe's home village but I did not know until some weeks later that some of his extended family actually worked at Libco.

The next morning I received a message to go into the bank and was met by one of the Liberian staff. I had been negotiating with American expatriate staff up until that point and they were nowhere to be seen. The bank official was very polite and handed me a letter stating that they accepted the offer that I had made to them.

'I am sorry that you had to wait so long, Mr Allen.'

I thanked him profusely and was driven back to the Contimba office to tell Victor the good news. Two weeks later, the expatriate staff disappeared overnight back to the States and the bank's affairs were in chaos. It was over a year before we could start paying our instalments to them. Serves them right, I say.

We spent two days with our lawyers sorting and checking all the necessary legal documents. This was quite a task for Liberian Law was American Law which was the only drawback to the whole project. Rudi Bunter flew out to sign the documents and I was able to accommodate him in my new flat about which I had not warned him. The flat was furnished so I had very few items to buy for it and it was all set up in a week. I was using one of Contimba's cars and was quite well set up.

By sheer luck, I had found a brilliant houseboy/cook, David who had walked into the office one morning looking for a job and I do mean that he was an amazing cook because he was as skilled as any chef and could just as easily cook for me or a party of ten, which he did quite often when I was entertaining timber buyers. The house was always clean and tidy and he could iron superbly. Rudi took to bringing out a second case full of his shirts which David would wash and iron immaculately because Rudi said that his wife could not iron his shirts properly. I am glad that David never met his wife.

I was also able to obtain my work permit, resident's permit and driving licence all in one morning. In Liberia, it was so simple. You filled out a form, gave them the fee and you received the documents. Cash is king.

I next had to spend a couple of days up at the mill to make an inventory, talk to the staff and workers and inform every one of developments. The road journey to the mill was over 350 miles with the roads deteriorating the further you travelled from Monrovia. The journey could take twelve hours so we invariably flew.

Wesua Airways operated an air charter company out of the Sinkor Airfield and the direct flight was 184 miles and cost about $400. The light aircraft could not take heavy loads, however, so we ran one lorry a week up to the site to carry spares and heavy goods. Flying up to the site one quickly started flying over sparse forest areas and my one abiding memory was the slash and burn fires that the locals started to clear forest areas to grow crops.

At ground level, these would not be so noticeable but from the air you suddenly realised their full extent as one could see dozens of them at height. This was to be one of my constant conundrums while I was logging. By necessity, we had to build roads into the bush so that we could extract the logs. The local people were delighted that the road was there, but, we should have destroyed it so that they could not move in for slash and burn for we never strip logged.

The Amazon destruction is very much due to strip logging. All the trees are felled in any given area and the land given over to cattle or plantations. We logged selectively only taking out about 10% of the trees in any given area. If left the forest would have recovered again but for slash and burn. The consequences of not allowing the people to carry out this primitive form of farming were all too clear especially during the tenure of the PDC. A ban was impossible.

Procedure for landing aircraft was to fly low over the mill to attract attention and then land at the airstrip about half a mile away and a pickup should be there by then to drive passengers down to the mill. I landed safely and jumped into the pickup with my bags and my food supply—no shops in the bush, but there was fruit and some vegetables.

Once when I was up at the mill the plane did a low pass and I drove up to the airfield to pick up the sawmill manager and his wife. I drew up at the end of the strip. No plane. I looked up. No plane. Just then the pilot and the sawmill manager and his wife walked out of the bush. The plane had burst a tyre on landing veered sharp right and gone in between two trees breaking both its wings off. No one was hurt. I had great fun with the owner of Wesua Airways because, when I suggested that his charter fees were too high he would quote a long list of expenses such as fuel costs, salaries, maintenance, difficulty of getting pilots, equipment, etc.

The tables were now turned for he wanted me to put the plane on one of my low loaders and transport it back to Monrovia. We gave him a high price and when he complained I gave him a long list of expenses such as fuel costs, salaries, maintenance, difficulty of getting drivers, equipment, etc. He had to agree.

The mill was quite large and had two production lines one with a band-saw and the other with a circular saw and they and the 'edgers' were in good condition and so was the mill building and storage facilities because sawn timber had to be kept relatively dry when cut. The generators had also been well maintained. Bush operations were in progress because Libco also exported round logs out through the port at Sinoe about a two hundred mile round trip which could take one and a half days and I agreed to go out to both operations on the following mornings.

Now, I knew that I had a problem because I was making an inventory not checking one. We had quickly established this problem in Monrovia when negotiating the investment. The staff at the mill were Italians and very resentful of the take-over package though they obviously had no other options it would seem but I am afraid that I did not trust them. To all intents and purposes they were good at their jobs but I doubted their sincerity and honesty. My suspicions were justified. I was taken out to both the bush operations the next morning and they were running slowly but thoroughly.

The bush equipment was in good order and the six lorries were making a sawmill run that day but I estimated, that, with planning they could make three journeys a day easily and I was told that they were taking two and a half days to do a port run. One and a half would have been sufficient even allowing for emergencies. What we seemed to have invested in was a sound workable sawmill and a slow bush operation, both of which would be capable of increased capacity and production but how was I to verify and ensure that we had received all the assets?

I found a solution that evening because there was a knock on the guest house door which was part of the managers' housing area at the top of a small hill. My visitor was a Ghanaian who was employed as the sawmill clerk. I gave him a beer and we talked about Ghana. His father had been born in Sunyani and we reminisced. He was an intelligent chap and I knew that I had found the key to establishing the true inventory. I showed him the simple inventory that I had compiled and one truck, a Swamp Skidder and a Jeep was missing, as well as a couple of big chain saws. The young man remained my eyes and ears for the whole time I was at the site. I had updated the inventory to include the missing vehicles and equipment. I was scheduled to have a meeting with the managers the next morning.

We all met on the veranda of the guest house and first of all I gave them a big thank you for continuing to work the mill during the last few months and would certainly be recommending a bonus for them to show our appreciation. I then carefully explained the whole process of the deal and asked for any questions. They were concerned for their future and I assured them that they could continue being employed subject to an interview and everything being handed over properly. I then asked them to check the inventory carefully and they confirmed that it was correct.

I explained my problem that I was experiencing difficulty in establishing the whereabouts of the items that I then listed. Looks passed between them. I merely explained that they had to be located before I could give them any guarantee of employment. I knew that the vehicles were close and I knew that I could find them if necessary. I think they had probably hidden them as insurance against the possible non-payment of their salaries but I simply told them that the deal could not possibly be concluded without them.

I met with the union reps and tried to reassure them that I was happy to work with them but they started to make outrageous demands for huge sums of money demanding compensation and making all kind of wild claims. During my stay, over three days, I then proceeded to talk with the foremen and charge hands, then small groups of the workers and got a completely different story so I risked a meeting with all the workers. I let the union reps speak first then moved in. I reiterated the details of the takeover, that money was tight, but we were confident that we could rescue the mill and be back to full work by the end of the month.

Furthermore, I was awarding them all an extra month's salary as a bonus and the company would in future be paying a thirteenth month's wages for the Christmas period and that I wanted to work out a bonus scheme for all workers with their input as to the best method. I was cheered and the union was defeated but I did see them again and they licked their wounds and agreed to cooperate, especially as I promised to double Labour Law.

That afternoon I was visited by the Chief Immigration Officer for Grand Gedah, who was polite and courteous but was obviously a mite too friendly with my managers because he visited them first before calling on me and then expressed concern for their futures. He was in their pockets and was probably concerned for himself as well. I gave him a beer and carefully told him what I had briefed the managers on that morning but that there was a hitch concerning the missing vehicles and equipment which could hold up the whole deal in which Sammy Doe was taking a personal interest as the mill was part of his development plans for his home area.

The Chief Immigration Officer was floored and he must have guessed that I had some connection with Sammy because he hurriedly left and scurried back to the managers. The vehicles and saws were back by the next morning and I said no more, departing back to Monrovia that afternoon. My priority now was to establish a responsible management team for the Mill and this was helped by the resignation of all but one of the Italian managers.

I was thus able to put in well experienced expatriates who worked well as a team and quickly brought everything under control. It was a pleasant place to work, with a close knit community and the managers enjoyed working there. I did get up to the mill on a regular basis but I also had to entertain log buyers and escort them day and night to check sawn timber and carry out logging receptions and give them a good time.

The Japanese were the most lively. They worked incredible hours in Japan and even worked their holidays for they were nervous they could lose their jobs, but while in Liberia they could let their hair down and were often quite wild, while the German buyers were very methodical and exact measuring every log down to the last millimetre. The French never got drunk on beer…only wine! Contimba thus moved from a buyer on the open market to a producer with a guaranteed production of logs which they could sell on to European buyers at source and by Christmas we had doubled the sales of round logs and sawn timber.

The sawn timber sales were actually more profitable and we were victims of our own success because our sawmill machinery was limited and we would need to rethink as to how we could increase production. Libco was a medium sized company with only about 150 workers and a turnover of about $1,500,000 per annum. We could increase the forest production of round logs for export but that would entail purchasing new trucks.

We discussed all options and my view was that we should give priority to increasing the round log sales at our mill with the simple purchase of new trucks to raise our income to $2,000,000 per annum, giving us breathing space to consider other options. Contimba maintained one of their staff at the Lemgo Head Office to source our spares unattainable in Liberia and deal with expatriate travel requirements so we put him on the quest to look for trucks. We were very lucky, for lying on the dockside at Bremen Harbour, were six Mercedes units that were part of a failed export order to Iran and we got them at a knock-down price. They were shipped within a week and landed in Monrovia shortly afterwards.

I have yet to introduce George who was our 'gofer' in the Monrovia Office. He was our 'Mr Fixit' and it was he who scuttled around the town paying our bills, searching for spares, dealing with any local problems and arranging for everything to be transported up to the mill. He had good contacts in the town and was very reliable in most respects. As soon as I had confirmation that the trucks were to be shipped I got on the mill radio to George in the office.

I talked him through the week's list of required items which we double checked and then I casually mentioned that I wanted six articulated logging trailers purchased within a deadline of three weeks. He asked me to repeat the order. I repeated it. Lorraine Fernandez the Contimba secretary told me later that George had almost fallen through the floor when he received the trailer order. I advised George to contact the other logging companies then go hunting in the industrial area of Bushrod Island in Monrovia and also gave him permission to travel to locate them.

Now Lorraine was another matter. She was an extremely able secretary, superbly efficient at her job and she could also speak four European languages fluently which was of immense value to the company. She was dictatorial and strong-minded and brooked no nonsense from unsolicited or unwanted callers. She protected our domain fiercely and was a valuable asset.

The one love of her life was her church where she sang in the choir. She thought Victor's office practices were quite slack and they constantly bickered in a friendly manner if they were both in the office together. He always called her Fernandez while I called her Lorraine. Ministers, officials, company managers and soldiers respected Lorraine and I always told them that she was really the boss and not me.

It is necessary at this stage to explain the 'dash', which I always represented as a mixture of a gift, a bribe with a tip and thank you attached. In Zambia it was unwise to offer minor officials and the police a small payment and you could be arrested for same, though payments were made to more important officials and government ministers but even they would suffer legal consequences in court if caught in the act. In Ghana, all officials expected the 'Dash' at all levels and it was very common in all aspects of Ghanaian life. So, if you were caught for speeding, you could give the policeman a couple of cedis and drive on, and you could arrange a meeting with a minister by 'dashing' his clerk. If you needed goods fast, you 'dashed' for prompt delivery.

Wages were low and I am afraid it was simply regarded as a way of life and is still prevalent in governmental contracts and business deals. It could also be used to escape crimes committed, avoidance of tax and for 'turning a blind eye' to illegal activity such as smuggling and drug dealing. I had to use it at a low level in Ghana occasionally but never used it for any other activity. In Liberia, however, they had developed it into a fine art and nothing happened or moved without the 'dash'!

If we had not paid up then our business would have shut down completely. It was a nightmare. You could pay all forestry dues, income tax, road tax, social security, port and harbour charges and surcharges, in full, and on time, but you would still have to pay the clerk to obtain the receipt. Every roadblock and police barrier demanded it in order for you to pass. Ministers and directors of government institutions took a monthly 'dash,' if they were involved with your business. The police would regularly stop motorists for false offences and you would have to pay to get off the charge and poor George could only move around in Monrovia efficiently if he 'oiled' the process. Everyone and everything was involved with no exceptions.

If the road tax discs were delayed at the printers, I had to send some of my staff down on to the main road to Sinoe to let the trucks through. The trucks could not move any forestry product without a Waybill signed by an FDA clerk. The forest manager had to give them 'lunch' money each day to ensure that the bills were issued. Due to a problem with aircraft, I once had to send some of my expatriate staff up to the mill by double cab pickup and I gave them the necessary 'dash' money for the barriers en-route with an instruction to use it. They did not do so and at the first barrier, the soldiers took all the spares off the truck, checked them against the waybill, allegedly found mistakes and wanted an even bigger dash.

I had to send George up with more cash for the 'dash' and a very curt note from me. The Liberian officials were completely open about the whole system quoting poor pay and conditions and several ministers frankly told me that they might not be in power for long and had to make as much as they could. Our biggest dread was the stop list issued by the Port and Harbour Authorities which blocked your exports if your dues, allegedly, were not up to date. Mine generally were but one could be caught out if the stop list was published early or late for I mostly paid every two weeks.

To avoid any problems, I had asked George to make an arrangement with the director to let us know of any such list 48 hours in advance. It worked perfectly until the director was on leave and his deputy deliberately issued an early list to 'line his pocket' and Libco was on it for the sum of $1.60; i.e., one dollar and sixty cents. I had paid a bill for over $30,000 the week before. I had to fly down from the mill and rush to his office, pay a dash and release my ships. I had two loading at that time in Sinoe. Revenge is sweet, however, because, after a huge future expansion, we advertised for an accountant and the deputy walked in as one of the candidates. He did not get the job.

My German employers had requested that everything be run properly and above board as per European standards right from our point of take-over. This, quite frankly, was a pipe dream and I had to perform a balancing act between what was feasible and what was not. Conditions in Liberia were rough and tough and the Liberians were not easy to deal with as individuals or in business and trade. They regarded a company as a cash cow and put enormous difficulties in our way.

It was, without a doubt, the most difficult job in which I had ever worked and one was under pressure on a daily basis. To avoid any disruption to my staff and workforce I took the flak quite deliberately so I was constantly under fire from the day I started as general manager. If there was a problem then I dealt with it allowing everyone else to get on with their work.

Our accounts were held in Germany but we had to file them in Monrovia as well and the problem of the 'dash' arose with our accountants Cooper and Lybrand stating that it could not be covered in our company accounts. I just simply asked how they accounted for it in their accounts? They assured us that they never 'dashed' and I snorted in sheer derision. Every company in Liberia 'dashed.' They found the correct method for us which was probably the same one they used and the meeting closed. Nothing or no one could operate in Liberia without it.

The FDA's regional forester was Bill Glay who was quite a character and lived in Zwedru. We often helped him with fuel and one or two minor favours and he did pop down for somewhat haphazard inspections now and again as he liked our hospitality and beer. It was his idea that we manage two reforestation projects based between our mill and Zwedru and I readily agreed because I knew it would help our relationship with the FDA. We would check the projects weekly and report any encroachments or damage but I did feel that it was akin to putting a very small plaster on to a big wound.

One evening, I received a radio message from the bush manager (I slept with a contact radio in my bedroom), but this was not a late call. I was informed that he had discovered many Anigre trees in a distant corner of one of our concessions. Anigre was a very valuable hardwood, light brownish in colour with a pinkish hue, and greatly valued by the veneer and furniture industry in Europe. I was more interested in its value and the high prices that I could obtain. Now we held 300,000 acres of forest concessions (smaller than the Anchor Ranch I managed in Zambia) and I knew the areas quite well but not in any detail, but I flew up to the mill and the bush manager, Mans Vroom, a very experienced man in the forest, drove me to the area where he said the Anigre was located.

There was Anigre there and in large amounts. I did realise that we were close, however, to a rival logging company's concession and so we sought out the boundary markers which indicated that the area was ours. I also specifically asked Mans to double-check all the co-ordinates. He again assured me that he had. I thus contacted head office and gave them the news and we started extracting the Anigre as soon as possible. It is not an easy wood to handle as it can easily split so it has to be bound with wire straps before transportation and we made good money from it.

While that bush operation was ongoing, I was informed by one of the expatriates passing through Monrovia, who was going on leave, that Mans had been behaving in a very peculiar manner recently and was drinking too much. Heavy drinking was a problem in the bush as there was little to occupy the expatriates and with all the machinery we used, it could also be very dangerous. I dropped everything and flew up to the mill late in the afternoon. Everything appeared as normal but Mans offered me a martini. A martini in the bush? I accepted the glass and took a tentative sip. It was pure wood alcohol which is very dangerous to drink and can damage the brain.

Mans had been on it for some time so no wonder he had behaved so unusually. I patched through a radio message to Victor and told him that I was bringing Mans down with me tomorrow and could he get George to buy a ticket for him to go straight back to Holland? Mans was a very great friend of Rudi Bunter and he had worked faithfully and well for Rudi for many years. Rudi was actually due into Robertsfield that evening so I collected him from the airport and explained the problem with Mans. He was sceptical about the incident and when he met Mans in my flat he did appear absolutely normal and we had a good dinner with wine and beer.

Rudi was just not convinced. That night Mans got up to go to the loo but went into Rudi's room instead, peed all over his boss and then fell sprawling all over his bed. Rudi was now convinced and took Mans back with him two days later. Unfortunately, this was not the end of the tale. I had always insisted on maintaining good relations with the other logging companies and Bashir Charrafidine popped into the Monrovia office for a coffee one morning. He informed me that Libco had trespassed into his concession and that most of the Anigre extracted was his. I listened carefully. I told him that I had checked the boundary markers carefully but he maintained that they had been moved.

He accepted my word that I would check as I was due to be back at the mill at the end of the week and check I did. I took the Liberian Bush Manager with me and checked the markers again. He thought that would have satisfied me but I slowly walked back looking for any signs of erased markers and we located two. Mans had moved the markers and now Libco faced a trespass case.

I contacted Bashir and he kindly asked me around for a Lebanese meal so we could discuss the problem. I informed him of my findings and told him all about Mans and his problem and then suggested that both our prospectors go in and agree on the number of trees extracted legally and illegally. Once this had been established, we could then decide on compensation. He agreed. Of the 100 Anigre trees we had extracted, seventy were from Bashir's concession thirty were from our concession. In compensation, I gave him the use of four of our trucks to clear a surfeit of his logs down to Sinoe during the period we would be shut down over Christmas. It was settled and even the truck drivers were happy because they received double wages for that period. Sorted.

The next complaint was not so easily settled. I was summoned urgently to the Executive Mansion one morning by Colonel Pino who was one of the volatile members of the PRC and when I was ushered into his office I was surprised to see the representative of Contimba's competitor sitting at his desk. Pino just shouted at me, 'YOU OWE THIS MAN MONEY. PAY HIM NOW!'

I tried to explain to Pino about the agreement but he was having none of it and ranted on at me. He demanded that I pay or I would have to go to jail. I told him that he had power over me to do that but I could not personally pay because it would break Liberian Law. Pino called his bodyguard.

'Put this man next door and keep him under arrest.'

I was escorted into the next room and the rep came in. I realised that he was new in Liberia but it was an absolute taboo for expatriates to complain about other expatriates to the PRC and I told him so. He was very arrogant and asked for the money. I simply stated that it would not be forthcoming unless it were under the terms of the legal agreement and so I waited. He tried again a little while later but I refused to talk to him. I stayed in the room for a couple of hours and then a high ranking policeman arrived and told me to follow him, which I did, slightly bemused, and walked down the stairs, went outside, where my driver was waiting, and we drove back to the office. How?

My driver had been worried and contacted George who, through his contacts, established what was happening and then drove down to Doctor Kassas to explain my difficulty. Doctor Kassas must have contacted someone in the President's office and I was immediately released. I refused to have any contact with the rep who had made the complaint and barred him from the office and many of my compatriots 'sent him to Coventry' as well. Rudi did meet up with him in Monrovia a few weeks later and blasted him but he was unrepentant. I would not wish any form of retribution to be exacted but he died in a car accident some months later.

Car accidents and fatalities were not unusual in Ghana and were normally drink-fuelled. While Victor was on leave, head office sent out one Walter Kuhns to cover the Contimba side of the business and this was about the time when head office also discovered a tax loophole regarding their office cars. If these were sent abroad within two or three years one did not have to pay the duty on them I think. The Monrovia office thus had two or three BMWs for its use at any one time which we could then sell on, but BMWs were no good for Liberia.

They were too sophisticated and complicated and useless in bush conditions. We needed basic, uncomplicated vehicles. We need not have worried. Walter was a bombastic and overpowering character who rubbed everyone up the wrong way, annoyed most people he met and was a very heavy drinker. He crashed one BMW coming back from Buchanan one evening and I had to go and fetch him. He was heavily under the influence when I picked him up and the BMW was a 'write off', which took some doing.

A week later, I got a phone call from him to state that he had had another accident at Mamba Point and he was at the Central Police Station about to be charged. He took some extracting as the police were angling for a big dash. I merely said it was too much and it would be better to leave him in the cells and pay the court fine in the morning as that would be cheaper and made to leave. They dropped their price and he was freed but the car was sold as spares.

Not three days later, he pranged the third car when he was totally drunk again and got himself into severe trouble by racially abusing the police. I was fed up and told the police to put him in the cage and I would come down the next morning and I did not hurry my breakfast. I collected George on the way. Walter was looking very sorry for himself in the main cage and I knew the feeling. I had been there and done that. George and I were ushered into an office. Who should be there but the officer who had kindly escorted me out of the executive mansion…Colonel Sam Cojolo.

He offered us refreshment which we kindly accepted and we discussed Walter. Sam explained that the charges were quite serious. They knew he was drunk but the man quite obviously disliked Africans and had been very rude and abusive. He was a racist. I agreed and asked what were his options. If he were charged he would face a court trial with much bad publicity and probably end up with a two or three year jail sentence or he could be released for a high price but he would not be given another chance. His case was to stay on file.

For my part, I had already decided that Walter Kuhns would be flying out of Monrovia that very evening back to Germany. He had wrecked three cars, upset Contimba's buyers and was a thorough nuisance who had caused endless trouble. I agreed $1,000 dollars for his release and Walter was wheeled into the office. Sam read out the charges and I explained the options. Walter was still trying to bluster and started to get very angry. I got up, shrugged my soldiers, and began to walk out with George. This brought him to his senses.

Sam carefully pointed out to him that if he was not on the plane that night he would be arrested and charged. I took him back to the house where he was staying and told him that I would collect him later to drive him out to the airport. He actually had the cheek to request his balance of salary and I told him most of it had been left at the Police Station. He left that night. I had already warned Rudi about his impending return. I for one was very glad to see him go.

I spent several days up at the mill before it closed for Christmas, scheduling the maintenance work and checking the security for the period. Two expatriates were always left to ensure all was well and look after everything. They had also agreed to tackle the ongoing problem of a sawdust fire. From the moment of my arrival, we had always burnt the waste sawdust or given it away to chicken farmers but the previous management had just dumped it around the site. A small fire had started and it was its nature to travel underground and then emerge twenty or thirty yards away, then duck down and keep on repeating the process so two Komatsu Bulldozers were going to clear it once and for all and dispose of it safely.

I visited the area where we had extracted the Anigre to reassure myself that the markers were now back in their proper place and was lucky enough to come across two forest elephants amid the trees. I was later to see one of the small forest leopards and they are beautiful creatures to watch. On my return to the mill all hell had been let loose. One of the fellers had been rushed into the clinic with a horrendous gash right the way down the whole length of his body from head to toe when the chain had snapped on the 6' blade of the chain saw he had been using. If you are squeamish, look away now for he was bleeding badly and was in deep shock. Night was falling and I could not call a plane and I suggested a pickup to transport him into Zwedru Hospital.

Prince, our dresser, was certain that the journey would kill him as he had lost much blood. By a sheer stroke of luck, again, Prince had retained details of the blood groups of all our workers who had agreed to join in a Malarial research project that a Monrovia organisation was conducting, and set up two blood transfusions from matching workers with makeshift apparatus. With no anaesthetic, Prince then proceeded to stitch up the wound which took him several hours. One hundred and sixty four stitches later, he had closed the enormous wound. The feller made a full recovery and was back at work in four weeks.

I gave Prince a handsome bonus as a reward. We issued every feller with complete protection—a helmet, with ear protectors, rubber upper body armour with reinforced elbow pads, crutch and leg protection and strong boots for which they signed. Every morning they would walk in proudly wearing all the gear and travel out to the bush in the same. They would then discard all the gear and cut down trees wearing shorts, a vest and flip-flops. I couldn't win. As a result of that accident, we did much improve our medical facilities and we even provided a small service for the local village.

I now led what was almost to become a double life. I had worked incredibly hard in Liberia and in recognition of that, I, and all expatriate staff, worked five months on and one month off on a regular basis. My German employers were very generous to me and the staff and I not only had my Monrovia salary in dollars but also my accommodation and most of my food was all found as were utility bills and house expenses. I was then paid an allowance in England and furthermore I received half a percent of all profits—generous indeed.

I worked hard and was rewarded well. Liberia was the hardest country in which I had ever worked but I had no real complaints. World events had nearly passed me by. Prior to my arrival in Liberia, the Shah had fled, Amin had staged his coup, Maggie was voted in and I was now 34 years old. I came back to England for a month's leave via Lemgo in Germany for a review of progress so far.

The holding company, Holtz Import, seemed very satisfied with progress and developments to date so I left to enjoy my Christmas in Bristol. Now, however, I had completely lost the timeline on my cars. I did have a big white V8 Rover saloon that was very powerful and I used it to go up and down to London quite regularly but I exchanged that for an MG Midget, which was fun. The doorman at The Inn on the Park was quite used to parking my Rover for me and a whole collection of client's cars including a Pink Cadillac that was longer than a London Routemaster bus but incredibly easy to handle. He struggled to get into the MG because he was very tall.

I once turned up in a converted camper Mercedes coach to pick up a client. He was a wealthy German industrialist who wanted my help to locate a tug with a big aft deck for his diving submersible. We drove right around England clockwise and eventually found a suitable vessel in Lowestoft. If only we had gone anticlockwise. My parents were kind enough to let me leave the MG in their garage and an uncle and his grandson used to enjoy exercising it for me.

My parents and family were well. Roger had divorced his first wife and was now living in Portishead with his second wife, Bronwyn, and her son Tony and I think, Helen, my niece, might just have been born. Roger's divorce had been quite upsetting for all concerned because Sue had given them a grandson which Roger later disclaimed and there had been no contact since. My grandmother was still living with my parents but the low level strains were still obvious so I made sure I took Mum and Dad out alone as many times as possible and then Grandmother separately as well.

They liked Downend and enjoyed walking along to the shops and my father quite enjoyed watching the cricket in the summer as the Downend Pitch was directly opposite the house where W.G. Grace lived just up the road. I revisited friends, including Jim Pickup, and also toured all the relatives. That December was bitterly cold, however, and I still had to deal with faxes concerning the company. A huge snowfall stopped all traffic and I had to walk up daily from Downend to an office agency in Park Row for three days running. It was a long walk for one who had been used to having a driver. I used taxis for the rest of the week.

I spent a few days at The Metropole Hotel in Padstow as I was thinking of buying a house in, or near the town. It was somewhat quieter then than nowadays. I met up with Jackie Stanley, the estate agent, and she promised to send me details of any properties that matched my requirements. I toured around that part of Cornwall not realising that I was to get to know it very well indeed. I liked the Padstonians for their friendliness and hospitality and the town was most pleasant. I also liked The Metropole Hotel and stayed there several times.

Between Christmas and the New Year, Victor phoned me from Liberia informing me that Prime Timber Products (PTP) had shut down; actually, the management had run and left the company high and dry. It was possibly a golden opportunity for us especially as their site was at Zwedru only about twenty-five miles from our present site. I phoned Rudi Bunter, who seemed quite excited by the possibilities this would offer, and flew into Hanover on 1 January for two days of meetings, discussions and planning.

It was agreed that, as far as Liberia was concerned, we would use much the same format as we did for the acquisition of Libco but there should be a tentative checking of all details before we made any commitment. This was too big a project for Centra Holz Import alone and we realised that we would need strong financial partners. Research revealed that the German Development Bank, based in Dusseldorf, did not have any timber industry companies in its portfolio and could be looking for one, and I was to sound out the Liberian Bank for Development and Investment in Liberia. I would also have a confidential discussion with Emmanuel Emeh, my landlord, and Deputy Director of the FDA.

I flew back immediately to Monrovia. I handed over the running of the Libco site to the sawmill manager who was also competent in forestry work though I would continue to be in control from the Monrovia office. My first call was to Dr Kassas. I carefully explained what we were planning and emphasised that we could not make any commitment as yet. All our enquiries were tentative in nature until we could draw in other investors and truly established the finances and current situation of PTP. He obviously relayed this to Sammy Doe and I was invited to the executive mansion two days later for a meeting with the President.

He was under some considerable pressure from his home village in Grand Gedah to resolve the problem of PTP. As President his people expected much good fortune and development to come their way—this was generally the expectation in all the African countries with which I had come into contact. I again emphasised that we could give no commitment until we were sure of our investors and the state of the closed company.

He promised to give all necessary help to the project and appointed one of his staff to liaise with me and to whom I gave the courtesy of a briefing at the mansion on a weekly basis. I now had a power tool in my tool box. I had the backing of the President. This was to prove instrumental in achieving successful negotiations. This could and would open doors.

I tiptoed cautiously into my initial investigations and search for support and facts. Over a couple of beers, Emmanuel Emeh and I discussed the possibility of a take-over of PTP. He was surprised as it was a huge complex of two sawmills, a moulding shop, a furniture factory and had 1.3 million acres of concessions but he could see no objection from the FDA's point of view, indeed, they would positively encourage it. I approached the Liberian Bank for Development and Investment (LBDI). They were lukewarm about the idea even though they had no logging companies in their portfolio but I played neither of my two aces during the first meeting and left them to mull over the idea.

I made an initial trip to Zwedru to see Bill Glay, the regional forester, to bring him on side. He knew of my connections with Sammy Doe and was in full support anyway in the full expectation that there would be something in it for him even if it was only free fuel and beers on my veranda and I had a brief meeting with Geoffrey Saydee, the Personnel Manager of PTP, that I had surreptitiously arranged through Bill Glay. I invited him down to Monrovia for a couple of days and we discussed the whole of the project in full detail, in my flat, for I needed to prepare my ground very thoroughly.

The company had been running up until Christmas and the workers were thus up to date with pay because they had received their salaries for the down time over Christmas as well, but, Geoffrey thought they might claim for redundancy pay. Legally this would not apply as we would be taking over the existing company not formulating a new one. He and I were fairly sure that we had the union on our side as our employment procedures at Libco were exemplary and our record was good. The local labour force were very keen to work for us and with us. Geoffrey rated the employees as quite hard working and keen to revive the company.

I then picked his brains about the company as a whole and it appears that there was ill feeling between the expatriates and Liberian staff and workers and the business was not run efficiently because of the weak management. I obtained as many facts and figures as I could and was now well armed and prepared for the next stage. Geoffrey was concerned about stealing from the site by some of the workers. I suggested that he hold a meeting and explain what might be happening as any thefts or damage could hold up the whole process and this he did. The stealing was reduced but thieves attempted to steal one of the pickups one night and I reported my concerns to the staff liaison officer the following Friday at our weekly meeting.

The very next day, armed soldiers from the Zwedru barracks were at the gate and patrolling the compound and no one was allowed in except for Geoffrey Saydee. The President was capable of acting very quickly, it seems.

At this stage our accountants recommended that we take on the services of one Eden Reeves who had many high level contacts in Monrovia and could help us with the bid for PTP. I did ask to see references but there were none. I was persuaded, however, to take him on as we could do with help. Eden, however, loved to talk, and seemed unaware that we were pressed for time. He was keen to tell everyone about all the people he knew and his alleged achievements. He could talk the talk but not walk the walk.

We eventually side-lined him to smaller jobs, which, quite frankly, George, our Monrovia office man, could have done more efficiently and certainly much faster. We ditched him as soon as we could but this is not the end of the story. Eden fled Liberia when the second round of trouble started and ended up in New York. He talked his way into a top United Nations post and I bumped into him much later in the Golden Tulip Hotel in Accra where he was disporting with a group of young ladies in quite a raucous manner. I greeted him and he recognised me but was anxious not to talk and certainly not about Liberia. He has since died but his son is now a big businessman in Monrovia and looks exactly like his father.

Assured that the site and assets were now safe, Victor and I commenced our double act visiting PTP's creditors around the town negotiating this time for a six month moratorium, then 25% paid in instalments over one year. Most agreed and the ones that stuck longer eventually acquiesced when we threatened to walk. Two really stuck. One was their fuel supplier, and the other, an insignificant hotel bill of less than a thousand dollars. I got two of the biggest creditors to persuade the fuel supplier who stated that they could organise all the other creditors to boycott their company if they did not agree and the hotel went bankrupt. We were home and dry.

At about this time, I took a gamble. I put our prospector, Jock McCay, into the PTP concessions. He was a wonderful character from Scotland who was also a drunk but never while actually in the forest. He was very experienced and knew the forest well so I let him have his head to complete a quick survey in the areas he thought best. Prospectors looked for commercial trees and would spend a week or ten days in the forest on any one trip camping or staying in villages and then report back. After each trip Jock would spend two or three days totally drunk, was then normally picked up by one of our lorries or vehicles, having made his way to a main road, and report back for another trip.

Head office in Lemgo had been working on the German Development Bank. They seemed very interested in the project, and they, in turn had also approached the LBDI and received the same lukewarm response that I had faced. It was time to move in. I had received letters of support from the FDA. the Mayor of Zwedru, the Grand Gedah District Governor's Office and a signed letter from all the creditors which I presented to the Director of LBDI explaining that he was now the only stumbling block for the activation of the project but he was still hesitant in the face of that barrage.

I reported his reluctance to the executive mansion, expressing my surprise. Here was a massive investment for Liberia with all the potential of being one of its largest timber producers and guaranteeing the employment of 500 Liberian workers, a deal to which all creditors now agreed and everyone was ready to move forward, and…who was blocking it…a LIBERIAN bank. It was scandalous.

On the following Monday morning, Lorraine took an urgent call from LBDI. The director would like to see me very urgently please. I was taking some of staff out for lunch and asked Lorraine to make an appointment for me later in the afternoon, for which I was deliberately late. It would not hurt the poor man to stew in his own juice for a little while for I had received a copy of a letter earlier that morning from the executive mansion, addressed to the Director of LBDI, pointing out all of the above, signed by the President.

The director did not mention the letter, merely stating that the bank had reconsidered the proposal and upon reflection, it would be willing to be the major partner in the deal and he was happy to give me full cooperation in establishing the project. He did not know that I knew the letter had been personally delivered into his hands that morning by the member of Sammy's staff who was acting as my liaison with the President, with the verbal warning that his present short tenure at LBDI was possibly in danger.

There now followed a flurry and whirlwind of activity. I composed a dossier of the Liberian aspects of PTP including all the facts, statistics and memorandums of understanding to date, with all letters of support and all background checks which was greatly appreciated by all parties and proved to be invaluable to our discussions and future planning and I also wrote the feasibility study. The German Development Bank was most appreciative and stated that it was one of the best set of documents that had come through their bank. I flew to Germany to the GDB and met the officials formally.

They were delighted when I formally presented them with a letter from the President of the Republic of Liberia—Master Sergeant Samuel Doe, expressing his thanks and support for the project. It all helped and the formalities could now commence for the takeover. I flew back to Liberia and spent another day with lawyers defining the documentation required and we then presented this to the LBDI.

In essence, this was to be a joint development project involving the Liberian Bank for Development and Investment, the German Development Bank and Centra Holz Import of Germany as technical partner based on Liberian Law so that we not run afoul of complex German legal requirements though we did have to conform to those required by the GDB. It was all quite complicated and time consuming and everything had to be cross-checked with all the 'i's dotted and all the 't's crossed.

In anticipation of our success, we had already started assembling our staff which were mainly Dutch or German but we could not directly locate a workshop manager so we advertised in the UK. I chose to interview eight candidates at the Tara Hotel in London and two more in Torquay. Being an Irish establishment, I had reckoned that the Tara would avoid the current IRA bombing campaign. At 9 am the first candidate came into my room for his interview. He was Fred Sullivan and after only a few minutes, I realised that he was my man.

He had huge experience in the logging industry and had worked in Ghana for many years. He was both technically competent and practical while being very placid and well organised. His German and Dutch colleagues could never understand him because he was a Geordie form Newcastle but amazingly, the Liberians, who spoke a peculiar form of English that was only half enunciated, understood him and he was old. Liberians respected the old. Rudi Bunter always needed an interpreter when he spoke to Fred. I did proceed professionally with the other seven candidates but none of them had bush experience and were more technical rather than practical. It was the same with the other candidates I interviewed in Torquay.

Fred was appointed to the job. I flew back through Dusseldorf as the head of the bank had expressed a wish to meet me and was then driven down to Lemgo again where we reviewed and finalised all planning.

In just over two months, PTP was back up and running initially with two forest operations and the sawmills running. I had not yet found a competent manager for the moulding shop. With our sister company Libco we thus had three sawmills and four bush operations and Lemgo Contimba was very busy doing all the marketing of the sawn timber and round logs. It was, without a doubt, a Herculean task that had been achieved and we had many ongoing problems to overcome in the initial stages. I estimated that 5% of all traffic coming into Zwedru from Tappita was PTP bound.

We ran a shuttle vehicle down to Monrovia at least twice a week to keep the PTP camp running and charter flights were almost on a daily basis for one reason or another. I had little to do with the marketing though I did manage the quality control of all our products and I needed to constantly check all aspects of the two companies and still took on all the problems so that the staff and workers could get on in peace. My driver in Monrovia now became the office driver and escorted the expatriates around town and I had two vehicles and two drivers in Zwedru.

I worked long hours for I now controlled the Libco and PTP companies as managing director and I was very much 'hands on' and always problem solving or helping when needed. With the multiple bush operations, I reckoned to visit each one at least once every week and the staff and workers could never tell when. There was many a surprised look on staff and workers' faces when they saw my pickup quietly waiting for them in the early morning light or mid-day or late afternoon.

My drivers worked one day on and one day off and if it were a long day, I would often let them rest and take the wheel myself. They were also paid extra for the unsocial hours. Life was made much easier for me when the company purchased its own plane, a Cessna 147, and then I could rendezvous with my bush drivers at dirt air strips. This saved me much time especially as the plane was mainly based at Zwedru. It was a valuable asset because Wesua had constantly raised its charter prices and it became a simple cost effective solution in the end. We now had air transport for our expatriates, timber buyers, payrolls, medical emergencies, visitors and me, for I was the main user.

I did learn to land the plane and I did fly it once or twice but really I was only interested in landing safely in case anything happened to Jose the pilot. We nearly lost Rudi Bunter and the plane. He had come up to Zwedru on a 24-hour visit to look at progress while on his way back from the Contimba office in Abidjan from which we were able to collect him direct. Returning to Monrovia, the plane had been caught in a violent thunderstorm and torrential rain.

Jose had to run before it and eventually was lucky to make a rough landing at the airstrip in Sinoe, down on the coast. We had received word that the plane had not landed in Monrovia and of course we were terribly worried that the worst had happened. We had about twenty radios at our camps and concession sites as well as several mobiles and our radio operator contacted them all but there was no news of the plane. At eight o'clock that evening we received a call from our site office in the port of Sinoe to say that Jose had made an emergency landing at Sinoe and everything was alright. We thanked God.

I controlled everything to do with both companies except Jose the pilot. I made it absolutely clear to him that he was in control of the plane and to fly or not, for whatever reason, was his decision. I would never interfere with his decisions. Jose most enjoyed going off prospecting with Jock McCay for the plane made our prospections much easier. In a matter of a couple of hours, Jock could look out the areas with most potential that would have taken him days to reach on foot.

Some months later, Rudi Bunter, Jock and I were discussing the opening of a new concession and were poring over the maps. For some odd reason Rudi got it in his head that a certain area would be worth looking at but Jock stated that it had already been worked by PTP several years ago. Rudi insisted so I got Jose and Jock to do a flyover and it was very sparse and they turned around and came back. They had seen another possibility adjoining the area but Jock maintained it was not worth the effort and the trees were mainly small. Rudi insisted that he walk in and check. Jock came back with a big fat brown envelope and Rudi's eyes lit up. He opened it in front of him and it was full of elephant poo.

'Elephants? Yes! Trees? No!' remarked Jock.

Rudi never interfered with forest decisions again.

The new concession was opened up down on the banks of the Duobe River which was a fast flowing forest river perfect for a dip in the sweltering heat and no, that had no influence on our decision to start another forest operation. It was wonderful to strip off and jump into the cool water. I asked my driver why no one else ever had a dip and he was adamant that none of them could swim, even himself. As we extracted the trees we got closer to the Cote d'Ivoire border and I would sometimes pop across to get supplies. I would strip down, put my clothes into a waterproof bag, swim the river, dress and go and do my shopping. The border was porous but not my waterproof bag!

Contact was also maintained with the use of our many radios all of which could only be heard on our wavelength and sometimes the conversations could be quite funny. I had a French jack truck driver who was responsible for following the logging trucks down to Sinoe and back doing any necessary repairs and dealing with any emergencies as they arose. He would call me.

'Geoffroi! Geoffroi? Over.'

'Oui, Thiery! Over.'

'Geoffroi. Cameon Trios il est mort! Over.'

'Quel problem, Thiery? Over.'

All the staff thought it was better than live radio. Unfortunately, Thiery was killed while crossing a main road in Paris sometime later.

As a managing director, one of my perks was Black and White Ebony and Stickleback Sipo. The Sipo perk was a special grain and very rare. If any of these were found they were mine. I only ever received one Stickleback Sipo which we cut very carefully in the sawmill and I exported it to Germany where it was stolen at Dockside but I did receive a large amount on the insurance.

In my travels, we were often hassled by police and soldiers manning barriers and roadblocks and even had problems with some of the local inhabitants who were trying to demand money and favours. It all came to a head one sunny morning when a young army corporal and another soldier angrily demanded to see me. Geoffrey Saydee tried to reason with them but they interrupted a meeting in my office and stated that I was under arrest.

I asked under what charge and by whose authority but they bundled me into the Jeep with a rifle in my ribs and drove me off to the army barracks. I tried to reason with them but they were intoxicated and pushed me into a small office. I was not too worried because Geoffrey Saydee had followed the Jeep to keep an eye on me and he was even allowed into the barracks. They were largely unintelligible and I had to resort to their patois before I found out the problem. There was a young man in the corner who had been a former employee at PTP some little while ago and shouted that the company owed him compensation. I simply stated that if he wished to accompany me and Geoffrey Saydee back to the office, I would look into the matter.

The corporal insisted that I should pay two hundred dollars. I refused and showed them that I had no money on me anyway. They roughed me up a little bit and poked a rifle into my ribs, trying to bully me into paying about which I was not very pleased and tried to explain that they were heading for severe consequences and I refused to talk to them at all, demanding to be released immediately. They were having none of it and frogmarched me across to the cells where I was forced to remove my shirt and shoes and socks and was duly incarcerated and told that I would not be freed until I had paid two hundred dollars.

My staff had moved to the rescue, however, and must have notified every notable official in the town. Colonel Choloprey, Chief of Immigration, was the first to arrive and demanded the cell door be opened. The soldier had the good sense to do so and we sat chatting, until Bill Glay, the regional forester, also came to join me, then the mayor, and a very welcome Jack, who was Lebanese and ran the local store and came clutching a crate of cold beer and some snacks and then the local Chief of Police. It was quite a party, though I suppose the venue was unique, and that is how the Commanding Officer found us all about an hour later and wondered what on earth was going on.

I carefully explained what had happened and he became very angry. He curtly ordered the cell guard to go and fetch the corporal and soldier who had arrested me and bring the young man as well. They were brought quaking before us and the colonel demanded to know what had been going on. Geoffrey Saydee had joined us by this time and stated that the young man had never been in the employ of PTP anyway. I would have wished to stop the two soldiers being pistol whipped by the colonel but I could not interfere and I am afraid the young man was beaten. They were stripped to their underwear and thrown into the next cell.

I was sorry about the brutality but they had brought it upon themselves. The soldiers were sentenced to three months in the 'Glass House' with hard labour and the young man broke police bail and fled. I thanked the colonel who could not stop apologising so I suggested that we move to his office and finish the beer.

Quite unusually, I was asked to visit Bill Glay's house for a Liberian meal the next week. I say unusual because it was normally Bill Glay who ate my food and drank my beer. I arrived with half a case of beer which he gratefully accepted and then he added that there was another guest in the house whom I had met before. It was Sammy Doe. He was visiting Bill because he was staying in his village for a couple of days in talks with the elders, who, even as a Head of State he was bound to heed.

The village elders were very powerful and it was brave man who went against them. Sammy thanked me for starting up PTP because he was able to now boast that he had helped his people in some way and he was grateful. He expressed concerns, however, about all the interference and hassle to which I was often being subjected and about which he was not happy. He wanted the project to be a success. Bill then explained that Sammy was going to give me two of his bodyguards for security which I could not refuse.

I was grateful for the offer and resolved to use them wisely and well. They only escorted me outside of Monrovia because I was well protected by Sam Cojolo, the Police Chief, while I was in town which was fairly rare anyway. My staff were a bit taken aback but soon realised that they were better protected as well because they never minded escorting the ladies and children into town or helping out. They were very pleasant and well mannered. They were given good accommodation and we also put them on the payroll so they received a double salary about which they were delighted.

Like my drivers, I always ensured that they were well looked after and cared for and had no hesitation in sharing my food with them. They were very good for the company and I was grateful for their duties. I insisted that they work one day on and one day off, like my drivers, and I know that they were very happy. I did not let on to Rudi Bunter that I now had security guards so he was very impressed the next time he travelled up to Zwedru to be met by two very smartly dressed soldiers clutching their automatic rifles and politely taking his baggage and escorting him to the vehicle.

There was no more hassle from anyone or anything and I could get on with the job and I hope that I was considerate to all members of my workforce, Liberians or not. They were working under good conditions, could achieve advancement and liked the bonus scheme and everyone received a thirteenth month payment. We had very few labour troubles and the union was quite happy with our terms and conditions. I must award praise where it is due, however, because I would never have been able to achieve the good relationships with the workers but for the efforts of Geoffrey Saydee. He was a good friend and helped me and the project enormously.

I am deeply ashamed to say that I was totally unaware that we had a company school on site. I knew about the managers' and staff and workers' houses because I had inspected them with Geoffrey Saydee and they were all part of a renovation scheme that I had immediately put in place for their immediate improvement and upgrading including improved sanitation and clean water supplies but it was not until I was out on one of my morning jogs that I discovered the school tucked away behind a line of tall shrubs and it was pretty ropey, I am sad to say.

I asked Geoffrey Saydee about it immediately after my discovery. He said it was not the company's responsibility but I had to inform him that it was. The previous management had deliberately misinformed him. I employed one carpenter to work full time on improving the school's structure and another to make desks, chairs and cupboards for the classrooms. I met with the headmaster and Geoffrey Saydee to discuss equipment and book needs. The headmaster, who looked very old indeed, had made a totally unrealistic demand for a $50,000 input which was out of the question so I made him sit down with Geoffrey and establish basic needs such as pencils, exercise books, chalk, etc.

The initial sum was a thousand dollars so I gladly donated that straight away but under the strict supervision of Geoffrey Saydee. We sold sawdust, odd lengths of timber and shavings at the gate which the previous expatriates had used as a beer fund so I instructed Geoffrey to channel that to the school as well. I was mindful of the fact that I was now responsible to development banks and could not be seen to be failing on our workers' health, housing and education, so, at the first board meeting I proposed that a fixed sum be set aside for housing, health, the clinic and the school. I am pleased to say that it was passed unanimously.

One condition of the deal involving the LBDI was that all our banking should now be conducted through them and I was not in agreement about this. My real reason was that the bank was not terribly efficient in its operations and I could see trouble ahead but I did make the point that LBDI could not make favourable commercial loans in the short term which may well be necessary in the fluctuations of the current timber market. I argued that I might need short bridging loans in a hurry. We agreed to maintain LBDI as our major bank but also maintained a commercial bank as well and my predictions were to prove correct. LBDI often ran out of cash and had constant difficulties in meeting our fortnightly payroll requirements so I always kept a cash reserve at our commercial bank for such contingencies.

One payroll descended into farce. The government issued a new one dollar coin which the Liberians hated and it was used more for making necklaces than anything else because it was quite attractive. When George arrived to collect the end of month cash payroll of $30,000, LBDI wanted to pay it in the new one dollar coins! We would have needed a Russian Antonov An-225 Mriya to transport it to Zwedru. The weight was phenomenal.

Our housing complex at Zwedru was pleasant and again, situated atop a small hill to catch breezes when they blew. They were quite well appointed and quite satisfactory for our needs. Mine was far too large for a bachelor so we converted the big lounge into a bar and equipped it with a pool and table tennis table. It could be used during the day and night. We also started a library and the wives often used it to socialise and hold coffee mornings. The bar was run on an honesty box which seemed to work well and was kept stocked by Jack from the local Lebanese store who was also kind enough to bring out goods from his shop when ordered.

Most Friday nights, I paid for the first two drinks for each person and the tradition was that you had to drink them both before you could criticise the company, me or anyone else. It worked quite well though the Liberian managers and wives were slightly reluctant at first but I valued everyone's comments and criticisms and they soon joined in. My expatriates were a good bunch and very competent and well qualified for the work.

The most trouble could be the Brits. Why do Brits have to moan? I had two. One was a forest manager who was good at his job but I already knew that he was using us as a platform to get employment at another company and he was critical of everything. If he had not gone, he would have been pushed. The other was an assistant workshop manager who was competent but had a difficult wife. She could be quite blunt and rude and was not ideal to be in the bush so she moved down to Monrovia.

Bill Glay and other officials would sometimes pop in and the company paid for their drinks. He must have informed Sammy about the club because I was approached by one of his bodyguard to ask if he might have a game of table tennis which he obviously liked. Thence started a rather odd tradition. Some time, normally just after midnight, there would be a gentle tap on the veranda glass doors of my bedroom and I would peer out, see one of the presidential guards and go around to open the door.

In would troop Sammy with two or three guards and they would play table tennis and pool for a couple of hours and then silently depart. This may appear rather strange to us but I realised very quickly that Sammy was frightened to sleep in one place for the whole of the night in case anyone made an attempt on his life. He was a very nervous president.

We did talk sometimes, though only at his instigation, and he did genuinely seem to want to hand over power in a democratic way. He was beset by the somewhat fanatical members of the PRC and genuine well-meaning advisors and bearing in mind that he was very young, he was finding the job very difficult indeed. One attempt at a counter-coup had already been thwarted and the insurgents had been executed by firing squad immediately as an example. Some of the power had gone to his head and it did not go down well that he was using the Presidential Jet for personal trips to New York from one of which I was to be a beneficiary. Sammy was immature and struggling to cope though his aims were noble and to be praised, he did not always receive the help he so definitely needed.

One night there was a somewhat earlier tap on my bedroom door and it was one of the guards who asked if I had a tie that Sammy could borrow as he had an important meeting with the Elders (I wondered what they were after?), I had a rather elderly relic in my wardrobe which I gladly handed over and thought no more of it until some weeks later when one of Sammy's guards delivered a package for me. It was a huge, luridly coloured tie that he had purchased in New York and was easily recognised from a distance of about three hundred yards. The problem was, that I had to wear the tie whenever I had a formal meeting with Sammy at the Executive Mansion.

By this time we had really withdrawn from Monrovia almost completely. We were legally bound to maintain an office in the town so we had a very small one tucked away off Randall Street and it was manned by Sam Cojolo's wife. She was not terribly efficient and we did not worry about that as George was very competent. As she was the wife of a senior police officer our office never suffered any hassle, threats or bother. She was brilliant at collecting outstanding monies as she would seize their cars with no help from her husband until they had paid.

Later in that year, we amalgamated Libco with PTP as an economy measure. The problem was that Rudi was on holiday on a remote island somewhere on the German/ Yugoslavia border and we needed his signatures on the legal documents to proceed. I still have no idea where I went but I flew to Zurich then caught a Swiss Crossair Flight, on which the co-pilot served refreshments, and landed at a small airport. Here I was met by Rudi and he drove me to the island where we quickly checked through the documents and then we did some serious drinking. I destroyed all opposition and Rudi even had a very sore head as he drove me back to the airport and I flew straight back to Monrovia.

I was now responsible for 1,500,000 acres of concessions, three sawmill production lines, and five forest operations. Our annual output was 70,000 cubic metres of round logs, 40,000 of which were exported while the balance was processed as sawn timber, mainly for export. Our annual turnover exceeded well over $12,000,000 and we ran at a profit two months after taking over PTP.

We employed a workforce of 350 workers, including ten expatriates. It was a mighty operation. I had tried to restart the moulding shop with a contract from Mothercare in the US but their quality control only allowed a 2mm differential from their specifications. My machines were up to the job but not my operators so a small team of carpenters used the shop to make the site's furniture and all that was required for the school.

I needed leave and head office was adamant that I must take it. Dida Rimker, Rudi Bunter's partner, who ran Contimba in the French-speaking countries, had invited me over to Abidjan for a weekend for some R&R where I had water-skied and attended a nightclub in Treshville, which, frankly, was quite boring, as the girls showed no emotion as they stripped. They were not even that pretty. This, however, was only a short weekend and now I was to go on one-month's leave. Rudi and Victor would take over while I was away and they needed another holiday when I returned to recover from the experience.

I flew straight back to the UK and spent a week relaxing with my parents then visiting relations before I drove down to Padstow where I had chosen a house and had agreed to complete the deal in a two-week time slot. It was a three-bedroomed bungalow at the top of the town in a new small estate called Rainyfields. It had been built as a self-build project of twelve bungalows. Mine was the second to be built and had used up 8% of the land available so I had a big garden. We quickly concluded the deal as there was no mortgage involved and the estate agent had pre-arranged all the legalities. I could just pay and move in.

The seller requested the full amount in cash. I knew he was trying to avoid his creditors and the taxman but that was up to him. He had some vague idea that the deal would be confidential. £32,000 in cash, withdrawn from Padstow's Lloyds Bank, that had specially opened on a Saturday morning so that I could collect the money. It was all over the town in minutes. He might just as well have announced it on Radio Cornwall.

I spent the rest of my time ordering furniture and buying everything I needed for the bungalow, all through Debenhams in Truro, but I bought essential furniture so that I could stay in it until I left for Liberia. I enjoyed having my own property at the age of 35 and it was all mine. I arranged for a neighbour as a cleaner and also employed a regular jobbing gardener to look after the lawn and plants. I met most of my neighbours and enjoyed myself in Padstow and its surrounds. I named my bungalow, Water-ma-Trout, which, in the Cornish dialect means, Wet my Throat.

Refreshed, I flew back to Liberia ready for the fray. We had a developing problem. The dollar was falling in value. Our sales were in European currencies. The exchange rate was becoming more and more unfavourable. My production costs were at $92 a cubic metre and I managed to pare this down to $85 but the slide was continuing. Contimba gained some breathing space by attracting American buyers but we were doing a fine balancing act and the profit forecast was not that healthy.

To be frank, I was also getting worried about the political climate. I sensed that there was unease in the air and in my talks with Dr Kassas, he was of the same opinion. There were further rumours of coup attempts and many senior officials were edgy which meant that their demands for 'dash' were rising. I shared my opinions with head office and Rudi flew out for a discussion. My overall view was that in the short term we were OK but at some time in the not too distant future there were going to be problems and it was better to be prepared.

We were very frank with the German Development Bank and I had written a brief synopsis for their perusal. I knew that Sammy had every chance of being elected democratically as President, that was a forgone conclusion, but it was the aftermath that worried me. Many of the Conga people were out to revenge the mass executions that Sammy Doe had ordered and could obtain financial backing. Remarkably, I also had another very unusual source. Opposite my flat in Paynesville was a big house and in it resided four members of the American CIA. I first met them when I was out jogging in the morning and we gradually became more friendly and sometimes had a beer together.

One of Donald Trump's election promises was that America would be withdrawing from its involvement in all foreign disputes. For the word 'involvement' I would prefer 'meddling' and, indeed, would also suggest that they often provoked the problem in the first place. I would dearly like to see the CIA files for 1975-2000 and their operations in West Africa for that period for I think they had 'many irons in the fire', which, they were stoking and were guilty of manipulation of the politics and affairs of Sierra Leone, Liberia, Cote d'Ivoire and even Ghana. The CIA agents were very clear that there were troubles ahead and I was to clearly understand that they would commence in the 1990s. Their prediction was spot on. How?

In essence, the German Development Bank was safe as it maintained substantial insurance policies for any event but we were not. Insurance in Liberia was not practical. To insure our whole operation would entail an annual premium of over $5,000,000 and this was only for the tangible assets. We could not insure against anything else.

I set to work wooing the LBDI Director and simply stated that Centra Holz Import were aiming for voluntary liquidation in the not too distant future as the directors wished to retire. I could not help but to mention that we were the only profitable company in their portfolio at present and perhaps they should take over our shares in the future. They mulled over the proposition for a little while and then came back with their demands. They would want full control of the business, including the management team, they would be the sole bank through which all company funds were to be directed and the project was to be theirs and theirs alone. We asked for clarification on several points and I tried to haggle on several details but in truth, we had jumped at the chance.

They did insist, however, that a full report on our business and operations be conducted by an international assessment team which duly arrived with their clipboards and cameras. I had merely told my staff that LBDI were conducting a survey and please to co-operate. I did receive a complaint, during the process about, Fred Sullivan and his lack of cooperation.

When I tackled Fred about his reticence to provide the necessary answers, he retorted, 'Answers? Answers? I can't understand the bloody questions!'

I had encountered Fred's bluntness in a previous incident. One of the front-end loaders had been brought in with a damaged 'prong' and Fred had spent much time late into the evening welding and repairing it. The operator had arrived the next morning and to test the prong, he banged it on the concrete forecourt. It had cracked. The operator was upset because Fred had allegedly called him stupid. This word is taboo throughout Africa and its use is forbidden.

I had to deal with the problem immediately and called Fred into my office. The expatriates called Fred Popeye because he had a very round face and seemed always to be smoking a pipe. When I asked Fred about the incident, he denied using the word stupid, and then added, 'I called him "effing stupid"!'

The situation was a little difficult as the operator was the son of Colonel Choloprey, Head of Immigration. I invited him over for a beer and recounted the story. He had the good sense to laugh it off and then blasted his son for annoying Fred who was an old man and entitled to respect.

The 'experts' were with us for two days but their only expertise was with their clipboards. They literally ticked boxes and left us very satisfied with their findings. They knew nothing about logging or the timber industry and charged the LBDI $50,000 dollars for conducting the survey. Please be wary of experts.

We handed over the company on January 1st 1984 and I spent the whole of the day signing documents until my fingers were totally cramped. We had paid off our expatriate staff handsomely but I was hit by Liberian Law! I had to stay in Liberia for two months to ensure that all was as we had sold. I knew about this but I objected. What about my living accommodation, flat, domestic staff, salary and transport and driver, business class ticket return to the UK, repatriation allowance and my small percentage of the profits?

The LBDI said that they would cover them but I knew that I could be left high and dry so they paid me an amount in cash of $15,000 which I gratefully accepted. I had offered to help PTP if the new manager wished but I was told that I was not wanted. I realised that the LBDI Director thought that he had got his revenge on me for my association with Sammy Doe and that I was out of a job, but I am afraid it was I who had the last laugh; for the company, under LBDI management, spiralled into terminal decline over a period of eighteen months and closed. The LBDI Director was fired. He had assumed control because he thought it would be a cash cow. He could appoint his cronies and friends to jobs at the plant, most of whom knew nothing about logging or timber production and also gain monies from the company as well.

No one had thought to remove the radio from my bedroom and I awoke early the next morning to 'trouble at mill'. The whole operation had ground to a halt. Evidently, the rather bombastic and dogmatic new managing director had addressed the workers the day before in what could only be described as a haranguing manner more suited to a school assembly than anything else and foolishly, adamantly condemned the 'dash' which he was not going to pay.

Geoffrey Saydee came on the radio and pleaded for my intervention. I felt very sorry for him but I could not interfere in any way. The sorry saga continued all that week and the workers went on strike with the union requesting that I be reinstated. This was not possible and some sort of deal was patched together and the project limped into action. I was very sorry, of course, that I could not help but the decision was the right one and I had rescued Centra Holtz Import from a very tricky situation.

I became a beach bum. I practically lived at the beach for two months and ate wonderful seafood. I also rescued a young boy from drowning who was a weak swimmer and had gone out of his depth. I wined and dined with friends but also ensured that David, my cook, was well set up. He had received a job offer from the British High Commissioner which he had turned down as his wish was to go farming. I gave him the $2,500 to buy the farm and he was delighted. I was very grateful to him for all his service to me.

I paid courtesy calls on Sammy Doe and Dr Kassas to say my farewells and the President was kind enough to express his wish that I would return to Liberia soon. I flew back through Lemgo/Head Office. In terms of salary and renumeration, the company had been very generous but I was surprised when Rudi gave me another very large cash sum as a bonus, for he was mightily pleased that I had managed to extract them from the Liberian situation. Returning to the UK, I was bronzed and fit, extremely well off and wondering what to do next.

As an addendum to this chapter, I have to add the following. Sammy Doe was democratically elected to the presidency in 1985, amid strong rumours of electoral fraud, and his time in power was noted for favouritism, corruption and the gradual withdrawal of American support. He became unpopular and in 1987, Charles Taylor invaded Liberia from Cote d'Ivoire with the aim of capturing Doe. This triggered a civil war and by mid-1990, most of Liberia was controlled by rebel bands each with its own warlord.

On 9 September 1990, Sammy Doe was requested to attend a meeting at ECOMOG Headquarters by General Quinoo who gave him a guarantee of safety. Sammy attended and left his bodyguards outside. Prince Johnson, a warlord, had found out about the meeting and rushed to the ECOMOG HQ. His men shot all of Sammy Doe's bodyguards and wounded Sammy Doe in the leg.

He was taken to Prince Johnson's base and interrogated, tortured and executed. His mutilated naked body was paraded through the streets of Monrovia. The rest is history. Charles Taylor was elected president in 1997 but was accused of war crimes in Sierra Leone, resigned and is now serving a fifty-year sentence in a UK prison at Frankland, County Durham.

Prince Johnson is now a Senator for Nimba County and Ellen Sirleaf was elected as the first lady president in Africa and has just ended her tenure of that post. Without naming names, the CIA agents in Monrovia had 'predicted' this sequence of events with alarming accuracy. One must also be aware of the problems in Sierra Leone and Cote d'Ivoire throughout this period.

Chapter Twenty
Water-Ma-Trout

During the previous months, when I had been battling away and then lazing on the beaches of Liberia, Water-ma-Trout had been occupied. My parents had sold up in Bristol and were intent on retiring in Cornwall so I suggested that they use the bungalow as a base to look around while they were house-hunting. The decision had caused problems because sheltered housing had to be found for my grandmother who was not at all happy about the idea, but no other family members were willing to take her in, so that was that.

I think it did cause a rift between my mother and my grandmother. At least my grandmother now had two houses that she could visit in Cornwall for short holidays and that was some compensation and she was taken out very regularly in Bristol by her other children and visited by thirteen grandchildren so she had plenty to do. My parents moved in, but alas, the furniture I ordered had not arrived. It was at least two months late so I vehemently complained to Debenhams and gave them a 48-hour deadline to deliver, berating them on their tardiness and bad customer service.

My neighbours were amused at the number of delivery vans and lorries that appeared from different parts of the country in the next two days and I received compensation. In one of those little intimacies of family life, my parents would not use the big double bed that I had purchased for my room but slept in the guest room with two single beds. My mother seemed embarrassed by this and was anxious I did not tell anyone. Now I have blown it.

I think my father would have probably wished to live in Padstow. He made good friends while he was there and played snooker at the Institute whenever he could. I think it reminded him, in many ways, of Barnstaple where he had been brought up as a boy. Unfortunately my mother was a townie. Padstow was too small for her and their eventual choice was a bungalow in Newquay into which they had just moved before my return. My mother had to leave her admirer behind. Eddie Murt, who had an artificial leg, was one of the local fishermen, had taken quite a shine to my mother and always greeted her warmly whenever they met. My father was not amused.

It was very pleasant to be back in Cornwall and I now had time to enjoy its life and scenery. I became a regular at The London Inn though I did also give my custom to all the pubs in Padstow. I was at the bar in The London Inn one evening when one of my neighbours mentioned that now I was unemployed (I suddenly realised that I WAS unemployed) I should be entitled to benefits. A quick phone call next morning revealed that I could not claim Unemployment Benefit because I had not kept my National Insurance up to date. I did try to point out that I had paid for fourteen years and that surely I was entitled to something. Surprisingly, I was not. They suggested that I contact the Social Security Office in Bodmin.

That very afternoon, at 2 pm I was visited by a very nice man from Social Security who filled in an enormous form. I gave him all the details of my salary as an MD and allowances and bonuses and gladly gave him the other facts that he required including my deposit account. He phoned me the next morning having calculated that I was entitled to a weekly sum of £300, vouchers for electricity and water, a big reduction in my rates and a Course for Executives in Truro for two days a week for which I would receive £36 a day and petrol expenses.

I was delighted. That was the good news. The bad news was that my deposit account held too much money and I would get nothing. I asked him if I could phone him the next day which I did and informed him that I needed to adjust the balance on my deposit account for the purpose of his records and amazing as it may seem, the account had less than the upper restricting limit. I must have been one of the very few people in England to actually phone a double glazing salesman and invite him around to conclude a deal in two hours and pay him on the spot. The salesman was delighted and so was I because I would receive far more from Social Security than the cost of the double glazing.

1984 was the year of the Miners' Strike and the terrible famine in Ethiopia and my conscience stirred. I made a few phone calls and then contacted Oxfam, offering my services and those of Fred Sullivan and Barry Gale to voluntarily organise the running and maintenance of trucks from Addis Ababa into the famine areas, based on our skills and experience of getting 35 logging trucks to port over dirt roads three times a week. They were not interested and I did try one or two other sources and again, met with a blank. We did try. I believe that we could have made some difference.

Padstow

Water ma Trout.

Padstow's history is said to go back as far as 2500 BC as there was a defined Celtic route from Brittany to Ireland which the Saint's Way, across Cornwall, celebrates. The town was founded by St Petroc in 520 AD and there was a monastery built in its distant past. In the eleventh century, Athelstan, king of Wessex, gave the town Right of Sanctuary which meant that fugitives did not have to reach the church to claim sanctuary. If they were within the town's limits, they would be safe.

The monastery was destroyed by King Henry and the Prideaux Family took over as lords of the manor from Prideaux Place which is still worth a visit. Sir Walter Raleigh lived in the town for some time and his original house is still on the quay. The port was always busy exporting tin, copper ore and slate and many emigrants left the quay for the colonies in America. Its major, consistent activity has always been fishing, however, even though that has had mixed fortunes throughout the passage of time. With the introduction of railways Padstow entered a boom period and that was the origin of its flourishing tourist trade for its population swells tenfold during the summer months.

The Atlantic Express ran directly from Waterloo to Padstow and brought much business and activity to the town and two goods trains left Padstow every day with fish for the London markets. The direct link was very valuable but the rail journey from Bodmin to Padstow was also stunningly beautiful with its wonderful scenery of woods and fields and sweeping vistas of the estuary. This is now the very famous Camel Trail which can be walked or cycled and is also a great ride on horseback. The Padstow Secondary School pupils attended school at Wadebridge and many travelled up and down by train every day.

The local trains had no corridors and there developed the unusual prank of stripping some unfortunate pupil and then passing his clothes along outside the train by use of the windows. The unfortunate thus had to emerge on to the platform at Padstow stark naked to be greeted by cheers and catcalls before he could reclaim his clothing. However, it came to an abrupt halt, so I am told, when a girl was stripped by her girlfriends and subjected to the same treatment. Fierce punishments were threatened and the practice stopped but I have often wondered why the practice was accepted when it was confined to boys.

Padstow has much more serious traditions than stripping boys on trains, however. It is known, may I say, world-wide, for its May Day celebrations on the 1st May every year, without fail, which is said to be linked with the advent of spring and ancient fertility rites. It is essentially a day for Padstonians for they celebrate it with much passion and energy throughout the whole of the day and their preparations are meticulous, thorough and very traditional and start some months before. It is a very serious affair which they also much enjoy.

Few people are aware, however, that the celebration commences the night before when the 'night singing' takes place. The old Cornish carols and songs are sang outside the older residents' houses especially the infirm and house-bound as Padstow is very good at looking after its old people. The next morning, very early, the Children's Horse parades through the streets with a crowd of youngsters in attendance. But what are these horses or 'Obby Osses' as they are referred to in Padstow?

They are costumes which basically consist of a very wide circular skirt with a solid circle of wood on which is hung the cloth and it is all painted black. The top half of the costume is a brightly coloured costume and the whole ensemble is completed by the wearing of a mask. It is quite heavy and the participants have to practise hard to get things right because the 'Oss must prance and dance. It has to swirl and roll and dip and sway to the sound of the drums and accordions of the accompanying players. I understand that it is exhausting work. There are two 'Osses' in Padstow—the Red 'Oss' and the Blue 'Oss'. The Blue 'Oss's stable is The Institute and it collects for charity. The Red 'Oss is the original traditional 'Oss, and is stabled in The Red Lion, and did not collect for charity during my sojourn in the town but I had to be a Red 'Oss follower as my cleaner and near neighbours were major players in the day's events.

It is said that the costume originated in an attempt to scare away foreign invaders and the creature danced on the cliffs until they turned tail and fled. I like the story even though it may not be true. I have already mentioned that I often prefer the apocryphal to the truth—it seems so much more interesting. You are also aware that I get this tingling sensation and a feeling of great awe when I am faced with living history and I used to get this feeling at the appearance of the 'Osses. A great crowd would gather, packing the streets and suddenly the horse would leap out to the cry of 'OSS! OSS WEE 'OSS!' and immediately start its prancing.

Each horse had a Teaser who would dance around with the horse to the accompaniment of the traditional May Day Song. It was believed that if a young lady was touched by one of the 'Oss's black skirts then she could expect a baby very soon. It was magic and it evoked a pagan atmosphere of the past. Thus the 'Osses would parade throughout the town all day watched and followed by thousands of people until they both met in the centre of town by the maypole where they were put to bed with a traditional song and I have to tell you that this was a genuinely sad moment for Padstonians and many would be in tears with sentimental affection.

My first May Day was fascinating and interesting and I really enjoyed the celebration. In the pub that night everyone was very happy and even the police popped in for a drink. They were needed for crowd control and isolated small problems but there had been few major incidents in past celebrations and had nothing to report on this, until…one of the radios crackled into life. There had been a stabbing. In these days of stabbings and knife crime, we are all aware of youngsters 'tooling' themselves up for nights out which can get out of hand, are we not? In Padstow, it was nothing new.

Stabbings did occur two or three times a year, almost invariably at family functions such as weddings, birthday parties and even anniversaries because there was a tradition of family feuds and bad feeling which could end in tears. I cannot recall a fatality but a stabbing injury was often inflicted. This particular stabbing was not true to form. A wife had stabbed her husband. She had been preparing her husband's crib (packed lunch) when he came back from the pub slightly inebriated. He became very sentimental and tried to give her a kiss but she chased him off to the bottom of the stairs, where he turned and avowed his love for her. She feigned to chase him up the stairs with the bread knife in her hand. He slipped backwards and the wife inadvertently stabbed him in…the buttocks! (No puns please.) It took him some time to live that down.

Another tradition was 'Darkie Days' when Padstonians 'Blacked Up' and toured the pubs singing and playing for charity. This was often led by a group called The Merry Makers who did magnificent work for charity. My parents were amused to have a smudge of black put on their faces when we were having a Boxing Day drink in The London Inn. Naturally the very term and the tradition fell afoul of the politically correct in London and it hit the national press in quite a big way, everyone having missed the point. It was nothing to do with racial discrimination or an insult. The local people had strongly objected to the slave trade when the port was used for transportation ships and blacked up as a sign of solidarity for the slaves and a wish to do away with the trade.

There was a mini tradition about which few people knew and I was only aware of it because I became a participant in the strange ritual. Having established myself as a regular in The London Inn I was often invited to stay after hours which could only be achieved by picking up ones beer glass, walking out through the front door and back in through a private entrance then into the bar again. The police did not seem to worry too much about late night drinking and I did get into conversation one night, well after hours, with one of the local bobbies.

A couple whom we knew as Clive and Mammie ran the pub. Clive and Kathy, the barmaid, were front of house while Mammie was on duty in the kitchen. We all watched carefully because Mammie's hand and glass would appear around the corner, reaching for the optic, put two gins into the glass, slide the hand back and we would all cheer. Clive and several regulars drank White Shield that required a careful knack of pouring but was not to my taste.

Kathy was amazed when I introduced her to my grandmother one evening, who ordered a Black Velvet, i.e., a Guinness and Babycham because she had not served one for many years. She was also surprised at the number my grandmother drank. Several years later, after Clive and Mammie had died, I was asked by one of the Directors of St Austell Brewery whether I could recommend their son took over the pub, but as he was a very young lad still going around on his skateboard, I thought a delay would be essential.

There were good restaurants in Padstow, though I thought Rick Stein's Seafood Restaurant was slightly overrated and his cheese board was quite shocking. Everyone jokingly calls Padstow 'Padstein' now but he has brought enormous benefits to the town. I used to take friends quite often to Rojano's near the Ship Inn (now a fast food outlet) where the meals were excellent and the Shipwrights Inn was also good.

Coleen, my cleaner, worked there as a waitress and when I had finished my first meal there, she presented the bill and I put my credit card on the tray. Two minutes later she was back apologising that they could not accept my card and did I have another? American Express would be fine. I quietly pointed out that it was an American Express Gold Card and she screamed.

She had not seen one before as they were rare in Padstow then. Nowadays, I would imagine that everyone had them. I had the same reaction in Wadebridge when I purchased a washing machine and tumble-drier. Wadebridge was a pleasant town and had most of the shops that I required for my needs. It was my habit to call in to Chatterbox, which was a coffee bar at the far end of the arcade, the whole area having been taken over by Boots now. Chatterbox was the home of Wadebridge gossip and it was run by two lovely ladies one of whom ran 'Bloomers' on the High St. Her erection of a sign depicting a pair of knickers had caused much consternation in the sleepy town of Wadebridge for a florist she was not.

The council had objected and given her much hassle and she was ripe for revenge. Opposite Chatterbox was a shoe shop which closed down and so we hatched the plot to apply for a licence to open a private shop selling marital aids. Our application was outrageous and was clearly dated 1st April. It should have fooled no one. All hell was let loose and it was debated in the council chamber and the press much enjoyed reporting the coverage. APRIL FOOL! We tried a similar trick with Padstow Town Council the following year, seeking permission to introduce camel trekking to the Camel Trail and submitted a spoof feasibility study.

They caught the date and replied that permission could be granted if we would pay the cost of raising the height of the five or six bridges on the route. It hit the press and

TV as a fun spoof and everyone enjoyed the joke, though I bet the camel world was disappointed. I do understand now, however, that you can go camel trekking in Cornwall?

There was no Polo in Cornwall so I rode out at Tresallyn Stables. There were set rides and sometimes we were able to go down to Harlyn Bay and gallop along the beach. The stables were owned by Rex Henry Trenouth and run by 'Mrs Coley', Jane, and there was always fun to be had there. I never fully understood the relationships between Rex, his wife, Esther, and Jane. Esther was always portrayed as a dragon but she was far from it. She was the district nurse and a very efficient and good one at that. I stayed on her good side. When the nearest hospital is 26 miles away she often helped in an emergency and very kindly did so when I was to take school groups to Padstow later. I was very pleased when she was awarded the OBE.

It was a glorious summer. I never had the hood up on the MG until September and I was travelling around Cornwall quite extensively. I attended the Executive Course in Truro that was held at the St John's Ambulance Headquarters I timed to coincide with my Truro shopping trips so I could kill two birds with one stone. The course leader was a very nice chap, especially as he never questioned my expense claims, but he seemed to have no experience of working in an executive position. He was a civil servant with a course handbook.

There were three other 'executives' on the course but they were traders and small business men who had hit an unemployment patch like me, not executives, but I did not mind. We half-heartedly looked at new inventions like a tractor to work on slopes and a new type of fish hook machine as well as visiting tourist businesses. We enjoyed expensive lunches and always ended up in a pub. I generally had a congenial and pleasant time and I could do my shopping.

Boggio was a very expensive men's outfitters in Truro and I took to buying my clothes there as money was no real problem to me. I was caught out one day by the owner who knew that I was looking for a leather jacket and insisted that I try on a new one that had just been delivered. It was superb and fitted me warmly and snugly while being supremely comfortable.

'How much?' I asked tentatively.

'One thousand three hundred pounds,' was the reply.

'Put it back!' I said.

I was not that well off.

I decided that I was going to have a house-warming party. Acting upon my neighbours' advice, we decided on a Saturday lunchtime as the optimum time for neighbours and friends. I had arranged everything and I thought the amounts I purchased were probably overkill but how wrong could I be? I had over thirty guests and we had a wow of a time with everyone tucking in and drinking as if there was no tomorrow, and the party moved out partially into the garden. I had a long hedge down one side and the older Padstonians got quite excited. One of them gave me a sly nudge.

'I 'ad my furst gal under that there 'edge,' he smiled.

There were nods all around and I reckon that many a young maid had lost her virginity under that long hedge.

I had a yard of ale and we spent Saturday evening in a drinking competition but the bravest lad was the one who drank a yard of Padstow's water. It was not long after the water pollution incident at Camelford Water Treatment Works when everyone's hair had turned green. Padstow's water always came out of the tap a milky colour due to its high aluminium content anyway so we were not bothered. By 9 pm, food and drink were running low so I phoned up Keith, who ran the local shop, to deliver more supplies and drink.

By midnight we were on the port, a very popular drink in Padstow, and this disappeared so quickly that we had to knock Keith up again for more bottles. At 2 am there was a knock on my front door. It was the police. Well, actually it was a policemen who had travelled out from Wadebridge in his Panda. There had been a complaint from someone somewhere but we were not making any noise at all.

Everyone knew Fred and he was invited in and joined the party and stayed with us until he received an irate call from the station demanding to know his whereabouts. Never let it be said that he was drunk when he left us but he was quite unsteady. The party finished at about 11 am on the Sunday with everyone demolishing a full English breakfast. It had been quite some do but I did not listen to anyone else's advice again about the best time to hold a party in Padstow. Mine had taken nearly 24 hours!

There was always something interesting going on in Padstow. At the town's Regatta Day I took part in the Birdman Competition as Prince Charles on his polo pony and achieved a watery jump of five feet from the harbour wall.

One of my neighbours had decided that fishing was scat (no good) for the short term and opened up his own milk round. He begged me to accompany him to a special camp to which he had been requested to deliver, full of naturists, because he was too shy. I assured him there was no need for him to strip off to make a milk delivery but I had to escort him and he found it very difficult going around the individual tents and kept his eyes fixed firmly on the ground. He received an unexpected call from a visiting tax inspector one late afternoon enquiring about his failure to send in a tax return.

My friend pleaded that he was illiterate and too embarrassed to contact the tax office. A special clerk arrived to help and a tax return was eventually made. Six months later my friend was in a Truro bookshop perusing the novels and he received a very friendly tap on his shoulder from the tax inspector who enquired after his progress. I also drove his milk float around the Padstow Carnival Route but I did get my own back a couple of years later. How? I had a party of Mount House boys at the house and we decorated the Land Rover and cage-like canoe trailer as a spoof vehicle for Padstow Safari Park with the porter's telephone number emblazoned on it. Sid and Coleen were fun. They provided me with endless tales of their activities.

One evening at the Buff's Lodge they decided that they would like a Chinese take-away and Coleen and a friend drove over to Newquay to get it. While they were waiting there was much activity up the street and the Police and Fire Brigade arrived to attend to a burning car. It was Colleen's and Sid would not believe her when she phoned up and told him. One of my immediate neighbours was a postman and he seemed to spend a lot of time in the loft of his bungalow. His wife was most annoyed to discover that it was full of Playboy magazines, hence his interest in the loft.

Early every morning I would go for a swim in the sea either at Harlyn Bay or round to St George's Bay and I would also often walk early on the many cliff paths in glorious coastal scenery. Between Padstow and Newquay there were seven bays and they were advertised on tourist posters as 'Seven Bays for Seven Days'. The tourists really did not worry me. I was out early and my bungalow was above the town away from the day trippers and holiday makers. It retains its popularity and is even busier than before. I used the bank and Buckingham's the grocer which was very traditional and sold quality groceries.

My father, an ex-grocer, loved walking around the shop and reviving his memories.

My parents were now well settled in Newquay but I am afraid that I never liked the town. It was a pleasant family resort when we visited on holidays but to my mind, it had become somewhat of a monster with ill-assorted fast food shops, cheap tourist attractions and the inevitable worse aspects of surfing. To me it was cheap and nasty. My parents liked it, however, and became involved with the Cancer Research Book Stall in the main street on which they served for many years. Their bungalow was bright and airy with a small garden at the back.

On their first night in the bungalow, my father was awoken by a loud noise. It was a roaring sound between a grunt and a cough then there was another loud roar. My father woke my mother and she told him it had probably been a dream based on me in Africa. Then came a louder roar and she realised that it was a lion. They had not realised that their back garden was only a road's width away from Newquay Zoo and the house was close to the lions' enclosure. The first day my parents moved in, The Sun newspaper published the banner headline about Newquay. SUN, SEA AND SEX CITY with reference to the amorous antics of the surfers and young visitors. Now I knew why my parents wanted to move there!

Having mentioned amorous antics, I once escorted a lady friend over to Rock for a swim and some sunbathing. It was interesting because this particular lady friend had arrived and wore clothes that were more appropriate for a London boutique. She was always overdressed. On this occasion, however, it was the opposite because she took off her bikini top and sunbathed topless. Ten minutes later, Geoffrey, the beach master arrived, puffing on his pipe. 'Geoffrey, I am sorry to have to tell you that I have received a complaint about this here young lady sunbathing topless on the beach. I have to say that I can find nothing in the by-laws about such a prohibition, so carry on. There's a grand view from where I am standing.'

My grandmother came down to stay with my parents for a few days before I took her over to Padstow for a weekend and she liked being taken out on to the road leading to the Headland Hotel. Why? She could see all the young surfers stripping off and changing and thought it was disgusting as she changed her reading glasses for her long distance ones and peered intently at them.

My parents were certainly more relaxed in their retirement though their marriage was certainly not a 'bed of roses'. My father could be quite jealous at times and was still easily exasperated with quite small things. My mother had always expressed a wish to own an expensive set of cast iron saucepans and I arrived to drive her over to Truro to choose a set, but she had changed her mind. My father had been difficult that morning and my mother simply said that if she had the cast iron saucepans, she might well murder my father so I took her out for lunch at the Headland Hotel instead.

In many ways, my little MG Midget was a godsend for I could only take out one passenger at a time so I was able to have quality time and individual talks with my parents, my grandmother and my brother Roger. He was now SWEB's agricultural advisor for Devon and Cornwall and lived with his wife and two children in a nice house in Perranporth where I was able to stay once or twice. I liked his house, I thought the town was grotty and loved the huge expanse and extensive length of the beach out of which Roger got little pleasure.

He rarely visited it at all. He was fairly stable and seemed to be going through a reasonable patch though there were links to him and one of his young lady helpers in the Cub Pack he was running. I took them all out for meals and invited them over to Padstow for water-skiing and a barbecue. Everything seemed to be quite stable though there had been quite a blip in the not too distant past. I met up with him at the Royal Cornwall Show and we had a riotous evening over at the Carlyon Bay Hotel with him and the other reps.

I felt duty-bound to take the ferry to Rock (Kensington-on-Sea) to visit the Reverend Mr George K Booth, my college lecturer, who had one of his present students cutting his large Leylandii hedges. We sat outside in the sunshine drinking tea and demolishing a packet of custard creams I had brought over with me. He looked well and seemed in good health. I thought that I had escaped any request for help but suddenly I saw his eyes narrow and he looked at me intently; a request was about to be made. He wanted to go over to Caerhays Castle Gardens on the South Cornish Coast to see their selection of camellias in which he was passionately interested. Could I drive him over?

I had to agree and the next week we made the trip. I knew George well enough to know that when he said he was interested in looking at plants, he would also be swiping cuttings and such was the case and this trip was no exception. He raided every camellia in the garden and stuffed the cuttings into his sports jacket pockets. As a priest, I only hoped that his confessor was a gardener.

In case you were wondering what I was doing about my status as an unemployed executive, I had sent off several desultory job applications and regularly took the Times Educational Supplement to peruse its columns. I had thought about setting up a business but in a fairly half-hearted sort of way. If I had known then, what I know now, then I would have had no hesitation in starting up a rocket farm. Not the space type but the plant type. Every chef, every cook, every menu, every meal, every sandwich and every salad garnish uses blooming rocket. Why? Why this mass introduction and injection of the wretched plant into every culinary corner of our lives?

I could be a multi-millionaire by now with rocket farms all over England. Hindsight is a splendid thing. I had also signed up with Gabbitas-Tring, the educational job finder, but was not impressed with any of their offers to date and quite frankly, I was quite well off as I was. I estimated my benefits as being £1,500 per month and I was certainly enjoying life in sunny Cornwall. There had been no rain for months. I think someone must have been shuffling around and organising my future life because I suddenly faced a combination of small opportunities that I would not have considered individually.

Firstly a friend was keen to try indoor/paddock polo on his farm near Alton, in his biggest barn, during the winter months, and wondered if I might be interested to help on two evenings a week? Rangitikei were also keen to sign me up for a selection of summer matches at Windsor because I had dropped down the handicap to a minus two (worth two goals). Neither of these offers were viable until I was phoned by the Head of Cheam School who was desperate to find a short-term member of staff for one year only to teach Geography and RE with all found.

Added to this, I had not indulged in any form of culture for four years and was keen to catch up on shows and exhibitions in London. I accepted the job for one year only on condition that I could play and carry out my polo commitments and did reduced duties. I accepted Cheam's conditions and they accepted mine but my MG would be unsuitable for all the travelling and I had another stroke of luck. My Uncle Dennis and his wife, my Auntie Violet, loved a day at the races and we attended the Newbury Racecourse one afternoon. Now, I do not bet but my uncle did and he asked my opinion which horse I fancied as the winner of the first race as they were being paraded around the ring.

I chose a grey, of course. It was on its toes and eager to go, eyes alert and strong in the hindquarters. It won. In the next race, I chose a complete outsider but it looked good to me…and it won. I chose four winners, another that came second and my choice came third in the last race.

Unbeknown to me, my uncle had backed every one of my choices on some sort of accumulator bet and won over five hundred pounds. We travelled back to Bristol in his quite luxurious Opal Commodore car which I coveted and it was not that old. He took my auntie and I out for dinner at Thornbury Castle that night which provided a splendid ending for the day and it was a fine celebration. Some weeks later he phoned me to say that he was putting the Opal up for sale and was I interested? I was and I purchased it at a very reasonable price so now had a very fine set of wheels, on the condition that I would escort him on another race day. I agreed.

Very sadly, it was not to be. He contracted cancer and died after a few months. An amazing thing happened on the day that he died. I always called in on my way back up to Cheam and when I arrived, he was busy with an insurance salesman. He had told the salesman that he was terminally ill with lung cancer and the salesman had stated that he could still sell him a policy there and then. My uncle asked me to witness both statements which I did and he signed and paid for the policy. The salesman left and I drove off sometime later, not realising that this would be my last goodbye. As soon as I arrived back at Cheam, my mother phoned to inform me that Dennis had died and the insurance policy paid up a cool £30,000.

Cheam was a very old-fashioned prep school with few modern buildings and was really a big country house with outbuildings attached. Its headmaster was Michael Wheeler, who was very keen on keeping parents and staff separate, and he had a splendid wife, Hilary, to whom I took an immediate liking. She had her own hunter, hunted regularly and was very sensible and good with the boys. Prince Charles had attended the school and absolutely hated it and the parents were very much of the 'Gin and Jag' set.

The staff was friendly, though quite traditional, and there were some splendid characters on board. I was given the usual detailed instructions about my teaching, that is to say, none, so I got on with teaching my English, Geography and RE. I was under a little pressure as regards RE because it was a Common Entrance form but all I really needed to do was a revision of the main Old Testament stories and a quick flick through the parables. I made it as much fun as possible and dare to say that I made RE interesting. It was well known that the head always gave the RE papers the quick once-over and warned the boys what was coming up anyway.

The English was with the 10-11 year age group and it was a valuable exercise for me to get back into such teaching as I had been out of the profession for nearly eight years. I liked that age group anyway so it was no trial. I did a full day's duty during the day on a Monday (no polo on Mondays) and a dorm duty on a Thursday. We were not expected to patrol the dormitories for everything was left to the matrons. We were really only on call for any emergencies which I thought was a wonderful arrangement.

If I was free on a weekend, I took the boys canoeing on the Kennet and Avon Canal and I also ran a week's canal cruise on the Oxford Canal for fathers and sons. The school had no camping gear so we could not camp or go on expeditions though we hiked in Savernak Forest and high up on the Marlborough Downs, but it did have woods and a splendid ornamental lake that was used as a swimming pool. I was not expected to take games so I introduced many of the boys to woodcraft, building bivouacs and simple cooking which all ages much enjoyed.

I taught canoe rolling on the ornamental lake as well. We sometimes received a royal spectator. If the queen Mother was racing at Newbury, she would drop in with her packed lunch and enjoy our antics though she was mainly interested in the ornate Italian garden which she had discovered when visiting Prince Charles as a pupil. She was always accompanied by her chauffeur but I never saw any security guards ever. One lunchtime one of the boys was showing off and stayed under the canoe for some time before he rolled back up. The chauffeur thought he was stuck and jumped into the lake just when the boy flipped up again. The queen Mother laughed and Hilary had to dry the poor man out. I told the boy that he should put that in his CV when he was applying for jobs.

I was able to lead quite a social life. Many of my polo friends were quite close and so I was wined and dined, I attended and helped with the indoor polo at Alton on two nights a week and I was able to catch quite a few shows and exhibitions in London. Later on I played five or six weekend matches at Windsor during the summer term. It was far from the hectic life that most prep schools required.

A couple of weeks after I started at Cheam, Michael wanted a quiet word to inform that he, and only he, dealt with the parents and he would rather I did not wine and dine with parents. I carefully explained that as far as I was concerned, they were not parents but polo players many of whom I knew already and was I expected to turn down their invitations? In fact should I turn down the invitation, I had to dine with the Irish Guards at a mess dinner in London that very evening as one of the fathers would be there. Sometimes, he could be quite silly.

The staff was very traditional indeed and you would not find such a staffroom nowadays. On my first evening at the school, the staff were sitting down to what was quite a formal meal when there was a hammering at the door. It was ignored. Conversation continued. More hammering at the door. It was explained that no interruptions could be made to staff supper. Hammering. Pounding. Thumping. The history master stalked to the door, flung it open and shouted: 'Blasted boy. What do you want?

'Please sir, is there supposed to be a fire in Dormitory Four?'

The history master rode a moped for transport and one night his worried wife telephoned the school to say that he had not arrived home. I offered to drive along his route directed by Mike Kidd, the PE and Games Master. After ten miles of unlit, winding main road we found him pushing his broken down moped along the side of the road with no lights and he had removed his shirt because it was such hard work. He was also quite deaf.

One lunchtime, there was an almighty roar and he burst out, 'Did you call me bastard, boy?'

He was holding up one very frightened junior boy by the scruff of his collar and shaking him like a rat. The poor chap had merely asked him to pass the mustard. The PE Master was also quite a character who lived in a small cottage at the end of the school lane at the side of the school and he brewed beer. It was awful because he never had the patience to wait for the full process to be completed and I could never drink it. He was a quiet and sociable chap but was a Jekyll and Hyde character if any altercation of dispute started in a pub. He just had to join in.

We always had to bundle him out to avoid him fighting. The Science Master was a kind gentleman whom the boys respected and liked but one senior boy tested his patience when he was in the changing rooms lining up for lunch so was whacked three times with a dap that turned out to be a running spike. The boy's father merely said that it had served his son right.

I had two accidents at Cheam. One was a very bad cut across my mouth from a bungee that snapped while I was loading canoes. It needed eight stitches, six outside the mouth and then two inside the gum, that the doctor said did not need to be numbed. After he had painfully stitched the two, I said that I should have disagreed with him because it hurt like hell.

I was then whiplashed in an accident on the road outside the school while a passenger in the backseat of the car and suffered a big haematoma on my left leg. I had an X-ray the next day and the hospital said it would heal but would I like some physio to help it along? A very attractive young lady physiotherapist proceeded to massage my leg for some twenty minutes and then asked if I would like to return in two days' time? Would I? I eagerly returned for the next appointment to discover a middle-aged man was giving me the treatment and there was no sign of the young lady. I bet they did that to all the male patients.

The school was very close to the American Base at Greenham Common which was being besieged by a large group of ladies protesting about the presence of American cruise missiles in the UK. I accept the right for people to protest in a civilised manner but some of the ladies were quite fanatical and individual bad behaviour caused problems as they were often abusive and very rude. Unfortunately, most were living in very poor and unhygienic conditions with inadequate sanitation or washing facilities. Groups of them would often turn up at pubs and many customers left, not because of any strong views, but purely because of the terrible smells. It was pretty awful I have to say and I had been used to the slums of African cities.

Prior to my appointment, the senior master at Cheam had been appointed to the Headship of Mount House School in Tavistock in Devon. Word reached me that he could be interested in employing me in the following year so we had several telephone discussions in the Easter term before I was invited down to meet him in the summer term. I only had a one-year contract at Cheam so there was no problem with giving a term's notice or the normal procedures when one resigns. A post was offered and I accepted and so began my longest period of employment ever. I will, however, leave you with two stories about Cheam boys.

In my RE class, I taught the grandson of a bishop who was heading for Eton and during the exam marking period, I received a fairly late-night call from the chaplain at Eton. There had been a question on The Last Supper on the CE paper and my candidate had written an extensive answer completely mixing and blending The Last Supper with Belshazzar's Feast. It was highly amusing and the chaplain was going to use it in his sermon in the main school chapel on Sunday. That was really going to cheer the Etonians up.

When back in Padstow, I received a phone call from the bishop who had a holiday home in Trebetherick. He had heard about his grandson's RE essay and would like me to come over for dinner which I duly attended, hoping that there would be no theological discussions over the meal. What they really wanted me for was to tell the boys to 'GO TO BED NOW' so the adults could have some peace and quiet. Both boys went to bed and I had an enjoyable supper from which I suffered no theological indigestion.

The last incident was a classic. We had a pupil who was the son of a famous racing driver but please let him stay anonymous. He was full of mischief which I like in my pupils. The boys' bathing facilities at Cheam were all in one big room underneath a flat roof with skylights. When the boys had finished, the matrons used the facilities either at night or in the early morning.

Three of the young matrons were using the baths one evening just soaking themselves and relaxing when one of them looked up at the skylight. There was the round face of the reprobate peering down at them in the middle of their ablutions. They screamed and fled. The boy was given six of the best but he reckoned it was worth it.

Chapter Twenty-One
Mount House

My interview at Mount House with Charles Price, the new headmaster, in his wonderful study with its magnificent views of the grounds and Tavistock below us in the valley of the River Tavy, followed the true lines of the prep school interview 'classic' model variety. Brief mention was made of teaching I think but Charles knew what he did not want. He did not want pupils stuck in storm drains, or pupils behaving like savages in the school grounds, or even near-drownings on the school lake. He required a thoroughly well organised activity programme for the junior and senior boys (no girls…yet) that was interesting, constructive, structured and safe.

Looking out over the school's grounds, its woods and river, the small lake nestling at the foot of the slope and distant views of Dartmoor, I knew I would be available to achieve that. The site and situation was perfect for the 'adventure challenge' that I was to devise for the school to run on Tuesdays for the Juniors, on Thursdays for the Seniors and many weekends and holiday activities. The 'adventure challenge' was to play an important role at Mount House for many years.

There was, however, one problem. Charles was offering me middle school English. I enjoyed teaching English and unlike many prep school English teachers, I had been trained in the mechanics and technicalities of the teaching of the subject. I was very qualified to take that age group but I had already been a Head of an English Department and I hesitated to take the post. I was very keen and eager to run the adventure programme and I liked the school and the staff to whom I had been introduced but I did not want to be in the position so prevalent in prep schools of waiting again for 'dead men's shoes'.

I thought it best to mull things over for a while. I would never have stayed on at Cheam but the only other offer I had received to date was an invitation from the President's Office in Liberia to become an educational advisor to the government. I sent a letter declining the invitation, thanking Sammy for the honour he wished to bestow on me but pleaded family commitments. He did include me on his Christmas card list until the year he met his ghastly death.

In the meantime, there seemed to be some glimmer of hope for me at Mount House. It was all rather vague somehow and perhaps, I should have requested a more formal response? I was led to understand that the present head of English, who was also the senior master, would be retiring in the foreseeable future when the post would become available and the post could be mine. That was my firm understanding.

I accepted the post at Mount House and moved into a spacious flat above the headmaster's study with even better views than he had on the ground floor. I had spent another glorious summer at Water-ma-Trout in Padstow and I did devote much time to the formulation and development of the adventure challenge programme. I was very clear about the various elements of the challenge in that it needed a practical series of interesting activities, carefully graded to ensure progress, and capable of being largely conducted within the school environs and above all, it had to be fun.

The challenge was integral to the various tests and the large variety of adventure activities that would be organised. I borrowed heavily from the Cub and Scout programmes and was able to present Charles Price with my format and proposals by late August. He liked the scheme and we went to print immediately. Then followed the arduous task of gathering all the practical equipment we would need and the provision of a budget which was a slight problem as I had missed the April deadline.

The bursar, Major Bob, baulked, but Charles pushed everything through. This was to be an ongoing problem throughout that year with the adventure challenge and my English teaching and I strived hard to establish my preliminary budgets and an expensive one for the following April, then forecasts for the next three years. Major Bob was impressed and surprised. This was unheard of at Mount House.

In essence, for the challenge, the juniors had a series of twelve tests to complete in each of the three sections of Bronze, Silver and Gold. They included basic house chores such as ironing and making a hot drink, developing into simple cooking, indoor and out; making simple kites and modelling things out of junk craft; tree climbing; grass sledging; simple tree and plant recognition; building a den; survival swimming; simple first aid; care of a bicycle; a short hike; a visit to a place of interest; dressing up and community service.

The seniors had a more advanced series of activities and tests such as advanced first aid; canoeing; building bivouacs and backwoods cooking; fire-lighting by different methods; survival swimming in cold water; hashing and letter-boxing on the moor; pot jumping; river running and the 'gun run'.

There were very few schools that had the natural surroundings of Mount House. It was perfect in nearly every respect especially as I was able to build, repair and alter the existing facilities on site. Mount House was set on a high hill overlooking the town and a series of plateaus descended to the athletics track and cricket square at the bottom of the site near the old stone bridge over the stream that was the outlet channel for the small lake. The River Tavy flanked the whole north side of the 42 acre site and was ours to use at will.

Tucked away in the north-east corner was Rowden Field, large enough for several football pitches, and more importantly, where the boys could let off steam as loudly as they liked. On the opposite edge of the field was a large area of woodland that was on a fairly steep bank leading up to near the back gates. The slopes were perfect for our grass sledges and turf skateboards with long continuous downhill rides, up which the boys then had to drag the sledges. They became quite fit. The juniors, and also some of the seniors, when they thought no adults were watching, liked to roly-poly on the steeper slopes and usually ended up swaying almost drunkenly dizzy when they tried to stand up at the bottom.

The lake was used for boating and canoeing almost throughout the year. We had what one might describe as a multifarious diverse and assorted fleet. We had proper canoes and small sailing dinghies, rowing boats, and the old type of sit-ons with a hump on the back. From my boyhood I had remembered learning to row and paddle on Trenance Lake in Newquay and I wanted the boys to experience the sheer fun of mucking about on boats. All safety rules were strictly applied and these did not inhibit the fun we had on the lake. I also held survival swimming sessions in the lake and I taught canoe rolling in the adjoining swimming pool. We held regattas and slaloms and the boys loved raft building and we held a couple of Kontiki expeditions when the boys took it in turns to sleep two hour shifts on the raft overnight, on the water, while the others slept out under the stars on the lakeside.

The River Tavy is the second fastest flowing river in England and our half mile stretch had a weir, small rapids, many rocky obstructions and a pool 150 yards long in the lower ends of our grounds. In flood it could not be approached and no one was even allowed on the river path in such conditions but its normal flow provided us with fun and sport. The weir had a salmon ladder with deep sections which we used for pot jumping. For all river activities the boys would wear a wet suit, helmet, trainers and a buoyancy aid and I would carry safety lines and a first aid kit.

Swimmers could leap into the deep water from each section then run back up along the bank to repeat the process finishing off with a swim in the river. The big, lower pool was used by the more experienced canoeists to get them used to a river current and it was a swimming spot for the school anyway so quite safe. If a boy capsized he received a badge stating, 'I can do Half an Eskimo Roll.' The main river sport was river running which the boys thought was fantastic and of which they never tired. Dressed as above the boys would each have a car inner tube on which they would shoot the rapids and float down the river. We even had to run the activity for staff and parents. As they say nowadays, it was so fun.

Rowden Field was often used as a site to introduce camping skills and many a boy, and later girls, spent their first night under canvas down there. If they chatted into the night it did not matter for there was no one to disturb. We could also play rough games down there such as British Bulldog and it was also safe in darkness to play night games. The boys liked to be frightened and could be nervous in the dark but were always excited when we played Will-o'-the-wisp, Flag raiding and Raiding the fort. It was a good way to tire them out as well.

All age groups enjoyed the woods where they could build bivouacs, light fires, try backwoods cooking (with mixed results) and even sleep out in summer. There is an atmosphere almost mystical and magical when you sleep in woodland especially if you have been told a ghost story before turning in. We would creep along to the badger set at the far end to watch the antics of these nocturnal creatures and there were several steep slides for the foolhardy.

I discovered an old set of metal cart wheels complete with their axles in the long grass at the far end of Rowden. I borrowed them and with the help of the welder who had constructed the climbing wall, we built a cannon which could be easily dismantled and assembled and thus started the now famous Mount House Gun Run. Not the easy one that the naval teams compete in but a cross-country obstacle course around the school grounds which was fiercely competitive. It was a sight to watch as a team of eight boys manhandled the gun all the way around the school and provided much entertainment, though one or two visiting parents became quite startled as the cannon lumbered towards them.

During my first year, I built an adventure playground at the top of the drive, not the sterile and boring ones you see in parks nowadays, a proper wooden one with scramble nets, rope walks, solid walkways, ropes to swing on and tubes to crawl down. It was a huge success and was popular for all age groups. It was to be gradually strangled sometime later by health and safety. First, we had to lay down a protective surface of wood bark which was worse than the soft ground on which it had been built. Then consecutive sections had to be removed because they were too high or too low and the concrete tubes had to be replaced by heavy plastic.

Towards the end of my time, the whole thing had to be replaced by a professional company and it was so sanitised and risk adverse, it became unloved and unused. I am afraid that I am one of those people who stoutly and constantly deplore this eradication of risk in all our lives. Children must learn to deal with risk in the outdoors as it an important aspect of their development. In this high-tech age, I suggest that all the children of today would love to try all the activities listed above and thoroughly enjoy them.

The school extended on to Dartmoor. Thirty minutes of walking or a five-minute drive and you could be on the 365 square miles of moorland that was Dartmoor so it was possible for our pupils to go out on it even during a games period, and we did. This is where we letter boxed, hashed and hiked.

Letter boxing started in a very small way when the more fit and able Victorians and Edwardians reached Cranmore Pool which is on the North Moor, ten miles from Okehampton, and is one of the bleakest, soggiest and marsh-ridden areas I have ever visited. It was considered a challenge to get to this god forbidden place and a box was provided for the visiting cards of the intrepid traveller who succeeded. From this developed letter boxing. Small containers are hidden all over Dartmoor (and in all other areas of the UK now) with a rubber stamp and a small book. By following clues in a pamphlet, issued by the Dartmoor Letter Boxing Club, you can locate the boxes.

Once found, you can 'collect' a copy of the stamp and put your stamp into the book. Thousands of people do it and some people have many thousands of such stamps. The member of staff who first took the boys out letter boxing admitted that he thought it was interesting for the boys but was half-hearted himself…until he caught the bug. He went on to be one of the 'big game' hunters and collected several thousand.

For those readers who have never hashed, it is really a run following a pre-set trail of sawdust markings. I would go up on the moor in the early morning and lay a two-mile trail, which the boys and a member of staff would follow in the afternoon. We would also do it on camps and expeditions.

I suppose that one might suggest that I was spending all my time on the adventure challenge? It was not so, as my teaching was keeping me fairly occupied as well. I had been issued with the usual detailed instructions by the head of department for the handling of same. i.e. none. It would appear that I was just to keep them busy and the word 'English' never even entered the conversation. I had to dig down into the pre-prep and first forms again to locate reading and handwriting schemes on which to establish my teaching.

I was fascinated by the air of surprise expressed by the junior school staff as no such enquiry had been made before. School children appreciate consistency when being taught and most of the boys I now taught were competent with their use of English, reading and writing and were able to build on that already taught. New boys, that is to say, those arriving in the second form and above, came from widely different teaching backgrounds, however, so I needed to treat them on a more individual basis.

I never forced them or attempted to bring them into line with the mainstream pupils. They could carry on using the methods by which they had been taught. I would only take the pupils back to stage one if they were dyslexic or had major problems. I was also mindful in my teaching of different rates of learning dimly remembering my time at primary school where I had to initially sit and listen to slow readers until I was allowed to go on and read at my pace. I heard all reading individually but let the faster readers surge ahead at their fast speed and bothered them but rarely.

I tested reading regularly so that I could establish progress and I had one boy who was not only making no progress but actually regressing. I asked Matron if his eyesight could be checked and it was very poor. Glasses were fitted and he shot back up the reading scores again. As an English teacher, I was always under attack about spelling and what was I doing about it? To me spelling is an important English skill but I was absolutely convinced that constant spelling tests were not the answer.

My classes did spelling corrections and all my pupils had a list of the words generally misspelt pasted into their English exercise books and everyday one of the class would bring an obscure word and the class tried to define its meaning. We also followed my usual technique of alternating different types of comprehension with story writing every week. I also had vocabulary charts displayed of good words that had appeared in their stories. Again, I tried to inject interest and fun into the lessons and made the classroom an interesting learning environment with charts and pictures and the pupils' work displayed. I still had enormous enthusiasm for teaching and had very clear views on educational methods and systems.

As a resident assistant housemaster, I was expected to do dormitory as well as school duties. The school day was quite long and you were on for twelve hours, and they were busy hours, as you were still teaching and doing activities. Two other members of staff were on with you and one always hoped that you would be with the hard working ones and not one of the shirkers for one or two of the staff spent most of their duty day talking about how hard they were working rather than fulfilling their duties. If only two staff were effective then the duty day was even harder.

For many years, I suffered when supervising the dining room for lunch and tea. There was a cumbersome system whereby the pupils were lined up outside the dining room and then filed in, sat down and then got up again to receive the first course, sat down to eat it, then got up again if they wanted seconds sat down again, then got up again for their sweet, sat down and ate it and then got up again to file out, table by table. They could thus get up five times and sit down five times in any one meal. That took up to three quarters of an hour.

The thinking was that the staff could sit down and eat their meal supervising their table in a civilised manner except that the non-duty staff were up serving food, no one was supervising the tables and there were forever admonishments about the noise. It continued long after I arrived until, hallelujah! self-service was introduced. I had already been through this experience at St Bede's where self-service was the only practical solution to a dining room too small to take expanding numbers and it worked like a dream, as it did at Mount House. It cut out all the hard work, hassle and noise and meal times became quite pleasant and very civilised.

On the day duty, prep had to be supervised after tea and the school looked after them until 8.30 pm but there were the usual bunch of parents who arrived late to pick up the day boys dragging the duty on until 9 pm and were most upset should we mention that 8.30 pm was the latest pick-up time. Boarding duties were normally from 7 pm until 9.30 pm when the prefects had 'lights out' and one patrolled the dorms, putting lights out according to the age group, chatting with the boys and ensuring all was well.

When I arrived, the first housemaster was unsuited to the job. He was slightly temperamental and not really suitable for residential care. He used the slipper with a vengeance and must have worn out several pairs. He was also a little too fond of the bottle and often smelled of drink. He was rather like my grandfather because he would walk into a pub and offer drinks all around yet he left the school with many debts including over £600 owing to the milkman. I used the slipper but once and it was in anger and I have never hit a child since.

Benbow Dormitory was for the oldest boys and right next door to my flat. After lights out I heard unusual noises from Benbow and investigated. The dormitory had a huge bay window with curved glass that was highly valuable which three of the Seniors had opened and climbed out on to a veranda glass roof. It was a very stupid and dangerous act. I carefully got them in and stood them out in the corridor in their pyjamas and gave each one of them six of the best and made it hurt. I informed Charles and he added his wrath to mine.

Unlike Cheam, it was the duty master's job to patrol the dorms and maintain discipline but it was also a time when the boys could relax. Those of you who have families and struggle at bedtime, just imagine what it was like to put eighty boys to bed. The school conducted regular fire practices every term with the whole school during the day, and another for the boarders at night. There was a senior prefect in each dorm which was very good for the younger boarders.

The problem with the night alarms was that the boys had duvets and if a boy was well snuggled down inside it, you could miss him on the dormitory check. During one famous week, there was a recurring fault with the fire alarm system and we had to evacuate the building nine or ten times. The local fire brigade had to attend every one and they were delighted at £35 a call out. The boarders were whacked, however, and in one lesson, I let the day boys read while the boarders slept.

Charles Price was easy-going though he could sometimes get annoyed with staff, me included, and he would often act on unsound advice but he was labouring under one fairly major handicap. The retiring head remained in the old headmaster's house halfway up the main drive, seemed determined to adopt a Mr Chips attitude to the school and was very reluctant to let go the reins. This was grossly unfair on Charles, especially as this was his first headship. The interference was not welcome and I know it was a big bone of contention.

At governor's meetings, individual governors would visit the ex-head for consultations and advice. I liked all the governors I met but this was unprofessional and a grave discourtesy to Charles. It was not a good situation. It was also unprofessional on the ex-head's part. Retired heads do not interfere with their previous schools. The Chair of Governors should have sorted all this out but did not.

Unfortunately, the ex-head was a sailor and owned a Drascombe Lugger in which he took the boys sailing out of Plymouth in adult life-jackets. He was then quite elderly and infirm, should not have risked the boys in such a way and I did express my concerns to Charles. The final straw for me was when the Adventure Store was opened one Sunday, while I was off caving with another group, using the spare key from the Bursar's Office, and he took four canoeing buoyancy aids to take a group out. It was thick fog, they became disorientated (lost) out by the breakwater and they had to be towed back. I hit the roof and sent a stinging letter to the ex-head, Charles and the Chair of Governors and I changed the locks on the Adventure Store. There were no more sailing trips.

By the end of the first year, I had built up the adventure activities to include caving, climbing, canoeing and hill-walking, the latter mainly on Dartmoor, converted my old classroom and an adjoining loft and workshop into the Adventure Store where the boys could change, the gear could be stowed and where I could conduct maintenance and repairs. The downstairs changing area could also be used for drying the tents which, invariably came back wet.

Beside this area was an old shire horse stable complete with fittings. In the good old days pupils had used it to stable their ponies which they rode to school and one of the fathers remembered that very clearly. We converted it into a climbing wall. We did have a plan, honestly, but the floor was cobbled and very uneven. The huge metal extra wall we welded and placed at an angle took some fitting and then there was the drilling out and screwing on of holds to complete.

It was called the Giraffe House because of the very high hay grill for the shire horses. I think it is still there. It had much use during my time and generations of Mount House boys have climbed, Jumared and abseiled in it. This was the second climbing wall that I had built and there was yet to be a third.

Our initial caving was at Pridhamsleigh on the other side of Dartmoor, near Buckfastleigh. It is best described as a multi levelled muddy maze and our boys, then girls, enjoyed crawling and wriggling through its myriad of tunnels and passageways down to a subterranean lake and then back by another route. De rigueur for caving was a light wetsuit, covered by the boys' boiler suits wellington boots and helmets. Our initial lighting was carbide lamps which were not perfect by any means and could give you a nasty mini burn. We later modernised to battery lights which needed to be recharged after every trip.

I was initially ignorant of the staining powers of Pridhamsleigh mud but the matrons soon informed me in no uncertain terms and we had to invest in canvas boiler suits for caving. Upon exiting the cave, we would wash off in a nearby stream but everyone needed a shower on their return to school. Pridhamsleigh was on private land owned by a farmer, who not only made his own cider but also drank large quantities of it. He made quite an income out of the entrance fee for the cave which I suppose was entered on his annual tax return?

He waged a constant ongoing battle with the local constabulary as many of his agricultural implements and machinery were always parked on the grass verges on the side of the narrow road. He was very cantankerous and treated all in authority with contempt. Thus he was jailed for three months for being in 'contempt' of a court order. He was incarcerated in Channings Wood Prison near Newton Abbott and he loved it. He thought the food was good, there was plenty of TV and entertainment and he enjoyed talking to all the inmates. He reckoned it was a good holiday. His sons were less happy. They were left to do all the work on the farm and grumbled every time I knocked on the farmhouse door to pay the caving fees.

There was another cave at Buckfastleigh not far from the church, which, incidentally has a fibre glass steeple, called Bakers Pit. Its use is not encouraged for general caving and access is restricted by the William Pengelly Caving Centre nearby. Its entrance is a wide vertical 40ft shaft which goes down through the remains of rubbish and debris from when the area was a rubbish dump to a large pot like chamber from which passages lead out to more passages. Reed's Cave is connected to this system and is fascinating because it has a very rare stalactite/stalagmite formation called the 'Little Man'.

It is only six inches in height but looks like a miniature gentleman in a top hat and it appears to have outstretched arms caused by side draughts. It is directly under the family tomb of the Cabell Family in the church. I was able to get two groups down to see it but it is now fiercely protected and the entrance from Bakers Pit has been concreted up.

I organised several regular events at Mount House. I ran sixteen ski trips while I was there to add to my collection of six that I had organised at Wycliffe and St Bede's. We visited Andorra, twice, which I did not rate highly as we had poor snow conditions and the resorts were too small even for school groups. On the first trip there, I was at the bottom of the chair lift and another colleague was at the top and I gradually became aware that the boys were beginning to act quite strangely when their group checked in.

Upon fuller investigation, it appears that the boys were buying hot chocolate drinks at the top station…laced with rum. The second trip to Andorra was odd as the resort rep was not expecting a school party and had put my group in rooms over five floors which

I insisted had to be changed immediately after the first night, and one boy's suitcase was missing. I was told that I could buy him new gear which I did in the local town and fitted him out with new ski gear, smart casual attire, a warm jacket, wash gear, towels and underwear. We returned to the hotel to find his suitcase waiting in the foyer and that young man had two sets of clothes for the trip and his mother thanked me profusely on our return.

We eventually found a resort that suited us and I wish I could say that it was I who discovered it. We found it by force of circumstance. We had booked at Sauze d'Oulx, a resort that I knew well and where I was well known. Two months before we were due to fly out the company phoned to say that there was a problem with the hotel and we would have to change to a French resort called Les Orres. I immediately phoned my friends at the Palace Hotel and enquired about the problem. There was no problem.

The Ski Company was rationalising their bookings and moving groups into their hotels which were under booked. I contacted the ski company and complained very strongly about the change and was visited by an area manager. He did not have a leg to stand on, especially as I was very friendly with the hotel owners in Sauze d'Oulx. It was a fix. I agreed that we would try Les Orres at a substantial discount for each party member and subject to an inspection visit. It was agreed and the next weekend I flew out to Grenoble and a taxi took me to Les Orres. The hotel was nothing special but the resort was very good and excellent for schools.

Furthermore, at the time of my visit, snow conditions were superb and I was able to ski for two days and dine and wine in French restaurants. Thus we found Les Orres and we visited it many times building up a relationship with the hotel, the resort and many of the local people. This made running a ski trip much easier and it is perhaps at this point I should mention that, contrary to many peoples' opinions, running a school trip is not a holiday for the teachers and organising a ski trip may be the hardest of all.

I always took the view that the pupils were my absolute priority and were there to ski properly and have fun and I could not have run trips without the staff who very kindly accompanied each trip and provided superb support. There is also the view that staff should never be in mountain restaurants while the pupils are skiing which is daft. A normal day for our school party was ensuring the group was up and about, had a good breakfast and then taken over to their ski school classes. At each stage, staff would supervise this. One getting the boys up, another already supervising breakfast, another at the ski store checking that everyone had their correct skis and boots, hats and gloves, etc., then two members of staff would walk the group over to the ski school where we would all assemble and do a double check.

Lessons were all morning with instructors during which the staff split into two shifts. One shift could ski until mid-morning and the other shift would man the restaurant at the top of the mountain, another at the midway point and I would be at the bottom. The shift changed halfway through the morning. Everyone, including the instructors, knew this system and if there was a problem, it could easily be solved by contacting one of the staff.

In the afternoon, the more advanced skiers would go off with staff while I supervised the beginners on the nursery slopes. Group skiing was also used. The boys would be given a defined small chair lift and could go up and down in groups of four checking in as a group at the bottom each time. When possible, I, or a member of staff, 'ghosted' the groups under instruction to check on the standard of teaching with the British and French instructors.

I had handled this quite diplomatically with the head of the ski school over a couple of glasses of wine and valued my input into the handling of classes, especially as I never criticised the competence of the instructors, only teaching techniques. Much of my time was largely devoted to the daily organisation of the trip keeping one step ahead of all skiing and the evening activity programme.

On our first trip, we experienced one or two problems with the ski company staff. They employed young people from the UK as waitresses and domestic staff. Unfortunately, one or two had a problem with independent schools and showed it. This was unacceptable. The company also employed British ski instructors and it was decreed that we could not use the bar before supper as they gave priority to the instructors who needed time to relax. Fortunately the ski company manager also resided at the hotel and I complained that both attitudes were totally unacceptable and the 'attitude' and bar prohibition disappeared. There was, in my experience, for many ski companies to treat school clients as second-class customers, an attitude that I would never accept and I always insisted that we receive proper service and civility.

It was the time when the French ski schools were running their closed shop policy and put constant obstacles in the way of fully qualified foreign instructors in direct contravention of employment law. On one trip to Les Orres all our instructors were arrested and taken down to Gap, the nearest municipal town. The ski company then faced a huge dilemma as we had no instructors and I did have every sympathy for them so we took over the boys for the whole of one day while they sorted out the problem. They hired in French nationals from all over the area for the remaining few days, even some from posh Chamonix.

I did, however, protest to the mayor of Les Orres, with whom we had become friendly, stating that, it was unforgiveable for the ski school to disrupt our group's skiing and very bad publicity for the resort. If it was not sorted out then we would be unable to return. I believe the point was taken. At least, there were no more problems.

Skiing was the only activity where I was prepared and expected an injury. We had two serious non-injuries. One was a leg injury that the local doctor plastered which was immediately removed on return to the UK. The other was a back injury which appeared to be serious. The boy concerned had to be taken down to the main hospital at Gap and admitted for several days. We were obviously concerned and I hired a car to shuttle staff to and from Gap for a member of staff stayed with him day and night.

It was a dark blue Renault Megane and at that time I also had a dark blue ski jacket. I was initially mystified by the courtesy of the local gendarmes who waved me on and through police barriers and speed traps with a smart salute which I always acknowledged with a wave of thanks. I was enlightened by Lydi in the ski shop, from which we hired our skis, who laughed and then explained that they thought I was a police officer.

The ski shop became my office and a base for everyone for Lydi loved children and she and the staff were brilliant at sorting out problems ski based or anything else. More importantly, she provided early morning coffee which was my reviver at any time and as she lived in Gap, she popped in to see our patient daily with sweets and goodies. The Head of French, Hugh Walkington, known as Huge Hugh because he was very tall, mainly consulted with the medical staff.

He spoke impeccable French and committed a slight faux pas with the senior consultant by suggesting that the patient be moved to a specialist hospital for back injuries. He was assured that he was already in the most specialised hospital already. Thoughts of the boy being medevacked back to England were dispelled when he started walking. He flew back with the rest of the group and was running around after a couple of days.

My deuxieme amour, Lydi, married a handsome gendarme with whom she is still living very happily in Gap.

The ski trips were gradually attended by girl pupils, families and friends and I began to baulk at the sheer physical organisation of travelling by coach to airports, boarding planes, finding luggage and then boarding another coach to the resort with a large party, so, as an experiment, we tried going out by coach all the way. The adults would have preferred to travel by air but then they were not supervising fifty children en-route. I was delighted.

As far as I was concerned, it was brilliant. Everyone got on the coach at the start of the journey and got off at the end. My only task was to check the group off and on at rest stops. We used the Channel Tunnel and even I was surprised when one of the staff suggested that the children let off steam in the adjoining huge compartment and even gave them a football to play with.

On our return, a different official was horrified at the suggestion. I was surprised to discover that apart from our double-decker coach, there was only one lorry and a solitary car on the whole of the train and they were right up at the front. The other advantage with a coach was that if it took you out then it was sure to bring you back. If a ski company collapsed while you were in a resort and you had travelled by air then you could be in trouble. We stayed with coach travel.

Parents introduced a new dynamic to ski trips. Our pupils were well behaved by anyone's standards and were good at following instructions and warnings. Some parents were not. We had several scares with parents losing their way on the pistes and ignoring safety instructions. I became increasingly wary of adults in such circumstances and often took an extra member of staff to cope with the extra responsibility. Parents are often a mite too quick to criticise teachers for lack of good supervision while being offhand with theirs and I came across this on many occasions at home and abroad. Parents also underestimated the high ability and standards of many of the pupils especially in adventure activities.

Every year I organised the sponsored walk in aid of specific charities with the proviso that all money raised was used for specific projects, equipment and designated groups. Our donations were not to be used for administration purposes. I would plan out a route of about fifteen miles either on a circular route, taking in a chunk of Dartmoor, or organise buses to a dropping off point at the same distance.

I was helped on the day by the Dartmoor Rescue Group that patrolled and helped with the route and ensured all was well and that all dogs were under control. We provided refreshment stations and a lunch stopover and a minibus service for the collection of cars if required. The staff were always keen to help but then that might have had something to do with the fact that they then did not have to walk? Unaccompanied pupils were actually escorted by staff so that they were not left out. I would patrol around in whichever Land Rover the school had provided for me to help with the adventure activities bearing in mind that I was often towing trailers as a necessity for large amounts of kit.

Our first Land Rover was quite elderly and cantankerous so I named him George after my college tutor, the second was a slightly more updated model which was faster and slimmer. She was obviously a lady and we called her Georgina. She was superseded by an almost new Land Rover that the pupils then called Georgie. Seventeen walks were organised and we raised well over £50,000 for local charities. The children were amazing, for even the youngest completed the walk and to everyone's amazement, many of them would then play football or tennis after their return while the adults collapsed into the dining room for refreshments though all the boarders slept well that night.

One of the walks commenced at Whiteworks, south of Princetown, with its infamous prison. One of the parents was the doctor for the prison and had left his car there and driving the Land Rover out to collect the cars left at Whiteworks, he asked to be dropped off at the main gate of the prison. As I drove on, I suddenly realised that there was a stunned silence in the back of the Land Rover and I just could not resist saying, 'Oh yes! A very sad case I am afraid but they do let him out on parole occasionally.' The doctor was quite amused by the story.

Another annual event were the Junior and Senior Barbecues which were held on the 'beach' by the side of the River Tavy when all the juniors on a Tuesday and then all the seniors on a Thursday, helped by a small army of staff and parents, cooked a meal over an open fire. The results were varied but we had much fun and it was a good adventure for all.

By tradition, there was a Leavers' Camp which had taken place in the school grounds marked by much disturbance, ill-discipline and outrageous behaviour, I understand. This had always been held when the rest of the school was taking their summer exams, the leavers having finished their Common Entrance and Scholarship Papers. I was asked to take this on and decided, immediately, that it should be off site, so the venue became The River Dart Country Park and furthermore, it would be adventure-based. The park at Holne was perfect for the camp as it was set in acres of woodland on the banks of the River Dart on the southern edge of Dartmoor.

The boys cooked, camped and looked after themselves, after careful instruction, and caved, climbed, canoed, letter-boxed, and visited a place of interest, in rotation, during the day then completed an evening programme of the blindfold sisal trail, orienteering, a river walk and a parachute jump. The initial formula was a success and continued for many years at Holne Park, Padstow and in the Brecon Beacons. The staff at the park were very good to us, especially David Rowe, the senior instructor, though the campsite warden was a nightmare and complained about everyone and everything.

One peg was an inch off one of the boys' sites and he blew up, another boy had dropped a teaspoon in the washing-up area which led to an apoplectic fit and his final rant about baked beans clogging up the washbasins was the last straw as our group had not had baked beans that morning. I complained to the park manager and he did not bother us again.

Our site at Brecon was at a very smart caravan site—Pencelli Castle, which I found in slightly unusual circumstances. Sometime before, I had run a small overnight camp in the Brecon Beacons so that we could make an early start to go caving in Porth-yr-Ogof Cavern over at Ystraefeldte in the Black Mountains. Upon arrival at the site I had booked it became clear that its advertising had been well overstated. It was long tussocky grass in a field full of scrap metal and the ablution block looked as if it had not been cleaned since the Second World War.

It was dirty and insanitary so I drove on through Talybont-on-Usk and out on to the Brecon Road. At Pencelli there was a very smart caravan site and I very cheekily drove in to see whether they had a pitch for one night. The owners were marvellous immediately offering us a site and thus began my very long friendship with Gerwyn and Liz Rees and Craig and Morgan, their two sons, who are now grown up and run the site with their parents.

On the first night, we held a Leavers' Camp where all 30 of the boys disappeared when they went over for a shower. One of the staff walked over to investigate. They were ALL in the farmhouse kitchen drinking hot chocolate with Gerwyn and Liz who were charmed by their liveliness and good manners, so much so that they wished the same for their boys. We helped get both of them into Christ College in Brecon—it was no great problem because both boys were good at rugby.

Logistics were always a problem as we captured all the school's minibus fleet and initially had all our catering supplies delivered daily from the school. Staffing was not a problem as some teachers came with me and I hired specialist instructors. There also developed a tradition of taking a military parent with us and/or a friendly marine who held the occasional inspection in true military fashion and would take the whole camp on a route march/ run early in the morning and scrutinise the washing-up minutely.

One marine, Marty, was a colour sergeant from Norton Barracks in Taunton, who arrived on a brand new Kawasaki 1000cc motorbike. When we were turning in he announced that he was going off for a ride. He came back the next morning having been to Newcastle upon Tyne and back. When he left the marines he ended up 'baby-sitting' geologists in Alaska. All staff were generous with their time and the camp and expedition programme that I ran could never have been run without them. They were good humoured and long suffering, I thought.

At the Padstow Leavers' Camp, we always stayed at Dennis Cove which was run by Barry and Anne who were charming and kind but Barry was always nervous about his campsite and facilities which led to minor tensions. We were on activities all day—cycling on the Camel Trail, cliff climbing, orienteering in Cardinham Woods, canoeing, tombstoning and walking on the coastal footpath and Barry came up with the idea of rotating the boys' showers throughout the day, to save the wear and tear on the shower block, which was not possible or practical.

Barry and Anne's concerns were always for the care of their site so we did try to help as much possible and they were not like some campsite owners who were like martinets insisting on daily inspections, raising groundsheets and banning barbecues. Prior to one camp a parent had donated a tent to the school which he thought was so complicated that he should go down and put it up for us the evening before. Now I have spent several years of my life under canvas and had dealt with a huge variety of tents but I accepted his offer.

I received a worried phone call from Barry who was concerned that it was getting dark and the parent was struggling and swearing in obvious difficulty. Barry was kind enough to help him which cost me a bottle of wine. When I arrived on the foreshore site the next morning there was this huge frame tent with six or seven rooms which the boys immediately called, 'The Dorchester.' Upon my departure from Mount House the Leavers' Camp continued but at a pukka site with chalet tents and all facilities in the middle of a holiday park in Cornwall.

The boys always camped using an extended version of the Blacks Vango Mountain Tent which provided them with a porch shelter in the event of bad weather. We first used them on Lundy where the warden was insistent that we site the tents almost up against the Linhay which I thought was a very bad idea as a westerly gale had been forecast. We were blown away and the tents were shattered and all over the field. The ridge poles had all broken at the point of their extension and thus began a long drawn out battle with Blacks which was eventually decided with a duel. They were adamant that we were erecting the tents incorrectly so after many months, they agreed to send one of their 'experts' down to Devon to check.

We met at D-Sails at Saltash with Don, the owner, as referee and witness. The 'expert' arrived and asked me to put up the tent. I agreed and also gave him a tent, instructing him to erect it on the other side of the building so we were out of sight of each other. I had mine up in ten minutes. The 'expert' had not really started and clearly had no idea what he was about. I threatened legal action and suggested that the press would be very interested in the duel. They paid up with handsome compensation. I had already solved the technical problem by fitting an extra sleeve over the point of extension and was kind enough to inform them of the adaption.

The Family Camp was a big event when the whole of the school grounds were taken over by our parents and offspring. Every plateau and flat piece of ground was used with the most fantastic variety of tents, caravans, campers, horse boxes, lorries and bivouacs. It looked like a cross between Glastonbury Festival and a refugee camp. The town of Tavistock was stripped bare of all groceries and provisions and off licences were depleted.

The owner of Creber's, the famous delicatessen, phoned the school to enquire what on earth was going on as he had run out of nearly everything. A full programme of events was held including a barbecue and donkey derby and there was much socialising and merriment until the early hours. The lake was open to all comers and we had a tug of war competition which ended in a grand match between mothers and fathers. The mothers won. There was a service on the cricket field on Sunday morning and more activities and competitions until lunch after which came a general wind down. We estimated that over five hundred campers enjoyed each annual camp.

Another annual task I had, though not in the summer, was to help let off the fireworks in the school's brilliant November 5th celebrations for which there was also a popular 'Guy' competition so we normally ended up burning about twelve 'Guys' atop a huge bonfire carefully constructed by the ground staff. The fireworks were spectacular and the display was superbly organised by one of the parents whose brother produced the fireworks professionally except one or two experimental ones that were normally included. I found myself starting with three inch mortars, then four inch, then five inch, then six inch, but when the seven inch was unwrapped, I declined. It was someone else's time to help out.

One winter half term, I had planned to take a small party mountain biking in Normandy. Unfortunately, there was a flu bug around which knocked out the small group one by one and I faced cancellation until Geoffrey Whaley and his wife, Jane, and their two boys, Tom and Jo, stepped into the breach. We were staying in quite a large modern farmhouse with wood fires and I am afraid that we depleted the store of firewood very quickly and had to make raids on the main farmhouse for more fully loading up the Land Rover.

It was agreed that the family would sightsee while I was dropped off for mountain biking and then picked up again at the end of my route. I really enjoyed this as I rarely had the chance to mountain bike alone without a school group so I had three glorious days of mountain biking around Normandy free and alone. On the night before our scheduled return, it snowed and it was the first snowfall in that region for fifteen years. We had the four-wheel drive Land Rover but I was towing a trailer so our exit out to the main road was difficult to say the least.

We were blocked at one point by a little Citroen that had slid into a ditch. Here I must mention that my fellow Geoffrey was of large muscular build and had putted the shot for England. I grabbed one side of the car and he the other while the driver reversed the car. It slid out quite gracefully except that Geoffrey was still in the ditch clutching the front mudguard of the vehicle in his hands. The main road was littered with abandoned lorries and vehicles through which I carefully threaded until I came to a hill (there are one or two in Normandy) on which vehicles were stuck but I thought I could see my way through so chose a crawler gear and edged my way up and through. Jane informed me afterwards that she and the boys had kept their eyes firmly shut while clutching onto the backseats.

The second half of the Summer term was very busy for me for I not only ran the sponsored walk and the Leavers' and Family Camps but I also had to set and mark exams and write academic and adventure challenge reports. As a form master, I then had to check all my form's reports and finish off with a form master's report before they were handed in to the headmaster. In those days, they were all written by hand and we had no opportunity to write them online.

Now, it will be pointed out that all staff did this but I am afraid the division of labour was unequal and it follows that this applied to remuneration as well. Many staff worked long and hard hours and gave up much of their time to the school, but some did not, yet they received the same pay without putting in the time or extra effort. One did receive free accommodation and food if you were resident which was always reckoned as part of the package but you paid quite a high price for this in loss of free time. I am a great believer in rewarding staff for work and effort and the use of differentials and incentives. This never applied at Mount House. I received no reward for my extra efforts especially when three major hurdles presented themselves.

The Head of English and Senior Master announced his retirement and I waited for the word to become Head of English. None was forthcoming so I broached the subject with the Headmaster who assured me that there was a slight delay. Unbeknown to me, however, there were other schemes afoot. A local prep school was closing and one or two Mount House governors were on its board as well as ours. The first hint that I had that all was not right was when a young temporary Head of English was appointed who had no teaching experience.

I sought immediate clarification and discovered the truth. The new Head of English would be arriving when the other prep school closed in the following year and he was also to be the Senior Master. There had been no advertisements for either posts and it is my very strong belief that Mount House could have done much better if they had done so. It was a retrograde step to appoint a traditional Senior Master keeping the school in the past. Sour grapes? No.

I would have been quite happy to apply for the Head of English under a full and proper interview process and there were other staff who would have welcomed the chance to apply for the post of Senior Master. The whole affair was unprofessional and, may I say, underhand. Mount House deserved a professional and younger Senior Master and not a relic from the past. I now found faced the prospect again of working under a Head of Department who solely did his own teaching with the top forms and never had any interest in the English department as such.

He was not a head of department—he was a senior English teacher. Period. As a Senior Master, he was staid and quite boring. Prep schools were changing with the introduction of professional and dynamic staff to whom the pupils could relate and our new man did not fit this mould. Unfortunately for me, he was also to take over as housemaster in the main school building and relations were quite strained right from the outset. A good married couple would have been the perfect answer as house parents, not a traditional Mr Chips.

I was now in the horns of a dilemma. There was apparently no hope of becoming Head of English at Mount House and as I explained to the head, I would never have accepted the post at Mount House but for that promise. I was to be faced with another problem, however, of which I had been gradually aware had been building up since my arrival at Mount House and was brought into sharp focus by the canoeing tragedy at Lyme Bay. There was the question whether the canoeists should have gone out at all that day bearing in mind the sea state, imperfect weather and inexperience of the instructors but the whole tragedy could have been averted if a support boat had accompanied them.

The British Canoe Union quite clearly discourages powerboats being close to canoes and they are right. Powerboats and canoes do not mix. Accompanying paddlers and expeditions at a distance, however, is a different matter. If a boat had been 'standing off', the deaths would never have happened. That one incident led to stringent conditions being applied to the adventure activity world in Great Britain. Up until that point, I had run all my expeditions to high safety standards and with tight controls but the only qualification I held was as a canoe instructor.

Like hundreds of other instructors I now had to formally qualify to conduct the adventure activities at the school. This was not a problem in itself but it did require another huge time commitment on my part as I now had to qualify in climbing, caving, orienteering, hill walking, mountain biking, first aid, life-saving, powerboating and cycle proficiency. I completed courses in Devon, Somerset, the Brecon Beacons and North Wales and within a year had full instructor certificates for ten activities.

As a school, we were not classed as an adventure centre and did not have to be legally inspected but we underwent an annual adventure audit conducted by staff from Plas y Brenin, The National Mountain Centre in North Wales. One or two of our activities flummoxed the inspectors. River running, pot jumping and the gun run caused them much amusement and bemusement but my life-saving qualifications and canoeing certificates covered the first two leaving the gun run to be left under the umbrella of first aid.

Many of these had to be renewed on a three-year cycle and it was quite a task to maintain them. Chaos was caused by the introduction of the child protection safeguards and protocols. No one could possibly object to their introduction but each and every organisation tried to insist that we had to take their course. That would have entailed me attending what was basically the same course, ten times. It was daft and reason eventually prevailed. I still consider the safeguarding checks as very important but when, oh when, is the government going to carry out its promise for each individual to hold one CRB that covers all. At time of writing I hold eight such checks which are very expensive. It is a ridiculous situation. Going off on so many courses made me the butt of much staffroom humour, I am afraid, but it had to be done.

Now arose a technical hitch. I had, since the time of my first appointment, requested my contract of employment and it had never materialised. A new bursar was appointed and he panicked to get the contracts issued and quick. The problem was that it was different from the terms that I had been told especially as regards to retirement age. I was told 65. The new contract stated 60. I would not agree to its change.

Now at that time, I was producing The Royal Hunt of the Sun, again, involving the whole school. 'A cast of thousands'—well, over 100. Just as I was about to start the dress rehearsal, I was called into the head's study. Charles had taken 'advice' from a somewhat dogmatic and ill-tempered Latin master who was a retired headmaster and had stated that Charles should insist the contract be signed or the member of staff should leave.

Unfortunately, Charles put this simple proposition to me and bearing in mind all of the above, I said that I would leave, thanked him very much and walked up to the hall to conduct the dress rehearsal. In the middle of the 'Climbing of the Andes', Charles furtively appeared at my side. He had panicked and phoned the Chair of Governors. He assured me that I could stay as long as I liked.

I produced many plays at Mount House both large scale and small one-act plays with my English forms and they were a riot normally based on the dramas written by Jonnie Ball. I liked producing them in the front hall where we could use the balconies and stairs but I also did another production of The Golden Masque of Agamemnon. My Head of English appeared to have no interest in drama.

I was, however, considering my position especially as I was asked to take on even more work. The school had four houses which were used in school competitions. I was involved with Hawks and when the master in charge of that house retired Charles assumed that I would take it over. His assumption was overturned. I point-blankly refused. I did not wish to add to the already incredible workload that I completed.

Other staff were not doing half the hours that I did. NO. I was not a willing horse to be flogged. The time had come to draw the line. I was not angry or annoyed because that rarely gets one anywhere. I merely did a thorough internal review and audit. I discussed this with Charles in an amicable way to let him know my feelings. I could not work under the present Head of English though I had now been given a Common Entrance form to teach and acted as their form master. The boys were charming but not of high academic ability and one was really teaching at a remedial level to get them even a low English mark while the Head of English was teaching 'la crème de la crème' with the scholars and high ability Common Entrance classes.

I was rarely asked to take games because of my commitment to the adventure programme but when I did I instituted the game of footballs (note the 's'). We would start a match with one ball and then I would throw in another ball, and another, every two minutes…until there were a dozen balls on the pitch and then let the game continue. It was huge fun and every player was able to join in. That is how I felt with the job at Mount House. More balls were being thrown at me all the time and more kept coming. In the middle of my deliberations another 'ball' was tossed on to the field.

A parent approached Charles Price during the school sports and placed a cheque for £14,000 into his hand. It was a deposit for seven daughters to join the school the following term. Now I am reliably informed that every Headmaster faces his conversion on the Road to Damascus at some time and it is recorded that Charles was not quite ready for his…yet.

'Over my dead body!' he retorted.

The next term, however, a girl did join the school and she was the first girl to attend Mount House, though she was such a tomboy, she immediately became one of the boys. That opened the floodgates and the following year the school became fully co-educational and thank heavens, the Headmaster was still alive. It was an interesting fact that, at that time, I was the only member of staff with experience of teaching in a co-educational environment.

It was also interesting to have more ladies in the staffroom which prompted a much more gentle air of civility and politeness. Now Charles knew this and was keen to draw on my experience but I had already informed him that I would be resigning at the end of the year as a sign to indicate that I was serious about going. One governor has informed me since then that a mild panic set in for neither Charles or they wanted to lose the adventure challenge which was not only so valuable for our pupils but also brought much kudos to the school.

Again, my guardian angel decided to get off his backside and help me out. I am not sure if it was a reluctant or divine intervention at the time. The Head of Geography announced his retirement so that he could concentrate on his farm and lead a more relaxing lifestyle. I did think at the time that relaxation and farming would not sit well together. I had a chat with Charles. I proposed that I took over the geography department and gave up house and dormitory duties, thus being able to concentrate on the ever growing adventure challenge and also hand over play productions to have more free time. He agreed. I stayed.

It took nearly a year to achieve but all went to plan. Charles thought that I might not have liked having to be form master of nine-year-olds as the geography room was situated near the junior classrooms. On the contrary I liked that age group and we got on well and had lots of fun. I also enjoyed teaching throughout the school from the third forms right up to the scholarship form.

My world lightened and brightened and I was now able to organise a department properly and well but this was not through some need for power or prominence. It just made everything so much easier. I was my own boss. The English Department had no overall structure, organisation or guidelines and my teaching in it had been akin to working in a vacuum, whereas now I could formulate everything and aim for a completely progressive and ordered learning programme.

The geography room needed a make-over and had one. It was quite large, had plenty of display space and was fully equipped with a TV for videos and an overhead projector and abundant shelves and cupboards. All were spruced up and much old material cleared out to the gubbins. Gubbins is a Devon word and originates from the names of a vagabond family group of outlaws whose hide-out was deep in the depths of Lydford Gorge. It is now a word used to describe rubbish as were the outlaws.

The dustbin area at Mount House was thus called 'The Gubbins'. For much of one summer, I worked on making charts and diagrams related to the geographical topics that I was to teach which were displayed in rotation according to the topics we were studying and the room was always alive with the charts, the pupils' work and displays of one sort or another. I also maintained a board-games cupboard for the use of my class and a geographical library with books and magazines. One parent kindly donated a set of geographical transparencies for the overhead projector which showed every kind of geographical feature known to man and it became a very valuable teaching aid indeed. I was under no pressure as regards the junior forms but there was a formal exam syllabus for Common Entrance to which I had to teach and the Scholarship Geography was of a very high standard.

As the geography department developed and expanded, so did the adventure challenge. The basic programme operating for the whole school continued on Tuesdays and Thursdays and never went stale during the whole seventeen years that I organised it. All the camps, expeditions and weekend activities flourished and developed to very high standards while never forgetting that we were organising them for fun and enjoyment as well.

Firstly, however, we had now taken on sailing as an activity which had a stuttering start. We tried sailing at Weir Quay which was only partially successful and I was never that comfortable with the safety standards there and then we moved on to the Devon Schools' Sailing Association which operated from lock-ups on the Barbican in the centre of Plymouth. Here our pupils really learned to sail well under the instructorships of Mike Duffield and Barry Wilkinson in Oppies to begin with then advancing to faster dinghies and on to Lasers. The teaching was superb and we sailed summer and winter, weather permitting, and many of our pupils became very skilled in all conditions. Some even did powerboat courses to quite a high standard.

Plymouth was busy and the pupils had to be alert at all times with a variety of craft around them. It was a tough area in which to learn and everyone learned quickly. Several of that group have now gone on to the big sailing races such as the Fastnet and Sydney to Hobart. Three of them have their Master's Certificate and are now in charge of charter yachts in sunny climes. Sailors were bred tough at Mount House then.

I have tried to calculate the number of camps, expeditions and trips that I ran from Mount House—and failed. There was normally a Sunday trip or weekend expedition every weekend of the term except for my two school duty weekends and major expeditions during half terms and in school holidays and well as activity days towards the end of the holidays. These were conducted in Plymouth, on Dartmoor, other areas of Devon and also Cornwall and then we extended into Somerset and up into the Brecon Beacons and the Black Mountains, the Wye Valley and North Wales including Anglesey.

We travelled abroad to France, Andorra and Spain and even reached Lundy Island in our travels. We caved in Pridhamsleigh, on the other side of the moor, and also in Goatchurch Cavern, Swildon's Hole, Stoke Lane Slocker, Browne's Hole, and GB under the Mendips, many of the caves under the Llangattock Escarpment and Porth yr Ogof near Brecon. Unusual caves we visited were 'The Earthquake' on Lundy and Le Cueva les Ardales in Spain to see ancient rock art and the spectacular sea caves at Morgat on the Crozon Peninsula in Brittany. We climbed on many rock outcrops on Dartmoor, including the famous Dewerstone (where it is said that the devil's hounds chased their victim over the crag to meet their death below), several Cornish cliffs, Lundy, the Llangattock Escarpment and in North Wales as well as the red cliffs on Anglesey.

The most spectacular climbing was on the cliff faces of La Pointe de Pen-hir near Camaret with its stunning views though they did get me into trouble on one trip. We were camping at Tregarvan on the River Aulne where we were canoeing, sailing and yachting at the kind invitation of one of our parents. On the Sunday it was suggested that the boys and I attend the Matins service at Landevennec Abbey but I was not keen. I had no objection to attending the service in normal circumstances but the service was noted for its length, some three and a half hours. I did give it consideration but declined, jokingly adding that God would be on La Pointe de Pen-hir, climbing instead.

Unfortunately, one of the boys' father was a priest who was staying with our host and reported my blasphemy to the headmaster which I found quite amusing, I am afraid. More is the pity, for the local priest in the village thoroughly approved of all our activities, especially the mud rugby that we played at low tide on the banks of the Aulne bewailing the fact that the French children did not play such games anymore. We mountain biked on Dartmoor, where allowed; many cliff and inland areas of Cornwall, including Goon Gumpas, the Camel Trail and Cardinham Woods; the Black Mountains and the Brecon Beacons, North Wales; Brittany and big expeditions to Spain in the area of the Alhaurin el Grande including the notorious Barra Blanca and the El Chorro Gorge.

The Barra Blanca was a huge caldera like geographical basin surrounded by high mountains with watch towers on every peak for this was where many Nazis fled after the Second World War to be under the protection of Franco and the whole area was heavily guarded and protected. The night Franco died, the Nazis disappeared and fled to South

America. It had good mountain biking tracks, an original heated Roman Bath, in which we all bathed, and we also swam in the pool from the Timotei Advert, though the water was rather shallow.

In both France and Spain, we were regarded with much interest for the locals were delighted that young boys were up for such activity and very pleased to see us and admired the difficult terrain in which we were biking. We did, however, take parents on these trips and by chance we normally had our own dentist 'dads' with us. We did take one mother with us who ignored all my advice and arrived with a (here, I quote my young cousins again) a kind of, sort of, like…district nurse's bicycle complete with the obligatory front basket.

We had to hire another more suitable bike from the local bike shop but I am sure her bike enjoyed the flight to Spain and back. In Brittany, we mainly mountain biked around the various peninsulas sometimes diving down to a beach for a swim en-route or for the sheer fun of riding through the shallow water. We were halfway across the huge expanse of La Palue when one parent called out to stop for he had spotted nudists ahead. I am afraid we carried on wishing them all a loud 'BONJOUR' and the boys wanted to ride back again but when I asked them whether they wanted to go swimming, they declined.

We canoed on mainly local waters though never on the Tavy outside of our grounds and I did 'The Horseshoe' on the Dart in low water conditions with three competent paddlers then we also canoed on the Exe, on the Tamar and in Plymouth Sound if and when conditions and tides allowed. We often canoed off Gurrow Point on the Dart, courtesy of another parent and paddled up and down river. We also took a week canoeing down the Wye for which I experienced difficulty finding another member of staff to drive the support staff until I hit on the brilliant idea of inviting the History Master to explore all the Border castles in Wales and the Marches. He was delighted. We canoed and he did all things historical. We also canoed in Brittany at Camaret including around the Ile des Mortes in the Baie de Roscanvel. Sea canoeing only occurred twice in perfect conditions at Dawlish and Harlyn Bay.

Orienteering was mainly conducted in the school grounds as I could easily set out a course but we also competed in competitions in around Devon and Cornwall and on several fixed courses.

Readers will perhaps be aware by now that we had a special relationship with Camaret-sur-Mer where we always camped at Trez Rouz belonging to Claude Jean Hugret who liked having the boys on site. It was close to the beach, very handy for mountain biking and canoeing and the boys liked exploring the ancient and Second World War forts for this was 'Hornblower' territory with the Goulet de Brest and the Pointe de Grand Gouin, the latter providing a superb grandstand for the huge firework displays at Brest. We camped in what I termed a 'time slip' because summer nights were so light that events never started until very late so we stayed up well past midnight and got up late in the morning.

The boys were particularly keen on Fort Capucins as one had to descend to a causeway to get out to the island on which it was situated and deep in its bowels were some of the old flying bombs that guarded Brest. Fort Fraternite was also popular as they could explore in pairs and was brilliant for wide games. The Second World War fortifications were vast and extensive and one had to be careful of not getting lost underground.

They were built using forced labour and indeed, quite a few innocent residents of Crozon were executed in the town square as a reprisal for the activities of the French Resistance so even during our camps there the local people disliked all things German, especially as U-boat pens were cited in the harbour. We explored far and wide in the area visiting the historical and the interesting. The groups enjoyed a French lesson in the old school at Tregarvan, which is now a museum, climbed Menez-Hom to see the wonderful views and were astounded by the ultraviolet light displays of the Maison de Mineral as the rocks seemed to be floating in space.

Naturally, as a geographer, I could not resist pointing out every geographical feature that we passed and the boys certainly knew a great deal about coastal erosion on their return.

We visited Plas y Brenin, the National Mountain Centre at Capel Curig in North Wales, every year for the week before Christmas and enjoyed full use of their facilities. We met many famous names in climbing and the boys enjoyed the exciting slide presentations that they offered in the evenings. It was traditional for us to climb Snowdon with one of the centre's instructors. Once we were turned back by bad snow conditions but the next year we made it up the Snowdon Ranger Path to the top where the boys were disappointed that the café was closed. We used their ski slope and also climbed indoors and out.

We were very friendly with the owner of the Capel Curig post office who was a Roman centurion with a cohort of Roman soldiers in a branch of the Sealed Knot and it was he who led his cohort from Brecon to Merthyr, up the old Roman road over the Brecon Beacons and down the other side but he could not walk for several days afterwards. The boys liked visiting Jo Brown's Climbing Shop in the village especially if he was there. He was getting on slightly and now earned big money by retrieving bodies from mountains worldwide.

I quite liked a stretcher type patient bag that I thought would be handy for rescue exercises in the school grounds but he announced that he was using it to bring down a body from the South Col of Everest. Sometime later he phoned me and asked if I was still interested. I declined as the thought of the boys using a used body bag was now rather macabre (If you are at all squeamish you should skip this bit). He assured me that it had been up to the South Col but had not been used. The body was so badly decomposed that they had used dustbin liners instead. He had a great sense of humour. One of our parents was a member of the Dartmoor Rescue Team and was considering the purchase of a new pair of very expensive boots. Jo took in his fleece emblazoned with Dartmoor Rescue badges and suggested that a pair of wellies would be better for Dartmoor.

As part of the Adventure Challenge, I considered it important that our pupils should use and learn about as much of Dartmoor as they could and they were taken to the stone rows at Merrivale, Fogintor Quarry, the Museum/Centre at Princetown, Lydford Castle and Gorge, Dartmoor Prison Museum, Wistman's Wood, Buckland Abbey and Crownhill Fort—all within easy reach of the school. When we were on expeditions and camps further afield, we would always take groups to places of interest that were cultural, scenic and interesting.

During the winter months as well, groups would visit the town of Tavistock to complete the town trail to learn about its history for it was a stannary town for tin assaying and an important market for miles around its pannier market dating back to 1105. The ruins of its old Benedictine Abbey can still be located and Francis Drake had been born south of Tavistock. It was also famous for its Goose Fair. Above all it was a pleasant and friendly town which the pupils enjoyed visiting.

Dartmoor holds many ghosts, tales and legends and everyone enjoyed these tales especially if they were told around a campfire at night with dark shadows as a perfect backdrop. Mount House pupils knew all about the Ghost of the Hairy Hand at Powderham Mills, the Golden Dagger at Soussons, the Lady with the Golden Leg and of course, the Devil's Pack of Hounds that hunted over the moor. They were also aware of the terrible legends about Lydford Law and its horrendous castle, suicide graves and the Lych Way for all bodies had to be transported across the moor to Lydford for burial there and nowhere else.

Life at Mount House was interesting, varied, frenetic, fun and sometimes, plain daft. I had two bugbears. One was the usual complaint of lesson disruptions mainly because of individual music lessons constantly programmed throughout the day disturbing the academic programme and lessons. Most lessons would have one or two pupils missing for same and in any week an average of 25-30 children would miss my lessons for music. The standard of music was superb but all the academic staff and their subjects suffered greatly from these interruptions and nothing was ever done to alleviate it.

At the time of writing, members of the senior management team of the Roman Ridge School are in the UK visiting senior independent schools in which this practice continues and were amazed. It does not happen in Ghana and the parents would never allow it. Here in the UK parents are paying high academic fees and another set of fees for music that take place at the same time. How? I could not write a geography report for one student as her combination of instruments meant that she could never attend my lessons and the headmaster suggested that I make one up.

Staff meetings were a trifle odd because we only discussed the top three and bottom three pupils in any form and the middle group of ten were never discussed unless one insisted. It never made sense to me and there was always a rushed few minutes for any other topics at the end of the meeting and it was considered taboo to mention anything controversial which meant that the head did not get to sound out the views of his staff at any length. To be fair, however, one could always ask to see him for any points that you wished to raise.

There were peculiar practices that no one thought to question such as the rigmarole of lunch which I have already mentioned and was so long-winded and unnecessary and breakfast and tea was not much better. There was also the unhealthy option of fried bread served at every breakfast and rice pudding available at every lunch both in an attempt to 'fill the boys up.' Then there was the Sunday Service which was pleasant and well attended by parents.

There was no problem with that but it entailed collecting all the chairs from the classrooms before the service and replacing them afterwards which was a dreadful task and left to the poor boarders who were not going out. Chairs should have been purchased for the service that were close to hand so that they could be assembled quickly. With the heads that I have inducted in the not too distant past I always suggested that they invite a relative stranger in to look around and question procedures and out-of-date practices. A 'third eye' is always handy in such cases. Let them ask WHY!

There were odd and eccentric moments and incidents such as the young temporary head of English, who was resident in the lodge down at the main gate, who was enamoured of a female Israeli Army sniper and ran up a phone bill of £1,700 in his nightly calls to her. A temporary science master who had a glass eye and would take it out and leave it on the bench at the front of the class in the science lab while he popped out, stating that he could still 'keep an eye on the class'. The annual arrival of a truck from the Royal Marines with over a 1,000 metres of black abseiling rope which was brand new but now 'unfit for purpose' being over one-year-old and a Navy Lorry with one of the mooring ropes from the Ark Royal.

The proposed disposal of allegedly unfit furniture which I suggested was very valuable and it was. The secret phone call from Major Bob (bursar) to come and collect him from Tavistock Hospital because a fishing fly had been caught in his right ear and of course I told no one (not). The pupils telling off the new bursar for rushing onto the playground with a stretcher (army style) to move a pupil who the boys had already checked over and put into the recovery position and treated for shock, waiting for the matron to arrive.

The poor painter who crashed through a glass roof adjoining the front hall when an assembly was in progress (he did recover). Returning to school after a North Wales trip just before Christmas to the sound of rushing water pouring into the surgery upstairs from a burst water tank and the bursar suggesting that we leave it until after the weekend. The whole school assembled at the front of the main building for the traditional school photograph and the art teacher driving up into frame to ask what was happening.

Finally, there was the battle of the washing machines. At previous schools, we used a commercial laundry who took 24 hours to return your washing but at Mount House we used washing machines except that they were obviously very busy and it was nigh impossible to find a free machine so the matrons imposed a schedule. My slot was 3 pm on a Sunday afternoon just as I was emerging from a cave or finishing a climbing session or racking the mountain bikes on to the trailer ready to get back to school for tea at 5 pm. How daft was that?

Evidently, the head matron insisted that that was the only time and she knew full well that I could not make it so I ordered a washing machine and tumble-drier from the town and had them fitted in my flat above the headmaster's study, which was not a popular move but it did solve my laundry problem. Pedantry was often present in many of the prep schools in which I taught and small issues could be blown out of all proportion, especially towards the end of term when everyone was tired and overwrought. Please do not think that all my seventeen years were spent on the adventure challenge and the teaching of English and Geography.

I was forever busy but I did maintain a private life which is contained two chapters hence because I did write stories during my time at the school which have now been included to divert and entertain you. Towards the end of those seventeen years, there was a gradual realisation that the century was drawing to a close and was I ready to stay at Mount House for another ten or so years before gradually sinking into retirement in the delightful West Country?

It was not for me, for Africa was calling again and I was about to embark upon the biggest undertaking of my life which involved a huge gamble with my personal life and finances. It was to be the biggest challenge of my life and would it succeed?

Chapter Twenty-Two
Gown and Out

Soft music, pastel-coloured walls, comfortable armchairs, up-to-date glossy magazines and…coffee. Was I in some new five-star hotel? No—I was actually in the new hospital's X-ray department awaiting an outpatient appointment. What a wonderful service the NHS now supplies, I thought, sipping my coffee and looking casually out to the courtyard where the sunlight was reflecting off happily splashing fountains. My reverie was gently disturbed by a lady's voice calling my name. I stood up and turned around to meet a very attractive nurse. She had a most pleasant smile and looked absolutely charming.

'Follow me, please,' she requested and we set off through a labyrinth of passages and corridors.

Modern the new hospital might be, I thought, but the architects and planners must have based their plans on the maze at Hampton Court. I frantically tried to keep my bearings so that I could get out after my appointment.

Eventually, we arrived at an inner reception area which had blue chairs along one wall and a row of cubicles against the other. A middle-aged lady and a young man dressed in hospital gowns and their ordinary shoes were perched self-consciously and uncomfortably on the chairs awaiting their X-ray.

'Please go into one of the cubicles and undress,' requested the nurse. 'I'm afraid you have to take everything off. Your clothes can go into one of the plastic bags and then you can put on one of the hospital gowns from the basket. I'll be back in a minute.'

Now, I must confess, a vague unease was at the back of my mind at this point, but I could not establish its cause so I quickly put it to the back of my mind and proceeded to undress. I put all my clothes in the white plastic bag provided and labelled Patient's Personal Property—carefully remembering to keep my shoes out—and reached into the basket for a gown. There were none. I looked behind the door. There were none. I looked under the bench seat. Again, no gown. Now I knew why I had been so uneasy.

My mind raced back over thirty years to my student days. The matron at my college had been steadfastly ignoring my persistent stomach pains for three days. Ever anxious to greet the first avocet on the estuary, she had dosed me with aspirin and told me to go away. She was a fanatical ornithologist and was obviously determined that I should not disturb her activities. By the fourth day, however, I was in much pain and insisted on seeing a doctor.

A very disgruntled matron escorted me to his surgery muttering about the beautiful weather and the pending low tide. I had no appointment so we had to wait for well over an hour until the doctor agreed to see me. The doctor was infamous at college because he forced you to strip even if you had a wart on your nose but he did give me a very thorough examination no part of which seemed to have any bearing on my stomach pains. Eventually, I was asked to lie down on the examination couch and he checked my stomach.

The left side was alright but when he touched the right, daggers of pain raced through my side. I screamed. 'Is that tender?' he asked.

Having broken into a sweat and breathing heavily, I assured him that it was extremely tender. I was allowed to dress and told to sit down in the waiting room again.

Three minutes later, out stormed an angry matron.

'You have to go to hospital. You may have appendicitis.'

I recorded her obvious irritation but by this time, I was beyond caring and feeling decidedly groggy. The pain was very bad. At least, I thought, the ambulance would come soon and we would be at the hospital very quickly.

'Come on!' snapped Matron. 'We have to catch the bus.'

The hospital then was an old Victorian monstrosity, a red brick building near the centre of the city. Feeling more dead than alive now, I was ushered into casualty and matron fired all my details at a surly receptionist through a little glass window. I collapsed onto a hard chair, feeling nauseous and dizzy and it hardly registered when matron bade me farewell and disappeared. The call of the avocets was very strong!

The waiting room had seen better days. Paint was peeling from the walls, there were cracks in the plaster of the ceiling and the furniture was very chipped and battered. The latest magazines were well over two years old but I did not care. I must have dozed off because I found myself suddenly being shaken very roughly and a loud voice was shouting my name.

This was no attractive nurse, for the lady shaking me, and making a good attempt to

dislocate my shoulders, was the casualty matron. She was big. She was awesome. Her hair was swept straight back and held down tightly beneath her cap, her features were sharp and her uniform crackled with starch. Imposing and stern she towered over me and I think she probably starched her voice as well.

'Follow me!' she commanded and led the way to a small, dingy and drab cubicle, the curtain of which she pulled back and I nervously followed her in. 'Take all your clothes off,' she demanded, 'and I mean e-v-e-r-y-t-h-i-n-g!'

Very shakily, I complied, aware of her stern gaze upon me, while I stripped for the second time that day until I stood stark naked in front of her, trembling with cold.

'Lie down on the examination couch while I get a nurse to bring you a hospital gown. I'll put your clothes in a safe place.'

She swept out of the cubicle, closed the curtain and I never saw her again.

With some difficulty, I crawled up on to the black upholstered examination couch and surveyed my surroundings. There was little to survey. The floor was bare. The walls were bare. The ceiling was bare. I was bare. I was by now erupting with goose-pimples on the outside but burning hot on the inside and my appendix kept on sending repeated sharp pains to my right-hand side. I could not stop shivering. The cubicle could easily have doubled as a deep freeze in a frozen food factory.

Some ten minutes later, I heard footsteps approaching the cubicle. At last, I thought, my gown and some action. A middle-aged lady popped her head around the curtain.

'Whoops! Soreeee,' she laughed and giggled off up the corridor.

I lost all track of time but after what seemed ages, a friendly male nurse arrived to take all my details and though I begged him for a gown, he proceeded to ask me personal details and seemed to have no problems at all about dealing with a shivering nude student. With a flash of inspiration, I asked if I could have a blanket as gowns seemed in short supply. He promised to bring me one straight away and…disappeared.

The next arrival was a doctor who looked about the same age as me and was extremely sympathetic when I explained my predicament. He called a nurse and asked her to bring a gown and a blanket then continued with a full examination of my stomach and a rather painful rectal examination, the details of which are best left out of this account but it did lead to action.

'My God!' he exclaimed. 'That's very nasty!' and he ran out of the cubicle.

Needless to say, the nurse had not appeared with either a gown or blanket and I was sure, by now, that I was going to die of hypothermia. I suppose I must have become delirious because I remember shouting out and nearly rolling off the couch. I curled up into a tight ball and evidently became unconscious. Then…nothing. I came around a few hours later, though I did not realise it at the time. As I opened my eyes, I was aware of a throbbing pain in my stomach and there were all sorts of drips and machines around me. I groaned. A charge nurse, who could have doubled as a heavyweight boxer, carefully explained that my appendix had burst just as it was extracted and they had put me intensive care for the night to keep an eye on me.

The gadgetry around me was impressive. There seemed to be a huge array of paraphernalia leading to and from my body. I raised my head slightly to see the operation scar. To my acute embarrassment, I was still naked. The charge nurse noticed my embarrassment.

'Don't worry,' he said, 'we've sent downstairs to find a gown for you. By the way, did you come into hospital with any clothes? You don't seem to have any!'

If I could have laughed then, I would.

'Are you alright in there?' asked a concerned voice and the young, charming nurse walked into the cubicle. 'Oh—I beg your pardon,' she blushed.

'I'm fine,' I replied, 'but do you think, after some thirty years of waiting in this hospital, I could be provided with a gown please—PLEASE?'

Chapter Twenty-Three
They Shoot Horses, Don't They?

It is my firm belief that Hitler and the German invasion would never have succeeded in conquering France if the French had left their autres directiones signs in place for they were totally frustrating my attempts to drive through the town of Rodez. Loyal Frenchmen who destroyed these signs in the misguided view that they were baffling their oppressors were wrong. They should have left them in place! Scientists and 'boffins' should have cancelled all their experiments and pooled resources to produce the autres directiones sign much earlier. This one sign alone would have led the crack German Infantry, Motor and Panzer Divisions to mill and grope around the French countryside, totally confused and bewildered for months on end. It could have changed the whole course of history!

Such were the thoughts in my mind as I circled the town of Rodez at nine o'clock in the evening on a dark night in October searching for a road sign to Sévérac-le-Château which was close to my destination…the Cévennes. Every roundabout, however, merely signposted autres directiones and not Sévérac. On my third circuit of the town, I spread my net wider. I was beginning to understand that French road signs rarely reveal the whereabouts of adjoining towns and villages but could indicate a town at some distance. I studied the map carefully, to no avail. Rodez had me in its grip and was not keen to let go. It was defying my attempts to escape.

A small bar hove in sight and I thought it was a good idea, at the time, to stop for a coffee and establish my bearings. Unfortunately for me, the bar was packed with the players and supporters of the local rugby team who were celebrating a victory in grand style. It was very loud and noisy. I managed to order a coffee, my panacea for all problems, and in my fractured French, asked the barman if he could direct me to Sévérac-le-Château.

He consulted with the three nearest rugby players, who in turn asked two drunken fans further down the bar and so on. My problem was soon known by everyone and eight of them took me outside, gabbling and arguing volubly proceeding to point in four different directions while one of them held my road map upside down. A huge argument followed between several of the players and supporters and I decided to beat a hasty retreat before the pushing and shoving turned into a scuffle or fight.

In sheer frustration, I extracted a compass from the glove compartment of the van, took an easterly bearing, and followed roads out of the town in that general direction. It worked.

By 10 pm, I was well on my way to Florac in the Cévennes musing on the large mileage I had already covered that day. The plan had been to potter down through France on a nostalgic tour of some of my old haunts but my memories had somewhat deceived me and I had been disappointed with places like Poitiers and Limoges. The countryside and landscape had done little to raise my enthusiasm so I had decided to drive on within striking distance of Florac which I was using as a mountain biking base for a few days.

Ten miles outside of Rodez, I was to be thwarted again as I came up behind a blue fully loaded horse lorry. It was slowly swaying and meandering along the narrow, twisting road which threaded its way through a series of low hills and tight bends so I had to play a cat and mouse game for another seven or eight miles before I could eventually overtake it. Thankfully, I drove around the next bend to a straight section of road to be greeted by a scene from Agamemnon.

A little way up the road there were bright lights, smoke, steam, shadowy shapes and scurrying figures all combined in one horrific picture. A blurred figure was running away from this holocaust towards me frantically waving and shouting. All this I had taken on in an instant but, as always, everything seemed to happen in slow motion. I knew that I was well within my stopping distance but a glance in my mirrors revealed the horse lorry rumbling towards me.

All livestock containers, especially those carrying valuable horses, require a special type of driving. You need to offer your passengers a smooth ride and this entails careful anticipation of any events or problems ahead and braking, steering and changing gear cannot be abrupt or carried out quickly. Emergency braking could cause very severe injury to your load. This was now my problem. I knew that if I braked hard I was doubtful whether the horse lorry behind could, or would, pull up safely and was likely to give me one enormous shunt at my back end. I also have to admit that I was also extremely worried about my brand new mountain bike strapped on the carrier on the back door!

The fates were with me, for, mercifully, at that moment, a small lay-by appeared on the right hand side and whilst braking carefully, I steered into it and rolled to a halt. The horse lorry driver had sensibly taken the softer option and I breathed a huge sigh of relief as he slid by me and gently stopped barely ten yards from the scene of the accident. But for that lay-by I would certainly have become one of the casualties.

I peered carefully through the windscreen to look at the accident ahead, especially as I was not keen to rush into the chaos of the scene. My involvement with rescue teams— mountain, cave and surf had taught me to check out the situation first. Indeed two of the teams with which I worked always insisted on drinking a cup of tea or coffee a little way from any incident just to look at the overall picture before starting a rescue. It was difficult to see the full situation. It would seem that horses had broken out of a field and had been hit by a vehicle or vehicles on the main road. No one was assuming control and there was a great deal of confused shouting and aimless running about by three or four figures.

I will always remember one Emergency Incident Manager Trainer stating that the biggest problem to overcome was whether you felt able to assume control of an incident and I do clearly recall assessing the situation very carefully. The biggest drawback, of course, was my limited French. During a good French meal in a good restaurant and after several glasses of claret, my French was pretty good but this was obviously not the situation now. I was stone cold sober and looking at a complicated situation.

The nearest horse made the decision for me. It had terrible injuries, was whinnying pitifully and struggling to gain its feet. I confess that I am not an animal lover in the British sense of the phrase. I am sceptical about the keeping of domestic pets, particularly in confined conditions and feel very uncomfortable about zoos. I do not like to see animals suffering, however, so decided to help if I could. But was I safe in making an approach? The answer was no.

Approaching vehicles from both sides of the accident could still drive into the horses and helpers. I reached into the back of the van and found two Petzl headlamps and a torch. Carefully I clambered out of the van and stood on the grass verge. It took a little while before I could persuade two of the bystanders to go off in opposite directions, at least five hundred yards up the road, to warn oncoming drivers and that the headlamps worked by swivelling the headset. I vowed that I would learn the French for opposite at the earliest opportunity (enface—it was easy really). It would also be a help if they put bushy branches blocking the road.

While I waited for the two volunteers to make their way up the road in either direction, I tried to explain to the lorry driver that he should back up away from the accident to a safe distance. Horses can sense injury and death by the smell of blood and his were already becoming very restless. Unfortunately, he was in obvious shock, however, and was in no condition to drive at all and judging by the smell of alcohol on his breath, this was probably just as well.

Are there guardian angels? Do they exist? If I had been asked this question prior to my trip, I would have jokingly dismissed such a possibility but I can now state, quite categorically, that I do have one. He appeared to me in the disguise of a Turkish lorry driver, short, very short, dressed in grubby jeans and a Marlboro T-shirt and flip-flops and smoking a strong Turkish cigarette. He seemed not to have showered or shaved since he had left Istanbul (Do angels shower?) and judging by the state of his blackened teeth revealed by a wide smile, his dentist could not have been a rich man.

His name was Fawaz and he was the most helpful angel I have ever met. He was a kindred soul as he spoke French very badly and it stretched all my linguistic skills in French and Arabic to suggest that he should put out the cigarette drooping from one side of his mouth as there was a strong smell of petrol everywhere. At first he thought that I was asking for a cigarette but when I mimed an explosion, he was about to throw the lighted cigarette away.

I grabbed his hand quickly, took the cigarette away from him and ground the butt out on the side of the horse lorry. Having done so, he swung up into the cab and with me directing from the side, still worried about oncoming vehicles, Fawaz deftly reversed the lorry a good hundred yards back up the road behind his own thirty-five ton lorry which he said contained chou-fleur. Chou-fleur, I pondered. What were they? Of course… cauliflowers!

Even in all the excitement of dealing with the accident, I do remember struggling to find a reason why a Turkish lorry driver would be delivering cauliflowers to Brittany (for that was his destination), where, as far as I knew, Brittany grew millions of cauliflowers; indeed, I had been led to understand that Brittany Ferries instigated the Roscoff/Plymouth crossing to flood the British market with tons of cauliflowers. I never did find the answer and most people in England and France thought I was mad when I posed the question to them.

But I digress. Though everything seemed to have taken place in slow motion it had only taken a matter of minutes and we were now able to move on to the scene of the accident. We did so very warily. Two large horses, both over sixteen hands, one a dark bay and the other a chestnut, were sprawled in the middle of the road with horrible injuries. My first impressions of the accident had been correct and the steam, blood and injuries still stick in my mind to this day. The horses' injuries were very bad and their pitiful whinnies were terrible to listen to. I looked around carefully. A third animal, a small grey pony was lying on the opposite verge about fifteen metres away, dead, without a mark or injury anywhere.

Fawaz and I peered all the way around the scene again. Where was the vehicle involved? Having lived abroad for many years, I regret that I had attended and observed many accidents involving humans and animals. These accidents were nearly always fatal for the poor creatures involved, the drivers were often badly injured and the vehicles a write-off. At Livingstone, near the Victoria Falls, one driver and his passenger were killed when they hit a sleeping hippo on the tarmac road. The hippos used to like the warmth of the tarmac and were difficult to see because of their grey colour against the tarmac. The car was very badly damaged but the hippo just got up and lumbered back into the Zambesi! I was fairly sure that no vehicle could have continued after hitting a group of horses. We were baffled.

By this time we had been joined by two other Frenchmen, a lady and amazingly, a doctor wearing a dinner jacket, who spoke very good English. I explained the problem. In my opinion, the vehicle could not have driven on and I was worried about the driver. We agreed to call a vet, the police and the fire brigade (les pompiers) to cope with the immediate problems and while the doctor made these calls the lady and the two Frenchmen agreed to search opposite sides of the road for any signs of the vehicle or driver.

Fawaz and I returned to the horses. Thankfully, the bay had died but the chestnut was still in great distress and pain and I tried to get close but it kept on kicking its legs and an approach was not wise or safe. It was then that I noticed a wide, faint set of tyre marks stopping at the horse's stomach. I ran back to my van, found my 'Rabbit Lamp' and standing between the tyre marks, pointed along the tyre tracks in a line along both directions. There was nothing to the rear, but, over a hundred yards away and to the right, in a field at the bottom of a grassy slope, was a large Luton van on its side. It had no lights and there was no sign of life.

The doctor grabbed his medical bag from his car, we called to the others, and we all scrambled through a small ditch and over a low hedge to the field where we ran down to the vehicle. Fawaz started to clamber up on to the Luton to get to the cab door but we realised that the windscreen had been smashed and this provided easier access. We flashed our torches into the interior—up, down, around, under the steering column, behind the seats—no driver. We searched around the vehicle—no driver. We double checked again—no driver.

This was worrying. Had the driver been crushed when the vehicle fell on to its side? Fortunately, the vehicle had toppled on to a small ridge so we were able to check carefully underneath. No driver! Where was he? If he had wandered off, dazed, he could be at some distance. I suggested that the doctor call the police again to request more manpower and while he was doing this we did a thorough spiral search to include the whole of the field. Still no driver. Nothing—absolutely nothing.

Now while the doctor was phoning the police again, it did occur to me that none of the emergency services had yet arrived. We were in a rural area but it had been a good twenty minutes since the doctor had phoned them. This was confirmed by the doctor who started a very heated discussion with the gendarme on the other end of his phone. I learned many new French swear-words that night as the doctor was obviously not pleased with the delay.

The Gendarmes evidently claimed that they had not received the initial call. Still cursing, the doctor and the rest of us heaved ourselves over the small hedge down into the small ditch on the other side. Fawaz attempted to jump the ditch, which I felt was a bit like a Shetland Pony trying to get over Becher's Brook on the Grand National course, landed awkwardly and fell back in the ditch. He rolled on to something soft which gave a faint groan. We had found our driver.

Providing a circle of light, the young man was quickly checked over by the doctor in a very efficient manner. Now conscious, he was able to answer some of the doctor's questions. Remarkably, he was relatively unscathed. He had obviously been knocked out and the doctor suspected a couple of broken ribs. Apart from some deep scratches on his face, he seemed to have escaped very lightly. An ambulance was summoned.

Though the driver was somewhat confused, we managed to piece together what had happened. He had been rushing back to his depot and had not stood a chance when the three horses had galloped onto the road ahead of him. He had braked hard. He remembered hitting them and the windscreen smashing before he was catapulted forward, incredibly, ahead of his truck, into the ditch, before his vehicle leapfrogged over the horses and the ditch and hedge before careering down the field to its resting place. He had not been wearing a seatbelt. Make of this what you will! It was certainly a lucky escape in all respects.

Leaving the driver in the doctor's capable hands, Fawaz and I returned to the poor chestnut horse which was still struggling feebly. The vet had not arrived and there was little we could do. Fawaz was unclear about what would happen to the horse so I tried to explain about the vet and mimed a pistol at my head. He walked off and I thought he was too upset to stay.

His place was taken by the doctor who now, like me, was extremely worried about the non-appearance of any of the emergency services. It had been ten minutes since he made the request for an ambulance, fifteen minutes since his heated exchange with the police and the initial call had been made well over half an hour ago. The thought did pass through my mind at this juncture, that, perhaps, the pompiers, who were volunteers, might well have been the rugby players celebrating back in Rodez, in which case, there would be no hope of a response. The doctor shrugged his shoulders, thanked me for my help and stated that he would take the driver back to Rodez hospital himself. We both helped the driver into his car. I gave the good doctor all my details in case I was needed as a witness and he drove off.

I returned to the horse where all the other helpers were standing. I was pondering over the remarkable events of the night but I was mostly very concerned for the horse and the lack of specialised help. I knew it must die but horses have remarkable stamina and staying power and this one was not giving up yet. Despite its terrible injuries, death might still take some time.

It was then that Fawaz reappeared. He tapped me on the shoulder and handed me a somewhat battered Colt 45 and a bullet. My scepticism concerning guardian angels now focussed on the possible existence of divine intervention by a higher authority involved in organising one's life according to a grand plan, because, by luck or brilliant management, someone had certainly arranged everything very well indeed that October night some ten miles out from Rodez. A guardian angel in the disguise of a Turkish lorry driver toting a Colt 45, encountering me in the middle of the French countryside, perhaps the only teacher in England who had direct experience of humanely killing horses and cattle with the self-same gun.

I knew instinctively that it was right to put the horse out of its misery and I did briefly consider the legal implications but one look at the suffering animal and I knew it had to be done. I asked Fawaz if he had any more bullets as it would be safer that way. He rushed off and brought me another two, his complete stock I think. As I loaded the gun, I became aware of a woman sobbing at the side of the road. In retrospect, I suppose I should have made enquiries but I was anxious to get the job over and done with. I asked Fawaz to hold the horse's neck lower down, well away from the Colt 45 and very carefully found the centre of the cross mark between the horse's head and ears. I pressed the barrel of the gun on to that point, steadied my stance and started to squeeze the trigger.

At that moment, there was a horrendous, high-pitched hysterical scream that jarred every nerve in my body and I was engulfed in a powerful bear hug from behind. I was jerked up and around and my arms and the gun pinned to my side, pointing to my right foot.

Now I would hate to malign all lady owners of stables and riding schools. All I can say is that, in my experience, they do tend to be large in body and voice, immensely strong and extremely determined. Though my observations and experience had been with the British variety, the French appeared to be no exception, for I was, at that moment, being attacked by the lady riding school owner (the owner of the chestnut).

Suffice to say that I have been in some very frightening situations during my varied life being a veteran of some five African coups but was I frightened now? No, I was terrified. The lady in question could easily have gone twelve rounds with a heavyweight boxer and I was no match. Whilst begging her to stop, assisted by her husband trying to drag her off and Fawaz also trying to break her vice-like grip, I decided that drastic and perhaps, ungentlemanly conduct was in order before I, or someone else, was shot by the bouncing gun.

I was only wearing light boots but I managed to scrape my left foot down her shin and stamp on her toe...hard. Her loud shouting and screaming turned to a genuine scream of pain and she let go. She was then bustled and pushed to the side of the road by several spectators who, I suddenly realised, had done nothing to help. The husband signalled me to go ahead and shoot the horse while he led his wife away. Taking deep breaths to quieten my throbbing heart and trembling, we approached the horse again. To our relief, it had died and no further action was necessary.

It was now nearing 11 pm. Incredibly, none of the emergency services had arrived. I made a mental note to drive more carefully in France for the rest of the trip to avoid accidents. What would have happened if there had been serious casualties? A local farmer arrived with a tractor and we dragged the horses to the side of the road which was washed down using buckets of water from a nearby horse trough. There was nothing more I could do. I had had enough anyway. I thanked everyone and said my goodbyes, leaving my details again with the local farmer and also gave them to Fawaz, who gave me a big hug. I suggested he move on quickly in case the police did arrive and wanted to express an interest in his Colt 45. He drove off with alacrity before I drove out of the layby.

Twenty minutes later, I had stopped near Sévérac-le-Château on the N9E11 and was amazed to see the local chateau fully floodlit. I decided to cross the route Nationale and find somewhere to park up for the night. Just then I heard sirens approaching and could see blue, red and yellow flashing lights. Two police cars, a large Renault estate (the vet?), an ambulance and two fire engines all sped by in convoy. I estimated their response time as being one hour and thirty minutes—and they complain about twenty minutes in the UK!

WANTED
DEAD OR ALIVE

ROBIN

WANTED FOR VANDALISM,
EXPLOITATION AND THE
THEFT OF EVERY GIRL ON
CAMPUSES HEART.

Chapter Twenty-Four
'I Bet…'

…You Wouldn't Pinch that Woman's Bottom!

Here we go again, I thought. Robin and Charlie were constantly involved in a never ending round of wagers and bets, most of which led to problems, trouble or mayhem. All three of us were comfortably ensconced in the bar of a rather smart yacht club, somewhere on the south coast, after a mediocre meal in a local Chinese restaurant. We were making a serious attack on the cocktail list as Robin had bet Charlie, who was well known for holding his liquor, that he could not drink his way through the ten listed cocktails, specialities of the bar. Charlie, though somewhat the worse for wear, was making serious headway while Robin and I were trailing far behind.

Earlier that week, I had been persuaded, or rather, hijacked, into joining Robin and Charlie on a shakedown cruise on Robin's splendid new motor yacht which he intended to put into the Caribbean charter business later that year. Knowing him as I did, I should have regarded his invitation to come down to his house for a few days with a great deal more suspicion. With hindsight, I now know he had set me up and I had fallen for it hook, line and sinker.

I had known Robin since our college days. He was tall, fair-haired and strikingly handsome. He had broken the heart of every girl on campus. He could never refuse a challenge or a dare even in those days and we had several brushes with the college authorities about our outrageous exploits which included sending out false National Service call-up papers to every male student, climbing the bell tower to leave the traditional article of feminine underwear up the flagpole and re-spraying the principal's car a lurid yellow.

Rumour had it that the matron in charge of the local nurse's home had 'wanted posters' of Robin pinned in every corridor. He was notorious but fun so it had come as a huge surprise one morning to receive an invitation to attend his wedding. I was due back abroad and I could not attend so I dutifully posted off a gift and my congratulations.

Later visits confirmed, however, that he had settled down in a beautiful house overlooking a scenic estuary with his beautiful wife, Vicky, who was vivacious, attractive and sensible. She was fully aware of what she had taken on and was adept at curbing Robin's wild activities. They had two lively children—Ben, aged ten, and Susie, aged eight—who were bounding all over the lawn with several dogs as I made my arrival.

The children had been invited to a party that evening (Could this have been engineered?) so Robin suggested that we eat out at the local pub. The Black Swan (known as the Mucky Duck of course) was an attractive, thatched hostelry that retained its unspoiled rural charms including an open fire, gleaming horse brasses and low beams. I did notice, however, that the landlord kept a wary eye on Robin. He had been banned on several occasions and though he was too good a customer to make this a permanent arrangement he was obviously wise to stay alert for the hint of any problems.

We had eaten an excellent meal and Robin and I wandered into the bar for a pint or two and it was there he mentioned that his friend, Charlie, might pop in to join us. Charlie evidently helped Robin with his boating projects and they needed to plan their next trip. Amazingly, I still did not catch on.

From outside I heard the roar of a powerful motorbike, a sudden squeal of brakes, shouting, then the crunch of gravel as the bike drew up. After a short time, a tall and thin leather clad figure stooped through the doorway and peered around the bar. He was too tall to stand erect and he looked like an unemployed Mexican bandit as he had long black hair, a drooping moustache surrounding a narrow mouth, a hawk-like nose and steely glinting eyes. He saw Robin, waved in acknowledgement and limped on to a short flight of stairs leading down into the bar area.

On the second step down, he made the mistake of straightening his lanky frame. His head cracked on a low beam, he lost his balance and catapulted down the steps into a table occupied by three regulars playing dominoes. Tankards, dominoes, chairs and players capsized, tipped and scattered all over the floor. In the middle of this chaotic heap lay the leather clad figure. For a moment, there was stunned silence. Even the wary landlord froze.

'Hell!' exclaimed the biker. 'My bloody leg's come off!'

To my amazement, he hopped up on one foot, took down his leather trousers then proceeded to re-strap his artificial limb back on to the stump of his right leg. Having tightened the adjustable straps to his total satisfaction, he pulled up his trousers, zipped up the flies, fastened the belt, looked around and smiled. 'Hey. Sorry, you guys. Let me buy you another drink.'

By this time, the landlord had regained his wits and was busy setting the furniture and everything and everyone to rights. Luckily, there was no damage and the biker was true to his word and refilled everyone's tankard.

'And two more over here,' called Robin, 'we are suffering from deep shock.' The biker limped over, clutching three pints of beer in his big hands.

'Geoffrey—this is Charlie,' smiled Robin. 'As you have already noted, he is slightly accident prone.'

We shook hands and sat down to our pints. The conversation naturally turned to boats and Jamie carefully explained that he had recently purchased a motor yacht and he was keen to take her out on a shakedown cruise. Charlie had agreed to crew.

Beware boating talk. Beware beer. Now on my fourth pint I launched into my sailing and boating stories. I had once owned a small cathedral hulled cabin cruiser moored at Shoreham. She had twin 60hp outboard engines and was very fast and stable. I often crossed the channel in her using a rather unique navigational method. Outside Shoreham, I turned left along the coast to Newhaven then waited for the Sealink Ferry, Senlac, and followed her across. I knew the captain and this was a fairly regular arrangement! I could have managed on my own but this method was far more reliable.

I had also skippered fairly large motor boats on the huge Kariba Lake between Zambia and what is now Zimbabwe helping numerous friends with their fishing trips. Then, a friend's small cabin cruiser had struck a submerged log on the Kafue River and we had all beaten the Olympic 100 metre swimming record to reach the shore pursued by a family of hippo.

In Mazabuka, in Zambia, a local farmer started water-skiing on his farm's dambo (a large lake). I had joined in to help drive the speedboats and most Sunday afternoons were spent at the lakeside or on the water towing rather ungainly water-skiers up and down. This activity suddenly ceased, however, when, at a barbecue one evening, on the shore of the dambo, one of the guests casually threw a couple of chicken bones from his plate into the water. There was a sudden swirl in the water as two crocodiles lunged for the leftovers. No more skiing.

I had been so absorbed in my yarns that I missed the knowing look between Robin and Charlie. Without hesitation Robin invited me to join them on the shakedown cruise the next day. I made the usual lame excuses all of which were firmly rebuffed. The bottom line was that I did owe Robin a favour and he was calling it in. My fate was sealed so we had another drink to celebrate before we lurched outside.

'Bet we get back to the house before you!' shouted Robin, leaping for his car and dragging me with him as well, while Charlie was still struggling with his crash helmet.

We jumped into Robin's black Daimler and sped off with Charlie roaring along behind. Every time Charlie tried to pass, Robin moved out to block the move and then infuriated Charlie by driving at 15mph in the narrow lanes. The last leg of the journey was on a track passing through fields and here Robin accelerated causing a huge cloud of dust to billow out completely obscuring Charlie in our wake, so, in the second field, Charlie threw caution to the wind and moved out on to the grass to overtake us. It was his undoing.

The powerful bike pitched, bumped, jumped and…Charlie lost control. He slid gracefully into a hawthorn hedge. We got out to check that he was OK. He was unscathed. We asked if he was alright and he nodded then jumped on his bike. As he roared off, he shouted: 'I'm going to beat you now anyway!'

Robin stomped back to the car, realising that he had been tricked and it was then that I noticed that the front offside wing was rather the worse for wear. It had a huge dent and horrendous scratches spread right back to the door. Evidently, this had happened the previous evening when another bet had led to an old garage door being hitched to the back of the car and then towed around the field at speed with a person on it. The object was to see who could stay on the door for the longest period of time. Robin had not fared well at this exploit so was very keen to shake Charlie off as quickly as possible. He had made the mistake of twisting around in the driving seat while towing Charlie and had forgotten about the granite horse trough in one corner of the field.

We arrived back at the house and met Charlie, minus his leathers and helmet, making his way down on to Robin's slipway. As we gazed out over the estuary, Robin explained that Charlie had sold up all his property in England and was hoping to buy a chateau in France. Meanwhile, he and his family were living on a yacht downriver and Charlie was now busy trying to get his dory engine started to get to the boat. We could hear Charlie muttering and cursing as he yanked the starter chord several times. It was obvious that the little outboard was not going to start.

'I bet you're out of fuel again!' shouted Robin.

'Bet I'm not!' retorted Charlie.

A few minutes later, however, we heard him clumping up the steps towards us and then he appeared carrying a petrol can.

'Told you so,' smirked Robin.

We made for the garage where Robin kept spare fuel. They could not locate a funnel so improvised with an old newspaper. They were making a pig's ear of the job and fuel was slopping all over the garage floor. It was then that I realised that they were both smoking cigars. I carefully edged back without them noticing. Eventually the task was complete and Charlie strode back across the lawn spilling fuel all over the place down on to the pontoon. I helped Charlie connect the tank, primed the fuel line and pulled the starter cord. The engine spluttered into life. We wished Charlie bon voyage and he disappeared into the darkness upstream.

'Robin?' I asked. 'Where did you say Charlie's boat was moored?'

'Downriver,' responded Robin.

'Charlie has just gone off upriver,' I remarked.

'Daft bugger,' said Robin.

We walked across the lawn.

'Strong smell of petrol,' I observed.

'Nothing to worry about,' said Robin, turning around to look at the lights on the estuary and casually flicking his cigar butt on the ground.

There was a loud whoosh and a ribbon of fire streaked across the lawn, down the steps on to the pontoon. Robin froze.

'I think I need a stiff drink,' he stammered and led the way back to his lounge where he poured two very generous brandies.

'Is Charlie always so accident prone?' I asked.

'Oh, this is a good day. He can be much worse than this. Did you hear about his accident?'

'Was that when he lost his leg?'

'No. He lost that in a plane crash!'

My mind boggled at the thought of Charlie in charge of a plane.

'This was recently,' continued Robin, 'when he was on his motorbike in France.'

Evidently, Charlie had taken his motorbike across the channel for a short weekend to start looking for his dream chateau. He had been notified of several possibilities so was keen to see them all before he returned. They were quite a distance apart, unfortunately, and he had been short of time. He ended up racing from one to the other. This was alright on the main roads but very risky in the narrow winding lanes of Normandy. The inevitable happened.

Negotiating a blind bend in haste, Charlie met an old French farmer driving his battered Citroen van to market. The farmer swerved. Charlie swerved. The bike ended up in a ditch and Charlie slid and got completely stuck under the van. The poor farmer was horrified and tried to drag Charlie out from under his vehicle but to no avail. His cries for help were answered by two burly farm labourers who rushed to the scene.

One of the labourers gradually lifted the van while the farmer and the other labourer tugged and pulled at Charlie. With a mighty effort, Charlie was dragged out from under the vehicle but was minus his artificial leg! The farmer suspected the worse and fainted in deep shock. When the ambulance arrived, it was the poor farmer who was lifted in and taken to hospital for a check-up.

The gendarmes evidently were very amused by the whole incident and Charlie was able to refit his leg, rescue his bike and proceed on his way.

I had planned to stay in bed late the next morning. I knew Robin had to be up early as he was supervising the workers on an extension that he was having built. Though he grumbled all the time he quite enjoyed lending a hand though Vicky now kept him under strict surveillance since Charlie had bet Robin that he could not operate the JCB on site. He lost the bet when he put the JCB's front bucket through the patio doors!

At 8 am there was a loud knocking on the door. It was Robin. Charlie had failed to return to his boat last night and his wife was very worried. We would have to go out and look for him. It was a beautiful summer morning and the sunlight was sparkling and dancing on the water. I quickly showered, dried myself on one of the huge fluffy bath towels, slipped on a pair of shorts and a T-shirt and made my way to the kitchen. There was no time for breakfast but I insisted on a cup of coffee despite Robin's impatience to be off. Vicky was about to start a full scale enquiry about the scorch marks on the lawn and Robin did not want to be a witness for the defence.

On the pontoon, we both jumped into a large dory with a powerful outboard and headed off upriver in the direction that Charlie had taken the previous night.

Surprisingly, we found him after only about fifteen minutes. He was still in his dinghy, fast asleep, about seventy metres up inside a little creek. Unfortunately, the tide was ebbing fast and he was firmly stuck on the mud. The first task was to wake him, and the second, to get him off the mud. Neither proved easy. We hollered, yelled and screamed from the dory. It took all of five minutes to rouse Charlie then another five to explain that we were going to throw him a line which he should fasten to the bow of his dinghy so that we could pull him off.

Robin nosed the dory in, being careful not to run aground so we were able to close the gap between us. At the third attempt I hit the dinghy with the line and Charlie made fast. As he was facing the bank, Robin slowly put the engine into reverse and gradually took the strain on the line which tautened so Robin increased the revs. At full power, the dinghy did not budge! After several attempts, we had to rethink.

'One of us,' explained Robin, peering at me closely, 'is going to have to swim over to Charlie, help him remove his clothes, get him out of the boat, then push it from the rear while I pull with the dory.'

I removed my shoes, they would not have stood a chance in the mud, and stripped down to my underpants to preserve my dignity. I lowered myself over the transom into the cold water. I carefully swam towards Charlie and his dinghy. My feet touched the mud about halfway there. It was glutinous and squelchy and I had quite a job to extract my feet. Using a flat breaststroke, I managed to get within ten metres. I tentatively put my feet down again. But I could make no progress.

Using the towing line between the boats, I hauled and dragged myself along hand over hand. Charlie thought this was very amusing, especially as the cloying mud kept on dragging my underpants down. Every metre or so, I had to stop and pull them up. By the time I reached the dinghy, Charlie was in fits of hysterical laughter but he quickly stopped when I explained that he was going to have to strip off and join me in the mud. It was now Robin's turn to laugh as Charlie stripped down to his underwear.

He was sensible enough to take off his artificial leg before I helped him overboard to join me in the mud. As he could not swim, I suggested he station himself at the front of the boat so that he could use the line for support. I would push from the stern. When we were in position, I shouted to Robin to take the strain. The dinghy began to move.

What happened next is still not very clear to me. Once on the move, the dinghy took off at a tremendous rate. I could not keep up and fell flat on my face in the mud. Charlie was so petrified that he hung on to the line with a vice-like grip and protesting and screaming, between great mouthfuls of water, he was dragged to the dory. I was left floundering in the mud. My attempts to swim on were of no use as I was stuck.

'Throw me a line!' I shouted.

It seemed like ages before this was organised and they took seven tries before I could eventually grab the line. I put several knots in it to prevent my hands from slipping and once again, the dory took the strain. Slowly and steadily, I was pulled over the mud. Slowly and steadily, my underpants slid away from my bottom but I dared not let go of the knotted line. Another metre and they were at my knees. I thrust my legs apart to halt their sliding progress. It was no good. Their progress was inexorable.

Soon they were at my ankles and at an attempt to avoid a branch, they slid off one foot. I made one last desperate effort to hook them on a toe to retain my ownership but as I hit deep water, they were washed off and sank to the bottom. I crawled over the transom of the dory stark naked and covered from head to toe in mud amid gales of laughter.

On our return to the house, the children were very amused at our bedraggled appearance and laughed all the time that Vicky was hosing us down. I had not bothered donning the T-shirt and shorts because of the mud so Vicky rushed out with towels for me and insisted she take me upstairs to supervise my showering personally, much to Robin's chagrin. Charlie was left to fend for himself and had the task of phoning his wife to explain his overnight absence. It was a long phone call.

With all the excitement, I had completely forgotten that we were starting the shakedown cruise that afternoon; though after the events of the last twenty-four hours, I was very apprehensive. Some of my clothing was unsuitable for sailing so Vicky raided Robin's wardrobe for me and I was fully kitted out. Now I really looked the part.

After a delicious cooked breakfast, the whole family, Robin, Charlie and me, three dogs and all our bags, piled into the dory. One of the workmen had helped Robin hose out all the mud so we all sat back to enjoy the trip down river, Charlie's dinghy in tow, with Vicky explaining the interesting features en-route.

Charlie's yacht looked a little the worse for wear and was very cluttered and untidy below decks. We were entertained to coffee and biscuits on board while Charlie packed a bag. Robin had much pleasure, of course, in relating the events of the morning and the hysterical loss of my underpants so I was the centre of the conversation.

We all clambered into the dory once again and headed out of the marina. All I knew at this stage was that Robin owned a new motor yacht. I had visions of her being a rather smart 'gin palace' sort of craft, with a rakish flying bridge that you always see permanently parked in marinas throughout the land. It had not even occurred to me why she was not moored up river opposite Robin's house. I was not expecting the Katrina. She was big—well over a hundred feet long—and painted white all over. From the tip of her bow to the sun deck at her stern, she was every inch an elegant lady. I was impressed.

Above and below decks, she was luxuriously appointed and no expense had been spared in her fitting out. I was given a full tour, starting at the bridge, where everything seemed to be electronic, through the crew's quarters, all well-appointed, the eight spacious cabins, the cocktail bar, the extensive lounge and a Jacuzzi and sauna. She was magnificent. My puny cabin cruiser at Shoreham was peanuts compared to this. Suddenly, however, I thought of accident-prone Charlie. With him on board, I mused upon all the possible problems and accidents we might have before we reached our first port of call.

Amazingly, there were no problems. The trip was smooth and totally uneventful. Katrina was a wonderful vessel and we made good time so now we were in this very smart sailing club bar attacking the cocktail list with gusto. Charlie was drinking the last cocktail on the list and was obviously going to win the bet and it was then that he issued the challenge to Robin.

'I bet you wouldn't pinch that woman's bottom.'

Now, at this point, I must interject that this incident happened some time ago when it was usual for men to pinch a lady's bottom as a sign of affection. Nowadays, of course, it is very non-PC.

However, there was, somewhat incongruously, I thought, a very large, well-built, almost buxom lady, sitting on one of the padded bar stools to the left of the main bar area. She was dressed smartly and spoke in an extremely forthright and very loud North Country accent and…she knew it all. She had been boring and irritating everyone in the bar for over two hours. She was with a very small, dark-haired man, perhaps her husband, because he looked very hen-pecked and agreed with everything she said. The couple with them had tried to excuse themselves on numerous occasions but the lady's hectoring voice seemed to have transfixed them in position and they seemed too scared to move.

Robin and I had not kept pace with Charlie on the cocktail list but Robin was quite inebriated and though I tried to stop him, he rose shakily to his feet, fumbled for our glasses and walked unsteadily to the bar. He sidled in beside the lady's ample posterior, ordered another round of drinks, and then, winking at Charlie, he raised his hand and gently pinched her bottom. She hardly flinched, but she did stop talking and carefully swivelled around on her stool and eyed Robin carefully.

'Did you just pinch my bottom, young man?' she asked in her thick North Country accent.

'Yes I did,' replied Robin bravely.

That was the last thing that Robin could remember for the next few minutes because the lady swung her fist straight at his face with a punch of which Frank Bruno would have been proud, making a good connection with his chin. It pole-axed him and he collapsed to the floor. The little man then started to kick Robin hard.

Charlie and I ran over to help Robin. Charlie, who was really quite strong, caught hold of the little man and held him aloft kicking and struggling all the while. I started to apologise all around in a vain attempt to cool tempers. I am not an aggressive person and my main line of defence is flight. By this time Charlie had released the little man thinking it was safe to do so, but to my utter astonishment, he stepped up to me and kicked me hard on the shin.

I once had a friend who was both kind and charming and would not harm a fly in the normal course of events, but if a fight started anywhere near him, he felt compelled to join in. Many was the time that his wife, or friends, and I bundled him out through a back door to avoid him charging into a brawl. He could not explain it. Something just seemed to snap inside him.

Well, so it was with me when the little man kicked me on the shin. Something snapped inside me and I let fly with a flurry of punches to his face and stomach. He toppled then joined Robin in a crumpled heap on the floor while Charlie had to restrain the lady from attacking me.

The barman, very bravely, I thought, appeared on our side of the bar and stood between the two warring factions. There was, I realised a great deal of tut-tutting and other expressions of disapproval from the other members. We were approached by an official of the club. He was the archetypal secretary dressed in a blue blazer and white flannels. His red face and white moustache twitched in anger and I thought he was going to have an apoplectic fit as he asked us to leave the premises. We dragged Robin to his feet. Discretion was the better part of valour—a dignified retreat was sensible in the circumstances.

'Quite raht too,' shouted the woman.

But hers was not the victory, for the secretary was adamant that she and her partner should also leave. She stalked out with the little man trotting along behind her.

I wondered if we should take Robin to the local hospital for a check-up but he point-blank refused. I checked him over as far as I was able. He seemed alright. I particularly checked his eyes in case of head injury and the pupils seemed of equal size and normal. I obviously failed to notice the gleam in his eyes, however, as he plotted his revenge on Charlie.

We walked back to the Zodiac, tied near the slipway, loaded the few supplies we had purchased, and cast off. The Zodiac had a powerful 80hp outboard on the back and was controlled from an open central driving position halfway down the boat. Robin was insistent that he would drive despite his aching jaw. We made our way out to the Katrina at her mooring.

As we came alongside, Robin asked me to get off first and take the stores from Charlie. This I carefully did and started climbing up the steps to the deck thus I did not see what happened next. Robin must have thrust the throttle full open and leapt for the gangway steps intending to leave Charlie on board. Charlie was slow to react, befuddled by too many cocktails, though it did not really matter because as he lurched forward to grab the controls, the accelerating Zodiac hit the mooring buoy and Charlie fell overboard.

Now, you might perhaps remember that Charlie could not swim so I ran up the deck of Katrina, shedding my clothes as I did so, stripped down to my underpants, again, and jumped in to help Charlie. I managed to grab him and we both held on to the mooring buoy. Meanwhile, the Zodiac, with its outboard running at full throttle and the steering hard over, started to make wild, bucking circuits around the moorings, bumping into and bouncing off boats, yachts and cruisers. Lights started to come on in some of these craft and a series of torches and spotlights started to illuminate the wild antics of the Zodiac.

Robin threw a lifebelt over the side and placing it carefully around Charlie, I towed him to the steps. Robin helped us out and the two of us stood cold, wet and shivering, while Robin pondered what to do. His plan to give Charlie a shock had backfired and he was now looking worried. The Zodiac was on its third circuit by now and had resisted every attempt to catch it by the other boat owners with boat-hooks and grappling lines. All seemed to no avail and now lights were appearing along the quay as well. This was turning into a major crisis.

Suddenly, Robin seemed to find inspiration and disappeared below deck to re-emerge, a couple of minutes later, with a shotgun and cartridges. Now I know Robin was an excellent shot as he had won several cups and held regular shoots on his land. He was, however, drunk and possibly concussed so I did cast some doubts on this drastic measure. I could not think of any other plan and waited, with bated breath, for the Zodiac's next appearance. It plunged out of the darkness, on the port side. Robin took careful aim and fired with both barrels. We thought he had missed at first but then the Zodiac sagged down on one side, straightened its course and shot straight up a mud bank at the side of the estuary. The engine cut automatically.

SILENCE.

We shouted our apologies over the water to all and then Robin suggested that we should lower the other tender so we could retrieve the Zodiac. At this point his crew mutinied and explained, in no uncertain terms, that we were somewhat unwilling to go crawling around another mud bank at that time of night. The Zodiac could wait until the morning. Robin reluctantly agreed and even made us a hot cup of chocolate, liberally laced with rum, while Charlie and I took a hot shower.

In the dreary light of morning we quickly retrieved the Zodiac. She was gently floating at the top of the tide with the prop shaft holding her fast. Her starboard tank was completely deflated but we managed to get her to Katrina and then swing her on board. She lay, like a dead whale, on the aft deck. We were famished by this time and trooped off to the galley. Robin seemed very contrite and even agreed to cook the breakfast. He had hardly started before we were hailed from the water.

'I bet that's the Harbour Master,' I suggested sarcastically.

'Don't be silly!' retorted Robin.

But it was and he looked very grim-faced. We could be in serious trouble, I thought, but Robin was at his most charming. He invited the Harbour Master aboard and immediately started to apologise profusely for the disturbance of the night before. It had been, he explained carefully, an accident. His friend, pointing at Charlie, had had the great misfortune to have lost a leg in a plane crash and now managed with an artificial limb.

Unfortunately, while trying to hand up the stores last night, he had stumbled and hit the throttle. The Zodiac had zoomed off, tumbling poor Charlie into the water. At this point Charlie very carefully rolled up his trouser leg to show his artificial leg. Though he did not like the idea of shooting at night, Robin explained that he was concerned for the safety of the other boat owners and took the drastic action that he did, being careful to shoot away from the other vessels, of course.

The Harbour Master immediately became very apologetic and sympathetic, expressing concern for Charlie and his leg, wondering if the water had damaged either? He came below to partake of a coffee and a bacon butty. He was given a tour of the boat and was most impressed, then returned to his launch all smiles and thanks hoping that we would visit again sometime. He even offered to take the Zodiac to a local yard for repair but this was declined as Robin owned an insurance company.

Half an hour later, we sat down to a full English breakfast and Robin opened a bottle of vintage champagne to celebrate.

'Here's to a very successful shakedown cruise,' toasted Robin.

'Shakedown!' I exclaimed. 'More like a shake-up cruise. I'm going to jump ship at the earliest opportunity.'

'You've done more than enough jumping off ships already,' retorted Robin, 'with and without your underpants! What do you think, Charlie?'

'I think you have both been pulling my leg for the last two days,' smiled Charlie and tapped his artificial leg. We all burst out laughing.

Chapter Twenty-Five
Polo

The king of Games and the Game of Kings

From 1985 until 2001, I was working at Mount House School in Tavistock, Devon. During that period, the world was marching onwards at home and abroad. The Berlin Wall was demolished, Maggie Thatcher resigned, the Gulf War took place, apartheid ended in South Africa and Nelson Mandela became President while one year later Jerry Rawlings was elected as President of Ghana, Princess Diana was killed in a car crash in Paris, the iPad was invented and the Twin Towers terrorist attack took place in New York. Though working in a prep school we were not cut off from the outside world and I can clearly remember the impact of these events especially as many of our parents were in the military and served in the Falklands.

The Gulf War did have an impact on many pupils in the school concerned for fathers fighting in the Gulf and there were quite a few. I vividly recall the announcement of Princess Diana's death as I was due to take a trip to Camaret not long after. On the campsite, however, it was obvious that the French were possibly more affected by the loss than many people in England and we were bombarded by commiserations and condolences which rather confused the boys.

There was a lady, a retired French Army major who even invited us to watch the funeral for the whole of the day in her caravan which I had to politely decline. Again, it was the fault of a perfect climbing day on the Pointe de Pen-Hir. The news of the Twin Towers attack reached us in the middle of a junior barbecue down on the beach by the Tavy and certainly came as a huge shock.

I still had my bungalow in Padstow where I spent what little free time I had during the term, normally on my short day off and during the holidays and could relax and enjoy the town and its environs. I was also able to pop over and see my parents in Newquay on a regular basis and they enjoyed visiting me in Padstow. There were initial objections about spending a night away from the school on my day off which I thought was rather odd as I regularly spent nights away on trips, camps and expeditions with no objections. So I swotted up on labour law discovering that the school was in contravention of employment regulations concerning staff 'on call' and presented my findings. The subject was never raised again. I was getting to know the Padstow people and my neighbours a little better and often joined The Merrymakers in their efforts to raise money for charity for the town and was often invited out on the fishing boats for the odd trip.

The Bristol branch of the family continued to thrive though there were the usual clutch of minor problems in which I had no need to be involved.

Sliding along the timeline, my brother, whose depression and anger continued to affect us all in many ways, moved out of a house high up above Perrancoombe to a big house further down towards the town having taken early retirement from SWEB due to stress-related illnesses. He had initially opened an animal feed shop in Perranporth which seemed to be a good business while his wife ran a pet shop in the centre of the town that did well.

For some reason, Roger gave up his feed business and took over an art shop which I thought was odd. Sorry Perranporth, but the town was not famed for its art and culture.

The business was not a resounding success and Roger's character was such that his customer relations were not of the best nor his association with the other retailers in the town. Next door to him was a Chinese take-away with whom Roger had skirmishes and arguments, especially, believe it or not, about a life-size RSPCA charity dog that Roger reckoned encroached on his frontage.

The takeaway never seemed to have any dustbins collected on the weekly rounds and this went on for a couple of years until Roger decided to sell the shop. No one would buy a property in Perranporth without a mine survey because the whole area was mined for tin but Roger did. A potential buyer commissioned one and discovered that there was a huge mine shaft between Roger's property and the takeaway next door. The owner of the takeaway obviously knew all about it for that was where he had been throwing his rubbish for years. He was fined very heavily.

Roger's children seemed to enjoy being in Perranporth and their schooling was excellent in so many ways as it was part of the community. They were mature enough to enjoy their secondary schooling in Newquay and Truro as well, though it involved the usual bus travel so common for children in Cornwall. Roger's pattern of life seemed to be cyclical. He would show an interest in a job or charity work and fully embrace it for two or three years and be initially highly popular and good at what he was doing, but then criticisms and arguments and rows would develop and he would withdraw, always on the basis that it was never his fault.

He ran the Perranporth Cub Pack successfully for some time and left in anger under a cloud, and it was the same with the Perranporth Band. He gave up the art shop and ended up in a legal wrangle with the purchaser and in debt for some reason or other.

I had made attempts to help in many ways. I had invited him to Gilwell Park when I was working there after the break-up of his first marriage and he seemed to enjoy that week, and he also attended one of the Mount House Leavers' Camps at Padstow when he was having another bad patch. He experienced difficulty with his relationships with the boys, being quite tough and aggressive on occasions, a flashback to his nickname 'Hitler' when he was running the Scout troop in Kingswood.

When his financial affairs became of some concern, I introduced him to one of my friends who was an insolvency practitioner and also let him use my accountant for well over a year but Roger fell out with both, nearly leading my accountant into serious legal problems after an enormous dispute with his bank. Paying the fees for the practitioner and my accountant for him also led me to cut back and maintain a tighter control on my finances.

This was at a time when he had found employment on a caravan site in Devon, as its manager, where his usual 'cycle' of work occurred and it all led to tears and a fiasco when he walked off the job. He was always angry and depressed, very quick to flare up and take offence and could not maintain any real form of social contact. He was a problem and a source of concern to my parents in particular.

After many happy years at Padstow, I decided to go into the holiday let business and purchased a bungalow at Harlyn Bay that was good for all year round lets. It was quite modern though it did have a slightly unusual configuration with the bathroom well away from the bedrooms. It earned a cracking income for me and I was able to let it to friends and took many school groups there as well. The beach was superb and the surfing normally quite good. It was looked after by a letting agent but the staff and owners of the Harlyn Inn also kept a weather eye on it for me.

It was decorated and maintained by one of the characters of the district, Dave Hollis, who swore that the bungalow was haunted and it was cleaned by a lovely lady, Francis, who, it must be admitted, was of somewhat generous build, and with whom no one crossed. She sat on a drunken US Marine one night, for he was drunk and had broken one of my windows, and refused to let him go until he had emptied his wallet to pay for the damage…and a few drinks, of course.

The best Lets were when Maggie and Dennis Thatcher visited Sir David Wolfson's house in Constantine Bay and her security guards stayed in my bungalow at very reasonable prices (not). Sir David's daughters were never well pleased with the annual event as there were sniffer dogs and security guards all over the house. Even the loos had to be checked.

Mounir Captan and his family stayed in my bungalow for a couple of weeks on holiday from Ghana and they liked the place enormously. They loved the Cornish greeting of, 'Alright then?' and Kareem, his son, adored cream teas. He ate so many that the Cornish were thinking of calling them 'Kareem' Teas. While they were in residence, I had arranged for a couple of workmen to build a small side gate and Mounir and his family were astounded at their sheer lack of progress.

They gave them copious cups of tea and biscuits and they kept on disappearing over to the pub. Mounir was worried that I was paying them by the hour but I had only forked out £35 for material and labour. Alas, three weeks later, the wooden gate was ripped from its hinges and used to fuel a barbecue on the beach. Mounir loved driving and has the unusual distinction of driving all the way up to London and back to Harlyn in one day on an August Bank Holiday.

The letting business made good money but you could have bad tenants sometimes. I had been persuaded to allow one of the inn's chefs to have the bungalow for a long winter let but I received an urgent phone call to come down quickly as he was abusing the place badly. What I found was heart-breaking. They had big dogs and about eight puppies and allowed them free use of the house. They had dragged out the duvets and pillows onto the lawn and let the dogs play and rip them to bits and the whole place was filthy and covered with dirt and grime.

I simply said, 'OUT. NOW!' and they left but it cost me about £2,000 to put everything right.

One day, at Mount House, my lesson was interrupted by Charles who said that I had a visitor. It was the sheriff from Bodmin Court who stated that I was in breach of a court summons and he had come to take me to court. He had no idea of the details so I drove down to Bodmin and was brought up before a magistrate immediately upon arrival. I explained to the magistrate that I had no idea what all this was about.

An official from Bodmin Council then stood up and stated that I had not paid my rates on Number 4 Dennis Cove, Number 4 Hawker's Cove and Number 4 Porthilly Cove. Could I explain why? That was easy. I did not own any of them and told them so. Could I prove it? I turned to the magistrate and simply asked how on earth one could prove one did not own a property.

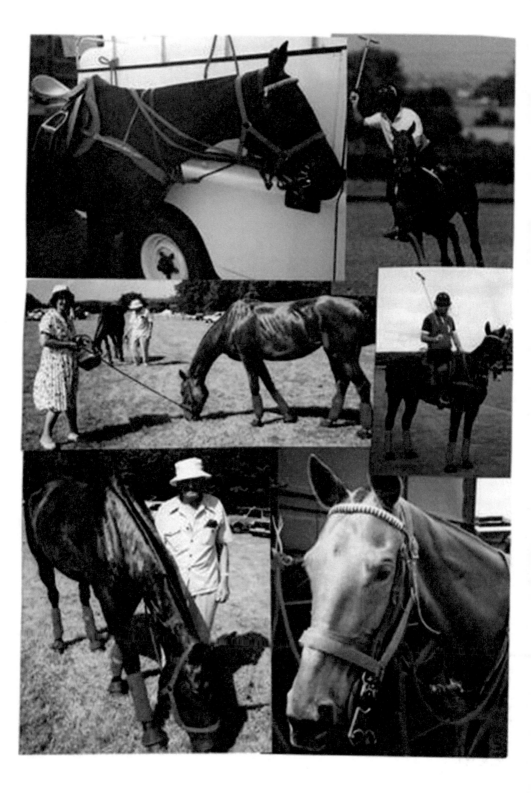

I am afraid that I let rip at the official for totally wasting my time and not having the common sense to proceed with such a ridiculous accusation. The magistrate took my side and gave the official a broadside that was quite succinct and savage. The case was dismissed but I had not finished for I demanded a formal apology for the wasted time and the slur on my character. Compensation was agreed at £200.

In the writing of this memoir, I have found it very difficult to keep track on the vehicles that I had owned. When I commenced my employment at Mount House, I do know I had the Opel Commodore and my red MG Midget. The Opel Commodore disappeared and I cannot think when, or to whom I sold it, so I only had the MG for some time but I was then driving the school Land Rover and minibuses much of the time anyway. I pranged the MG on black ice late at night and it was a 'write-off', as they say, and then ensued a battle with the insurers because I had deliberately, and with their permission, paid a higher premium, for I knew the car was increasing in value and not depreciating.

The insurers quoted the depreciation value and the dispute was settled in my favour by my insurance broker over a period of a couple of months with a threat of legal action. There was at that time a classic car dealer on the main road through Horrabridge and I noticed that he had an E-Type Jaguar in the showroom. I had always wanted an E-Type and made enquiries only to discover that it was an Alfa Romeo Duetto Spyder. I road tested it, liked it and bought it on the same day. It remained my personal mode of transport for well over a year and it was well serviced by one Jan Organo, an Alfa Romeo specialist, on the road to Cape Cornwall north of Penzance who had arrived as an Italian prisoner of war and never left.

Well over a year later, I was in contact with a friend that I had not met for some years. Dick had been the Scout leader of the Southampton Scout Troop with which Jim Pickup and I had been in contact, many years before. He was the professional recorder player and expert on fungi, and now, he was the editor, I think, but please do not quote me, of the Classic Car Magazine. I told him about the demise of my MG and my replacement Alfa. He became quite excited and wanted to know whether it had a 'blip' on the bonnet.

You will understand that I have no technical skills or ability whatsoever and whereas I can deal with petrol and oil and that sort of thing I had no idea what a 'blip' was in that context. It was evidently an indentation for the badge and I discovered that I did have one as he made me go downstairs on a very wet Dartmoor night and find out. He asked how much I had paid for the car and it was exactly £1,300 as I had managed to haggle it down from £1,500. He assured me that it was worth well over £10,000. Wow! Did I want to sell it? Did I? You bet. I drove it to Bristol Parkway and received a cash sum of £13,000 for the vehicle.

On 22 October 1989, my father died in Treliske Hospital. He had been experiencing a bowel problem for several days which suddenly became very serious and he was rushed to Truro from my parents' bungalow in Newquay. My father was scared stiff of hospitals and it must have been an alarming experience for him. It was a half-term holiday and unusually for me, I was not on a camp or expedition, but staying with friends in London, visiting Nour's mother who was having an operation at Cromwell Road Hospital at that time.

I travelled down by train the next morning to Plymouth then drove immediately on to Perranporth to be greeted with the news of his death by my brother at his front door. He had died of a heart attack during the night mainly of an acute left ventricular failure and a recent myocardial infarction which was clearly stated on his death certificate. The reason I mention the details above is because some years later I was to experience a heart problem and the same details appeared on my diagnosis. Like father—like son.

I stayed down with my mother for a few days, trying to be as cheerful as possible in the circumstances, and Roger and I helped out with the funeral details which were mainly handled by the undertaker anyway so there was not much to do but I had the feeling that my mother was not mourning greatly, as it were. Now, when my grandfather died, my grandmother did a little mourning for the sake of family decency, decorum and respect but there was no doubt that she was glad that she could now enjoy her life. My mother did not say anything but there was a small element of relief in her mourning that was very noticeable.

Surprisingly, on the night after my father's death, she asked to go out for a Chinese meal never having eaten one before, especially as my father hated foreign food and she wanted to go out shopping on the next day at the supermarket as usual and we also visited Truro.

I had to pop back to school one morning to get clothes for the funeral and sort out one or two things. When I returned my mother had placed about a dozen black plastic sacks by the front door, full of my father's clothes, and had phoned The Salvation Army to come and collect them. I asked her if she had checked the pockets as my father collected pens and he had some very fine specimens.

She had not so we carefully unpacked the jackets and suits (my father was a careful dresser) and not only did we find a pen in every inside pocket but £200 in cash as well. When I found the first £200, my mother explained that it was probably for their golden wedding anniversary which had been scheduled for the following weekend at the Headland Hotel but she could not explain the rest of the £2,000. Her comment was not very apt in the circumstances for she stamped her foot and said, 'Oh! I could kill him!'

There was, of course, a family funeral and several of my father's friends and acquaintances were in attendance at the Truro Crematorium to which the chief mourners had travelled in a rather elderly Cornish hearse that squeaked and protested on every corner and bend. All people's last wishes must be respected but I do find funeral services rather uniform and too sad. I have determined that I do not want a funeral service because I have donated my body to the Bristol Medical School to do with as they wish, after which my body parts may be pickled for public exhibition so maybe my relatives will still be able to see parts of me long after I die.

I would like a happy and cheerful remembrance service if my friends would care to attend and I have even thought of recording a dying speech for the event rather like the pirate chief in Peter Pan, just to wake everybody up at the commencement of such a service. If I am in Ghana then I am in trouble though I am sure their medical school would appreciate a 'Bruni' to practice on.

Now Ghana is a country that knows how to 'do' funerals and that could be quite exciting. I would be very sorry to miss it, I believe it was Ghana that first started the tradition of 'fantasy coffins' and one such coffin maker calls himself 'Papa Doc. Six Foot Under Enterprise'. He creates coffins to match the deceased. A preacher would have a bible, a pilot, an aircraft, a fisherman, a canoe, a baker, a loaf, and me? Surely a canoe or a polo pony?

My mother was probably lonely after the death of my father but she did have many friends and neighbours and she was not one to mope. At that time she could still drive so visited my brother and often came up to Tavistock for a few days. She was a good driver and even visited her relatives in Bristol on a fairly regular basis. She continued to do volunteer work for Cancer Research mainly in the form of manning the weekly book stall and enjoyed their meetings and functions though she did not approve of the minor 'politics' within the committee.

Looking at my medical records for this period, I was rather taken aback by the length of the list. During the summer, before my appointment to Mount House, I had been experiencing headaches and arranged, through my Padstow doctor, to have an MRI scan at Treliske. I duly attended for my appointment one sunny afternoon and was fed into the torpedo-like tube twice, the second time having been injected with a special dye so that I would appear in glorious technicolour. I was wearing one of my favourite silk shirts, and after I had thanked the staff, I walked out through the modern carpeted reception area.

A receptionist screamed at me, 'Blood! Blood! You're bleeding all over the carpet!'

I looked around and saw that I was bleeding quite badly from the point where the staff had injected my arm. The receptionist rushed over and pushed me none too gently into a tiled area of the floor then proceeded to deal…with the carpet, scattering tissues all over the bloodstains. This was surely an innovative form of first aid. I gripped my arm with my handkerchief and walked back into the scanner room where I was dealt with efficiently and well.

I could not but resist a jibe at the receptionist on the way out, suggesting that next time I had a stain on my carpet I would call an ambulance and ask her to deal with it. I did make an interesting spectacle, however, as I walked along the corridors and out to my car in a very bloodstained shirt.

I then had several minor operations for blood clots at Derriford but they did not appear to be too serious or debilitate me in any way.

I had urinary problems and my specialist suggested that I have a prostate operation, which for some reason, took place in the new hospital in Exeter. I can only think that I was anxious not to be off work and I could be operated on during one holiday there. I had undergone pre-op several weeks before but I must say that I found preparations for the actual operation rather casual. I was allowed to wander around the hospital at will and the people performing my operation sought me out, even my surgeon, Mr Pocock (I kid you not).

I was informed, after the operation, that he had removed most of the prostate but I was just put into a bed and abandoned. No fluids. No food. No care. I rang the buzzer several times and the staff told me that they would come back but they never did. When the night sister arrived, she was furious. She peeled back the bedding and it was like the aftermath of an evening's soiree with the Marquis de Sade, with blood and fluids everywhere, and I had to be cleaned up and moved to another bed.

Bless her. She found me food and drink. She told me not to make a complaint because she had already done so and I fell into an exhausted sleep. I had to stay in hospital for two more days and I was told to shout loud and clear if I required anything. I was surprised to hear that the Hospital Car Service could drive me back to Tavistock so ensured that they received a handsome donation; then, I made a mistake. I never told my parents when I was going into hospital for anything.

My routine was to phone them after the event to inform them that I was back from hospital and all was well. My mother was now living on her own, however, so when I cheerfully phoned to say all was well, she drove up to see me in a panic and insisted that she stay with me to ensure I was OK. Now I do appreciate it when people visit me when I am ill but I hate visits to last any longer than half an hour and cannot stand permanent attendance at my bedside for hours on end. It wears me out and I get edgy. I like to recover mostly alone, thank you. With great difficulty, I persuaded her to go back the next day, but mothers being mothers, she subjected me to the same treatment in the following year.

I woke up in the middle of the night in my flat at school and was in agonising pain all over, barely able to move and I could do nothing about it. I was crying out with the pain and endured this for several hours before I was able to bang on the floor of my bedroom when I heard Charles go into his study. He ignored it at first but then realised that something was wrong and galloped up. The school doctor was called and an emergency ambulance arrived to rush me into Derriford Hospital.

The ambulance had definitely seen better days and I felt very unsafe travelling in it as he rocked and rolled around corners. Upon arrival, I was immediately shunted into a mixed reception area, the curtains drawn, and apparently, and amazingly, forgotten. I must have been moaning and groaning in my pain and the lady beside me told me to shut up twice. Several hours later, the school had phoned in to ask how I was and were told that I was not in the hospital. There was no such patient.

I could not give you a precise statistic for the number of Mount House parents who worked at the hospital. It was probably well over fifty and the school sent out a general alert to most of them. The 'Find Wally' game was converted to 'Find Geoffrey'. I was eventually discovered by one of our nursing parents and I must say, all hell was let loose.

I was all in, totally exhausted and was even unable to answer any questions. They gave me a precursory check, painkillers and something to make me sleep and I knew no more until the middle of the night when I awoke in great pain again. I pressed the buzzer and there was immediate action. I was rushed down to ITU, I think, and connected up with all sorts of wires, gadgets and monitors and I was now in safe hands as it was run by one of the parents who, as a group, now seemed to take over my care completely as I never saw another non-Mount House doctor or specialist again during the two weeks I was in hospital.

I lost count of the number of tests that I had and they tired me out more than anything for I think I must have undergone every conceivable medical test known to modern science. I also had to wait at X-ray and other test centres until one of my doctors suggested that I travel in my bed then I could receive priority. I wish I had known that before. I was eventually placed in a ward for chest problems as the general theory seemed to be that I had suffered from a pulmonary thrombosis but I was unable to sleep because of the noises from the various aids to breathing in the ward.

Most nights I was wheeled down to the intensive care unit to sleep if they were not busy and wheeled back for breakfast in the morning courtesy of Mount House. My mother arrived and camped by my bed. I was delighted to see her but not for too long. I persuaded one of the doctors to tell her that I needed complete rest and half an hour a day would be appropriate for a visit. On the night of the annual fireworks display, I was kindly visited quite late by a barrister parent whose father had been a famous surgeon at Derriford. It was nothing more than a social visit but the staff were under the impression that I was taking legal advice and I detected a slight improvement in my care.

I was in limbo for the medical profession was baffled. The tests had come up with nothing. My only symptoms now were an all prevailing tiredness and I could hardly walk. I spoke to the general ward doctor one morning about my condition and he told me that I was likely to be in hospital for many weeks. Two hours later a senior consultant, who was not a Mount House parent, told me that a virus was the problem and I could go home. I could barely walk and had to crawl up the stairs to my flat where I chilled out for a whole week and willed myself to recover.

The school doctor was very kind and supportive but he gave me a drug that scared me to death. It caused hallucinations. I had truly alarming dreams about being savaged by animals and falling over cliffs and they seemed real. I stopped them immediately. I managed to complete most of the term but only by going steadily and slowly. My doctor suggested that I had ME. Anything was possible I suppose but I fought the lassitude and weakness and gradually got back to normal.

I broke my right wrist two times while I was at Mount House. Both were falling off without permission. One was from my mountain bike in Cardinham Woods and the other was a polo accident on Plaster Down. My wrist was plastered on a Sunday afternoon and I had to attend the Fracture Clinic at Derriford again on the Monday morning. The orthopaedic surgeon was Patrick Loxdale who was not a Mount House parent but knew my connections with polo and Mount House.

The clinic was packed and I was called first after he had announced to everyone that he was dealing with his first polo injury ever and that I was probably going up to see Prince Charles immediately after. He had the wrist X-rayed again and said that it would need a new plaster and warned me about breaking anything on a weekend as the hospital was left to the porters and the cleaners to run. I was told not to ride or play polo in the meantime.

I did have a bit of fun in the plaster room. One young lad had been waiting anxiously outside to have a plaster fitted and as he seemed a bit nervous I chatted to him about his favourite football team which was Manchester United and he was determined to have a red plaster put on his arm in their support. As I came out with my blue plaster neatly fitted, I called back, 'Thank you. Sorry, you only have pink plaster left now though.'

The poor boy looked horrified so I assured him that I was joking.

The following Saturday I was out exercising my polo ponies with Harold, my groom, on Yelverton Golf Course. (We have to catch up on the polo scene next.) One is not allowed to play polo left-handed. The stick is always played in the right hand. Your left hand holds the reins so I had reasoned that I could still ride because I was not using my right hand at all. It was a beautiful summer's morning so Harold and I opened up the horses into a flat out gallop across the diagonal bridle path that cuts across the course. There was a roar.

'I TOLD YOU NO RIDING!' yelled Mr Patrick Loxdale.

We stopped both horses on a sixpence and cantered back towards him but by this time he was laughing and shaking his head. He was amazed that we could control the horses with one hand on a loose rein and I invited him to have a go some time.

'Lasers do not damage the eyes,' said the specialist at the Plymouth Eye Hospital but one had certainly upset mine somehow. I had walked into the staff room at Mount House after a session with the pupils down on the lake and a young member of staff had been playing and fiddling with a laser pen flashing it around the staff room. When I came through the door, he shone it directly at me and it caught me in the eyes. I had an immediate reaction and had light flashes and blurred vision which did not clear and I had to sit down but there was no improvement. I got back up to my room but I still had major problems so phoned a taxi to take me up to Tavistock Hospital, leaving a note for Charles.

They would not deal with it and organised a community ambulance to take me down to the eye hospital. There was no obvious damage to the eyes which they established after a couple of hours but there was obviously a problem with my sight because I could read hardly any of the letters, big or small, on the eye chart. Now, I have always been very short-sighted in my right eye ever since childhood and I have always had to be overprotective of my left eye.

For the moment, everything was blurred in both. They asked to see me the next morning again but I had to travel into Plymouth and out again by taxi as it would have been totally unsafe to drive and the school was very worried as the accident could lead to all sorts of complications. Another specialist examined me in the morning and confirmed again that there was no sign of obvious damage but agreed that the laser had obviously affected my sight. It was over a week before my sight returned to normal and thank god it did. I am still very wary of eye injuries to this day. So lasers may not damage your sight but they can affect it.

There was a major alteration in my lifestyle at Mount House and a large number of factors combined to bring it about. Firstly I cannot say that I did not get on with the new senior master as it would have been unprofessional not to, but I did not enjoy working with him and I really had done my fair share and more of residential duties which were a constant tie and responsibility.

My workload had eased when I had taken over the geography department and I had now completed the scramble for qualifications for the adventure challenge. I did not have the responsibility of producing plays anymore and the adventure store was very well organised and efficient. It had taken much hard work to achieve this so now I could pause and reflect and this was linked with a fairly quiet time in the family's affairs.

I sold up the bungalow at Harlyn Bay and purchased a flat in Yelverton above the Devon Tors Inn run by Pete and Kathy Gibbs with their son, Little Pete. It was a two-bedroomed flat on the third floor and was very comfortable indeed. It was also about the time that I had sold my Alfa Romeo at a very handsome profit so I was in 'funds', as they say. The one thing I missed in life was polo, so I put on my 'thinking cap' and thought that it would be a good idea if I could start a polo club in the area, and eventually teamed up with the Spooner's Hunt who were also keen to have a go at the game of kings—the king of games.

I purchased a Land Rover and trailer and bought a horse from Mike Biddick down in Cornwall called Jack that seemed a very light ride. I rode him and he seemed quite sound and biddable and brought him up to Tavistock. Our first attempt at polo was in an indoor riding school and the paddock polo was interesting to say the least and enormous fun and funny. Jack nearly had me off twice and achieved it on his third try so I let one of the Dartmoor farmers try him. Four men jumped on his back and Jack had them off.

One of the 'horsey' ladies tried him and he behaved like a lamb. Any of the girls could ride him but not the men. I phoned Mike the next morning and took Jack back to Cornwall the next day. Sorry Jack but you obviously needed a new home in a harem. Mike was very good about it and returned the money immediately. I next met up with him at a steeplechase at the showgrounds when we were trying to load one of the Mount House parent's horse into a lorry. Julian returned to the cab to find a rope to help coax the horse up the ramp and in.

Just then, Mike Biddick happened to pass by and saw the difficulty. Now Mike had enormous hands and I stood to one side as he smote the horse with 'a great rout up his ass', as Mike put it and the horse shot into the box.

When Julian reappeared, the horse was quietly eating its hay in the lorry as if nothing had happened. Julian's son was at Mount House and he had a love of horses. He trained under Martin Pipe, is now a trainer and you will often see the name of Warren Greatrex appearing on the race cards.

I phoned around my UK polo friends to size up the market and was invited up to Cowdrey to meet up with Terry Hanlon who had several for sale. He was a bit of a dealer and as soon as I walked into his flat at Ambersham he asked my views on Bute so I knew what to expect when he showed me several horses in a muddy field. My search was not going well but I did bump into Julian Hipwood whom I had not seen for ages and he was then one of the top players in the country. He had just received a new batch of polo ponies from Argentina all destined for High Goal Polo so I knew that they were out of my league and pocket.

He did let me try one and it was like driving the horsey equivalent of a Ferrari that also could spin and turn at speed. It was sheer pleasure to ride. Ghana came to my rescue. Nour Captan phoned me to say that Ross Farge had left some of his horses with another friend of ours, Roger White, who played polo at Rhinefield, I think and they were for sale. I phoned Roger and arranged to go up to Midhurst to try them and play them a chukka each at Cowdrey having no idea that Roger had differences with the club. I had merely phoned the lovely Sarah and she had been kind enough to offer me a couple of chukkas. I duly tried Ace and Lady and agreed to buy Ace who was a wonderful pony and agreed a price.

Unfortunately, Ace went lame and Roger came up with the idea of buying them both, taking Lady first then Ace when fit again and passed by a vet, and he was offering both at a discount. I accepted and I had my first English polo ponies despite the fact that they were Canadian. I brought Lady down first having already arranged for her to be stabled with Harold Cole in Tavistock just above the town, where she seemed quite happy, but ten days later I received a phone call from Roger White who stated that Ace was pining and could I collect her soon.

I drove up on my day off and collected Ace and a very strange thing happened. I was on the A30 just before the Sourton Junction when there was a commotion from the borrowed horse box. I pulled into a lay-by and Ace was whinnying and stamping her feet. She seemed OK so I drove on to the accompaniment of Ace's whinnies and banging. She knew (how?) that she was meeting up with Lady.

Upon my arrival at the stables, Lady was weaving backwards and forwards in her excitement and Harold told me that she had been circling the paddock madly for the last two hours. She had known as well (how?). They were overjoyed to see each other and cantered around the paddock with each other for half an hour.

I can only describe our attempts to start a polo club in Devon as interesting, lively and fun and our local matches were played with great enthusiasm, vigour and shouting, for it must be noted, rather oddly, that all the individual novice players were firmly of the opinion that they already knew all the rules while the other players did not. My experience, having played polo for well over twenty years, is that this is a common phenomenon in all clubs everywhere.

It is part of the fun of polo and I believe it even applies to the Royal Princes. Heaven forbid. The fun was greatly improved with the arrival of that famous Devon character, Art Cole and his family, who, I believe, held the record at that time, for being the largest family in Tavistock and all his family seemed to be excellent riders though their mounts were somewhat varied and raw. To hell with the skill of the game I loved the fun and amusement.

I was not, however, improving my polo so I 'looked up the line' to the Taunton Vale Polo Club who were kind enough to invite me up for chukkas. There was a problem. I had a Land Rover but no horse box so immediately purchased a suitable one for both horses. You will appreciate and understand, that, as far as my polo was concerned, I was a 'Colonial Boy.' In Zambia and Ghana our grooms looked after the care of the horses, tacking up and bandaging and to see all was well with our mounts. It was a foolish rider, however, who did not check everything was safe and well and this also applied in the UK.

In Ghana, many of the players had never tacked up a horse in their lives and knew nothing about horse care whatsoever screaming at their grooms if a leather broke or a bandage became undone. I was no novice at completing these tasks and could demonstrate this to new grooms if necessary but I must say, that, the realisation that I would now have to load and unload my horses, tack them up and do everything for myself was a sobering reality check for me and I did not find it easy to begin with especially as there was little time between chukkas. I managed, and Harold, who looked after Ace and Lady for me in Tavistock was a great help and did me and the horses well.

He had a love/hate relationship with Lady, however, who had a sense of humour bordering on the sadistic mostly directed at Harold. If he put a brush down when grooming, Lady would pick it up and throw it out of the box, then deliberately stand on his toe, or when he was trying to clean out her hooves, Lady would gradually lean on him more and more and almost push him over. She was also a thief and tried to steal the other horses' feeds. She was a 'horsey delinquent' and loved teasing Harold.

While playing polo at full gallop, she would suddenly miss a beat giving you the impression that she was about to fall but she always carried straight on and she played good polo. Ace was the better of the two and I admit that she was my favourite. I enjoyed playing her and she gave me many years of good polo. During the summer months I would often ride out early morning and if I was on Ace she would not pass the greengrocer's in Tavistock unless she was given a carrot and I would often ride her around the grounds of the school or up on to the moor.

Ace was bombproof and you could do anything and go anywhere with her. She did a publicity stunt at Avon Farmers walking around the aisles of their central store and she became quite well known in Tavistock. It was wonderful to take both horses up onto the moor and we had wonderful rides, though Lady, true to form, would shy if she saw sunlight glinting on the granite. On a family camp one Sunday morning, one of the parents and I dressed up in the full regalia of highwaymen and 'held up' the cars as they came in demanding money for charity. We raised well over a hundred pounds, but there were no golden guineas.

Harold was 'arrested' one morning by Simon Dell, the local policeman. If I could not exercise the horses, Harold would take both horses out leading one or the other on a leading rein. He would trot down the Old Exeter Road and into Bedford Square before making his way up Down Road to the golf course. Again, it became a common sight in the town. Simon informed Harold that he was breaking an ancient law forbidding the leading of horses through the streets of the town. Harold saw the joke and merely said, 'This is no horse. This is Lady and she be a polo pony.' Case dismissed.

Taunton Vale Polo Club is situated not far from the Taunton Racecourse at Orchard Portman and the club's stables are directly behind the course. When I first joined, it had one polo field and a small stick and ball area at the far end. The club house was traditional in that it was quite an elderly wooden building with adequate facilities. It was small, friendly and pleasant and I liked it very much indeed. I was playing off a nought again as a returnee from foreign lands and not having played for several years. The chukkas were well organised and I was also able to play in several matches during my first summer there. It had a good variety of players and professionals would turn up from time to time to join in the higher goal matches.

The captain was one Jeremy Barbour who seemed uneasy with me somehow and I never fathomed why. I was very friendly with Karen and Kevin James who really befriended me and helped me out quite a lot and with whom I spent much time especially during the long summer holiday much of which I could devote to the club. I was not overly ambitious and just enjoyed playing my polo at a low-goal level with friends and club members for the first year.

I purchased a small caravan and a Renault car to tow it. This came about by chance as I had popped into a local garage near Launceston for petrol and noticed the car on the forecourt and it had a tow hook. I looked inside and it was covered with dog hairs which put me off but I did test it and it seemed very good. The garage owner let me have it at a knock-down price and a few Mars Bars and five boys spring-cleaned the inside to perfection. The caravan I also bought cheaply near Taunton.

I often stayed up on the field to do odd jobs and complete bigger projects with Kevin. I became quite good at painting, some of which ended up on me. I was, however, laying plans for the next season, for Ace and Lady were now comfortably ensconced in the polo stables at Orchard Portman under the care of Karen James and the polo grooms. It had happened by chance. Ace had received a small cut in one match which required stitching and Karen suggested that I leave her up at Taunton for the week which meant that I also had to leave Lady, as the two were inseparable.

I drove back to tell Harold that the horses would be away for the week and he took some sort of umbrage and told me not to bother about bringing them back and he was adamant. It was most odd. Now this might have been connected with an incident that had occurred recently when I had arrived to pay the farrier, which I always did promptly and in cash. I caught him waving his hammer around Lady's head and objected.

I knew Lady had been doing her usual 'lean' but many horses do this and the cure was not to threaten her with a hammer. I phoned Karen and told her what had happened and that Ace and Lady were now homeless and she took them both in straight away. When I arrived for polo the next weekend, Ace and Lady were standing in the 'lines', beautifully turned out and ready for me to play. 'Colonial Boy' had won again.

Lady continued her tricks and was considered the 'in-house' comic of the stables. She was always lean looking and we could never get weight on her but she ate like… well, you know. One night the grooms had set up all the feeds ready for an early departure the next morning. Lady broke out of her box and ate the lot. After every match, Lady's legs would puff up and then, the next morning they would be fine.

There could be no separation from Ace ever. She was left behind in the stables one afternoon because of some minor ailment. Thirty minutes later, she got out of her box, trotted up the road to the polo ground and stood in her place in the lines, looking very pleased with herself. The lorry could carry eight horses which I used occasionally for hunting in the winter that I bought it. I think you could call us 'social' hunters because we never caught a fox and were mainly interested in the gallops across the fields and drinking in the pub afterwards. We always attended the Boxing Day Meet at Wiveliscombe every year which had a big turnout and was most enjoyable.

In January, however, our thoughts turned to polo for Kevin and I had conceived a plan to increase the number of horses we needed and to make money to cover our polo expenses. We headed for the Ascot Sales with my lorry, a limited amount of capital and a small wad of fivers. We always arrived early, had a good breakfast then scouted along the stables of the horses that were being sold, having previously studied the catalogue to pick out any likely contenders.

These were, of course, ex-racehorses and we were taking a chance on all of them but a surreptitious fiver to the groom often elicited the information we required before we headed for the sale ring. All bidding was in guineas and our limit for each animal was one thousand. We normally bought between two and four horses with the idea of training them up for polo early in the year and then selling them off at the end at a profit. If a horse did not make the grade we would sell it off for hunting or as a hack. The system worked well for several years.

On one visit to Ascot, I was interested in a stunning grey that would have been ideal for our purposes but as I have a soft spot for greys, I asked Kevin and Karen to double check. They concurred. It was being sold purely as the owner had gone into debt. It was outstanding and it had no reserve. In the ring the auctioneer could only start the bidding at one hundred guineas and no one seemed interested so I slowly bid up to five hundred. Twice the hammer started to go down to me for six then eight hundred and I stayed with the bidding up until one thousand five hundred. It sold for thirty-two thousand.

We were always on the lookout for horses and bought several locally to add to our pool of ponies and one of those was Singsing. She was well bred and out of a famous Derby winner. She was incredibly fast but would not race and she was not the most friendly horse I had ever owned. The problem was that she was being offered to us by Jeremy Barbour whose wary attitude to me made me proceed with care. We agreed a low price and I was insured because I had already received another higher price for her by a Mount House School parent.

She was not easy in the box, would not take the stick, indoors or out, and was a nightmare if you tried to stick and ball her. It was kill or cure. We tried her in a slow practice chukka having warned all players that there could be an explosion and she played like a dream machine. She was so fast that I missed the ball several times and she could turn and outrun any horse on the field. She was a polo machine and she had huge stamina, and…she could hardly be distinguished from Lady. She was a huge asset to our team as we had now formed one. I suppose you could now say that I was a Half Patron as I paid half the expenses of Armitage Shanks, our ad hoc Polo Team and that Kevin was our voluntary hired assassin.

We then had a small group of friends that played with us whenever they could. The name had come about at a club tournament when the secretary had asked me the name of my team when I was taking a pee in the Gents. There before me on the urinal was ARMITAGE SHANKS. I did write to the company telling them what I had done and they sent me a cheque for one hundred pounds which I thought was inordinately kind of them.

Kevin and Karen would bring on the horses and I would help whenever I could and my favourite time was taking them over to Brean Sands which offered a superb huge flat area for training the ponies to stick and ball ending in a flat out gallop along the sands which Singsing easily won. My horses were polo ponies but we liked to enjoy them as well and Ace could even jump though I never allowed her to do so when hunting.

Even at that time, polo ponies were selling for three thousand pounds and it was vital that they stay well for the sport. They were valuable and needed to be well looked after. Our scheme worked fairly well and we never lost money though we did have to sell out those horses that did not make polo at cost with Karen always ensuring that they went to good homes. One was actually used for shopping in Wiveliscombe which was a charming life for the horse.

We did not travel great distances for matches and games. We often played at the newly formed Dulverton Club at the Craven Arms on Exmoor. It was a very small and delightful club with none of the swagger or organisation of the bigger clubs up country. Their field was on the banks of a river in a sheltered valley and we often camped up there for weekend tournaments. I still had the caravan but Kevin and Karen and their young family commandeered it so I was sleeping in a mountain tent.

One weekend I had accepted an invitation to a ball at Helligan Barton, near Wadebridge, which was owned by Mount House parents. In fact, I had visited the property before because we had played an exhibition match at a Pencarrow Game Fair and stabled my horses at Helligan Barton, except, that, on arrival, we could not locate the stables. We were directed to a magnificent building that looked even grander than the main house in which were fabulous huge Victorian stables. It was like a five-star hotel for horses and I think Ace and Lady were most reluctant to leave the next morning.

I drove down to the ball where I was escorting a lady and had a good time dancing the night away yet I did not have one alcoholic drink. Returning in the early hours of the morning, I was stopped and breathalysed by the police three times and I reckoned it was a British record.

Matches at Dulverton were played with much enthusiasm and we quickly learned 'to play to the whistle', as the umpiring was fairly erratic. After a match we rode the ponies down to the river and they, we and the grooms, had a fine time splashing about to cool off. It was almost idyllic. The club still retained a connection with India for this area was the initial venue for all the Maharajahs to warm up their polo before departing to London for the Hurlingham Season. Most were formally polite and friendly but several still retained a very old fashioned colonial attitude.

I had to rescue one groom when one of the Indian players had fallen off as he mounted and he was hurtling abuse at her and threatening her with his whip. He was obviously attempting to dole out the same treatment to the poor young lady as if she was one of his syces back in India. It really was a nasty incident. He later turned his venom on me as I happened to be umpiring when he fell off and knocked himself out cold for at least three or four minutes. He was administered first aid and was helped from the field but tried to insist that he should play on.

Unfortunately for him, the Hurlingham Polo Association (HPA), the governing body for polo had introduced a new rule that year stating that if anyone was knocked out for more than one minute then they could not play on and had to be replaced. I knew all about the rule because at that time, I was a member of the HPA representing Ghana, and I had helped formulate and pass the rule. The Indian player was, however, most abusive towards me as the umpire and I had to report him to the HPA and he received a one-month ban. I am sorry that he had to abuse his hospitality and behave in that way.

Each year Rollo Clifford organised a polo match on his estate at Frampton on Severn. The match was played on the best polo field on which I had ever played between the main house and an ornamental lake, except that it had a large oak tree in one corner. Local rules applied, however, and you were permitted to bounce the ball off the tree to outwit your opponents. The very first time I played there I was coming down from London so agreed to meet Kevin and Karen and the horses at Frampton.

I generously gave Kevin my road atlas with the simple instruction to drive up the M5 and turn off at Junction 13. It was dead easy. When I arrived there was no lorry and no

Kevin and they were very late for the start of the match. They had lost their way. How? My road atlas was so old that the motorway was not even marked. Whoops!

The ground was great to play on and I only wish that the umpiring could have the adjective applied and it was a case of 'playing to the whistle' again though I did fear for my ponies. The most dangerous foul in polo is a 'cross' where someone gallops directly in front of you or into your side. It is dangerous and there can be injury to man and horse. I am afraid that the umpires at Frampton had not heard of the rule and we had to insist on providing one of our own umpires to play safe.

The chukka bell was the big bell above the eighteenth century house, Frampton Court, connected by about one hundred yards of rope. After the game there was always a slap-up tea. It was all very friendly and enjoyable. I bumped into Rollo only last year when I revisited Frampton with the friends of the Bristol Museum. As I stepped off the coach Rollo was greeting the guests and he recognised me and I am afraid that we spent most of the afternoon talking polo though I did get a quick tour of the fabulously renovated sixteenth century wool barn.

Kevin and I took Armitage Shanks to a tournament at a newly formed polo club in Berkshire (it was not the Royal Berkshire) where we had been invited to play in a weekend tournament. It was not that great a success as the promised accommodation was fairly ropey, the evening function was a drink in a local pub and we had to buy all our own food and drink. It was not the proper way to treat an invited team.

Despite that, we were in the final on the following day. It was fairly late in the evening when I met the captain of the other team in the final who had organised the tournament to celebrate his wife's birthday. It would be a wonderful gift, would it not, he suggested, if her team won tomorrow's match. I responded by wishing her all the best while hoping that she did win. He then offered me a 'bung' to throw the match, in an envelope containing a sizeable sum of money. I declined politely and walked away. I could not help 'winding-up' my team, however, explaining what had happened up to the point where I had been offered the 'bung'.

'Should I have accepted it,' I asked. 'Five hundred pounds each and another five hundred for the grooms.'

They were furious that such an offer had been made and though we were very civil and polite at the throw-in at the start of the match, we played hard and took no prisoners. They had an Argentinian 'assassin' playing for them who obviously thought that he was so good he was bound to win. With Ace and Singsing, I was able to mark him out of the game. Singsing could destroy his every move and he eventually lost his temper with me and the umpire awarded us with a Number One which entailed us going another two goals up, for Kevin had been busy scoring goals all the while. We won handsomely but I am afraid that the opposition were bad losers.

Before the match, there had been a handsome display of silver on a table outside the clubhouse evidently donated by the losing team in anticipation of their win. It had been suddenly replaced by wooden cups and fruit bowls. I could not resist commenting on the silver's sudden disappearance and several of the club members had the decency to look embarrassed. Strangely enough, we were never invited again and I cannot think why!

I also played in a Taunton team at Rhinefield down in the New Forest. I was staggered at the amount of traffic racing around the New Forest now and upon arrival at the ground it started to rain and the whole two day tournament was played in it. It was horrible. I was so miserable that I cannot remember what we did and whether we won or not. Rhinefield also has a local rule concerning the forest ponies. If they appear out of the forest on to the field then the game has to stop and I am sure one group was popping in and out of cover just to annoy us. I also discovered that Singsing did not like the rain and she was a little flat in both matches. I was very glad to get back to Taunton. It had been a miserable and expensive weekend.

Taunton Vale Polo Club had two major associations. When Hong Kong was handed, over the polo club there transferred everything to Taunton and its association exists to this day. It hosts one of the major tournaments of the polo season and it is generally well attended with much pomp and ceremony involved. If you were not playing then all members were expected to help out in a variety of roles to make the weekend a success and I can say that I did my share and had fun doing so. I was timekeeper for one match and the referee and I were away from the crowd on the far side of the field, by design, to officiate during the match.

Now polo, I have to say, was years ahead of all other sport associations. At time of writing football has just started experimenting with the third official and rugby has had it for some time but polo has had it for years and years. Officials in polo are two mounted umpires, two goal judges, a timekeeper and an unmounted referee to whom the umpires can refer. All other sports have a lot of catching up to do I tell you. The referee on this day was a retired Major of Artillery and I was able to wind him up about Moncrief Carriages, Big Bertha, the Beaufort Gun and a host of historical artillery facts about Bristol and its defences. He was astounded at my knowledge wondering how I knew so much about the artillery. Many old residents of Hong Kong attended the weekend and it was very jolly and pleasant. The tradition continues.

Polo was also much enhanced at Taunton by the arrival of visiting players and my favourite was Paul Clarkin from New Zealand. He was a very good player and was very patient with his help, advice and coaching. I learned much from him on and off the field. He was friendly, pleasant and kind and a good example of a polo player par excellence. He was a brilliant rider but had very bony knees to deter the 'ride-offs' I think. He was very amusing because he seemed to think that everything printed in the British press was the gospel truth and his favourite paper was the Sunday Independent which was always full of bizarre stories, all of which he believed.

He, like me, always thought of the welfare of his ponies on and off the field. The one piece of wisdom he gave me, as I nearly rode into his backhand shot, in a tournament, was never to do that for he did not take the shot in case he damaged my pony. I thanked him for that after the game. Lesson learned. Paul was killed in a polo accident at Cirencester some years later and was greatly missed by the polo world.

When I was a member, the Royal Naval Polo Club was based at Taunton and they contributed much to the atmosphere and enjoyment of the game and with my new secret weapon, Singsing, I often played against them at the club. Kevin was nearly always my captain and I tried to follow his instructions to the letter. Prior to that his instructions on Match Play could be difficult when I only had two ponies. Singsing could outrun most horses on the field and sometimes I would be called up from the back to take a player out at the front.

One Navy Commander completely lost his cool when I was marking him and said that I was picking on him all the time. Alas, the Navy eventually moved to Tidworth and combined with the Joint Services Association I think. I played only once at Tidworth when I received an emergency call from Rangitikei, the New Zealand team for whom I had played at Cowdrey. They were a man down and was I free for a friendly charity match? I took four of my horses up with the lorry and a couple of ad hoc grooms. There were then two fields at Tidworth and indeed it might well be the same now. One was at the bottom of a hill and the other at the top.

Upon arrival, I was directed to go up the hill to the top ground where many lorries were parked and there was one solitary smart box on the lower ground which I recognised immediately. The Prince of Wales was playing in the charity match. His security had decreed that he should be down on the lower ground alone and secure while everyone else had to march to the top of the hill and where was the match being played? On the lower ground, of course.

Evidently, princes cannot climb hills. I jumped out of the cab of my lorry and strolled over to Prince Charles apologising for disturbing him and asked whether he was aware of the security arrangements. He vaguely remembered me and when I mentioned the red sports car his mind clicked into gear. I jokingly accused him of gamesmanship making all the other players ride up and down the hill between chukkas, pointing out that none of us were members of the Duke of York's Regiment.

He laughed and invited everyone back down the hill and my somewhat shabby lorry was parked right next to the Prince's smart outfit. He wondered why I was playing for Rangitikei and I told him that my grandparents and several uncles and aunts lived in Christchurch which seemed to satisfy him. At that time, Charles was playing off a 'Four' handicap. He was not worth that. He was a good 'Two' and a possible 'Three' on a good day, for he could play well and hard. His presence at a charity match was a big draw for the crowds, however, and this was important for the organisers of the event. At charity matches, we always took this into account and considered the entertainment value and display of good polo to be important. It would not have been good to defeat the Prince's team hands down so we played the first chukka cautiously but well, offering the thrill of the game at speed.

I was playing at Number One in attack and the Prince was thus marking me. Even with his superior play and polo knowledge, he could not outwit Singsing so I had to go easy when riding her but he was a tough opponent when I was mounted on Ace and Lady and difficult to catch out. Peter Grace, my captain, could not believe Singsing and on several long runs she was cheered on by the crowds. Her speed was fantastic.

Prince Charles was fascinated by her for she was a handsome chestnut and a wonderful horse. He presumed that I had paid a high price for her and was astounded to hear that I bought her for £500 and told him her story. He was initially sceptical of the fact that we bought ponies at the Ascot Sales for a maximum of £1,000 and turned them into polo ponies. He had just paid an enormous sum for an Argentinian horse that he did not like. I suggested that he send Ronnie Ferguson, his polo manager, down to the next Ascot Sales. At the next HPA meeting in the Cavalry Club on Piccadilly I was accosted by Ronnie who stated that I had got him into trouble when the Prince asked why he could not get horses like Singsing. Was she for sale? Not on your life.

Kevin always had a hankering to be a farrier and would always be around when shoeing was in progress. The working life of a farrier can be very short as they quickly develop back troubles but Kevin watched, observed and helped with the easy tasks such as trimming and somehow, he obtained official permission to 'shoe' under supervision and he was good. For some obscure reason he could not buy the proper tools in England because of regulations so we travelled to Brittany for a weekend to find some, helped by my campsite friend, Jean Claude. We located the tools and purchased them to Kevin's exact requirements then returned to our hotel in Camaret. Kevin was a lager drinker and drank pints in the UK.

In France, the lager is served in small glasses not much larger than a wine glass. Kevin could not cope with this and the barman was amazed as Kevin downed glass after glass in quick succession. Kevin breathed a huge sigh of relief when we reached Roscoff to catch the ferry and lager was once again served in large glasses. I always looked after my Taunton farrier very well and always left cash for Karen to pay him when he did my horses. Unfortunately, some of the other players did not settle their bills so promptly but could then never get emergency shoeing for their horses.

Taunton did not play high-goal polo but they did have a big medium-goal tournament every year which was well attended by six or eight teams and was hard-fought. One year I met up with Simon Kesayo from Kenya who was the hired assassin for the Macarthy brothers all of whom were playing in the tournament. Simon was not entirely happy with his supplied mounts and I let him borrow Singsing for the tournament. He was a brilliant rider and was good with horses so I had no concerns. Singsing excelled for him and he rated her as a high-goal pony. The Macarthy brothers made me a ridiculously low offer for her which I totally rejected.

Many of my polo friends were clear that I lacked the devil when I was playing, that I was not aggressive enough in matches and tournaments. They were probably right but I enjoyed my polo at Taunton and deep aggression is not really embedded in my soul. I am not by nature an aggressive person and I really saw no reason to change. I knew my limitations as a player and realised that I would never play medium or high-goal and I also much enjoyed the social side of the game. I really enjoyed my time at Taunton and made many good friends though I have failed miserably in my duties as godfather to Karen and Kevin's children.

The 'winds of change' started a blowing. I was happy with my polo, Mount House School and my bungalow at Harlyn Bay. I was working hard and enjoying a social life but there was the dawning realisation that I was sliding into what Freud called, a 'contented cow' attitude. I was getting on with fewer and fewer years to my retirement and I am not sure if I wanted to fall into those twilight years without making a mark or taking on a challenge. I do not wish to be at all critical or rude to my colleagues and friends but I would not be happy with a secure and safe retirement as a kind of rebellious Mr Chips. The thought of a safe and comfortable retirement was not for me. I think Africa was calling me again perhaps.

My complacency was shattered by the death of one of my very good friends in Ghana in a polo accident out at Ningo. The funeral was booked for the next weekend and promised to be huge even by Ghana standards. I think Charles was somewhat surprised that I wanted to go out to Ghana just for a weekend but I think he had probably become accustomed to my unusual activities and requests. I flew out on a Thursday afternoon and was back by 9 am on the Tuesday. The funeral took place at Ningo, I still visit the grave on my visits now, and we drove out from Accra in a convoy of over one hundred cars.

There were well over five hundred people at the funeral and he was given a good send off by all his friends. We had been good friends and drinking companions and had enjoyed much fun together. I was able to have a long discussion with Nour and the Captan family during my stay and they were anxious that I move out to Ghana again so I agreed to come out over the next Easter Holiday (no polo) to look around. It seemed that they were pleased to see me and I realised that I was missing all my Ghanaian friends during that trip but employment was the problem.

I had dropped in for a coffee with a very good banking friend, Frank Adu, who was a polo player, and also the director of Cal. Merchant Bank which he had established and developed by force of his strong personality. There and then we hatched the idea of starting a school and the idea grew and grew from a dream into reality. There were many obstacles, problems, setbacks and difficulties to overcome but we did make the decision there and then. It was helped by Nour's suggestion that I play my polo in the holidays in Ghana and that idea did appeal so we will stay with the polo theme briefly.

I played one more season at Taunton and I also informed Karen and Kevin of my possible future plans as I did not want to let them down suddenly. I am not sure if they understood my reasons for possibly bailing out but increasing polo expenses that year helped in my decision. It was tough playing polo on a teacher's salary even though we were making our own ponies. Playing polo in Ghana would be of enormous relief to my bank manager though we were not really on good terms as he had turned down a loan on a business project in which Kevin and I were interested. A pub had come up for sale at Thorn Falcon and Kevin and I had visited it and checked it out thoroughly.

Financially, as a business, it did not look too sound but the pub was opposite a very upmarket traveller's site and though he was reluctant to admit it, the landlord took considerable sums of cash over the bar at the rear of the premises and the income was further enhanced if there was a wedding or a funeral. None of this could I explain to my bank manager but the simple fact was that if all else failed, the building could easily be converted into three cottage properties at a handsome profit and that is where I pitched my loan application. Unfortunately, my regular manager was taken seriously ill and a locum was in place who was adverse to all loans and risks. We lost the opportunity.

I retired from polo in the UK in 1998 but I did not retire from polo because from that date onwards, I played polo in Ghana during every holiday. I maintained all my teaching, school responsibilities and the adventure programme during that time including camps and expeditions after which I became a member of the Accra Polo and Hunt Club and played on borrowed ponies. I helped Nour with preparing his ponies for polo though he did not approve of my methods and I played chukkas and games. It was a much larger club than when I had joined some fifteen years before and new development now surrounded the improved club house and facilities.

I was now playing with the new generation of players many of whom were the sons of my friends. Arguments and disputes were still common but they did not affect me so I was just able to enjoy myself though I was doing much preparation for the school during the day. It was nice to enjoy social times with the large group of friends that I had and I was also able to help out Nour and Mounir with their salt project whenever they needed my help. I was around when they were building the new salt refinery at Prampram and it often fell to me to entertain the Spanish expatriates in the evenings as they expected to go out nearly every night.

That was fine but my problem was that Nour had quite a large collection of cars which were not dissimilar and I tended to forget which one I was driving so I always had to 'daqh' the watchman at the various restaurants and hotels to remind me which was my car. Mounir suffered a very bad injury to his hand one morning and I offered to go out and help with the refinery in which I did for a couple of weeks. I was also involved in arranging a funeral for a friend of Nour who had died of alcohol poisoning, details of which appear in the next chapter.

Nour and Mounir were fantastic hosts and there were always people staying in the house whether it was me, relatives, friends, or of course, polo players from Nigeria, Sudan, UK, Canada, South Africa, Germany and Burkina Faso, many of whom were great characters, some larger than life. Nour often hired a wonderful character called Rod

Gutteridge, a South African, who played off a 'Four' handicap. He was good on the field and played professionally and well but he could also enjoy himself and party hard as I found to my cost on several nights.

The Accra Polo Club was famed for its hospitality and treating visiting players well, though quite a few returned to their own countries with sore heads. Ghana also played matches abroad in Nigeria, South Africa, UK, Canada, Zambia and Ireland and it was usual for the Ghana team to take a 'gift' from Ghana to present to their hosts. When the team visited Ireland, they took some traditional gifts with them and bottles of Guinness.

Bottled Guinness was brewed in Ghana with a high alcohol content and was much liked by the Ghanaians. Indeed, believe it or not, this Guinness is exported from Ghana into London at the rate of one container per month for the Ghanaian community in London. The Irish declared it as 'not bad'. I took a couple of bottles back for my grandmother in the UK but she found it too strong. It was a potent brew.

I holidayed on two occasions while at Mount House. On the first I took my mother to Spain in my converted white van. We travelled Brittany Ferries to Santander and then spent two whole days travelling through central Spain to Alhaurín el Grande to stay in an apartment. We had an agreement. I could mountain bike in the early mornings and then take my mother out for the rest of the day. It rained and rained and rained. All the orange groves had their oranges floating in the deep water.

We ate in restaurants at night and it was in a small tapas bar in Alhaurín el Grande that a distinguished retired bullfighter fell in love with my mother. He was enamoured of her and he showed her his museum of bulls that he had killed and all his costumes. She was flattered. We did manage to visit Malaga, Coin, Alora, Marbella, Fuengirola and Ronda and we spent Christmas with friends in Estepona before driving back up through western Spain back to Santander. The second trip was undertaken alone for mountain biking in the Cevennes during which the chapter, 'They Shoot Horses, Don't They?' was written.

On the vehicle front, I had now sold my Renault and lorry and purchased two pickups. How? I purchased a red Nissan Double Cab in the UK and a dark blue small Ford pickup in Ghana. In the late nineties, the Double Cab was fairly rare on British roads and only the Gas Board had a small fleet of them, but, of course, they were quite popular in Ghana. I enjoyed full use of mine in the UK for it was four-wheel drive and could go anywhere, so I did, but I used it for my mountain trips and other expeditions. It was the perfect vehicle for me.

As my plans to open a school in Ghana progressed, I thought it time to tell Charles my intentions two years before I actually left and I found it very interesting that no attempt to retain my services was made. This was, I consider, part and parcel of not rewarding staff for their contribution to the school. My efforts at Mount House were appreciated but never in financial terms, not even on an ad hoc basis like St Bede's in Eastbourne.

However, I was not able to fully commit, until the beginning of the summer term of the year that I was leaving and Charles did not seem happy. It was unfortunate that I could not resign earlier due to the huge risk I was taking. I had to be as sure as possible before I moved, I am afraid. I regret that I found the advert for my job in the TES risible and almost an insult:

Junior Master required for the Autumn Term. Help with games would be an advantage.

Prep schools can be the most peculiar of places with odd ideas. I enjoyed my seventeen years at Mount House and still remain in contact with a whole army of parents and pupils who have supported and helped me with my various fundraising activities and I am still in contact with many of the staff in their well-earned retirement.

To conclude this chapter, I had to visit my solicitor, who was also the Chairman of the Mount House Governors, to settle my affairs before moving abroad. He thought I was mad to embark upon such a foolish venture and told me so. He has since had the good grace to confess that he was wrong but then he did not know how close I came to proving him right.

Chapter Twenty-Six
The Funeral Pyre Fire

'Hi honey. I'm home!'

'Jeffers—Mahabara—que feck?'

After an uneventful night flight and a very early arrival at 5 am at Kotoko International Airport in Accra, I had quickly passed through the Health and Immigration formalities, then waited the statutory one hour for my baggage at a ponderously slow and creaking luggage carousel, checked with Customs, who were now boasting a Green Lane—Nothing to Declare—that I felt was extremely optimistic in the face of the mountains of luggage with which most of the Ghanaians travelled, walked outside the building, found driver Musa with his beloved Land Rover and made it to the house by 6.30 am. Where I shouted my usual greeting in the time-honoured way!

My best friend was there to greet me, drinking Lebanese coffee and watching his some three hundred channels on the satellite TV. We used to joke that he had one channel for each day of the year.

I quickly showered and changed, unpacked, sorted out the various articles that I had been requested to bring out with me; a remarkably small number for once, I recall (no bikes and no dogs this time) then wandered out onto the veranda where I was greeted by the 'pack'. This was a motley and somewhat unusual collection of dogs ranging from moth-eared mongrels to valuable and pedigree animals, all of whom were controlled and bullied by a diminutive miniature dachshund named Princess.

The veranda was a good place to watch the house waking up. Every morning, Kwasi, the Houseboy, would deliver my coffee and I would start on the usual morning ritual. After being greeted by the dogs I would sit and listen to the birds but these were nearly always rudely interrupted by the African grey parrot who could chatter quite well and greeted all guests with a 'Hello'. The drivers would gradually arrive, one by one.

Quiet George was usually first and busied himself with the pickup, followed by the brash and noisy Musa, who of course, prepared the Land Rover. Usually, much later, Suli arrived and prepared one of the saloon cars depending upon which one was running at the time. My friend seemed to collect cars somehow and there were usually six or seven in the compound many of which were silver coloured.

I recall that this had led to total confusion one evening. We had been entertaining a group of Spaniards working on a project at Prampram and it had fallen on me to entertain them in the evenings. Purely by chance I had transported them in a different car every evening and this had caused much amusement and hilarity. One evening, however, I actually forgot which car I had been using and we and about seven security guards searched the large car park of a four star hotel for about twenty minutes before my key fitted one of them!

Preparation of the vehicles at the house always involved the driver jamming his foot on the accelerator and revving the engine as hard as he could and this cacophony usually coincided with the arrival of the manager of the salt project who was famed for his loud voice and shouting, particularly at the drivers which would set the dogs off barking and the parrot screeching.

By this time, my friend was usually up and amid all the din began another daily ritual. Kwasi and/or Akuse, the cook, appeared and the day's menu was debated with much merriment, many suggestions and haggling over the money required for the shopping. In the middle of all this my friend would think the dogs too noisy and Kwasi would have to put them in their pen and this was usually when Romeo, a very large dachshund, would craftily slip under my chair to hide.

I enjoyed the mornings…they were never dull and I remember them with affection. They were fun! On this particular visit, however, my peace was to be disturbed. I was looking forward to a relaxing Easter break of three weeks riding, playing polo, mountain-biking, canoeing, swimming, drinking lots of cold beer and generally partying and having fun with all my friends. I had no inkling that my plans were about to be thwarted.

My friend casually announced, that, unfortunately, he had to make a quick trip to Lebanon for family business. I was disappointed, of course, but his younger brother was still around who was a lively socialite in Accra society, and his youngest brother was also visiting…who was not! Mention was also made of a mutual friend who was in the Nyaho Clinic and I was requested to ensure that Kwasi delivered food to him on a regular basis. This was normal practice in Ghana hospitals and I foresaw no problems.

Nyaho Clinic was just down the road and I knew it well as they had patched me up several times when I had resided in Ghana many years ago. Indeed, during my last trip, I had cut my head and required stitches after a mountain biking spill so promptly presented myself at reception. To my amazement they located all my records, this after a gap of some seventeen years! My friend's brother had driven me to the clinic and was supposed to ensure that the stitching was completed with clean needles but felt very queasy when they were being administered and had to go outside quickly.

This particular morning we all adjourned to the office in central Accra where we drank copious amounts of coffee and ate several packets of biscuits. I wandered across the busy street to call on other friends. Most of the traders and vendors knew me and called out loud greetings and remarks as I dodged the traffic, and I returned to the office in a happy state of mind, to be welcomed back by my friend.

'There is good news and bad news!' he said gleefully.

Mentioning the name of one of my former employees from Liberia, he explained that he had managed to escape from the upheavals in Liberia and was coming to Ghana to look for a job, and that he was bringing his wife with him.

'And you will have to look after them and entertain them while I am away!'

I greeted the news with some trepidation. It was a long story but five years of my life had been spent running a logging company in Liberia under very tough conditions. We had employed sixteen expatriates including the rather colourful character who was now coming to Ghana. I had heard little of him but I knew that after the civil war in Liberia, he and his wife had returned firstly to a rather hair-raising logging job in the interior where disturbances were still common, and then later had worked in Monrovia in a local factory and his wife had run a nightclub and taxis. Conditions in Liberia were very rough and tough at the time, but having not been paid for some months now, they had obviously decided to try their luck in Ghana again.

After a late lunch, we drove down to see our friend Victor in the Nyaho Clinic with his lunch and some fruit. Victor had lived in Ghana for many years and I clearly remember his fish and chip shop in Osu—the first in Ghana. It had initially prospered, but, alas, the demon drink had taken a hold, the business had struggled and Victor's problems just became too much for him to handle. His wife returned to England and the shop became a chop bar where he continued his drinking and was looked after by an assortment of girls.

After I had greeted many of the nurses that knew me, we were ushered into a side ward to find Victor sitting up in bed reading a magazine, looking quite bright and cheerful. He attacked his food with relish, was very talkative and stated that he was looking forward to his discharge from the hospital soon. My friend carefully explained that I would be organising the delivery of his food and transport him home on his discharge. We left Victor bantering with two of the nurses but on our way out, we were asked to see his doctor who explained that Victor had been diagnosed with cirrhosis of the liver and that there was little that they could do for him, especially as he had continued his drinking. Even at this stage, I was not that concerned and we returned to the house. This was very busy.

My friend proceeded to pack and this process was constantly interrupted by the arrival of his friends and relations to wish him well on his journey. This was traditional and because we had so many visitors Kwasi was assigned to deliver Victor's food with one of the drivers. Hospitality was in full swing for all the visitors and the other house staff were extremely busy supplying drinks and snacks. Kwasi returned very quickly with the news that Victor had taken a turn for the worse and we were to go to the hospital now.

We rushed there quickly to meet a somewhat chaotic scene. Victor was now in great pain and because he had been rolling around in the bed the nurses had placed him on a mattress on the floor surrounded by pillows. He was connected to tubes and drips and groaning and extremely restless. If there was ever an incentive to give up alcohol this was it. The whole situation was very sad indeed. We returned to the house very subdued.

It was no surprise, therefore, when we received the news the next morning that Victor had died two hours earlier and we sent a driver to his house to inform whoever was there.

It then fell to me to try and contact his wife in the UK but this proved very difficult, though I was able to talk to his daughter and broke the news as gently as I could. During all this time my friend was preparing to go to the airport and he wished me a hasty farewell amid my many telephone calls. His daughter knew the address of the wife and though there had been little contact between them she agreed to inform her.

I was aware of the many difficulties when expatriates died in a foreign country and had to ask whether she knew where Victor's will might be. Surprisingly it was in her possession and I breathed a huge sigh of relief, because, without a will, many of the assets would have passed to the state. The daughter was very matter of fact and confirmed that she would come to Accra as fast as she could and keep me informed of her travel arrangements. I thought all had gone well as we concluded our call but I was suddenly aware that she had just mentioned that, in his will, Victor wished to be cremated.

To us in England, of course, this is a usual request and would cause no problems, but in Ghana, there are no crematoriums and this would present a huge problem. Not only this but I suddenly realised that there was much to organise—removing the body to the mortuary, obtaining the death certificate, sorting out the funeral arrangements, etc.

One must also remember that a funeral in Ghana is a huge affair with large numbers of mourners and attendant guests and many traditions to observe. It was also going to cost lots of money.

I called Kwasi to get some coffee, forgetting that he had gone with the driver to show the way to Victor's house and I thought this was odd because Musa the driver already knew the way. My coffee was brought by Akuse, however, who laughed when I asked why Kwasi had gone to Victor's house. Akuse had arrived in my friend's house many years ago as a slim 14-year-old housemaid, often too shy to talk but now she was the cook and was very large indeed. Very outgoing and happy, she cooked wonderful Lebanese food and reigned supreme in her kitchen but she would not answer my question.

Apart from Victor's death, everything seemed surprisingly normal. The dogs gave me their usual greeting, the parrot was whistling away and the other drivers started to arrive but I was busy trying to plan my day to sort out everything that required doing. It was not going to be easy especially as getting many things done in Ghana requires the 'dash'.

My thinking was disturbed by the arrival of my friend's second brother who obviously noted that I looked worried. Just at that moment the telephone rang again and I went inside to answer it. It was one of the customs officers at the airport who I have known for many years and was a great friend.

'Geoffrey, there is a man here who says that he knows you. He has just come in on a United Nations charter flight from Liberia with his wife.'

This was not good timing but I spoke to my English friend and promised to send one of the drivers to the airport to pick them up. This had now compounded my problems and I returned to the veranda. I explained this further complication and then had a sudden brainwave.

'Martin Manu!' I exclaimed. 'Tell me, is he still around and more importantly, is he still a Mason?' Both answers were in the affirmative.

'And was Victor a member of the lodge as well?'

Affirmative.

'Would I not be right in thinking that the lodge normally conducts the funeral arrangements for their members?' Affirmative.

Martin Manu was a friend of many years standing. He had held top security posts in various government regimes and had often helped me with difficult problems in the past. He was also a leading figure in the Masons and I disappeared to the phone again to see if I could contact him. He was at his house and after catching up on his news I explained the problem. We agreed to meet at his office later that morning for me to give him all the necessary details.

By this time, my friends from Liberia had arrived and we sat on the veranda drinking more coffee. Both he and his wife looked very tired and they had obviously had a very bad time with the difficulties of getting out of Monrovia and a delayed flight, so I agreed to drop them at a cheap hotel so that they could rest. We all clambered into the Land Rover and I explained about Victor as we drove to the hotel. Having ensured that all was well there, we then proceeded to Martin Manu's office.

As we arrived he was unloading files from the car and we offered to help. Lifting the last of the files from the back of the boot, he revealed an AK-47 at the back. It was in good condition but looked well used.

I raised my eyebrows at Martin.

'Lots of car theft and house robberies around at present,' explained Martin. 'In fact, I had to use it on Saturday night.'

Evidently, he had been returning from a late-night function and noticed that he was being followed so stopped, got out and pointed the gun at the car which had come to a sudden stop. The occupants did not wait. They slammed their car into reverse and disappeared back up the road at high speed in a cloud of dust.

His office was very untidy and he had to call his PA in to move all his papers so that I could sit down. He knew Victor well and very efficiently took all the details down promising that his lodge would see to all the arrangements deciding that I would be better to stay out of all these on the grounds that charges tend to be higher for the European Community! I gladly concurred and agreed to stay in touch.

We returned to the hotel where my friends were anxious to go shopping for essentials so they dropped me at the house and I let them have the driver for a short while. I collapsed on to my chair on the veranda and called for Kwasi. He came very sheepishly. I asked for some small chop then grilled him about his visit to Victor's house that morning but I had to insist that he tell me.

Evidently, the old fish and chip shop owned by Victor in Osu was not only being run as a chop bar with local food and drink but also as a house of ill-repute and according to Kwasi, the young ladies were very beautiful and attractive. It was in fact…a brothel.

At the polo club that evening, several of Victor's friends approached me about his death and were anxious to pay their respects but this proved difficult. Firstly, there was no relative at the house to whom these condolences could be offered, and secondly, should condolences be offered in such an establishment? I emphasised that the daughter and wife were yet to arrive in Ghana and perhaps it would be better to delay any proposed visit until they did. This seemed to be acceptable. I was able to assure them, however, that all arrangements were in hand and that there should be a very good funeral.

Out on the veranda the next morning, I took several calls from the UK. The daughter and wife were going to the High Commission in London that morning to obtain visas and had planned to fly out the next day. I confirmed that all arrangements were in hand but that there was one major problem. There were no cremation facilities in Ghana so would they consider a normal burial? The daughter was very upset about this and the situation required careful handling but the problem seemed insurmountable.

Arriving in the office later that morning, I quickly completed the little business and shopping I needed to conduct, then wandered into another friend's office further up the block. Accra is very multi-racial and the young friend I was visiting was from the Asian community. Over yet another coffee, I explained everything that had happened and the problem of cremation.

'Have a word with Pops,' my friend said.

The Indian community has special permission to have funeral pyres in Osu cemetery.

I hastened along the corridor to his father's office and presented my problem, fairly safe in the knowledge that as a Mason, he would also help. Though the request was unusual he agreed to look into the matter and promised to get back to me as soon as possible. Having thanked him, I returned to his son's office to finish my coffee. Within five minutes, his father popped his head around the door and stated that everything had been arranged.

At this point it should perhaps be explained that a funeral pyre is really a big fire on which one burns the deceased in a coffin. As far as Ghana was concerned, big logs were stacked in long rows then doused with oil. It was then expected that the nearest relatives would set light to the pyre and wish their dead relative well on his final journey. I realised that this would need to be carefully explained to Victor's daughter and I decided to warn her by phone. She did not seem to worry about the nature of the cremation and I agreed to meet her at the airport the next evening and booked her hotel in advance.

Generally, I thought I had done well but when I returned to our house, two of the drivers were missing and I learnt that my friend's younger brother had sent two of them on leave while my friend was in Lebanon. This complicated matters and it was a delicate situation so I phoned around all my contacts and found a temporary driver. As luck would have it, I employed him regularly on my many trips to Ghana and I eventually gave him full-time employment when I returned to Ghana to live. He became a trusted employee and looked after me very well indeed.

At the polo club that evening several members declared their intention of going to the airport to meet Victor's daughter. I could not persuade them otherwise especially as one member had obtained a special pass to meet her at the aircraft. They were very keen to bring her to the club for a welcoming drink but they had obviously forgotten that Ghana hotels do not hold reservations until late at night. She was duly met and escorted to the club for a rather lengthy welcome and of course, her hotel room was let to someone else. We thus had to tour around six or seven more hotels to find a room for the night. Which we did at 1 am in the morning.

On the veranda the next day, I made rather a slow start and carefully sipped my coffee before sending one of the drivers to collect Victor's daughter. When she appeared, she had had a bad night. Evidently, there had been much noise in the hotel and much worse, she had been bitten by bedbugs. The decision was taken to invite her to stay in the house so we collected her things and installed her in one of the downstairs bedrooms. We then served her coffee and breakfast and she seemed much brighter and cheerful. During conversation, however, she dropped a bombshell by simply asking where her grandfather was!

No elderly relative had been mentioned until now but by carefully posed questions, it appeared that Granddad had been living with Victor for the last few years. I surreptitiously sneaked off the veranda and immediately despatched one of the drivers to the house at Osu to find out about the grandfather. Returning to the veranda, I made polite conversation until the driver returned. In whispered tones, he informed me that the grandfather was indeed at the house and had been very well looked after by the young ladies.

They had cared for him extremely well, bathing and washing him, feeding him, providing conversation and social interaction for him and generally attending to all his wants and needs. He was even given beer to drink twice a day and seemed perfectly happy and content even though his memory was failing him. All this I explained but the daughter was still not aware that her father and grandfather actually lived in a brothel.

This was a big hurdle to overcome in the very near future and I did not look forward to providing the details. Everything was further complicated by the fact that our transport pool was now depleted and I could have done with all four drivers and four vehicles instead, we only had two. Eid took one driver to the office in the morning which meant that I only had one driver to cope with Barry and Maggie, the house shopping and Joy.

Barry and Maggie were quite demanding (Maggie always was) but they often had to resort to taxis to get around while Joy and I dealt with the funeral arrangement. Akuse did the shopping by taxi as well so that was a help. Mounir had escaped all this by leaving for the salt works at Prampram with the manager, Omar, early in the morning. I was left to juggle with everything else. Joy was anxious to see her grandfather at Osu and also to check out Victor's possessions and documents probably before his ex-wife did, so I had to carefully explain the circumstances under which he was living and she was quite upset, imagining the worse. I cannot say that she was pleasantly surprised when we arrived at the 'chop bar' but it was very clean and tidy, the girls friendly and kind and they had been doing a very good job of looking after her grandfather who was well dressed and cared for in all his basic needs and requirements.

They had fed him well, given him his daily beers and ensured that he was safe. Unfortunately, he was suffering from dementia and had huge problems with his short-term memory. The girls said that he insisted on reading the same copy of 'The Daily Graphic' every day and had talked to him as much as they could. They made me and the driver some tea while Joy looked through all her father's rooms and possessions, realizing that she would have to spend a day there to sort everything out so I let her take one of the house girls down with her the next day to do the job properly.

I was chasing around trying to deal with Barry and Maggie and Joy who was now happily settled in the house, but thankfully, the Masons had taken over the funeral arrangements completely and I had nothing to do on that front. It was but a temporary lull. Joy announced that she wished to take her grandfather back to England immediately after the funeral so that he could be placed in an old peoples' home near her. This was a noble intention indeed. I was not sure if it was the correct decision. I was well aware then that some of the homes for the elderly were inadequate in the UK and I genuinely thought that the girls had looked after her grandfather very well indeed.

Given the choice, personally I would have stayed where I was. We discussed this frankly and she thought about it carefully. The decision was to take him back. That was the start of a major problem. Her grandfather's British passport was two years out-of-date and he had no other British documentation apart from proof of his pension which he was still receiving. I arranged an urgent meeting with the British High Commission which Joy and I attended.

It was like hitting a brick wall. His passport could not be renewed in Accra for it was long out-of-date and would have to be renewed in London which could take two months. He could not be considered for emergency repatriation as he obviously had financial means and residency in Accra and his granddaughter was now with him. Finally, the case would need to be thoroughly investigated because of the strange circumstances anyway and that would come at a high cost especially as Joy had no power of attorney. The bottom line was that here was a very old British subject, suffering from dementia and unfortunate circumstances by no fault of his own, being blocked in his wish to return to the UK.

I must add here that during the whole of my time to date in West Africa, I had found the British High Commissions uncooperative and hidebound in their approach to difficulties and problems or troubles and I was to face a further obstruction from them in the not too distant future. In Nigeria, they had been downright difficult when I had been attempting to extract an acquaintance from a Lagos jail and the British diplomats in Ghana and Liberia were frightened of their own shadows during the various coups and civil unrest.

Diplomacy took priority over assisting their nationals. I always found help within the Lebanese community along the West Coast and knew their ambassadors and staff. I had by now, of course, been out in Ghana for many years, on and off, and had built up a fairly extensive network of contacts, friends and regular helpers. Accra is a small city and I was known throughout many parts of the town and managed to formulate a plan to help Joy and her grandfather.

The funeral was a grand affair and Victor would have enjoyed it as many friends were there to see him off in the church before we all drove over to Osu Cemetery. This is a huge cemetery near the centre of Accra and it was quite a long walk for the mourners following the coffin to the funeral pyre at the top of the hill. The fire had been carefully prepared and the coffin was laid on top and both Joy and Victor's ex-wife touched it by tradition before the flames consumed poor old Victor in his coffin and that was the end of that.

The next day I rustled up my contacts and was assisted by the Mason's again. By the end of the afternoon, everything was arranged. A ticket for the grandfather was bought through a contact with KLM who eased the necessary documentation checks through, and Joy, on our instructions, booked her grandfather as an assisted passenger on the flight. We were met at the airport by a contact who dealt with all documentation speedily and well, and Joy and her grandfather boarded the flight. Joy was worried about the UK Immigration controls but we had gambled on the fact that they would never refuse such a case. Joy phoned us later on the following day to confirm their safe arrival. Job done.

I was quite pleased to get back to the UK for a rest from that trip.

Chapter Twenty-Seven
The Roman Ridge School

Mention has already been made that the Roman Ridge School was some time in the making. In my many trips out to play polo in the late 1990s and in the first two years of the new millennium, I had been very busy in its formulation and conception and there were certainly many obstacles, difficulties and frustrations to overcome before we opened in September 2002. Right from the outset, however, it must be noted that I was helped considerably by my many friends and contacts throughout Accra who gave me much support and direct assistance in my endeavours.

I believe the school to be unique in one unusual fact for it must surely be the only school in the world that was started by polo players. This would only be natural as many of my closest friends played polo and they were the first I approached for an investment in the project. I am very grateful for their faith in me and the school.

Some five years before the school started, the idea was suggested by our present chairman Mr Frank B Adu Jnr (now Dr) when I had been looking for a new challenge, possibly in Ghana, and I was able to take up the idea and run with it, so we both started to draw friends into the net. My biggest catch was Gordon Gopaldas who I caught on the veranda of the Ningo Polo Club, late one afternoon after a match, as I was very keen to have him as an administrator and he had three boys to be educated anyway.

It was one of the best decisions that I have made in all my life because Gordon 'caught' our proposed aims and ethos immediately. He became 'The Rock' for the school and is still running it superbly to this day. I was lucky that he was at a loose end at the time. Without Gordon, there would have been no school. He is a remarkable man in all respects though I have noted the increase of his grey hairs recently. He also saved my life in a scene that would have done justice to a Keystone Cops movie, though I am afraid you will have a little while to wait for that tale.

The initial board of directors was composed of Frank Adu, me, Gordon Gopaldas, Daniel and Kingsley Awuah Darko and Dr Nortey Omaboe joined it just after the school started. We then persuaded Fafa Awoonor to join us as deputy head, who did so enthusiastically and wholeheartedly. She was a skilled French teacher who had taught at Ghana International School but like me, was up for a new challenge. My titles were Headmaster and CEO.

Already I begin to hear queries in the air with all this talk of the board of directors and shareholders and CEOs, so to be clear. The school we were proposing was to be a fee-paying private school to be run at a profit with a dividend payable to its investors by a board of directors. I am not sure if such a school exists in the UK. I was nervous at first because I thought we may be going too far down the corporate business route but our chairman and board members were men of stature in the banking and business world and everything had to be above criticism and according to the letter of the law for we also had lawyers, a company secretary and auditors. OTT?

It proved not to be and the school has always been financially viable and runs at a profit. There was always the unwritten agreement and understanding that educational necessity overruled financial considerations on some occasions. The board of directors provided us with strong and stable advice and continue to do so until this day.

Our main requirement and biggest problem was the acquisition of a suitable premises for the school. I had big ideas but my chairman's advice was sound. Start small—grow big. He was right of course so I lowered my sights and embarked upon a hunt of the northern suburbs of Accra with every estate agent in the city. I was anxious to avoid starting in too small a premises which might have entailed moving to larger ones in the future. The Entente Cordiale came to our rescue when the French school moved to new premises on the Ring Road abandoning their old site in Onyasia Crescent in Roman Ridge, Airport Residence.

Roman Ridge was an area of Accra developed by Nkrumah as a diplomatic area and was a good situation for many parents. I made a tentative inspection of the old premises. It would be suitable for our initial requirements but it was in a state of serious disrepair. Gordon Gopaldas and I realised immediately that much work would be necessary before we could welcome pupils on to the site.

It had then twelve classrooms that were just sufficient but a building programme for the future was a must, which would place a strain on our original budget. It was a lucky choice because we were able to expand the site through the years by acquiring adjoining properties which we rented or purchased. I left Gordon to negotiate the terms of the lease with the agent who represented the family that owned the site, for it is not wise for expatriates to embark on this as prices tend to be set at a higher rate when we are involved in negotiation.

I had already discovered, from my earlier experiences in Ghana, that though the British had been a colonial power in the country, they were regarded with some respect and many Ghanaians admire the English way of life. Indeed, during my first three years of working in the Brong Ahafo, which was a time of upheaval and unrest in many respects, several high status officials and influential Ghanaians had secretly yearned for the return of British rule, a view which I have never heard since, I might add, for though Ghana's passage through independence and political turmoil was quite difficult, it is essentially a peace-loving country with a very friendly people who are keen to develop the country in all possible ways.

Both British and Ghanaian societies have their faults and failings and their good points and achievements. By opening a British school in Ghana, we were hoping to take the best from both cultures and mix them in a unique blend for the educational good of our future pupils.

Thus began our juggling act. Talking to investors, negotiating for the site, obtaining estimates for building works and repairs, sourcing materials, ordering furniture and supplies, talking to prospective parents, completing all the legal documents re the company and the school, and trying to arrange a major loan. This was a headache. I had prepared a practical feasibility study that was actually acclaimed when I presented it to my bank in the UK sometime later, for its clarity, depth and vision.

Not so the banks in Ghana, for each and every bank we approached were playing the pre-expenses game; namely, that to consider our application for a loan, they would need to rewrite the feasibility study to their guidelines, formulate our proposal and have a dedicated member of their staff to direct us in the process, all at a cost of $15,000-$20,000, whereas it would have been perfectly possible to act upon our feasibility study with no charge. We were at a disadvantage as we had no assets or security and at that stage, no cash flow. We had raised nearly $150,000 from investors.

Later, we were able to support any loan application with our strong cash flow figures for education is one of the few businesses that takes its income prior to giving the service—all fees being paid at the start of a term (mostly anyway). For this reason I always had to be very strict about collecting fees and it was my one hard line though we did have arrangements for half-term and monthly payments to assist our parents and we did have reductions for siblings. We tried to be as helpful as possible. Our fee base was also lower than other schools in Accra and nor did we distinguish by race or birth but our income was in dollars due to fluctuations with the local currency, the cedi. It was part of our policy to assist our parents as much as possible.

Cost was also a major factor in our selection of a uniform. We were initially opening with classes from Reception to Class Six and bearing in mind that our children would be active for parts of the day, we needed a practical uniform suitable for the intense heat of Ghana as well. I had to strongly protest at a founder's meeting when they suggested ties for all and skorts for the girls and I forced through a light blue short-sleeved blouse and skirts for the girls while the boys had light blue short-sleeved shirts and blue shorts. They all wore dark blue ankle socks and black shoes.

The games kit was also light and practical. This, initially, was made up by a local seamstress until our numbers grew so fast that she could not keep up with the demand. Perhaps it might be a lesson to many schools in England but our uniform code was strict from the outset and we would not allow any deviations or non-uniform items. We were also strict about haircuts and appearance.

In what I considered to be local research, Gordon did invite me down to the school where his boys were being educated for a PTA meeting. The school was run by nuns, and the headmistress was a very forceful character who harangued the children and the parents non-stop. There was a three-line whip at all meetings because one parent had to attend. It was compulsory with no excuses accepted. As for fund-raising, volunteering was taboo as you were designated to an appointed task. One of her complaints was late pick-ups and there I did sympathise, for I had suffered with this at Mount House having to be on duty all day then waiting for day boy parents to pick up their children late.

It was discourteous and high-handed. In any emergency, I would not have minded but I would not accept any other reason. The school never took any action on this and it was a constant niggle for duty staff. Her other complaint was a hoot and I had to stop myself laughing for she was complaining about the trend of modern trainers which then had reflectors or lights on their heels. Why, she asked, did her pupils need lights to go around the school in the daytime? The school did not conduct night classes. It was quite a rant, delivered in a loud voice with her Irish accent berating the parents. It was pure theatre.

I had already received reports about another PTA meeting at a school in Accra where the feeling ran so high that physical fighting broke out, and this was allegedly a reputable school. Needless to say, I approached the introduction of a PTA very warily and carefully as we had many things to establish before we could think of its formation.

The Roman Ridge School.

Here, I now enter upon the realms of conceit for I had learned much from my past educational experiences, the good and the bad and the old and the new, and I now held a wealth of experience not only in the world of education but also in business and finance. I was hoping to bring all that experience to bear on the opening of the new school. Furthermore, I knew exactly what I was doing and for what I was aiming. I have, in this volume, been critical of some aspects of the schools in which I taught but they also had many good and strong points on which I could build while discarding the pointless, unnecessary and weaknesses in turn. In effect, they have provided a strong foundation for the Roman Ridge School in Ghana probably without realising it.

I was also following the tradition of independent school headmasters in the UK who were highly involved and motivated with all aspects of their schools and controlled them with kindness, enthusiasm, compassion and passion. I hoped to aspire to their example whilst realising that I would have to stamp my authority on the school from the day it opened its doors and provide very strong leadership to my pupils and staff.

Here, I must now introduce another loyal member of the Roman Ridge School team who was probably the most important and valuable employee on site, my driver—Alidu—for he was my guardian, protector, security, advisor, mentor, comforter and driver all rolled into one and loved the school and its children. My memory is somewhat dulled but I think I employed him through my best friend, Nour Captan, who 'dashed' him to me for the school for which I am forever grateful for we developed a special bond, though I doubt Alidu would have thought this at first as I carefully instructed him in driving and the various responsibilities of working for me.

His first lesson came on the first day of his employment when I had sent him into town on an errand in my pickup which was distinctive and already quite well known in Accra. I received three phone calls from friends to say that the pickup had been seen driving at speed on the newly completed Kanda Highway. When challenged, he denied it but when I told him who had phoned, he realised that he was in trouble, for he had been observed by a police inspector, a deputy minister and our lawyer. It was a lesson he learned that the car was better known in Accra than me.

He settled down and was a very safe and careful driver from then on and really cared for and looked after me. We developed a double act with on-site contractors whereby Alidu would sit in on our discussions establishing the work that needed to be done and the contractor and Alidu would then go outside to establish the price with the severe warning from Alidu that I would not pay a cedi over the agreed amount. It worked well.

He also knew every fast food outlet within two miles of the school as he was responsible for buying Gordon's junk food lunches during the week and yet Gordon was always thin and slim. Me…? And I did not eat junk food. He was also the first school driver responsible for driving the children at which he was very good indeed, being very popular with the pupils.

In July 2002, I bid my final farewells to Mount House and to all my relatives. I spent a full week with my mother being able to assure her that I would be popping backwards and forwards to see her quite regularly then travelled out through Bristol Airport to Accra via Schiphol, Amsterdam. I was then fully committed and I would either sink or swim in the challenge that was to come. July and August were frenetic. Gordon and I had temporary offices in one of the back storerooms of the school and the builders and decorators were everywhere. Visiting prospective parents were dodging wet cement, scaffolding bars, piles of rubble and delivery trucks and lorries.

We held meetings for prospective parents and I much enjoyed meeting some of my new pupils and I was captivated by their good manners and gentle personalities. There is one major difference between Ghanaian children and many of their British counterparts. The Ghanaian children were keen to be in school and eager to learn. I had interviewed staff during the previous Easter holiday and the new appointees were now dropping in as well.

During the Easter holiday, I had visited the British High Commission on a courtesy call to inform them of our progress and I was ushered in to see a young lady in the commercial section who was obviously very new to the job having been only in post for six weeks. She told me quite bluntly that I could not open the school without the permission of the British High Commission. This came as news to me and I asked her under whose authority she had made that statement and could I please have their name.

She had no such authority but just thought it was the right and proper thing to do. She was well out of order and became quite nervous when I asked her if she was willing to inform the president that his grandchildren could not attend the school because of a lack of so-called 'permission' from the British and I wondered what the press might make of that? I then went on to give her details of the board of directors and the influence they could bring to bear on such a silly statement. She became very flustered and I brought the meeting to a close. We heard nothing more about this but the High Commission did try another high-handed attempt at interference not long after.

I cannot recall if it was by mutual consent or a unilateral decision by me. Perhaps it was both? I gave up polo on the basis that the school could not possibly manage without me for the first couple of years and I was conscious of the fact that many people were depending on me now for its success. It would be unwise to indulge in such a high risk sport for fear of an accident and I did not want to let anyone down. I still maintained contact with the club and all my friends. I did not ride again. I was sad and I missed the game with all its excitement and thrills. I had plenty to occupy me now, however, and was keen to make a start.

We brought all the staff in early for we had much training to complete and had been determined to recruit the best staff we could find but they needed much help with the concept of British independent schools and our methods. I tried to make the training as interesting as possible though it was difficult trying to explain our overall vision and concepts. I had formed the impression that Ghanaian teachers were generally sound teachers based on their historically excellent results but those days had long gone and we now needed to start from scratch for we were aiming to establish the British Independent School System in Ghana commencing with a preparatory school before developing into a full secondary school.

Our standards and quality needed to be comparable with the UK. We were not aiming to be an international school or a hybrid conglomeration of other models or methods. I rated the British Independent School System as one of the best in the world and I aimed to be part of that world in Ghana. I firmly believed in striving to achieve academic success, good pastoral care and a full programme of activities to develop our pupils' full potential.

On these three principles, the school was founded and I allowed no detraction or distraction from them, for they all carried equal weight and were irrevocably bound together. It was a message that I had to constantly expound, and still do. It was radical new thinking for Ghana for parents believed that the longer and harder their offspring studied, the better would be their results. Children were regarded almost like empty vessels to be filled with knowledge and rote learning was the norm.

My theory was not some remote educational and philosophic intellectual exercise based on vagueness and 'pie in the sky' thinking. It was a practical and down-to-earth method and it worked. Ask any independent school in the UK. It works and that is why independent schools are so successful. The three principles cannot be isolated, relegated, diluted or reformed. They are woven into the fabric of school life and form part and parcel of its ethos.

We also believed in the development of responsibility, discipline, excellence, morality, integrity and leadership in our pupils based on Christian values. From the outset we were a Christian school and avowed this much stronger than many schools in the UK today, though we welcomed all faiths and there is no friction within the school as we enjoy celebrating all their holidays in a spirit of understanding and respect.

Our mission, which was some target in Ghana, was to develop in our pupils belief in themselves, by nurturing and developing their talents, gifts, abilities and character to become confident individuals. Ghanaian children were expected to be polite, respectful and obedient and of course, we did not aim to change those aspects of their characters but we did not want them to be submissive and shy which many of them were. We aimed to incorporate the good points into an added mixture of assurance and openness, frankness and verve. By the end of our first term their transition was remarkable and our chairman was highly impressed with their new-found confidence.

We then set out the following promises to parents, many of which were innovative to Ghana:

- Small class sizes (from 16-20) to enable close care and attention.
- Individual pupil attention to enhance each child's development.
- Highly qualified and dedicated staff. Our recruited staff were experienced and qualified and were supported by experienced multinational staff.
- A full programme of sport and extra-curricular activities.
- Pupil, teacher, parent interaction which we regarded as vital to our success.
- Firm discipline with a system of merits and demerits and no corporal punishment.
- Good manners at all times, everywhere.
- A sound Christian framework, which has already been mentioned.
- A caring environment with emphasis on strong pastoral care.
- Educational career choices. SATs tests 1 and 2, Common Entrance and IGCSE and A Levels and an ability to assist with choice of schools and universities and colleges.
- A solid Ghanaian cultural base involving the children in their rich cultural heritage.
- Supervision from 7.55 am until 5 pm. At the request of many parents.

Now the above were all pledges that we made to our parents and pupils and this was unheard of in Ghana. Within the various schools in Accra, not one had made such a list. Ours was practical and possible and we kept to it. Parents in Ghana are normally kept at arm's length and are nervous of complaining in case of retaliation from teachers and heads but we recognised that the education of their children was a shared venture with the staff and welcomed parental support and involvement much to their total surprise and disbelief I think.

Within the above are several hidden factors about which a minority of parents were not happy. Firstly, our pupils did their prep in school supervised by their teachers. I was well aware that even in the UK, parents often 'did' their children's homework resulting in false marks and this applied in Ghana with some parents even employing tutors to help their children with homework. This defeated the whole idea of establishing whether work completed during the day was understood. With our system, the teachers had immediate feedback about the day's lessons.

One of the biggest difficulties with developing the school was the employment of tutors for extra work and this is a huge problem in the UK and other countries as well. Most tutors in Ghana were not well qualified and were working for the money with no real thought of the damage that they might be causing. We have been unable to fully assess that damage, especially bearing in mind that our pupils were being taught by fully interactive teaching by professionals, while many tutors were teaching by rote. We discouraged all extra coaching but always provided holiday courses and extra help within the school.

On my trip out to the school last year, I was accosted by an angry parent whose daughter had not gained good IGCSE results and blamed the school. I enquired whether he had employed outside tutors and the answer was affirmative. It was he who had ignored our advice and therefore his daughter's low results were almost certainly his fault—not the school's. It was totally taboo for any of our teachers to be paid to coach our pupils and resulted in instant dismissal if it happened.

My biggest battle was with the acceptance of special needs children. I had discussed this with the founders and we had agreed that we should accept one child per year with the proviso that the child was accompanied by an assistant at all times. Such children have extreme difficulty gaining a place in any school in Ghana and were generally shunned by all in the community as were the disabled and the mentally ill. Many parents were against this policy and complained to me about it. I merely turned the tables on them and asked what if it were their child that had Special needs?

I am immensely proud of the Roman Ridge School and I rate one of its best achievements as accepting these children who formed such a delightful part of our community. They were fully accepted by all the children and flourished and grew under the safety of our caring staff. Those children have now gone on to develop careers at home and abroad and have always expressed their gratitude to the school for the start that we gave them.

Three anecdotes must be included here. The Roman Ridge School acquired a well-earned reputation for the education of special needs children and we frequently had visitors arriving to see how we had achieved this. They were surprised that we did not isolate them in outbuildings and make no attempt to separate them from the main school pupils. They were absolutely free to join in the full life of the school with the assistance of their helper. There was no magic formula or restrictive care plan. They were allowed to develop at their own speed and ability and that was that.

It is possible that Asheshi University might never have opened its doors in Accra but for The Roman Ridge School. Its founder, Patrick Awuah, had an autistic son, Nana Yaw, for whom they could not find a school in Accra until his mother turned up in my office one morning and we accepted him. Patrick kindly acknowledged this in a major speech at the university a couple of years ago and Nana Yaw was one of the successful pupils who is now at college in the USA.

Finally, I must mention Marley, who was a nine-year-old girl who was physically disabled, though she could run and walk at some speed. She would run excitedly into school every morning and give you a huge hug, nearly knocking you over in the process. She came across me in Maxmart supermarket one day and ran at me at full speed from the other end of a long aisle. Three metres before she got to me, she slipped and demolished a huge display of cereal.

I was keen for the school to have a caring, stimulating and well-planned environment where the children could feel secure and find school enjoyable. I was aiming for a family-oriented school with kindness and care as an essential element.

Our school day was a long one by design. It suited most parents who were working as they could drop their children before work and collect them afterwards. Most Ghanaian children had little to amuse themselves with after school anyway and liked being with their friends for lessons and the activities on offer on campus. Mornings always started with an assembly then the morning consisted of five lessons with a twenty minute snack break, followed by lunch.

We did not have a dining hall so everyone ate outside under the big veranda roofs with everyone sitting at picnic tables. This still applies throughout the school. The children either brought in their own lunch or ate meals provided by an outside caterer. We encouraged the children in playground games to run off their energy and we introduced skipping and even conkers. To hell with health and safety.

In the afternoon, two more lessons were taught and after a short break, the children did an hour's prep/homework followed by activities before being collected. The younger children could be collected earlier but most wanted to stay. At the end of the day, the children were always reluctant to go home because they enjoyed school so much and there were many tears when we broke up for holidays. At our first half-term break, I had five 11-year-old boys sobbing their hearts out at the front gate because they had to be out of school for four days. In England, the children cried when they came to school, not when they left.

The school officially opened its doors in September 2002 with a minimal role of only 22 students, as many parents were wary of committing their children to a new school run by an elderly grey-haired revolutionary and 15% of our student role went by the name of Gopaldas. Last-minute preparations had been undertaken up until 11 pm the night before and I had even been forced to make the contractor who brought the sand for the adventure playground in the form of dirt go back and get sand, which was delivered at 4 am and I was there to check it. We had held induction days for the pupils and we had a family fun event on the Saturday before the start of term so that everyone could get to know each other—pupils, parents and staff.

My staff had worked tirelessly for the three weeks before we opened but here I must now mention my ancillary and admin staff, my lady cleaners, ground staff, security guards and drivers who gave us magnificent support and were absolutely amazing with the children. After some fifteen years many of the original staff are still employed at the school and normally have to die before they can leave. They were diligent, caring, good at their job and kind to everyone. They make up an essential element of our success and make a massive contribution to the school in many ways.

The first week was a success and I had already realised that success was in our grasp by the close of school on Saturday. Saturday? It was my wish, and the first pupils were very keen as well, that we run as many activities as possible on Saturdays. I ran swimming lessons at the Shangri-la Hotel Pool from 8 am to 11.30 am with the help of Auntie Hannah who was in charge of our domestic staff and good with the younger ones, and who later learned to swim herself. Congratulations.

I considered it vitally important to teach as many children as possible to swim and the programme was enormously successful. We were able to commence advanced lessons as well with some of the older pupils. With these sessions and indeed, with anything involving a starting time, the parents were not good at punctuality at all and I had to impose fairly strict measures in all aspects of the school until most parents realised that we could not wait for anything up to one hour for the arrival of their children. I would brook no delay and would start or leave on time except in cases of an emergency.

On Sundays, we often embarked on expeditions and trips and could not wait for latecomers. I often faced upset parents who had arrived late still expecting us to be waiting for them, and I merely pointed out that it was not polite to keep the rest of the group waiting, to which there was never a reply.

The Saturday programme included on-site activities such as football coaching, extra lessons, basketball, art 'attacks', climbing wall, abseiling, cricket nets and if wet, board games, of which the children had little experience. I had reconfigured the water tank tower for abseiling and had patiently and slowly built an easy climbing wall at the back of two of the junior classrooms with the help of a local mason, which provided 'serious fun'. Our chairman was doubtful about its construction and suggested that we get an expert to inspect it. I told him that this had already been done because there was only one expert in Accra. Me!

On Sundays, we roamed far and wide for many of our pupils had visited places abroad but not the sights of Ghana. The children generally explored Accra during school time with visits to the museum, cathedral, churches, mosques and temples and to the airport and fire brigade headquarters. Most of our children have visited the Kwame Nkrumah Memorial Park and Mausoleum and Independence Square.

At weekends we moved out of the city. Only twenty miles from Accra were the Aburi Botanical Gardens high in the Aburi hills, which were not only interesting to tour but also provided a good picnic site. The pupils could play in an old crashed helicopter on one of the lawns while the more adventurous climbed trees at my instigation; a simple activity that they had never done before. We also held small family events there with games and lots of fun including tug of war and many wide games.

It had the advantage of being cooler than Accra and it could be very cold up there in the mornings when I took the older boys mountain biking on a three mile downhill track to the plains below…at speed. Locally, we also visited my friends' salt site at Prampram where the children could 'walk on water'. The lagoon was mostly a couple of inches deep at some states of the tide and they could run over the water and they loved to splash and play in the deeper sections.

We visited Ghana's waterfalls at Boti, in the Yilo Krobo district in the Eastern Region with its Umbrella Rock, and Kintampo on the Pumpum River, which were quite impressive, but the Ghana Tourist Board was always short of money so the approaches and surrounds were nowhere near a tourist standard. Other places on our list were the Tafi Atome Monkey Sanctuary and the Akosombo Dam which was always worth a tour, though the guides did tend to talk down to our children which they accepted with good manners.

We did make one venture into the Shai Hills Game Park and were assured that the dirt road around was in good condition. It was not and Percival, our first school minibus, did well to get around and survive the tour. Perhaps it was our Basil Brush mascot that brought us out through the gates safely. We saw no game at all though we did visit one of the most ancient sites of Ghana which was a rock shelter, where, it is suggested, Ghana's first people lived. A popular trip was boarding the Dodi Princess large pleasure boat on Lake Volta to travel five kilometres up the lake to Dodi Island which was not very interesting but we could dance on the boat and they served reasonably good food. The parents liked to attend this trip.

We had good friends at Afrikiko on the shores of the River Volta, two miles below the dam which had its own pleasure boats, a swimming pool and games areas. It later became one of our bases for camping and canoeing. We visited Kakum National Park on many Sundays as the children liked the Kakum Canopy Walkway, which transported you at great height through the top of the trees. It was quite a challenge for some of our children who were nervous of heights. We would always pop into Hans's Cottage on the way back for refreshing drinks and crocodile watching.

Ghana has some 500 kilometres of coastline dotted with a myriad of ancient castles and forts, most built between 1482 and 1786, which played leading roles in the slave trade that commenced in the fifteenth century to provide labour for the New World colonies. We visited many of these.

The two most impressive examples are at Cape Coast and Elmina. To visit them is a sobering experience and I always felt a sense of foreboding when I entered them, just as I had many years before when I visited Bagamoya and Fort Jesus in Mombasa. It was important that our pupils visited such sites to learn of the terrible history of that trade. At time of writing the debate about Colston has been reactivated in Bristol. He was a philanthropist and was charitable and generous with his wealth—mostly the proceeds from the Slave Trade. It was not a 'normal' trade which had become acceptable.

Everyone in the trade knew what they were doing and ignored the barbarity of their inhuman activity. Many other groups and peoples were involved in the trade however, which is sometimes conveniently forgotten in an attempt to 'whitewash' history, for it was the Arabs who bought the slaves in the interior, many of whom had been captured in inter-tribal wars including those of Ghana, and brought them down to the coast where some of the local indigenous people were also involved in the trade.

At Cape Coast Castle, which to me has the most forbidding dungeons, up to 1,000 men and 500 women were kept separately shackled and crammed into the poorly ventilated and dank chambers for up to three months, sitting in their own filth and excrement for the whole of that time. The next time they saw light was when they passed through the 'door of no return' to be taken out to the ships for the ordeal of the Middle Passage. This would be the last sight they had of their country. Elmina was a bigger castle, built in 1483 by the Portuguese and situated on a magnificent headland overlooking the Gulf of Guinea. The Portuguese treated the slaves in the same cruel manner. It is estimated that over 6,000,000 people were transported from the West Coast during the trade and that between 60,000 to 90,000 perished on the voyage.

I had many friends with chalets along the seashore between Prampram and Ningo and I would take out small groups at the weekend for the novel experience of sleeping and living by the sea. It was a whole new world for many of them and it was a source of constant delight to me to see their nervousness, awe and excitement.

During our first year, we embarked on a five-day trip to the north which was quite a trip. We visited the famous crocodiles at Paga which are very tame and live in the nearby village. This was where I was able to sit on a crocodile. We toured the Pikworo Slave Camp where the slaves were gathered in their preparation for the journey to the coast, and the mudbrick-built cathedral at Navrongo before moving on to the famous Larabanga Mosque, the 'Mecca of West Africa' with its Sudanese architecture and big patterns adorning its walls. Finally, we rounded off the trip with a visit to the Mole National Park but saw few animals. I was used to the wonderful game parks of East Africa and Zambia. Mole was very disappointing.

With this plethora of trips and activities, my readers will now be wondering if we did any work at the school? We did, and lots of it, for our pupil base was mainly Ghanaian and the children were very eager to learn in class. Our lessons were based on the English National Curriculum and our work and results were compared directly with those of UK children through SATs One and Two initially, then, as the classes progressed up the school, the comparison included the Common Entrance Exam at 13+ in which the pupils were examined in English, Maths, Science, French, History, Geography, RE and IT.

The Common Entrance Exam was of a high standard and broad-based providing us with perfect guidance for IGCSEs later. As our teachers were new to these exams we monitored them through the marking and then had them moderated by professionals in the UK. I was also insistent that there should be no coaching or extra sessions for SATs for they have become increasingly unreliable in the UK as many heads deflate SATs One and inflate SATs Two to show false improvement and the invigilation of the exams are often suspect with many of the teachers helping pupils directly.

There are also a growing number of cases of head teachers tampering with the results. SATs Two, in my opinion, is also spoiling Year Six for many pupils as they constantly practice papers despite denial by heads, and ignore the normal lively curriculum. Our pupils complete two practice papers where individual problems and difficulties are ironed out and invigilation is rigidly controlled. We get good results without the hassle and bother that accompanies the exam in UK. Our Common Entrance results are really an in-house guide though can be presented to the UK schools by a few individuals who wish to adopt that route. It is also important as a target for Years Seven and Eight to aim for and as mentioned, is also the backbone for IGCSE.

We had clear handwriting and reading schemes and policies from the commencement of the school and built a library which became the most used facility in the school because our children's love of learning also extended to books. The Science Lab was also popular and much used though I had problems with the Secondary School Labs at first. I have to say, however, that the most used facility in the school was the Adventure Playground near the entrance on which the children would play for hours.

Our teaching staff was small and Fafa Awoonor and I taught as well during the first two years. I took class six English and Geography in years four, five and six and Fafa took French in all the classes. She had been a highly successful teacher in the Ghana International School and now she faced an even bigger and daunting task…French with the pre-prep—the five-year-olds. She came out of their first lesson exhausted and shattered, having never taught that age group before. She has never forgiven me.

I had my problems. We were embarking on a comprehension about Dolly the Cloned Sheep, when one boy put his hand up and said it was not possible because God made all creatures. Religion in Ghana is taken very seriously and several parents were fundamentalists and the boy would not be moved. We moved on to the next passage. It was the same boy, he was very bright, who asked me whether I was really his headmaster, which I confirmed.

In his previous school, he had never seen his headmaster on the premises. He had a very enquiring mind asking about the future secondary school and was I going to start a university? He introduced the idea, jokingly, that I was going to provide a 'womb to tomb' educational facility, which I thought was hilarious. Humour was not really encouraged in Ghanaian classrooms but I soon put a stop to that and had tremendous fun with my pupils in class.

If attention was lagging, I would toss an educational hand grenade into the room such as, 'Is lined paper heavier than plain paper?' or make them do a silly exercise. My staff were amazed when I told them that they should send their classes running around the school to waken them up. I was trying to establish interactive learning in every sense of the word and my staff were finding this concept difficult.

We held regular training seminars and a seminar is what every Ghanaian loved but our aims and ethos were not coming through loud and clear. We had contact with a UK junior school head in London, Liz Armah, who was later to take over as head of the junior school, who gave a two-day seminar on creative writing.

Two weeks later, I was conducting a book check in which I checked every pupil's exercise books and work folders to ensure that the teaching and presentation were correct and assessing of every child's potential was being fully maintained. There was a notable absence of creative writing in one junior class and I called the teacher in and informed him about its omission. He asked what it was. He had sat in an interactive seminar for two whole days on the subject and not taken anything in. He had no idea about creative writing at all.

I also queried one member of staff's lesson and exercise on the use of the apostrophe for possession. Now he knew that I was keen that grammar and punctuation could be taught easily by the use of learning from children's mistakes and he assured me that as the class had made so many mistakes he had taught the lesson. He was surprised when I pushed his class exercise books across my desk and challenged him to find one such mistake. There were none. It was an abstract lesson completely out of context.

It was the same teacher who I asked why he had a chart showing English hunting dogs on the classroom wall. We insisted that all wall displays should be relevant to the work in hand or be filled with children's work and he admitted that it was wildly irrelevant. The teacher is still with us and I only have to mention 'English hunting dogs' and he goes off into peals of laughter.

Earlier on, I caught a serious problem. On a book check I would compare the validity of marks given for any exercise with the half-termly order sheets as a cross-check. One member of staff was not only wildly and incorrectly marking but he was also completely fabricating his order sheets. He tried to lie his way out of the situation which only made matters worse and he had to be sacked, which was a pity because he was really quite a good teacher and I could never understand why he did it.

With the arrival of Liz Armah as the junior school head, we changed our tactics. We stopped the 'seminar' method and moved directly into the classrooms in a friendly and helpful way dealing with problems and difficulties as they arose. 'Nudging' the staff into better teaching. Both Liz and I were good classroom practitioners and we were able to bring the staff around.

In the very first term I started the tradition of making two formal lesson visits, one by invitation, and the other not, and continued this during the whole of my time at the Roman Ridge School. Another tradition of the school, which you will not be surprised to learn, is that there were no lesson interruptions for any reason whatsoever so that our pupils' lessons were not disturbed. It created a very good learning environment and concentration was much enhanced. My teachers then only had to help students who had been absent because of illness not because of a string of music lessons.

Generally, our children were very healthy and we encouraged good habits and lots of exercise but Malaria was a slight problem and amazingly, coughs and colds, especially in the rainy seasons. I suppose I could now pretend that it was all planned but it was a fact that our children did not suffer from as many coughs and colds as the other fee-paying schools. Why? We had limited air conditioning. The classrooms all had ceiling fans and only the specialist rooms had air conditioning such as the library and science labs.

The other upmarket schools were fully air conditioned and that caused lots of coughs and sneezes so I held out against the parents' demands for a fully air-conditioned school on those grounds for it would have entailed enormous expense. The end justifies the means because several years ago, electricity prices rocketed and the schools with full air conditioning struggled to pay their electricity bills while we were able to survive the crisis with our fans.

I have mentioned that while I was at St Bede's School, I encountered a strange case of child possession, which is still prevalent in Ghana and many other African countries to this day, and we came across the problem several times at the school. One family had three children who were absent quite regularly and sometimes exhibited strange behaviour. We enquired of the parent to ask what was happening but the children were whisked away and an aunt came hinting that they were 'possessed'.

Quite sophisticated and cultured sons of one parent were taken across to Benin to be treated for alleged behavioural problems. Benin is still a centre for Juju and other African 'magic'. We had no way of interfering or checking these actions as there was no one to whom you could report. A father arrived unexpectedly in my office from abroad one morning as his oldest son was giving us cause for concern both academically and socially. We called his son into my office. The boy was totally shocked to see his father and crawled across the floor to touch his father's feet with his head and then knelt before him while his father harangued him for his misdeeds. The father promised me that his son would return after the impending holiday a different boy and he did.

He was totally subdued and very docile. He had been taken back to his ancestral home in the north and had spent a week with a 'medicine man.' Now if I was in the UK I would, by law, have reported this to Social Services. In Ghana traditional medicine was still practised and accepted. We could really take little action.

By the end of the first term, our numbers were increasing and from 29, we grew to 60 in a few months. This growth was partially due to quick thinking on my part because a family of eight children arrived from Mali requesting admission. Their English was very poor so I immediately arranged a special class for them all and gave them a dedicated teacher who gradually inducted them into the school. The whole family grew and flourished under our care and we did well for them.

I had also promised to keep my class six to a limit of ten children for the first year. Numbers stood at nine when I had an application from a parent with twins so I opened another class six. Gordon had initially informed the mother of my limit on class six suggesting that she see me 'personally' to discuss the problem. She duly arrived and requested that I try to help her twins, sliding a brown envelope across the desk towards me.

'Brown envelopes' in Ghana were very common and many heads in schools accepted them as a matter of course to guarantee admission for children. I laughed and pushed the envelope back. Neither I nor the Roman Ridge School accepted or used the 'dash'.

Several investors had not come up with their promises which left us in a precarious financial position and I was even ploughing my modest salary straight back into the school to keep going. I was living in a small flat within the school to save cash anyway but we could not raise a bank loan and I needed a fairy godmother or intervention of my guardian angel, quick.

At Christmas I flew back to the UK ostensibly to spend Christmas with my mother, see relatives and buy resources for the school for many of our requirements had to be purchased in the UK. I also approached my bank for a loan which was declined because it was for a foreign purpose so I took the brave move of mortgaging my flat to raise a substantial sum of capital through one of my former business contacts, which the school agreed to pay back over five years. We were safe for the immediate future. It was a gamble that paid off and I was even able to move into a proper house not far from the school.

School and my life settled into an enjoyable routine especially as our numbers steadily mounted and we were acquiring an excellent reputation in Accra. We constantly monitored all aspects of the school including teachers and children, the activity programme expanded and thrived, and we were able to complete the refurbishment programme. During my tenure at the school we were able to complete building developments year after year and I am afraid I cannot recall a time when we had a rest from builders. Initially we built new classrooms alongside the road and then a two-storey classroom block on the site of our old original offices.

We mainly catered for Ghanaian pupils. Our reputation however was reaching the expatriate community and we were receiving enquiries from many diplomatic and business foreign families. Enter the British High Commission again. If British children were to attend the school then they would have to inspect it. It was good timing for me for we had just been inspected by a small team from the Independent Association of Preparatory Schools and we were recognised with distinction in the fastest time ever for any school worldwide of which I duly notified the British High Commission and also informed them that the school could only be 'inspected' by qualified professional inspectors.

They seemed unable to recognise the IAPS and insisted that they check so I refused. Now I had already interviewed several British families and their children who were very keen to be admitted and I did not want to disappoint them. At that time, Janet West, was our director of studies, and a very competent one too, and her husband was head of DFID in Ghana. We arranged for him to come around and have a coffee and I asked if he would like to look around the school. He enjoyed the visit though he was a little surprised to see British children in the classrooms already. All became well.

The British can be bloody pompous sometimes and the defence attaché at the High Commission was more pompous than most. Ex-Mount House naval parents had now been promoted to driving ships around the world and would often give me the courtesy of inviting our children on board if they were calling in at Takoradi or Tema. I received irate calls from the defence attaché saying that he was in charge of on board visits and he was obviously totally unaware of my friendships with the commanders or admirals and used to blow a gasket when I would phone him as a courtesy to say that we had been invited on board another ship.

We also had a slight tiff with the UN, a global organisation responsible for paying several of our pupils' fees. They wrote to inform us that they had changed their policy and would now pay fees at the end of each term not at the beginning. We wrote back to say that our fee paying policy had not changed and our fees were to be collected at the start of term as usual. We were obdurate and immovable. Eventually they gave in.

There were permanent problems with Accra's electricity and water supplies which were totally erratic and unreliable. I was in school most mornings by 6 am to check the water tanks and we had to order road tankers on many a day which was quite an exercise. To overcome the electricity 'outages' we had to install generators to cover for the lack of supply. With the high price of electricity in Ghana it is often cheaper to run on Generators.

When we felt the time was appropriate, Fafa Awoonor took over the task of starting up the PTA, of which she made an excellent job. She was rightly cautious in the initial stages as we wanted no problems to develop and we are now blessed with a very active, cooperative and highly functional PTA that has contributed much to the school and the children.

Fafa reckoned that I had mystical powers. She would sometimes catch me looking over our boundary walls and coveting the land of our neighbours. Powers or not, the plots surrounding the school gradually fell into our hands one by one. Our present façade is very misleading as the school seems very small but there is now a whole secondary school hidden out of sight, for each year the school grew in pupil numbers up to 13+ Common Entrance and then up to IGCSEs and A Levels eventually accommodating nearly five hundred pupils in all. New three storey classrooms were built with three science labs, a senior library, an art room, changing room and showers, a language lab and two IT rooms and this was followed by a sixth form facility with a covered atrium with classrooms, offices and staffrooms.

Having built three new science labs which were fully equipped to modern standards with lab technicians, I discovered that they were not being used. I was assured that they were but I used to patrol the school two or three times a day, when the pupils would put out a red alert, and I never saw them being used so I asked security to run checks. They were not being used so I threatened to turn them back into classrooms. They were used then. We also had good IT facilities which were the envy of many schools in England and Ghana.

There was a slight change in roles now as I was appointed Principal (still CEO), Fafa was appointed as head of the senior school and Liz Armah eventually handed over to Valerie Mainoo as head of the junior school. We had a somewhat top-heavy management team based solely on the fact that our teachers needed much guidance, help and support. I was increasingly aware of the problem of succession. There would have to come a time when I would have to hand the school over and I thus agreed to the management plan though I was not altogether happy with Fafa being given independent control of the senior school as I felt she needed help in that role before she could become successful. I would have preferred a much more regulated handing over and this proved to be correct. Fafa should have been mentored and helped in her role.

There were now two major developments at the school. Quite frankly I had hedged on appointing a board of governors as I wanted to fully establish the school and its aims and ethos in my own grey haired, revolutionary way, with no other input to ensure that my dream would become a reality, and I am absolutely convinced that this was the correct approach initially. My chairman, however, was 'champing at the bit' as he was often being approached by parents on educational matters about which he knew little so they came back through me anyway.

When the time was ripe, we did form a board of governors under the able leadership of Joyce Aryee and the members were helpful and supportive and offered good guidance and suggestions. They also dealt with disciplinary appeals involving pupils, staff and parents though they were few and far between. They could occasionally go off at a tangent and I went head to head with them if they tried to introduce modern American educational methods. This was a British school and that was that. They did pass a motion about introducing the International Baccalaureate (IB) exams which I knew was not practical in our circumstances though I was given a two year deadline to achieve it. At the next meeting, I presented a document with full costings. The IB would be massively expensive. I heard no more and the deadline is now twelve years on.

One of our greatest problems was establishing activities for our children. We provided many afternoon activities and the ones at weekends have already been covered. Our sport included football, basketball, netball, cricket, hockey, tennis and athletics and all the pupils did PE as well. We also provided traditional and modern dance, ballet, traditional drumming and karate. We offered drama and musical productions as well as limited music lessons, mainly on piano.

I was ever anxious to improve the quantity and quality of our activities. Firstly, I imported a small fleet of canoes and we canoed regularly at Afrikiko beneath the Akosombo Dam and many of the senior boys reached a very high standard of paddling and many pupils tried the sport and even, one or two staff. We welcomed a young lady, a Newcastle United professional player, for a term and she taught football to the boys and girls. Why are the British schoolgirls not taught football? Our senior boys were quite disgruntled when Lucy arrived to take them for their first session but they were polite and respectful, as was required, and off they went. They thought she was fantastic after the first session. A good lesson in feminism perhaps? Our present girls' team is undefeated at time of writing, which follows the tradition of our staff team as well.

We needed something more, however, and I hit on the idea of starting Scouts. I made contact with friends in the British Overseas Department in London and received their OK to start a British unit abroad. The Ghana Scout Association insisted that we should become part of their association instead and there was a political impasse. I was not keen to join the Ghana Association as many of the Scouts were involved with politics.

Indeed, I had encountered several Scouts accompanying soldiers on looting activity during coups so we joined up with the Baden-Powell Scout Association in the UK. This was formed when the UK Scouts modernised in 1967 and the Baden-Powell Scouts broke away to run on the original lines set out by BP. They were keen to have us and the Ghana Scout Association was left high and dry. They had no jurisdiction over BP Scouts. We formed a Beaver group of 36, two Cub Scout packs of 36 and a Scout troop of 48. 156 in all and we met every week. All the pupils could not wait to get to meetings.

The uniform was school uniform from the waist down and a khaki short-sleeved shirt with a neckerchief around the neck. By chance the BP programme really suited our children and they loved the activities and tests each week and the rewards they received. We purchased camping equipment and regularly camped at Afrikiko and Stone Lodge which were relatively safe areas and run by friends. We had troop and unit flags which impressed the parents and visitors no end and these were dedicated and blessed by a priest. The pupils took to Scouting like a duck takes to water and it was a highly successful programme which challenged and maintained a constant interest for our pupils. The ethos of Scouting also helped the ethos of the school.

After seven years, we were seen as highly successful and the school had a wonderful atmosphere which was noticeable as soon as you entered the gate. There was something magic about the school which still exists to this day. Our pupils are a constant delight and fun to be with, ever anxious to help each other. The older children help the younger ones and there is a real family spirit. From the day we opened I had a 'headmaster's corner' right outside my office over which loomed the head of a tin elephant and the younger pupils were always told that the back end of the elephant was in my office.

I would often sit with the children and hear them read or talk or they could come and tell me their troubles but mostly they chattered away on a very friendly basis and they loved to talk. They were happy. I had achieved what I had set out to achieve. Do I have any regrets? No. I wish that I could have built up better music and art departments. Our children loved singing and had weekly music lessons but we had no hope of offering a typical music programme such as in UK schools. It was purely because there was not the skill and expertise on offer to achieve this and nothing to do with lesson interruptions. The art department suffers from the same problem, though all our children paint and have fun.

My one abiding embarrassment was our small sports field across the road owned by the Museums' Board that they had earmarked for staff housing which was not to our liking or standard as it always looked shabby and rundown. It seems that maybe, just maybe, we have finally arranged to purchase the plot so that it can be developed properly. Hooray!

If you are prepared to risk everything then you can do anything and a Chinese proverb simply states, 'Fall seven times, get up eight'. I dreamt it, I believed in it and I had achieved it. There is no doubt that I am very proud of the Roman Ridge School but I could not have done it alone. I had enormous help and support and much appreciate that.

The school is thriving and still maintains its original ethos and aims.

As an appendix to my personal educational ideas and philosophy, I have often used the following poem, written by Dorothy Law Nott, with my students and staff as a practical guide to education.

CHILDREN LEARN AS THEY LIVE.

If a child lives with criticism, he learns to condemn.
If a child lives with hostility, he learns to fight.
If a child lives with ridicule, he learns to be shy.
If a child lives with shame, he learns to feel guilty.
If a child lives with tolerance, he learns to be patient.
If a child lives with encouragement, he learns confidence.
If a child lives with praise, he learns to appreciate.
If a child lives with fairness, he learns justice.
If a child lives with security, he learns to have faith.
If a child lives with approval, he learns to like himself.
If a child lives with acceptance, and friendship, he learns to give love to the world.

It is to this end that we strive.

Chapter Twenty-Eight
Acute Myocardial Infarction

Early in December 2006, I had returned from a headmasters' conference in the UK on a Saturday with my usual carrier KLM. Alidu, my driver, had been at Kotoko Airport to meet me and drove me to my house on Sir Arkuh Korsa Road. I asked George, my houseboy, to prepare a sandwich and a cup of tea before I dismissed him, with thanks for staying late, and retired to bed as normal.

On Sunday I felt unwell but Alidu drove me to the school so that I could catch up on my paperwork and prepare for Monday. Upon arrival we were welcomed by security and Alidu carried my briefcase into my office. I was perfectly capable of carrying my own case but he would have been deeply offended if I had, for this simple act showed that I trusted him and he performed the same task every time I came to the school gate.

I cleared my desk but I was now feeling quite ill and obviously looked unwell. Alidu was insistent that he drive me to the Lister Hospital where they checked me over, gave me various tests and an ECG. They said all was well and that I had probably contracted a small bout of food poisoning from the KLM flight so I went to bed early. The next morning I felt ghastly and drove gingerly into school where I vomited a great deal but I stayed at the gate until all the children were in then asked Fafa to take assembly.

Alidu took me down to see my Dr Kanda who had not had much business from me during the previous seven years and I spent more time dining out with him than consulting him professionally. He gave me another ECG. I had experienced a massive heart attack and he immediately asked Alidu to rush me over to the Lister Hospital again where I was admitted as an emergency into a private room.

A heart specialist was called and all hell began to be let loose. Gordon rushed to my bedside and the heart specialist recommended immediate medical evacuation back to the UK for specialist treatment. By lunch time Gordon informed me that the news was all over Accra and worried parents were jamming the school's telephone lines to enquire after me, and Nour and Mounir rushed to my bedside. Now while all this was happening, I was not really feeling that bad and visitors were quite surprised to find me quite alert and sitting up.

I had experienced none of the usual symptoms of a heart attack, for which I was very grateful, and even ate a sandwich for lunch and drank a glass of water. A worried Gordon, in the meantime, had contacted BUPA, under whom I was insured, and started to arrange my medical evacuation which proved very difficult as BUPA were prevaricating—they even started to phone me in my hospital bed to discuss details. The heart specialist came in and pushed everyone out of my room informing me, in no uncertain terms, that I was seriously ill and I was to lie flat on my back and rest, otherwise there could be consequences.

When BUPA phoned me, he took the call and informed them that I was in a critical condition and my case was urgent but I was not to be bothered with the details, so Gordon took over the negotiations. BUPA maintained that they could not arrange a flight back to the UK for three days but could arrange one late on the following day to Milpark Hospital in Johannesburg. I had no choice but to accept on the specialist's advice, for I needed treatment fast, though we did get written confirmation from them that I was entitled to be repatriated to the UK eventually.

My evacuation required that I undergo several procedures to facilitate drips and other monitoring devices and this could only be done at Korle Bu Hospital on the other side of Accra. Alidu was mostly in attendance during the day so I sent him back to the house to pack a suitcase for me. Another problem arose when Gordon could not locate an ambulance for the transfer to Korle Bu so we had to call on the army who arrived in the ambulance that was used to attend polo matches. I was strapped to a trolley with drips attached and was tucked in tightly with the sheets.

Lister Hospital is situated at the end of a very rough, unmade road and the journey to the main road was agonising in the extreme. I was being banged around and moaning in the process all the while and Gordon was steadying the trolley, which was a tough job. We made it to the main road and proceeded to Korle Bu in the dense traffic. The ambulance speeded up when we got on to the main road alongside the lagoon, where we hit a big bump.

The back doors of the ambulance flew open and the trolley and I headed for the door. Gordon made a wild grab for the trolley and managed to jam his foot under one wheel, thus preventing me from ending up in the lagoon. It was a very close call and Gordon saved me from a very bad accident and a possible bout of Typhus if I had been thrown into the lagoon. I must admit that I was shaking for some while.

There are three heart surgeons at Korle Bu and they operate in theatre every day except for Tuesdays. I had planned my arrival well. It was a Tuesday. All three were in attendance and they swiftly whipped off my shirt and put in various intravenous lines including a monstrosity of a device into the right hand side of my chest that seemed to have about six or seven outlets or inlets and looked like a small bagpipe. They were very pleasant and I thanked them for their care and kindness.

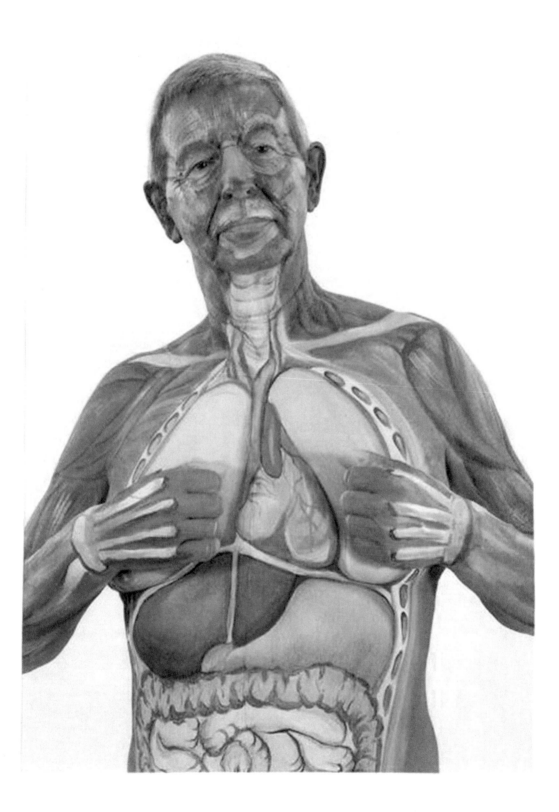

They accompanied me to the airport and we had an argument at the security gate as the guards thought they were hospital porters. A Lear jet was waiting on the tarmac and I was formally handed over to two medics who asked a myriad of questions after which I was strapped to a board-like device and hoisted into the aircraft, nearly dropping me in the process. I reckoned that the Lear jet was probably purchased third-hand during the Angolan War as it was spartan and not in the best of condition.

I was kept on the board and given instructions not to move. I could not anyway because of the straps. The two medics were efficient but not very friendly and I had noted their reserve with the Ghanaians at the airport. We took off for the six and a half hour flight to Johannesburg. The medics' attitude thawed somewhat and they offered me a snack and cool drink.

The flight was very boring and I tried to sleep but I was becoming very uncomfortable in my prone position. Two hours into the flight, I told the medics that I was going to die. They assured me that all would be well for Milpark was a fine hospital. I told them that I was not worried about my heart but the developing agony of being strapped to the board. I wanted to sit up. They had to obtain radio permission for me to do this and I was carefully guided into a comfortable seat.

It was an immense relief and I fell asleep at last only to be awoken a short time later when we landed in Luanda for refuelling. Soldiers boarded the plane and tried to insist that we had to get off and pass through formalities in the main building. The medics quite rightly refused as they were on a mercy flight. One soldier pointed a gun at me and ordered me to get up. I refused, pointing to all my tubes and drips. He shouted and I said nothing. If I moved, I would die anyway, I informed the soldier. An officer appeared and the medics strongly remonstrated with him. He ordered his men off the plane and we took off.

When we landed in Johannesburg some three hours later, it was like a scene from a movie. An ambulance drove up to the plane with lights flashing and I was strapped to the board again and handed over to two competent medics and a doctor, rushed into the ambulance and as we raced out of the airport with blue lights flashing and siren wailing, the doctor gave me an ECG, smiled and said it was OK. He assured me that all would be well. It was the middle of the night at Milpark Hospital and a crash team had been on standby for over six hours awaiting my arrival.

I was rushed through the outer doors and we sped down to ICU where I was met by Dr Dalby, a cardiologist, who was very good at explaining everything as his team proceeded. I was undressed, a sheet put over me and I was carefully attached to an array of machines and devices some of which were plugged into the device in my chest and others on my arms and legs. All sorts of electronic devices started to bleep and whirr and the team worked on and around for over an hour. They all looked much more concerned than I was and three of the team stayed with me for the rest of the night checking and making adjustments

I was under strict instructions not to move at all which was easier than being on the board because I was on a special mattress which was quite comfortable. I could not sleep because I suddenly realised that with all that had happened in the last 24 hours, and especially now being in ICU, I was in danger and in a serious condition. I was frightened that I was going to die and I wept.

One of the nurses noticed and kindly held my hand and wiped my eyes with a tissue. She understood that I was scared and she informed me that I was probably suffering from shock as well. It was all normal but I could not sleep still and they would not give me anything in case there was a reaction with the drugs with which they were treating me. A wry and slightly amusing thought did pass through my mind though. In previous hospitals, I had been left naked on two occasions with no clothes and here I was again, true to form, naked, though under a white sheet, many thousands of miles from the UK. Nothing changes.

The next morning, I was exhausted. I had not slept for over 24 hours and Dr Dalby was most concerned about me. There was a consultation in one corner and I was given something to make me sleep and I slept for the whole of that day and night. They had even fitted me up with a fluid drip and catheter and I had not awoken. I needed that long sleep and woke up ready to face the world again.

ICU was well organised. Every patient had a dedicated sister and two nurses in full attendance and I was not allowed to raise a finger. Everything, and I mean, everything, was done for me. I was fed, toileted, bathed, cleaned, checked and monitored by two lovely Basuto nurses who were jolly and nice and I enjoyed their banter because they talked to their patients all of the time. Their conversation was mostly in English but they did sometimes lapse into Chilapalapa which was a common language developed on the mines and farms of South and Central Africa as there were so many different tribal languages. It was a working language and I had a reasonably good knowledge of it from my time in Zambia.

Very often, the two nurses would use it when referring to my personal 'bits', saying that I had a small penis or josh with each other that I would make a good husband. I let it continue for a couple of days and then, when there was another reference to my personal 'bits', I said, very quietly, 'Icona. Me a hamba lappa and bawpa chai mningi!' (Stop it or I will come over and sort you out!)

They screamed. The sister looked up and eyed me carefully.

'You speak Chilapalapa?'

'A little,' I smiled.

'You didn't tell us?'

'No one asked,' I replied.

She was from Lebanon and asked if I spoke Arabic.

'Sabaah alkhayr. Keefik?'

She gave me a wry smile.

I had been in Milpark for three days before Dr Dalby and I discussed my condition. To date I had not undergone any tests to establish the cause of my heart attack and he was insistent that he would prefer me to rest as I was for a day or two before undertaking any. As it was, we waited three days as he was not altogether happy with my condition. I did manage to get messages out to my mother and the school that all was well. I was informed that Gordon was phoning twice a day to check on my progress.

I rate myself as intelligent but I was hard-pressed having to lie naked under a sheet connected to so many devices for a full five days. I was not even allowed to talk for long periods and all my friends will know how hard that would have been for me. During that period, there was always someone at my side day and night to look after my every need. What a service.

On the fifth day, they unplugged me and I tried to walk supported by my two nurses. I could not so they put me in a wheelchair to give me a bath which was sheer luxury and I wallowed for over half an hour. They moved me into another ward which still had twenty four hour care and I was connected to a smaller array of portable machines, but now I was allowed to sit up and move in the bed which seemed like bliss, and eventually

I could walk around with a nurse to prevent falls and chat with the other patients. I was even allowed to wear shorts and a T-shirt, thus ending my time as a nudist.

A couple of years before this, I had been treated for an abscess on my bottom which had developed into a fistula and I was now fitted with a Seton Suture in my bottom (apologies for the details). The hospital's colon rectal specialist paid me a visit. She was Russian with a completely unpronounceable name. She was dressed in colourful, flowing silks with lots of jewellery and gushed around me asking questions and examining my posterior in detail.

'Oh. You poor dear. You poor dear.'

She kept on repeating this and I became quite amused by her antics, especially when Dr Dalby attended, as she accused him of not looking after me properly as the suture site had become infected. He was used to her ways and merely laughed. I was ordered to take a bath every day to which was added a special liquid in which I soaked. This was slightly more complicated than it seemed as I still had a 'minder', for someone had to be with me all the time so I was actually bathed by the nurses during the daily ritual. Was I embarrassed? Did I mind? What do you think?

Dr Dalby started the tests and the first was a stress electrocardiography on a treadmill. I refused. I was scared stiff. Geoffrey Allen—caver, climber, canoeist, orienteer, sailor, mountain biker, wild water swimmer, polo player—was frightened. Dr Dalby himself had told me that I was not to put myself under any strain or stress and now he was asking me to get on a treadmill. He had to assure me and coax me on to the machine and I gingerly started walking.

It did not last long and I got down from the machine to be told that I was going straight in for an angiogram in one of the treatment rooms which he duly performed making his entry on the inside thigh of my right leg. I mention the detail of this because the next morning I had very extensive bruising in that area and while taking my bath, the colon rectal specialist came in.

As I could still not pronounce her name, we had agreed that I should call her Madame Derriere, which she thought highly amusing. She was horrified at the bruising and gave poor Dr Dalby a real telling-off for harming her patient. I enjoyed her visits as she was always asking me the meaning of British words and the use of vocabulary.

At the end of the first week, I had a long consultation with Dr Dalby about my heart attack. My right coronary artery was occluded close to its origin with a retrograde filling into a large distal vessel from the left system. In the left system the left main and circumflex showed significant disease but the left anterior artery (LAD) had a long section of diffuse narrowing in its central portion which appeared to be within an intra-myocardial segment. I needed a double heart bypass but this was complicated because of the long apparently intra-myocardial course and they could easily do the right coronary with distal grafting but could not find grafting points for the other.

There was a problem. I had no expectations or preconceived ideas before this consultation and strangely enough, the above news did not come as a shock or surprise. I knew the heart attack had been acute so I merely asked could we go forward, and if so, how? At that stage, the prognosis was not good. I should return to the UK and would probably lead the life of a semi-invalid until…?

Dr Dalby did insist, however, that I remain in hospital for another week to ensure that I was fully rested and stabilised before I left as he still had concerns. I agreed. I was totally in his hands.

The ward that I was in was for eight men but was quite noisy and I did find it quite difficult to rest and/or relax. Visiting wives were very kind to me realising that I had no visitors and came in with cakes and small titbits for me as well as their husbands and I thought this inordinately kind but they did tend to involve me in the noise, banter and disturbances of the ward. Added to this, a mad Greek gentleman was also making consistent attempts to leap over from his bed to mine in the middle of the night and also rampaged around the ward.

BUPA saved me. They were not in my good books at this time because of their prevarication about my flight, which I reckoned was pure fiction because treatment in South Africa was far cheaper than in the UK. The lady rep of BUPA actually visited me and immediately pointed out to the staff that I was some sort of advanced member of BUPA and was entitled to a single room. I was immediately transferred to one where I could now sleep and relax. She also consulted with Dr Dalby and commenced making arrangements for my repatriation a week hence.

I was now able to wander the hospital which was quite modern and had an atrium with shops and a coffee bar. I was under strict instructions not to put one foot out of the premises because of the high rates of crime in Johannesburg. I was able to phone home and reassure my mother that I would be home for Christmas but I could not contact Gordon for some reason. When I returned to my room Gordon was sat on the bed reading a newspaper. Our chairman was concerned and had told Gordon to fly down to see that all was well.

He was surprised to see me up and about and we had a long chat about the school which I had deliberately shut out of my mind so as not to worry. There were all sorts of cards and get well wishes from the children and staff and Gordon wanted to know the diagnosis and prognosis. This was a very serious question for me. I was aware that the prognosis was not good and felt it my duty to give Gordon the facts. I told him that he should not bank on my return to the school. This may have been a gloomy outlook but I did not want the school to bank on my services any more.

On the positive side, the school had now been thrown into a situation where they had to go it alone which may have been no bad thing. Now was the chance to see if the legacy could be maintained. Gordon stayed with me for a couple of days after which I insisted that he return as the school was going to rely much more on him now. I was still waiting for the all clear from Dr Dalby which had not yet materialised. I was undergoing a battery of tests every day including three blood tests.

I was never keen to have needles stuck in me so often yet the nurse who administered the tests was brilliant and I rarely felt a thing. I must say that I found all the nurses on the African Continent very competent which was not always true of UK nurses I am afraid. When I had my first operation on my abscess one nurse tried to get a line into my hand and still failed after thirteen attempts until I could take no more. A junior doctor arrived to berate me for being difficult and I merely showed him the bloody mess all over my hand and suggested that the nurse should give up medicine and try butchery instead. From that point on I always insisted that the anaesthetist put in lines prior to an operation.

I was bored. There was a TV in my room but it had very limited channels so I redecorated and did a makeover of my flat in the UK. I thought it all through step-by-step and amused myself for some days forgetting to take into account the blunderings of B&Q that were to occur. Dr Dalby brought in a book about logging in West Africa in which he thought I might be interested and I was amazed to find that it had been written by Jock McCay, my forest prospector in Liberia, of which there was no mention. I liked talking to the staff and they seemed to have time to talk to the patients. How different from the UK?

I had lost weight and I was ordered to eat more as my legs were very stick-like and I needed no bidding as the food was excellent, completely unlike any hospital in which I had been a patient before, though Derriford in Plymouth was to excel as well. I complimented the caterer when she came around with the menus one morning and half an hour later all the kitchen staff assembled in my room and I had to give a little speech truly thanking them all for such a fine menu. They all left beaming and wishing me good luck.

One week later, I was given permission to travel and the nurses helped me pack. I said many genuine thank yous before I was put in a taxi and driven off to O.R. Tambo Airport. It was odd. I did not know Johannesburg well and it was rather eerie as we drove through obviously affluent housing areas which were all very quiet with very few people and I rarely saw any black people until we got to the airport. I was well looked after as I was flying business class as an assisted passenger on Swiss Air to Zurich, then on to Heathrow. Someone stayed with me for the whole of the time and I was conducted to my seat before the other passengers boarded.

The cabin crew were most solicitous and seemed to know all about me. I had a snack and soft drink and was then given more blankets and slept through most of the twelve-hour flight though I was aware that I was being checked from time to time. At Zurich Airport, Swiss efficiency took over and I was well looked after until I was placed on a British Airways flight. At this point I should mention that I had no warm clothing and was, of course, in business class at the front of the aircraft.

I had timed my arrival with dense fog at Heathrow and the flight was delayed and they left the doors open. I begged for a blanket that took ages to come and was actually told off by one of the stewardesses for travelling without an overcoat, so I told her my circumstances and she questioned whether I should be travelling. I asked if she was a qualified cardiologist to judge my case and she walked off.

The plane was delayed by two hours and I was whacked. We did get into Heathrow but there was no wheelchair or anyone to assist me and the stewardess informed me I would have to walk to the baggage hall. I did not make it. It was the cabin crew who found me and they called medics to help me and they suggested I go to hospital but someone arrived to help me with my baggage and I limped out into the arrival hall. With that one incident I have never travelled through Heathrow again and never flown British Airways again either.

I had arranged for my nephew, Tony, to pick me up and I had visions of falling asleep on the back seat of his Audi as we drove down to Devon. He was quite shocked when he saw me and left me sitting on my suitcase while he went for the car. He was ages. He had forgotten where he had left the car and it was not his Audi. It was his partner's Nissan Micra. We turned the heater on full as soon as I got in and I slowly began to thaw out and talk a little.

I could see that he was worried but I did manage to converse for much of the journey. We stopped at Sedgemoor Services where I bought him lunch although I could only manage a hot chocolate. We reached my flat at Yelverton late in the afternoon. I turned on all the heating that I had, full blast, and collapsed into an arm chair while Tony made drinks for us both. We had bought essential foodstuffs at the services so I was alright until the next day.

Tony was reluctant to leave as he was worried about my condition. I persuaded him to go as he had yet to reach Cornwall, then had a quick shower and collapsed into my huge double bed falling asleep immediately. I went to bed at 6 pm and did not wake up until my mobile rang at 8 am.

'David Longdon here. Can you come over to see me?'

I was befuddled after my long sleep but I wondered why the managing director of Brittany Ferries was calling me at this hour of the morning to request that I pop in to see him, because that was the only David Longdon that I knew and, how did he know I was back in England anyway? I suppose it was quite amusing really as the conversation progressed and I explained that I had only flown in the night before from Johannesburg after a heart attack and could not meet with anyone at the moment.

'I know all that,' was the reply, 'but can you get over this morning to see me?'

I asked for clarification and he must have thought my brain had been addled by the heart attack. I was totally and utterly confused. The call was not from Brittany Ferries but the local doctors' surgery at Yelverton and my new doctor, whom I had never met, was also David Longdon. I apologised for the confusion.

In truth I did not feel that I was up to getting up, let alone walking over to the surgery, and asked if it would be alright to see him the next day. He was adamant that it would not be and that he was coming over to see me straight away. Within five minutes, he was at the door and gave me a thorough examination with which he was generally satisfied. He was concerned, however, that I was living alone and was kind enough to give me a personal number in case of emergency and arranged for carers and nurses to help me three times a day. He advised me to take things easy for a few days to get over the flight. I asked him how he had known about the time of my arrival and he said that Dr Dalby had phoned him direct from South Africa to inform him of the details of my case and ETA. What a kind thought.

It was just before Christmas, for which I was not really in the mood, and I seemed to be in limbo for the festivities so I made a late booking for my mother and I to celebrate it at the Carlyon Bay Hotel near St Austell for a few nights. Over the last couple of years, she had found much happiness with a male companion or perhaps, boyfriend, called Cyril, who was a 6'4" ex-marine and very much a gentleman. They were both very fond of each other and Cyril had given my mother much happiness especially as she enjoyed driving around Cornwall in his big Rover saloon car.

We invited him over for Christmas Day lunch and the three of us visited the Eden Project on Boxing Day. Now I had been warned by Dr Dalby that my memory might not be up to par and he was right. I had hired a car, which was my usual practice, but not only could I not remember which carpark I had parked it in at the Eden Project but I could not remember the type of car I had hired. I had to work my way through the car parks, pressing the key until I found it.

As far as they were both concerned, I was merely recovering from a heart attack and I had given no other details as I did not want to scare them with any details about the matter. I was feeling much better and was anxious to see the specialist at Derriford Hospital in the new year. I was taking one day at a time. We were also able to visit my brother who appeared to be going through a good patch at that time and my niece and her two children and my nephew, who had driven me down from Heathrow and was now able to tease me about losing a car in a carpark.

My brother, Roger, held financial power of attorney for my mother, promising faithfully to look after her in my absence. There were, however, already early signs that he was being dictatorial with the powers but I was in no state to discuss that problem then. I had already understood that he was suffering from depression which explained his sudden mood swings and anger.

Early in the new year, I visited my cardiologist, Guy Haywood, who happened to be an ex-Mount House parent and I had taught all four of his daughters who were very talented in music. He was also much the poorer for their talent as they required musical instruments of considerable value as they progressed to concert standard. I had seen Guy briefly before Christmas and he had concurred with my doctor that I should rest until after the holiday. Thus commenced another battery of tests, the main one being a stress echo test performed by Dr Gareth Morgan Hughes which was vital to the final outcome.

As I understood it, the hunt was on to find grafting points. Guy knew that I would never be happy to lead a life of reduced quality with the 'Sword of Damocles' hanging above my head. If I could escape the original prognosis it would be fantastic. The process took some time but by the beginning of February, there was good news and bad news.

They had details of possible grafting sections on the LAD but the stress test showed a flat blood pressure response and a possible dysfunction of the left ventricle. There was also a new and extensive ischemic dysfunction of the left ventricle and a new and extensive inferoposterior wall motion abnormality that had developed. I underwent a CT angiography of the heart. This was an all or nothing test and I was nervous about the result. It established that there were grafting points they could use for a coronary artery bypass graft (CABG). I could have the double bypass operation. It was an enormous relief to me and gave me renewed hope. It probably sounds silly that I could be looking forward to a heart operation but the alternative was not to be considered.

I had been dealing with BUPA all the while, as each test had to be approved by them and it was quite difficult obtaining the permission for the operation as the process was full of minor obstacles, and I experienced some delays. I gained permission and sat back and waited for the date of the operation. There was one complication. I would be unable to return to my flat to be alone after the procedure. I looked around nursing homes and other alternatives and I was becoming increasingly concerned when my guardian angel obviously helped me out. My flat, number seven, Devon Tors was above the Devon Tors Pub which was run by great friends Pete and Kathy Gibbs. They knew of my predicament and kindly offered to take me in after the operation in their house on the outskirts of Plymouth, which was a kind and generous offer indeed and for which I was very grateful bearing in mind that Kath was a brilliant cook. How could I possibly refuse?

They say that trouble comes in threes and I was now hit with another huge problem. My mother was rushed to Treliske Hospital with a severe stroke. She had been out shopping with Cyril the day before and her eyesight had gone funny. By the time she was back in the flat, she was not well at all and an emergency doctor was called but did not turn up. They decided to wait until the morning which was a bad mistake. They should have called me or Roger, or better still, an ambulance. I believe that immediate help might have alleviated the severity of her stroke.

I was not informed until the next morning and drove straight down to Treliske to find Mum in a casualty cubicle, comfortable, but the stroke had paralysed the whole of her right side. Her face was not drooping and she could talk but her speech was unclear. They were about to move her into an assessment ward in the main hospital and I stayed with her until she was settled in, then promised to return the next morning.

I drove back, saddened and disheartened and found an urgent note from my doctor to phone him immediately. His phone was picked up right away and he asked how I was. I started to tell him about my mother but he already knew through the grapevine from the pub, I suspect. He was sympathetic about my mother's stroke. He was, however, more concerned about me. I was not supposed to be put under any stress and he was worried that the news about my mother might trigger a reaction.

I was his first patient the next morning over at the surgery and he gave me an ECG and checked me over. All was well and bless him, he monitored me every two days after that. David Longdon could be gruff and straight forward. He was also remarkably considerate and kind for he knew that I had to visit my mother every day which I dutifully did. My mother was moved into Phoenix Ward, the specialist unit for stroke victims where attempts were made to help her improve with simple physio and exercises. Mother was then 88 and was not really expected to fully recover.

I was also worried about her care on the ward. The staff were always too busy and the ward was untidy and messy. I hated my mother being in that situation. I met other relatives who expressed the same concerns and many of them brought food with them for their relatives and helped them to eat it, for the nurses were run off their feet. I took to arriving before lunch as well and fed my mother regularly. I received a ticking off from the sister about this. I merely requested that she guarantee that my mother be assisted with her food. She could not. I carried on.

I noticed that the lady in the next bed was not capable of feeding herself and for three days in a row her lunch was put before her then whisked away about an hour later untouched, with the comment, 'Not hungry today then?' I mentioned this to the sister who blew her top and told me to stop interfering. I asked for an apology. She refused. I made a formal complaint which was upheld and she had to apologise to me in front of the disciplinary committee—I was also able to express my concerns about the management of the ward.

We considered it best if my mother was moved to the St Austell Community Hospital where she stayed for two months and where the care was better. We had to make plans for her future and our deliberations were somewhat hampered by Cyril who had the firm conviction that Roger and I were planning to put Mum in a home, about which he became quite belligerent and upset. I had a serious talk with him one afternoon and informed him that all options were up for discussion and the final decision was to be made by my mother. Unfortunately, the hospital that my mother was now in was where Cyril's wife had died and that had not helped matters. I had other things to worry about, for now my operation was looming but I did not tell her until after the event.

My surgeon for the operation was James Kuo and I had already popped into the hospital for all the necessary pre-op tests before I met him on 26th February. As far I was concerned, he was my lifesaver and he carefully explained the procedure. He also pointed out that there was a risk with the operation of which I should be aware and I clearly informed him that this was a chance that I was happy to take. He asked to examine my chest then my back and also wished to look at the seton suture in my bottom.

With my shirt off and trousers down, I lay on his couch face downwards when Guy walked into the room. I looked at Guy and whispered, 'Guy? James does know where my heart is, doesn't he?'

I was still having regular checks with David Longdon and the operation was scheduled for 14th March. Between visits to my mother and Derriford, I had been arranging the makeover for my flat and had agreed that B&Q did the work, based on the helpfulness of their surveyor and initial contact. I was to have a new kitchen, new bathroom and my hallway floored, the work to be scheduled while I was away in hospital with the keys left in the Devon Tors Pub. Immediately I had paid the deposit and made all arrangements, I received a letter to say that their Southampton Office was now in charge of the work.

This had not been mentioned in any discussion in relation to the job and I was not happy. Furthermore, they intended to make a chain of some fifteen deliveries to my flat of all the white furniture and fittings required. Firstly, I would not be there and secondly, it was a small flat and there would be no room. I phoned the local store who were offhand and unhelpful though they did agree to store everything locally at their store and move it all out in one go just before work commenced. I carefully made notes about all this and my concerns, which proved invaluable later.

During the whole of this period, I had been in constant contact with the Roman Ridge School mainly through Gordon who had become aware that my prognosis had now changed. There was now hope of a return to work. This was good news for our chairman who had convinced himself that the stress and strain of running the school had caused my heart attack, so perhaps we should put the record straight. My heart attack was not caused by a dissolute life style or excessive drinking or high blood pressure or strain or stress. There was a deformity in the heart and this had led to all my problems. I phoned the chairman personally on this point and he seemed relieved. If it had been caused by strain or stress, I would not have contemplated going back to work.

On 13th March, Peter Gibbs very kindly drove me to Derriford Hospital in preparation for my operation the next morning. He was pleased that as a private patient, I had free carpark passes and duly delivered me to Meavy, the private ward, wished me luck and said goodbye. I was given a gown and told to body shave and shower and then waited reading on my bed until the chef came in. It was a private ward and he asked me what I would like for supper.

I was surprised and said that I was due for a bypass the next morning enquiring if I was allowed to eat. I was. I ordered a mixed grill followed by ice-cream and had a pot of tea as well. It was excellent as was all the fare that I was, eventually, to receive. Thoughts of, 'the condemned man ate a hearty supper', did pass through my mind. I was nervous. Of that there was no doubt and I did stay awake for some little while to be woken at 6 am to take a shower, and then I waited.

I was due to go down at 7 am. No one came. I waited. At 8 am the sister came in with the news that my operating team had been up all night with an emergency and could not operate on me that day. It was quite a blow and I must admit to being deeply upset and quite shocked. They let me rest for a while and then I had to phone Peter to come and collect me.

'Bloody hell!' he exclaimed. 'That was quick.'

I explained what had happened and returned to his house in Woolwell for lunch and Kath also left out a cold supper for me as they were working at the pub.

The next morning I was not feeling that good and thought it was probably nerves or slight shock from missing the op. When Pete and Kath came down, I dashed to the toilet to vomit. Now they say troubles come in threes and the symptoms I was now experiencing were identical to those in Ghana when I had my heart attack. Peter put me into the car because we were very close to Derriford Hospital and he rushed me into casualty, explaining my symptoms to reception. I was put in a wheelchair and rushed into one of their crash rooms, I think, and again connected to monitors, gadgets and machines and observed carefully.

I was sick again which was neatly caught by one of the nurses and we all waited. Everything seemed alright so I was disconnected and put into cubicle with a nurse in attendance who also had to conduct a blood test. She could not find a vein properly and blood spattered all over her and me, leaving us both in an awful mess. They had to remove my clothes AGAIN and put me into a gown. The nurse had to go and clean herself off and a doctor conducted the blood test.

I was worried and sweating profusely through the whole of this episode. They decided to keep me in under observation for the night and I was put on to one of the wards. I did not sleep as I was now worried that they would not operate on me on the following Tuesday and I was becoming quite distressed. A doctor arrived, gave me an injection and I slept. I was discharged the next morning and Peter came to collect me again.

I slept all that day and most of the next until Peter dropped me at Derriford again on 18th March, my operation having been rescheduled for 19th. We followed the same procedure and this time I was wheeled down to theatre at 7 am. I knew nothing of the operation and I was told afterwards that it had been over eight hours because of the grafting difficulties.

Several of my friends asked if I had not been worried about dying. I was nervous about the operation obviously. In reality, however, you would never know whether you died, would you? You would know nothing about it. I had been warned that after the operation, I would be placed in a special unit with dim and subdued lighting and monitored for some time. One also had a morphine button to deaden any pain, which you pressed as necessary. I had no conception of time or space and drifted in and out of sleep or unconsciousness, I am not sure which.

I was aware of being checked regularly and then I awoke to a cacophony of noise and bustle in a hospital bed in an ordinary ward. I was thirsty with a very dry throat but everyone seemed to be rushing around and I could attract no ones' attention. I could see the operation dressing on my chest and I was wired up again to a couple of machines and monitors. No one paid me any attention whatsoever. No one checked me or spoke to me for a good hour until I caught a male nurse's eye. I could not speak and pointed to my throat.

He got me some water and while doing so, I heard a strange gurgling noise from my chest which persisted. I was near panic, sipped the water and managed to explain my concern. The nurse listened and dashed off. A doctor came in and explained that the noise was normal after an op. Everything was just getting back to normal. Relief. After two hours, I was helped into a chair and told that I had missed breakfast so had to stick to water.

I gradually gathered that the ward sister had forgotten to put in her budget requirements and everyone was busy checking stocks and ordering anew and running around like headless chickens. Guy arrived and was not amused that I had been unattended and abandoned. I was to be moved straight down to Meavy immediately except that the ward sister would not allow me to be transported on an NHS bed. One had to be brought up from Meavy. Now how petty was that?

It is remarkable that many of us think that private medical treatment is preferable to the NHS and I suppose that bearing in mind the above incident, that could be true. Many wards on private wings, however, depend on moonlighting staff from the NHS and in my experience their care can be erratic because it lacks continuity and this particularly applied to the ward sister during the day on Meavy. She was charming and kind but scatty and shirked the more difficult medical tasks. Her nurses were far better than her and more professionally competent. I arrived exhausted after the operation, glad to be back in Meavy Ward again and the nurses settled me into my room, which was very pleasant indeed, it being bright and airy. I was given several cups of tea and actually ate a light lunch having had nothing to eat for well over twenty-four hours.

Recovering from such a serious operation was not going to be easy and I listened to all instructions very carefully. I was allowed to sit up but also had to rest flat and I could not roll on to my side and this prohibition lasted for two months. I was not allowed to lift weights, not even a full kettle, and had to exercise by walking up and down the Meavy Ward corridor two or three times a day. Stairs, initially, were taboo and I was unsure of anything I was to ask immediately.

I obeyed all instructions to the letter. I was surprised when the sister came in and asked if I wanted a shower and amazingly, she left me alone to have one in the bathroom in my room. It spoiled the wound dressing and messed up some of my attachments. The doctor was not amused and told me off in no uncertain terms until I told him that the sister had offered it. I heard him berating her from the other end of the corridor and he and the nurses took an hour to dry me down and off, put on a new dressing and ensure all was OK.

My supper was superb and I was also able to order breakfast. I was told that I could eat whatever I liked for the first two weeks. That suited me fine because the chef was brilliant. Guy came in to see me later and was surprised that I had not been disconnected from various devices and aids. By this time a strict Scottish sister had come on duty whom I found quite fierce at first but she proved to have a heart of gold and we became quite friendly.

After Guy had gone she and another nurse pushed an ominous looking trolley into my presence, which I eyed warily. The sister and nurse were good and talked me through each process as they gradually disconnected the drips and catheter and several other tubes without hurting me a bit. I was then warned that they were going to remove a rather large tube from my neck and it could be a painful process. It was, and brought tears to my eyes. I was then cleaned up and told to sleep and I was allowed a tablet to help me do so.

The next morning I was checked over by the Registrar who had assisted James Kuo with the operation. He checked the chest wound and then the stitches in my legs where they had extracted veins to put into my heart. The 'seamstress' who had put in well over forty stitches in both legs, though mainly in the left, had not done a good job it seems and most had to be replaced. I had visions of a general anaesthetic. This was not to be the case. He wheeled in an even more ominous trolley than last night's and then gave me five or six injections to numb my legs. He then proceeded to take out the original stitches and put new ones in.

I could not watch and asked to be laid flat so that I could not see the process. While stitching, he was complaining about his daughter who had won a scholarship to Devonport Girls' School, with which he was delighted, and the shortness of her skirts, with which he was not. Skirt length at the Roman Ridge School was not a problem as we were very strict with uniform requirements and I suggested a transfer for his daughter to Accra.

Ex-Mount House parents were constantly popping in and out and bringing in goodies, especially the ones working at Derriford and the sister was beginning to wonder how many other complaints I had as the staff kept popping in from all departments. Peter and Kath were constant visitors and Peter enjoyed using his free parking pass, which, he alleged, was his only reason for visiting me. Free parking at Derriford was a rare achievement, he thought.

Surprisingly, my brother came to see me with Bron, his wife, and Tony. I was quite taken aback and he stayed for well over an hour. I ordered coffee for them that was willingly supplied by the Meavy kitchen. Nine months later, I received a bill for £3.36

for the coffees!

I had to make an agreement with the nurses with whom I bribed with boxes of chocolates. If a visitor was outstaying their welcome, I would press my buzzer and complain about the dressing. The nurses would shoo everyone out saying that they needed to check it carefully. I seemed to be making satisfactory progress about which I was delighted. Before I could be discharged, however, I had to attend a meeting of the five or six bypass patients who had been operated on that week, so I was bundled into a wheelchair and taken up to the NHS Heart Unit on one of the higher floors. The meeting was very long-winded and should have taken half the time though I did appreciate the necessity of it. The lady was rather boring in her delivery and I think I must have started to daydream.

My reverie was interrupted by the comment, 'You cannot fly for six months.'

I queried this. Two of my Ghanaian friends had undergone the same operation last year and flew home after four weeks and I knew others who had been operated on in India and flown back in the same time at the expense of the NHS. The lady was adamant but wrong, as I confirmed later. I had to complete one final test before I was released, sorry, discharged. I had to ascend and descend a flight of steps which I did the next day.

Now, again, if you are squeamish, then look the other way. My heart surgery had entailed sawing through my sternum and opening up the rib cage to gain access to the heart. This was a normal procedure and one had to be very careful when convalescing that one did not endanger that repair. For that reason you were not allowed to drive for six weeks or lift even a light weight. It was recommended, that, while being driven, you held a cushion in front of you over which the seat-belt could be fixed.

Peter kindly arrived with a cushion and having offered my profuse thanks to all the staff on Meavy Ward, he started to drive me back to his house in Woolwell. I had to ask him to drive slower and not swing around corners as I was being forced across the seat with the G-force sensation. It was really most uncomfortable.

Thus commenced a rather unusual lifestyle for me as I was now almost completely inactive other than the light exercise regime I had to adopt. Before the operation, I was ordered to take things quietly but I could get around and drive and socialise. I was further inhibited by the after-effects of surgery in that my eyesight was marred sometimes by clear prism-like structures so I could not read much and my memory was not good. In those first few weeks, I was quite forgetful and experienced difficulties with my concentration. I had warned Peter and Kath about this in case they had to deal with anything, and they did.

I locked them out of their house late one night by leaving the catch on and retired to bed. It took them ages to wake me up. They had told me about the catch. It was a memory slip on my part. Sorry. I had always been an early morning person so I would get up early and have a shave and take a half bath. I was not allowed to shower yet to protect the dressing then had my coffee and breakfast while watching the news.

Peter and Kath would come down much later, as they worked in the pub until quite late, and eat their breakfast, after which came the important ceremony of 'the putting on of the socks', for I could not get mine on since the op and Peter had not been able to do so for quite some time, so Kath very kindly did the honours for both of us.

We were then ushered out of the door for the first of my bi-daily walks with Peter accompanying me on the orders of Kath. Peter had COPD and she thought a walk would do him good. There were few flat areas where they lived and we quickly exhausted those. As our walks were extended we met hills with which I coped quite well. Peter did not and I would often leave him sitting at the bus stop to regain his breath while I ambled on a little way.

The day that I had been dreading arrived. A nurse turned up to remove the multiple stitches in my legs. Peter and Kath knew her, but then, they seemed to know everyone, and the preparations were made. I had to be a brave chap as Peter and Kath were present. Many came out perfectly and some did not. My legs had swollen slightly and these stitches had to be stretched to cut them. I winced once or twice (understatement). I was very glad when it was all over. To this day I still get little sharp pains on the wounds like sharp pins and needles. Later, I also suffered from extreme night cramps and was told to wear shorts as much as possible as a cure. It worked.

Two minor events became major events in my little world. Firstly I reached the Spar shop barely a quarter of a mile from the house. The only problem being that I could not gain entry. Another taboo for me was pushing heavy glass doors. I waved to Sue inside the shop and she waved back. I pointed at the door and she waved again. I could have been there all day but for the kindness of a mum pushing her baby in a pram with two toddlers at her side. She opened the door for me and I had to explain my embarrassment to her.

I gave Sue 'what for' and she always sprang to the door after that. I do not overcome just small targets, however, for now I was going for Tesco superstore at Woolwell more than half a mile distant.

I made it after ten days and I felt as if I had conquered Everest. Medically, I was OK and regaining my strength and faculties. I had a small setback (another slightly gory bit coming up, I am afraid). The chest wound was oozing pus and I had to attend the Yelverton Surgery every day to have it drained. The nurses there were full of sympathy and were very gentle with the treatment.

It was on one such visit that I decided to go up to my flat, across the road, to see the completed makeover. I had been on the phone to B&Q and they had assured me that all was well and everything was finished. There had been a few minor problems which they had fixed free of charge, they added. I tentatively opened the front door and peeped inside. There was chaos throughout the flat. The old bathroom suite had been dumped on my bed, with an assortment of pipes and fittings, the new bathroom fittings were not fitted, the floor was half-complete and the kitchen had not even been started.

My antique mahogany dining room table had been used as saw bench and scratched beyond repair and all the new kitchen fittings were still in their packaging in my lounge. Of the contractor there was no sign and when I did manage to get hold of him, he was on board a Brittany Ferry heading for a holiday to France.

I phoned B&Q in Southampton and was told that I had probably made a mistake. How? I asked. I was standing in the flat with the shambles directly in front of me. I realised that I could not cope with the situation and informed B&Q of my circumstances and would have to phone off. I was phoned about ten minutes later by another person who stated that the contractor would be able to finish it the next day so I simply asked how he was going to do that long-distance from France.

I phoned one of the local solicitors, who was an ex-Mount House parent, and asked if he could help. He swung into action straight away and took all my documentation and then visited the flat to take photographs of the mess. He banned me from worrying about it and told me to leave everything to him. B&Q initially proposed that they would finish the work and make good all damage within two weeks.

I was adamant that they should never enter the flat again. I never knew the real background to the solution of the problem because my solicitor deliberately kept me out of the picture. My only contribution was to give him a list of my regular contractors and confirm my agreement to have central heating fitted which was not part of the original plan. Within two weeks, everything was complete. The new kitchen and bathroom fitted, the hall floor laid, the damaged furniture restored and the flat deep cleaned. B&Q had even settled my solicitor's fees. My Mount House friends had done me proud.

By mid-May, I was able to thank Peter and Kath for their generous hospitality and move back into 7 Devon Tors where they were still able to keep an eye on me from the pub below. Their generosity and kindness will be remembered forever.

I still could not drive. I had travelled around mainly in taxis and I discovered that I had to be careful which type I booked. It had to be a saloon car and not a traditional cab or a 'van' type as I would be thrown around at the back. I had tried a bus and had to get off at the very next stop because the ride was too rough. I had managed to talk to my mother through relatives' mobiles and I then travelled down to see her by rail. It was difficult and I will admit that I got back completely exhausted.

She was happy to see me and I was now able to concentrate on her problems and wishes. She was much incapacitated and would need help for the rest of her life as she needed to be lifted by two carers for all her needs. She wanted to go back to her flat and we moved heaven and earth for her to do so but the arrangements were slow and complicated. It was achieved eventually and she was looked after by two carers visiting her four times a day.

Her flat had a marvellous view of the harbour at Newquay and the events on the Killacourt. We managed to receive a grant to convert her bathroom into a wet room so that she could shower every day, and everything it contained was specially adapted for her needs. The family bought her a small power chair so that she could be mobile in the flat but it was a bit risky taking her into town and she demolished several shop displays in various establishments.

I received the all-clear towards the end of May from my surgeon, James Kuo and Guy, my cardiologist. I would never be able to thank them and their teams enough for saving my life and my lifestyle. I was also very grateful for the help of David Longdon and the Yelverton surgery team who were magnificent in my support and I had many friends who also helped and assisted me in a thousand ways. I value the continuation of my life and I am eager to use it for the benefit of others however I possibly can. I have been given a chance and am determined to use it.

Having been cardiologically 'resurrected', I understood there to be an air of nervousness at the school. Firstly I was making an unexpected return, for the odds against such an event were very much against me before Christmas, and here I was bouncing back after a few months. Secondly, several of our board of directors and governors were concerned that I was going to pop my clogs at the first sign of strain, not realising that I was now a recycled teenager.

Fafa and I had entered upon an agreement from the very start of our friendship, that, even if we disagreed and fell out for any reason we would still remain friends. And this was tested. As an assisted passenger, I flew out KLM from Bristol Airport, whose assisted passenger systems and handling were very poor indeed, and landed in Accra to be welcomed and even clapped by customs and immigration staff. What a welcome.

Alidu was with George, my houseboy. They were both very pleased to see me and had even gained permission to bring the school minibus right up onto the concourse for me, for they thought I was too weak to walk. The following morning at 7 am, I was at my usual place, at the school gates, welcoming the children, parents and staff and I had to be careful not to be hugged too hard and my right hand became very tired shaking hands with all and sundry. We had a whole school assembly and we thanked God for my return safe and sound. My office had remained as it was and apparently nothing had changed within the school.

The children were about to take exams at all levels and everything appeared to have run smoothly. There had been one or two disciplinary problems involving staff and pupils and one member of staff had been sacked for inappropriate behaviour. I could still feel and knew my school extremely well and I was up to speed by the end of the first day. Some of the 'vibes' and atmosphere had gone. It was not sparkling. It had lost some of its special atmosphere. I knew that activities had been cut and there was an undercurrent of over-concentration on academics with a decline of overall pastoral care.

If you had asked me how I knew this, it was because I was sensitive to the needs and ethos of the school. It was not fundamentally flawed but certainly ragged at the edges. I could feel it. I often visit schools in the UK and I can sense the atmosphere of the school as I walk in through the front door. Schools do have atmospheres and they are almost tangible. Visitors had always remarked on the lovely atmosphere at Roman Ridge but it had slipped.

There were also two major concerns. Against all my principles the school now closed early on a Friday once a month for staff training and it was always in seminar form. I held, and still do hold, that staff training should never be in the children's school time but should be in staff time. I am deeply opposed to British schools holding inset days during term time as I think it is morally wrong, especially as schools are complaining about the children themselves being absent in term time.

As far as the Roman Ridge was concerned, we had already established the seminar system was not successful for the training of our staff, which led on to the next concern. The recruitment of qualified and dedicated staff was a constant problem which we recognised in a top-heavy senior management team to help and advise our new staff. Fafa, fairly desperate for teachers, had taken on known colleagues from the Ghana International School. They were not a success.

They were elderly, set in their ways, were 'talk and chalk teachers', had many personal problems and did not believe in our aims and ethos. They ranged from mediocre to very poor on our scale of teaching. I complained to Fafa about them but she had autonomy in the senior school and wished to keep them. As principal, I still visited all classes and gave Fafa feedback notes on ALL her staff clearly stating that I was happy to help within the classroom and with the training of staff. The offer was rejected. This ultimately led the school into problems later in our short life.

I returned to my work with vigour and energy and apart from the problems above, we regained our ethos and activities so that we became a happy thriving community again. I was, however, still concerned about the succession and gave it considerable thought. My experiences of the last year made me acutely aware that I was but a mere mortal and anything could happen. I discussed this openly with Gordon and the board of directors and founders. The obvious and planned route was for Fafa to take over as principal and that was no problem to me except that she needed help, advice and backup in her position. Who was to provide this?

At this stage in our development, our founders, directors, governors and administration were very strong and firmly believed in our aims and ethos. In particular, Gordon Gopaldas was the 'rock' to which all could hold and I am very thankful that I made the right choice in approaching him as a founder. I held the rather simplistic view that the school should be stronger than any individual and was well-equipped to face the future.

I made the simple decision to retire at the end of the academic year 2008. It was a difficult decision but I would not be severing my links with the school. I was still a founder and remained on the board of directors. As a founder, I could still act as a watchdog to ensure there was no deviation from our aims and ethos and as a director I was also able to ensure the future financial path. I hastily add that at this juncture, the board of directors were more capable than I of directing the finances of the school and I had no concerns whatsoever about that but I did, and still do, have concerns about future financial investment possibilities for the development of the school and the form that this should take. The board is actively exploring possibilities as I write and we await developments.

I also promised the school that should any emergency arise then I would come running and I have kept to that promise. There was a slight problem with several of the parents as they thought I had resigned and it had to be clarified that I was retiring and at my insistence, my departure at the end of term should be a joyous departure not a sad one and so it was. I left Ghana in July of that year and returned to the UK to start life afresh and for a medical check-up. It would appear that the world of medicine was anxious to keep hold of me for a little while.

My heart operation also had a more lasting effect on me because of BUPA, for when I returned for the summer holiday, a bill for £36,000 had been sent to me from the NHS. Upon enquiry they informed me that BUPA had declined payment. I phoned BUPA quoting my corporate membership number and the reference of their approval for the operation. Apologies all around. It was a clerical error and the invoice would be settled immediately.

On my return to the UK at Easter, the amount was still unpaid and the NHS were getting edgy though I must say that they were understanding and helpful. Another call to BUPA. They checked and double-checked. Profuse apologies this time. They would deal with it within forty-eight hours and phoned me to state that it had been paid.

Three months later, I was threatened with legal action by the NHS for the amount had not been settled and I phoned straight through to BUPA requesting that I speak to a supervisor or manager urgently. I carefully explained my problem quoting all necessary references and explained that I had logged all my calls to them over the last year about this matter. The manager asked me to hold while he checked and came back on the line in a few minutes.

He had found out why the invoice had not been settled. I had changed my corporate membership to a personal membership over the phone and the new membership did not cover my operation. I explained that this was not possible. As CEO of the school, I controlled the BUPA account and had made no such call or alteration but he was adamant that I had. When, I enquired, had I made this phone call? Was it logged? He assured me that I had made it on 19th March and gave me the time as 11.45 am.

'Just remind me,' I asked. 'What was the date of my operation please?'

'Let me see. Oh yes. It was 19th March,' he replied.

The penny had not dropped for the poor man.

'So, if I am right in thinking, you are saying that halfway through my open heart surgery, I phoned BUPA to change my membership details. Is that correct? I was in theatre from 8 am until late in the afternoon.'

The poor chap became very flustered. I informed him that he and BUPA would be hearing from my solicitor with immediate effect and that I was sure the national press would be very interested in the remarkable patient who had made a phone call halfway through a double heart bypass operation. Amazing—don't you think?

My solicitor was on holiday so I used the same ex-Mount House parent that had dealt with the flat. You do not often hear solicitors chortle with glee but mine did. I gave him the details and said that I would drop in all relevant documentation to him that afternoon and we went through one or two more details over the phone. He relished the case.

As soon as I had replaced the receiver, BUPA were on the line and a senior executive at that. He felt sure that it was all a terrible mistake and everything could be sorted out immediately but I merely informed him to contact my solicitor and gave him the number. My solicitor put BUPA through the shredder. They paid the bill immediately and paid compensation to me for stress and for damaging my reputation, which was no small sum, and paid all legal fees and interest to the NHS. To this day I have no idea whether BUPA were just downright inefficient, stupid, trying it on or indulging in sharp practice?

Chapter Twenty-Nine
I Grow Old, I Grow Old

I shall wear the bottom of my trousers rolled. (T.S. Eliot)

I had retired officially. Fortunately, I was still very active and had no intentions of sliding into a life of senile decay, growing geraniums or quietly sitting on a park bench watching life go by. I was determined to grow old disgracefully and set about enjoying life as much as I could. Now in this respect, I had two friends, also retired headmasters, whom I had known for many years and with whom I often met up for a lunch or pub meal, and they were of the same mind.

Rather unimaginatively, we called ourselves The Three Musketeers and at one lunchtime session, we drew up a list in which we challenged or dared each other to complete various tasks which we all had to complete. The loser was to treat his companions to a meal at Thornbury Castle in Bristol. Each challenge was scored out of ten, marked by mutual discussion with the final decision by majority if necessary. I am sure you will appreciate that my colleagues should remain anonymous unless they decide to publish. The list?

1. Achieve and gain a world record in the Guinness Book of Records.
2. Visit a previously unvisited country outside of Europe.
3. Take part in a play, film or pantomime.
4. Complete a sporting feat.
5. Serve on, or chair a committee, for at least a year.
6. Complete a charity project.
7. Write a book or anthology or a learned paper.
8. Prove an exercise regime.
9. Lose ten pounds.
10. Life model for an art class.

This added much spice to our retirement. How did I fare? Watch this space though I have to say that many of the items were ticked off almost automatically as my retirement progressed. I also held a private BID List which I kept on completing so had to keep on forming new ones.

My mother was still residing in her flat in Newquay, with carers visiting her four times a day which was not very exciting for her but better than being in a home. My brother and his wife visited her every week though Roger rarely stayed with her for any period of time and would go off into the town. He still held power of attorney for my mother and strains between her and Roger were beginning to show. She was also visited by friends from the Cancer Research Committee on which she had served and the Bristol relatives would drive down to see her on a regular basis. She also had wonderful neighbours across the landing who were very kind to her and helpful.

Mainly, however, she spent much time with Cyril who was with her most days and they were obviously very fond of one another. For my part I had maintained almost daily telephone contact with her and was able to visit her regularly in the holidays and take her out, for Roger would never do so. He never touched her and never took her out at all, ever. I was thus now able to spend much more time with her and we would go off on local trips with John the taxi driver for lunches and teas. She enjoyed shopping trips and meeting people and like her niece, she liked garden centres, a trait which I thought, perhaps, ran in the family?

The only problems we faced during that time was dealing with hospital transport and the hospital itself that I labelled 'Terrible Treliske'. They seemed totally incapable of doing anything properly or well and the hospital transport was a nightmare. Booked ambulances would turn up very late or not at all and my mother could be kept waiting for hours for the return journey and even, on one visit, totally abandoned. We eventually took to letting John, our taxi man, take us over and back. He was reliable and helpful unlike Treliske.

Unbelievably, Treliske struggled to cope with disabled patients on day visits and were not very good in the wards either. Third-world hospitals had better facilities than Treliske. My mother had to undergo a minor operation to remove a cyst. It was delayed for five hours and my mother, who was supposed to be a priority patient, then disappeared.

I had no news or contact whatsoever from any of the nursing staff…absolutely nothing. She could have been dead for all I knew. I insisted that I was staying until I received confirmation that everything was OK. It was 11.30 pm before I was told that this was so and she had been up in a ward for some six or seven hours. I was furious. I was back at the hospital early next morning and located her ward and we struggled all day to get her home for she had to travel by ambulance and not minibus. I was amazed just how blunderingly inefficient the hospital was and booked a private ambulance to take her back.

On another visit, she was seeing an eye specialist. The department, again, could not cope with wheelchairs and suggested that she move to an ordinary seat. How? I asked. They assumed right from the start of the appointment that my mother was to have a cataract operation and as I pushed her from room to room, we were given leaflets about the operation and all the tests seemed to be geared towards that fact, which I thought was very odd because we had not seen the specialist yet. When we eventually saw him, after some considerable delay, with my mother still supposed to be a priority patient, he informed her that he could not perform an operation successfully. I pushed all the paperwork and details about cataracts onto his desk.

'Why have we been told that she was to have one and received all this paperwork and leaflets about cataract operations?' I politely requested.

He thought it was probably standard practice, giving no thought to the fact that my mother had been worrying about it for over five hours.

I resorted to another very strong letter of complaint and I did receive a reply full of platitudes and regrets and the possibility of staff training, which I dismissed as totally unacceptable and was given the pleasure of appearing before another hospital committee, where I let rip. I did not take prisoners and did not give an inch in my concerns for the manner in which they had treated my mother, me, as concerned relative, and their complete lack of facilities for the disabled. My only wish was that I never had an accident in Cornwall and would prefer Derriford instead. I would be happy to pay for the ambulance myself.

Medically, I was now due to have an operation for a couple of hernias under my private healthcare scheme, and was duly booked into the Nuffield Hospital at Plymouth for their removal by laparoscopic/keyhole surgery. The ground floor staff at Nuffield are all regulars and I knew most of them and the specialists and surgeons through Mount

House School. To prove my point, I knew not one member of staff on the upper floor who were relatively efficient with a cold manner. They appeared to be all moonlighters, again which proved to be the case.

My operation was due for early afternoon. Unfortunately, it was late in the afternoon before I was taken to theatre and operated upon. I returned to my room quite late to be informed that I had missed my supper. I politely told the nurse that I had not missed my supper—the hospital had made me miss it and my request for a sandwich was rejected. I had also been informed that I could be collected at 10.30 am the next morning and Peter had agreed to do this before he opened the pub.

There were delays with my discharge for some reason and I approached the sister on duty to enquire about my discharge. She told me off and tried to insist that I return to my room immediately in a very abrupt and rude manner. I stood my ground and insisted that she apologise. She flounced off, saying she was busy. I had to phone and tell Peter not to come because I was yet to be discharged. At midday the sister came in with my discharge papers and was, again, most abrupt to the point of rudeness.

I quietly informed her that she had greatly inconvenienced me and my designated transport and that I would be making a formal complaint about her rudeness and treatment of me. She did have the decency to try and make an apology but my surgeon was adamant that I should complain, which I did. I was actually visited by the administrator of the hospital at my flat upon receipt of my complaint because I had also sent a copy to BUPA.

She apologised and stated that the sister was not working for them any longer and gave me a generous voucher for M&S. I also mentioned the point about general staffing and most private hospitals reliance on part-time staff providing no continuity for care. It was a serious flaw and she agreed. Was anything done?

Unfortunately, the operation was not a success and I had to return to the Nuffield for open surgery. My reception was well conducted, the operation was on time and I was given a supper, except that it was cold. I was on the end room of the meal delivery rota and the meal and drinks were nearly stone cold…and this in a modern private hospital. I immediately complained and a hot meal was duly delivered. Thus thanks. The next morning the night sister was handing over to the day staff and I heard her say that my operation was completed by keyhole surgery.

I politely interrupted to inform her it was performed as open surgery and I had stitches. She still argued so I lifted my gown to reveal stitches. There was no apology. I was, however, discharged on time. She was probably glad to get rid of me. Without a doubt, that was the most painful operation I have ever experienced as regards aftereffects and I include my open-heart surgery, for the whole of my lower abdomen was swollen as were my genitalia and everything was very painful and sore.

I had to phone an emergency doctor to provide pain relief and was in agony for several days. I could hardly walk and I was not comfortable either sitting or resting flat. I called the emergency doctor again at the weekend for more painkillers and something to make me sleep for Peter thought I looked awful and needed rest. On the Monday, I phoned the surgeon who asked me to come in but this was just not possible in my present state so I was visited by David Longdon, my doctor, instead. The swelling took a good ten days to subside before I was able to visit my surgeon. He apologised for all the problems but it sometimes happened and I was the recipient this time.

I liked Tavistock. It is my favourite town. I knew many of the traders and they knew me and I was always almost certain to meet ex-Mount House pupils and parents whenever I was shopping. Everyone was friendly though one or two shop owners could be grumpy at times. I liked the market and its different stalls that seemed to vary each day and I was now happy to wander around aimlessly enjoying conversations and imbibing coffees.

I thought I was being attacked one morning at Donella's Coffee Shop in Paddon Row when I was suddenly surrounded by a group of excited young ladies and kissed and hugged by them all. Ex-Mount House girls, of course, all extremely pretty and delighted to see me as I was to see them again. The ex-Mount House boys were always much more formal in their approach. It is interesting to note nowadays, however, that even the boys will give you a half-hug as a greeting. It was interesting to hear all their news and achievements and I am old enough now to even have met their sons and daughters. I was delighted for them all. My annual Christmas card list is heavily weighted towards Mount House and the parents and pupils were most generous in their support for my charities and fundraisers.

I was drawn into two more activities which took up more of my time. The leaseholder of the flats in which I lived had always been unreliable and cavalier towards his legal obligations and requirements, failing to consult his tenants as was our right. His idea of being a leaseholder did not conform to mine or that of the government. Redecorating and building repairs would be tackled by his builder with no estimates or information to us tenants in the block and the work charged on an hourly basis. We faced huge bills as a result. This had rumbled on for some time but now I had the time to bring him to heel.

We formed a tenants' association, for which I was chairman, and he slowly started to conform to his obligations at our insistence and legal requests. Unfortunately, however, he did not conform to the consultation or estimate process the next time the exterior was to be redecorated. We warned him but he went ahead and was adding extra work to the rising bill. We called in a surveyor who deemed some of the work unnecessary and also questioned the cost, especially for the demolition of the chimneys which needed repair, not demolition. The bill was huge and the tenants refused to pay as he had not conformed to correct procedure and we had not been consulted about the work.

We took the case to the Lands Tribunal. The landlord was told, in no uncertain terms, that he was going to lose the case and was advised to settle out of court. We met and he admitted that he had been told to settle and offered a measly sum to be taken off the huge bill. The secretary and I had already agreed tactics, taking the line that we would rather gain a settlement through the Lands Tribunal and suggested that we bring the meeting to a close. He settled at a very satisfactory figure for every tenant and we also insisted that the management of the block be handed over to a proper management company. Tuffins took over the reins.

During this rather long-winded process, I had also been appointed as chairman of the Devon Schools' Sailing Association which had taught so many of the Mount House pupils to sail properly and well. It was going through a bad patch, was about to lose some funding and was facing a high rent hike for its premises on the Barbican. Ultimately it could have faced closure. I set about gathering the facts and figures as no one really had a very clear idea of all the various facets of the association.

I personally visited all stakeholders to take their facts and opinions, and advice if it was given, but I did not necessarily act upon the latter. I listened carefully to all members of the association and visited schools to assess the DSSA's contribution to their school life and the value to them of that contribution. Personality clashes, personal agendas and Youth Service politics were all involved. I gave no opinion—I merely listened. I had no axe to grind in any particular form. I merely wanted to help the association to achieve a solid financial footing so that it could continue.

The one facet about which I was passionate was that the association taught sailing brilliantly and generations of school children had learned to sail safely and well under its auspices. All stakeholders agreed on this one simple fact which was a good starting point for a new beginning. The one thing I lacked was statistics about our members, age groups and sailing sessions undertaken and I could not get a proper handle on expenses until a new member emerged who took on the task of finding these figures. His was a valuable contribution.

Modelling

All stakeholders could see that I was not confrontational, critical or disparaging in any way and this built up a consensus of cooperation and help. The threat of eviction was lifted, the enormous rent hike postponed, we applied for as many grants as we could possibly complete, took on a part-time manager and also gained a contract from Alternate Complimentary Education (ACE) for regular sailing sessions throughout the year.

We started a Saturday Morning Club for children, teenagers and adults and also held well attended open days. As a money-maker, I suggested getting a fleet of sit-on canoes for which I raised the funds by doing a sponsored super triathlon. I leapt 60ft out of a tree to start then jogged down the Tamar Valley to Calstock where I cycled on to Greenway Quay. I then canoed down river with the intention of paddling around to the Barbican along the Plymouth Waterfront accompanied by a rescue boat. The one thing you do not want when paddling down the Tamar is a southerly gale-force wind blowing up the Tamar.

Conditions were very rough indeed and the rescue boat called it a day at Saltash, so I mounted my bike and cycled around the Hoe to the Barbican to the accompaniment of thunder and lightning and a torrential downpour. I arrived at the boathouse looking like a drowned rat. Due to the splendid generosity of my sponsors, especially the Mount House parents, I raised over £2,500 which was matched by the British Canoe Union and the association acquired a dozen sit-on canoes which have been very well used.

I gave the association one year of my time by agreement as a thank you for teaching the Mount House pupils their sailing skills but then left them to manage for themselves which I believe they have. It was also a great pleasure to hear that Mike Duffield, who has worked tirelessly and devotedly for the association received an OBE for his dedication to schools' sailing. It was long overdue and well earned. Generations of school children will be grateful for his instruction and that of his splendid team of instructors. I leave the DSSA with one anecdote that occurred when the Mount House pupils first started their sailing instruction down on the Barbican. Barry, one of the slightly older instructors, was showing them how to rig an Oppie and suddenly realised that he had lost my pupils' attention. They were staring over at the smart marina across the other side. Barry enquired why.

Four or five of them said: 'Oh! That's my father's boat over there.'

'And that's my grandparents' yacht with the blue flag,' said another.

'I think my uncle keeps his cabin cruiser in that marina,' added another.

By a quirk of circumstance I was also launched into a modelling career. A friend taught art at Plymouth College of Art and I approached her about the best way of completing my dare of life modelling for an art group. She arranged this for me and I thus posed naked for an adult group for over two hours one evening which I found rather liberating, uplifting and quite enjoyable and there was no embarrassment on anyone's' part. It was very amusing because one adult arrived late and it was a Mount House parent.

She did a double take and I just smiled. I had swotted up on what might be required and the lecturer was pleased with my professionalism and poses. She had a contact with an agency in Truro who phoned me with the suggestion that I go on their books. I then was in gradual demand for portrait modelling as I evidently have an interesting face to paint and draw and I was then offered a session as a model for body painting.

This is hard because a full body paint can take up to six hours and can be quite tedious. My first sessions were with Josephine Vanoer, who is of some repute, and later I have been painted several times by Helen Eyre who lives in Frampton Cotterell in Bristol. As a Christmas joke, I sent my cardiologist and heart surgeon a picture of me as an anatomical model with the ribcage open, revealing the heart.

I was also in demand for costume posing for clubs and societies and I have now built this up into an art show where I pose in costume for half an hour, having started the session with a monologue based on the character. The characters are varied and include Samuel 'Peeps', a Pompous Pharaoh, a Native American Chief, a Hungry Cannibal, the Sugar Plump Fairy with a Take Off Problem, the Big Bad Wolf, 'Rough Old Feathers' and 'The Oldest Swinger in Town'. I enjoy photo shoots most of all as I prefer the fast action and poses required and you can be indoors or out. You have to think fast and obey the photographer's instructions to the letter. It can become quite fast and furious and one needs a good stock of poses as you can work at over 200 poses in an hour.

I now hold thousands of photographs in stock from these sessions. The oddest photo shoot I attended was at the British Camera Club at Kennington in South London where a photographer was using Victorian cameras to take pictures the really old-fashioned way. I now also work in films mainly in London and mainly on documentaries that seem to be released abroad. Apart from the 'Rushes', I have never seen myself in a film to this date. Work with personalities must remain confidential I am afraid as part of my contracts but I have filmed in all the Royal Palaces and fine houses of London including Mansion House, Apsley House, St James's Palace, Kensington Palace, Kew Palace and Queen Charlotte's Cottage as well as in all the major art galleries.

I also filmed once at Highgate Cemetery which was very interesting. Film work pays better than ordinary modelling and I was once paid £250 for a two-minute shot of me walking through the rain in Soho clutching a coloured umbrella. I also do voice-overs for information and educational films. The wettest shoot I ever worked on was on Bodmin Moor on Carbilly Tor where the heavens opened and the sponsor wanted to cancel the shoot. We were modelling mountain and outdoor clothing.

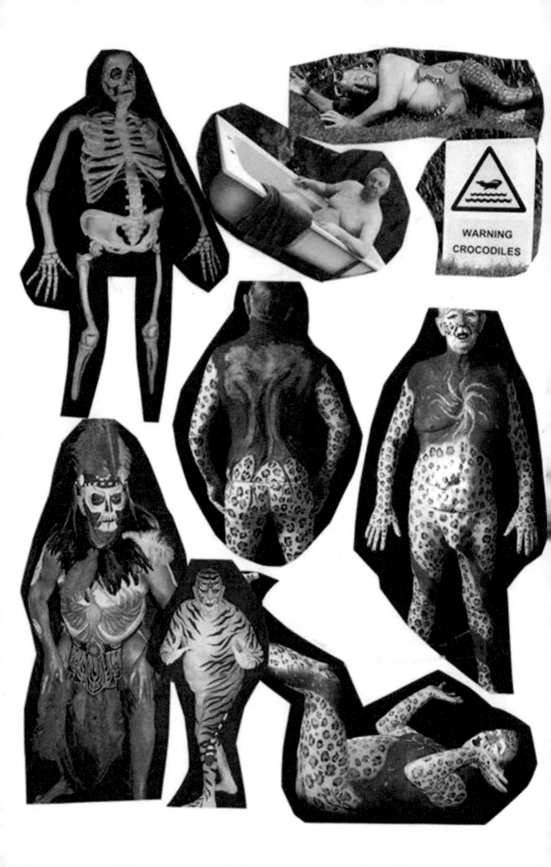

WARNING
CROCODILES

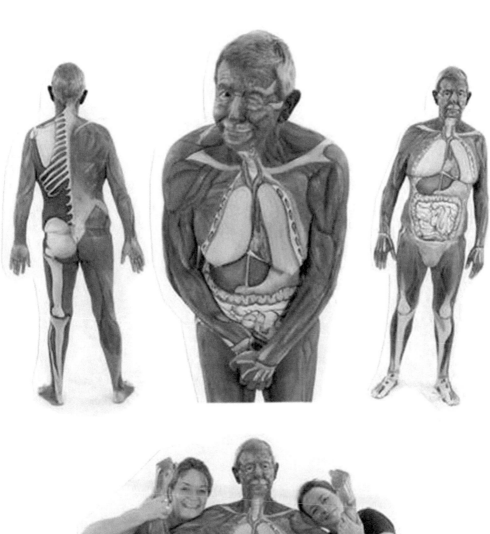

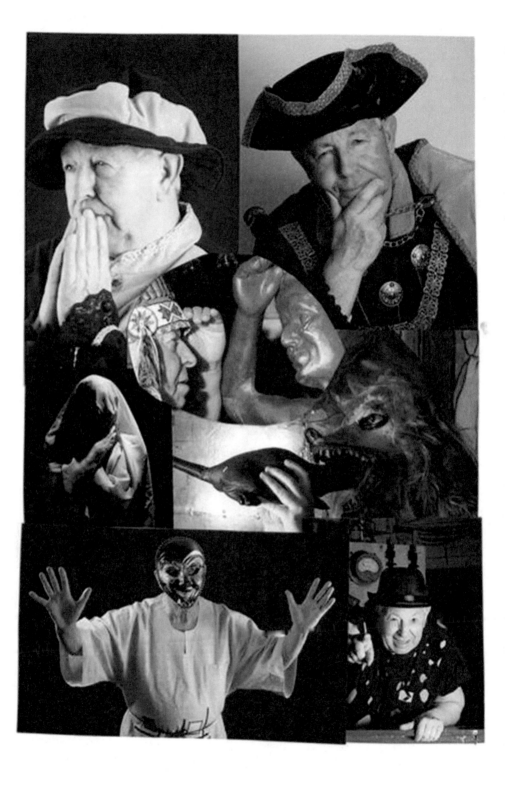

I am sometimes called upon to do specialised modelling for animation students where the poses are active and always on the move but in a repetitive pattern. and I also pose for sculpture classes because of my 'wonderful head'. Big head?

Are the pictures of me going to be worth thousands of pounds in the future? I was modelling portrait and costume at Cirencester a couple of years ago and I was dressed in a medieval velvet green outfit. A retired judge who was a pure portrait artist with some talent was in the class and grumbled about the costume which he had no need to paint as he could have concentrated on my head. By the end of the session, he had completed a very good painting of me and the costume.

We walked out together and he apologised for his grumpiness and I joked that it was probably an affliction associated with 'judging' as it were. A lady was quite taken by the picture and asked if it was for sale. The judge sold it for 300 guineas. Somewhere in the environs of Cirencester hangs a portrait of medieval me and I imagine it hanging in a great baronial hall peering down on a feast below. Then again, it could be locked away in someone's wardrobe.

This was not the record for a picture of me, however. A friend who owned a small art shop in Bristol was interested in photography and had set up a small studio above it on the second floor. He asked if I would come up one evening to model 'art nude' for him so that he could try out the studio and his equipment. For the layman 'art nude' is when you appear to be nude but you are not. You wear shorts or briefs to cover your 'bits', if you know what I mean.

He agreed that I could take the four best shots for my portfolio and he would take the rest. We worked for two hours and he was obviously pleased with everything. I then chose the four best shots and he groaned at each selection. The first one was a picture of me seated holding my knees and it was a beautiful shot cast in partial lighting. It was real quality and he hated parting with it. A year later he was writing a book on small studio photography and asked for the picture for the cover. The picture was under my copyright and I knew it was of high value. He had to pay a thousand pounds for the photo. We are still friends.

I am now on the books of Fusion Management in London though I still retain my original agent and I also work for some of the professional London photographers occasionally when they need an old grey haired model who is falling to bits!

I am mainly active in the Somerset and Devon areas where I am quite well known by local artists and tutors and continue modelling at art clubs, colleges and adult education groups as well as for individuals, and I find it interesting and challenging. This is the area where I perform my art show and have taken several bookings for children's groups recently.

I qualified as a senior appraiser while I was at The Roman Ridge and have used this qualification in the UK assessing senior management teams in business and education on a freelance basis. I have also been assisting the IAPS in the induction of new heads in preparatory schools in the South West. My main forte seems to be in the area of mentoring, however, and I have helped many students in many ways over the last few years including some from Mount House and The Roman Ridge School. I enjoy the work and am happy to assist in any way that I can.

I have had several health checks since I returned to the UK, with my cardiologist and with the BUPA Wellman Checks. All was well. I am still eternally grateful for the skills of my cardiologist, the surgeon and his team and the help of my local doctor without whom I would probably not be alive today. To mark my thanks I have, since that time never accepted presents or gifts from friends and family, preferring instead for them to donate to a charity of their choice as a practical thank you to the skill of the whole team at Derriford Hospital. It is a small token of my appreciation for the receipt of a great personal gift—my health.

Let us cast our minds back now to the great 'Headmasters' Challenge'. You may have noticed that I have been slowly ticking them off one by one. I had completed my charity work by being the chairman of the DSSA in Plymouth for one year and I had also served on the tenants' association committee as chairman. I had taken part in many films and also acted as a nude life model. I had also completed a sponsored triathlon for charity. Five down.

I was in the middle of compiling an anthology which was nearly complete and I had no problem with the exercise regime as I exercised most days by walking, swimming, cycling and canoeing. I was attending Weight Watchers and had already lost six pounds. I had two challenges to complete. Visit a foreign country not previously visited outside of Europe and achieve a world record. I suppose that I could have just booked a foreign holiday to complete the first but I felt that this might lose me marks. As luck would have it, I need not have worried.

Mounir Captan's daughter, Lara, was getting married in Lebanon and I was invited. Not only would it be a chance to catch up with my honorary brothers but it was going to be a big occasion and many of my friends would be there. It would also enable me to see something of Lebanon which I had not visited before and I was also scheduled to spend a few days in Tripoli, the Captan's hometown.

I arrived in Beirut late at night so had no impression of the city until the next morning when I was given a tour of the immediate coastline and the marina by two lovely ladies in a splendid Mercedes. I met Mounir for lunch which was typical Lebanese cuisine and very good. I had forgotten, however, that smoking was permitted in restaurants and that did slightly mar the taste. I walked back to the hotel to meet Lara's in-laws who were very hospitable and friendly and delighted that their son, Kareem, was marrying Lara. There was another big dinner that night and I realised that the weight loss that I had recently achieved was in great danger!

The next morning, at breakfast in my hotel, it was like a reunion for the Accra Polo Club with many friends having flown in for the wedding and it was quite late in the morning when I strode up to Mounir's hotel and met Lara, who was slightly nervous and excited at the same time, in one of the luxury suites on the top floor. She had met Kareem 'on-line' it would seem but the correct arrangements and protocols had all been properly observed after that and they were obviously in love.

What a wedding it was. There were hundreds of guests and the decorations were fabulous. Now Lara and Kareem had already had a formal and legal wedding a couple of days before and this was to be a celebratory occasion with a big party attached. Everyone was there and it was one grand reunion for me as I greeted friend after friend and we swopped news. Most had heard of my heart attack and were delighted to see me. I also met Nour. He had stayed in Tripoli until the last moment as Eid, his youngest brother, was suffering from dementia and he had driven down with him that evening to be at the celebration. Eid seemed quite well and mingled with all the guests and it was a pleasure to see him at the function.

Nour was also well and I was scheduled to join him in Tripoli for three or four days. The food was magnificent and there were cabarets, dancers and singers, all famous and very popular in Lebanon and we danced the night away. It was a fabulous celebration in every respect. I was aware of the cost of the wedding which was $800,000 because Lara had told me that she had questioned her future father-in-law about the amount prior to the event. He simply told her that she was not his daughter-in-law yet. After the wedding, she could then tell him off. Too late, of course.

The next afternoon, Mounir and I drove north to Tripoli up along the Corniche. I was slightly bemused to realise that the large expanses of what I took to be banana plants were actually hashish, grown quite openly on the side of the main road. It took us a little while to clear the suburbs north of Beirut especially as this was also the site of the port, before we drove into less populated areas and it only took an hour or so to arrive at Nour's house in Tripoli which was part of an apartment block owned by the Captan family.

I had always been aware that Nour belonged to a very large family indeed. Now, I thought that my family was quite large but compared to the Captans, we were a minor group. I had always been amused by Nour's attempts to explain his relationships to a particular relative or guest at the big house in Accra and he would go into long explanations about the person being a distant cousin on his mother's side who was partly related to another branch of the family who had also married his uncle's son, etc. Nour's family knowledge was encyclopaedic and he never seemed to forget who was who and their origins.

Members of the Captan family had served as ministers and high ranking officials in the various Lebanese governments in the past and seemed well connected in modern politics. Strange as it may seem, I knew Nour's sisters but had never met them as I had often talked to them on the phone trying to contact either Mounir or Nour when they were in Lebanon either in Tripoli or Beirut. Marianne, Leila and Jasmine were familiar to me from their voices only and it was a great pleasure to meet them in person. I was also able to meet their husbands and it was to them that Nour turned for advice about security.

Lebanon seemed fairly peaceful at the time of my visit but even I was fully aware of the 'no-go' areas in Beirut which also applied to the whole of the country south of the city controlled by Hezbollah, the Bekaa Valley and the area to the north of Tripoli. Personal security was always a concern and on several outings we were accompanied by armed security just in case.

Tripoli is well spread out but it is normally quite easy to walk around its environs. With a population of 250,000, it is the heartland of the Sunni Muslim Community and it is famed for its Mamluk period style of architecture. It was founded by the Phoenicians in the ninth century and became part of the coalition of trading centres including Sidon, Tyre and Arados—the three cities, hence the name Tripolis (Three Cities). It was conquered and occupied by the Assyrians, Persians, Greeks, Romans, Byzantines, Arabs, Crusaders and Ottomans. The city was badly damaged by the occupation of the Crusaders in 1109 and they burnt down the then famous Dar al-Ilm Library, though they left a magnificent castle. In 1289, Sultan al-Mansur Qalawun sacked the Knights and destroyed the city.

Modern-day Tripoli suffered a heavy Syrian bombardment in 1983 during the civil war and endured fighting between various Palestinian factions during that conflict. The two huge refugee camps of Beddawi and Nahr al-Bared are to the north of the city and the latter became a war zone as fighting broke out between the Lebanese Army and the militant group Fatah al-Islam. During my visit the city was full of refugees from the Syrian conflict and it was obvious that Lebanon was supporting a huge influx of them almost swamping their small population.

During my visit, I was able to visit the Hammams (Baths) of Hammam al-Abed with its beautiful domes, Hammam al-Nouri, and the largest baths of Hammam al-Jadid which was very spacious with fountains and courtyards. I was particularly interested in the Madrasa of Al-Quartawiyat with its stunning decoration. It was like no other 'school' that I had seen before, and Al-Tuwashiyat Madrasa with its plainer sandstone and less ornamentation, was much smaller but had a pleasant atmosphere. I was allowed to visit the Al-Mualla, the Burtasiya and the Great Mosque.

My favourite was the Taynal Mosque, however, which was quite magnificent with its courtyards and prayer halls. I could not go inside the Citadel of Raymonde de Saint-Gilles for security reasons but it was set up on a hill dominating the old city and looking down on to the Abu Ali River. We wandered the old parts of the city looking at khans and souks and I enjoyed the hustle and bustle of the narrow crowded streets with their wide selection of goods and trades.

On another morning, we walked along the seafront to the marina and drank coffee from a street seller. The port was active and working, far less grand than the marina in Beirut with its floating gin palaces and superyachts. I was surprised to see an almost derelict trade fair site on the foreshore. This was the Rachid Karami International Fair site designed by the Brazilian architect, Oscar Niemeyer, of Brasilia fame, and was intended as a major international commercial and exhibition space. It was a definite blot on the horizon now. Nour's sisters spoiled me terribly and I was really worried about my waistline now as they cooked delicious Lebanese dishes that were all far too tempting. The fruit and vegetables in Lebanon are the best that I have ever seen or tasted anywhere and it was very hard to resist or cut down. At least that is my excuse.

I also like seafood and my appetite for same was assuaged in the beautiful little port town of Byblos which we visited one day to meet Turkish friends driving up from Beirut. The guidebook on Lebanon quotes an old saying: 'The more things change, the more they stay the same.'

This certainly applies to Byblos. Its history dates back for 7,000 years and it is a contender for the oldest town in the world. The delightful port and surrounds still evoke the atmosphere of its previous inhabitants, namely the Phoenicians, Greeks, Romans and Crusaders. The port was important for its paper trade during the Greek period, Byblos being the Greek word for papyrus/paper. It is also famous for its gift to us of our alphabet which was then only 22 letters. As long ago as 3000BC Byblos was exporting Lebanese cedarwood to Egypt through the port and it has a long history of sea trade.

It is also important for its huge archaeological site which covers many acres and many historical periods of settlement, the finest of which is the Crusader Castle, Chateau de la Mer, which commands and domineers the approaches to the port. I spent an interesting hour climbing through its ruins before exiting to the Phoenician ramparts of the 2nd and 3rd Centuries which can still be clearly seen. There is a charming Roman theatre right on the edge of the coast and further Roman remains take the form of the Roman Colonnades and the Temple of Baalat Gebal dedicated to the goddess Astarte.

The site is enormous and contains much more such as the Temple of the Obelisks built over a predecessor, the L-Shaped Temple, the king's Spring and the Royal Necropolis where nine tombs have been uncovered so far. The site also contains prehistoric and medieval settlements and remains. The town is a veritable goldmine of history spreading over many thousands of years. The archaeology is superb and so was my seafood lunch which I felt that I had truly earned, having been historically saturated throughout the morning.

Drinking my after-lunch coffee at our table directly overlooking the harbour I did think that it might be almost the perfect place to retire as I imagined myself basking in the sun on the harbour wall looking out to sea. I really enjoyed my first trip to Lebanon and it is really quite easy to get to with no visas to worry about. The hospitality of the Lebanese is very generous and I would happily visit again. This visit meant that I had now ticked off the ninth item on the challenge.

The last was proving difficult. I had to gain a world record. It dropped into my lap. I was doing a sculpture pose for a private client and we had been discussing the challenge by way of conversation when she mentioned that she was attending the Midsummer Skinny Dip in South Wales, to raise funds for Marie Curie Cancer Care. It was to be a world record attempt to have the most number of people skinny-dipping at any one time. I was on and made all necessary arrangements to take part.

Thus, on 19 June 2011, 412 people and I turned up very early one Sunday morning at Llangennith, on the Gower, to shed our clothes and dash into the water. Actually, it was not as easy as that because it was all very well organised and the number of people had to be carefully counted. We also had to be naked in the water up to our waists at least for ten minutes.

It was chilly and the surf was quite high. I am now the proud possessor of a certificate from 'Guinness World Records' stating that I had participated in the greatest number of people skinny dipping at a single venue, which is 413, and was achieved by Midsummer Skinny Dip in Llangennith in South Wales, UK, on 19 June 2011. I had cracked it and I was the first to complete the challenge with a high score of 90%.

On 10th August this year, I was filming in London with our regular crew who were excited because they had all signed up for 'Streak for Tigers' at London Zoo. My producer bet me £300 that I would not join them. I accepted the bet only if he would agree to take us all out for dinner after the event. I streaked. He lost. The dinner was excellent.

Now I had also bared all for tigers thinking, perhaps, that if a tiger did see me naked, it would probably run away. My mother was coping in her flat in Newquay, though only just. Unfortunately, Cyril had been diagnosed with cancer and was receiving treatment and yet he still struggled over to see my mother whenever he could. There were very happy together until he became very weak and died in Treliske some months later.

My mother was heartbroken and we were all very sad. For some reason, she did not want to go to the funeral and I think she wanted to remember Cyril as he was, retaining her fond memories. Unfortunately, she now deteriorated and really became quite ill requiring night care as well as help during the day. We employed night nurses when we could but I would often have to drive down to stay with her during the night which was very wearing and the mileage on my car was huge. Her neighbours were very helpful and deserve our gratitude.

The situation could not continue and I was warned by my doctor to stop the night journeys as there could be consequences to my own health. My mother had to go into respite at Newquay Nursing Home for a while. I was not enamoured of the facilities and felt very guilty that she was in such a place. She did, however, start to make good progress and we were thinking in terms of her return to her beloved flat, until the company providing her care discontinued the contract, giving no reason or explanation, and we could not find another provider.

My mother was stuck and we had to arrange her long-term care at the home which entailed the sale of her flat, many possessions and no hope of a return. She would have to contribute to her care from the amount raised by the sale for a long time to come.

Unfortunately, my brother tried to rush the whole process and my mother needed a bit of time of time to adjust to everything, a bit of space to think things through and come to terms with what was happening. He had completely forgotten that his power of attorney did not confer dictatorial powers on him and that he was expected to consult with my mother and the family as clearly set out by the Office of the Public Guardian (OPG).

It was not a good situation and I had to rescue many of my mother's personal possessions and some of her more valuable furniture in his rush to strip the flat, which I later sold to give her a small cash supply, informing my brother of the amount in hand and pointing out that she would like to keep this to one side for emergencies. He immediately stopped withdrawing my mother's weekly petty cash from the bank against her explicit instructions.

He was also keen to give her motorised chair away which annoyed the family as they had bought it for my mother. My mother insisted that it be donated to the cancer charity shop where she had been a volunteer. The flat was stripped bare for interior repainting and even the curtains were removed. Now, I was still chairman of the Devon Tors Residents' Association and when a flat was empty, the curtains were left up so that people would think it was still occupied.

Roger insisted that the estate agent had given him the instruction, which was not so as I checked. All his anger and obstreperous behaviour was caused by his problems with depression but he would never accept help or advice from anyone and blundered on like a bull at a gate. My mother was deeply resentful and Roger's impetuous and unfeeling actions had hurt her a good deal.

Every time I visited her, she was weeping and sobbing, extremely upset and very emotional. One week she and Roger had a row and I think that was the straw that broke the camel's back, especially as one of my mother's friends was present and took Roger's side without knowing any of the facts. I tried to discuss the problem with him to no avail, declaring that it was none of my business.

The rest of his family tried to help and he lost his temper. Their weak relationship had broken down completely and Roger was refusing to follow my mother's wishes and instructions about her money. He was still withholding her cash and her emergency fund was fast vanishing. I had discussed the problem with friends and received the old adage of a reply that families always had their problems, you know. We tried to sort it out with the help of the family solicitor who had known Roger and his anger for years as he was a neighbour, who came to the rather odd conclusion that Roger and I were fighting each other because of bad blood and not talking to each other.

I wrote a reply to say that I was perfectly prepared to talk to Roger but Roger was not prepared to talk to me…or anyone. His legal opinion was slightly awry. I approached a local solicitor and explained my difficulty. I did not want to suggest anything directly to my mother in case anyone thought I was trying to influence her views. I was aware that she had options which she needed to discuss independently while being offered professional advice and she had to reach her own decision.

It was tough on her but she was absolutely, strongly and emphatically resolute to change the power of attorney into my name as soon as possible and she had already instructed the solicitor to proceed before informing me. I fully appreciated that my brother would be upset to the point of being incandescently angry about this and I was sorry that his depression had led my mother to take this step. I felt really sorry for him.

I knew that he could not help being what he was and how he was, but the depression from which he suffered had caused him much misery and isolation. Friends and family in the UK will perhaps be surprised to hear that I do pray every day, though my children and staff at the school appreciate that I do, and Roger is always included in them with the hope that he will allow himself to be helped. The barriers that he has erected could be overcome even now.

The legal process rolled on amid countless forms and procedures and I was already aware that Roger would try all in his power to block it so I had to parry his every move. He tried to force the rest of his family not to see my mother and actually forbade Tony from visiting his house because he had visited her. My mother was enraged when he approached her doctor and stated that she did not have the capacity to make the decision, and the doctor, without consulting anyone else in the family or my mother, arranged to have her capacity tested.

I thought this most unprofessional and told him so. The change of power of attorney was finally granted, then came the difficult task of retrieving my mother's accounts, documentation, bank details, etc. I am afraid that he had to be threatened with a solicitor's letter to hand them over or the matter would be put into the hands of the police as a criminal complaint. They were all dropped at my accountants. Since then, he has cut off all communications with my mother and me and all the family in Bristol. He is very fond of his grandchildren but otherwise leads a fairly solitary life in St Austell. I wish him well and my prayers are with him.

My mother was moved out of the Newquay Nursing Home several years ago. It was never altogether satisfactory and was often understaffed and quite cramped and dark. My mother was not thriving there and one of my cousins suggested that we bring her up to Bristol where she would be with the family and receive more visitors. I visited ten or eleven homes to the north of the city which generally left me quite depressed and I could not imagine myself in any of them. We eventually located Woodlands in the Kendleshire area which seemed quite good, came recommended, had space and light and where they ran activities for the residents. The food was also good.

I made no attempt to persuade my mother and I think her decision to move up to Bristol was probably mainly because she had become nervous of Roger and his anger some time ago and was happy to move away. It proved to be a good move in most respects and she has now been there for some years. The family would prefer, of course, that she not be in a home and I have experienced all the various guilty feelings of her being in one. She is safe, however, and my anxiety levels have been assuaged knowing that she is not at risk.

We brought up her small electric mobile chair that I had purchased for her in Newquay and this gave her some independence within the home though it must be admitted that she was not that a good a driver of the same and caused minor damage in the corridors and broke her foot in a doorway. On the last occasion she used it, one of the carers had wrapped my mother's buzzer cable around the arm. My mother decided that it was lunch time and just drove off, not realising that the cable was still attached. She was on the second floor and trundled along the corridor to the lift, opened the doors, closed them again and pressed the button for the ground floor with half the buzzer cable still on the second floor. She was stuck in the lift for a couple of hours. I am afraid that was the demise of the chair and we moved her down to a ground-floor room.

I also purchased 'Emily' for her which was a vehicle converted for wheelchair use and which she adores. I located Emily in deepest, darkest Cornwall going for a bargain price and drove her back to give my mother a big surprise. Emily was purchased out of frustration with local transport schemes and extortionate taxi fares for the use of their mobility vehicles. The local minibuses obviously do good work but they could be unreliable and it was very difficult to make bookings and the local taxi companies fleeced the disabled and immobile for large sums. For a trip of one mile over to Sainsbury's, they were charging £15 one way and on Christmas Day, they charged me over one hundred pounds for a drop-off and pick-up for less than two miles.

Their fares and service need investigating, especially as they did not offer a proper service at weekends. Emily is a character as she has foibles, tricks and idiosyncrasies for she is ever playing tricks. She will refuse to start some mornings and you have to lure her into ignition but she always starts when my mother is using her. Her doors will suddenly close on you and bang you on the backside and her lighting circuits go on and off at will. Her best trick is turning on her interior light during the night and running the battery low but she has enabled me to get my mother out and about and we are able to visit Bristol and the local area at our convenience, not relying on taxi cowboys.

She loved seeing the harbour area again and we even got her around the SS Great Britain. We also visited St Mary Redcliffe which she much enjoyed though was very disappointed that she could not get a coffee as their restaurant is not wheelchair-friendly. We often picnic on the sea walls, high up on the Downs, and she likes touring the University Botanical Gardens which does have a coffee bar.

Most Saturday mornings, I collect her and we go up to the Daisy Chain Coffee Bar in Frampton Cotterell whose owners, staff and customers all spoil her and she likes seeing the families and children. The noise and hustle and bustle seem not to worry her. We do sometimes visit the Kendleshire Golf Club where the manager, staff and golfers are always most hospitable. Sundays we are normally at her favourite nieces for lunch with the Helliker Family who are to be included in this tome. In Bristol, she is having a much better quality of life and there is plenty going on in the home. Every Monday she has her hair done and then there are pamper days, massage, film mornings, painting and craft, as well as Barbecues, fairs, singers and singing, bingo, pantomimes and nail bars.

At 98, she is suffering from cataracts and macular degeneration so is now registered as partially sighted and her short-term memory is not what it was. She is easily upset and very often has fixations about past events. She also nods off quite a lot but we all love her and are praying that she will stay with us long enough to receive the telegram from the queen on her 100th birthday. She is a very special person, as all mothers are, and I am amazed at her bravery and courage over the difficult years she has experienced since the stroke. She has endured much and I am one of her greatest fans.

Time condenses when you get to my age and I suppose that is to be expected but I do know that I have been out to Ghana three times since my retirement. Once for a personal trip to see friends, once for the celebration of the school's tenth anniversary and once by urgent request. When I left the Roman Ridge School, Fafa Awoonor was appointed principal and Valerie Mainoo was running the junior school and she was a safe pair of hands. Fafa had no experience as a CEO of the business side of the school so Gordon Gopaldas was running that extremely well.

Fafa had the hardest task of spearheading the growth of the senior school through IGCSEs to the A Level programme in the sixth form. Her struggle to find suitable, appropriate and qualified staff was extremely difficult and the academic challenge of developing all the educational programmes was a Herculean task. She had also lost a valuable director of studies who was joining her husband on another foreign posting for DYFID. Janet West had been very competent in all aspects of curriculum development and had established a very strong foundation for our academic success.

I regret that due to my heart attack, I had not been able to thank her properly for her contribution to the school. I do not think that the school realised that Fafa was under so much pressure and stress and things came to a head concerning a difference of opinion between the governors and Fafa. At this point it was thought best that I make an appearance in case any emergency arose so I dropped everything and flew out as a possible backup, hoping that my appearance might diffuse a difficult situation.

I was not sure of that as Fafa was a mighty feisty and determined lady who maintained strong principles. We always met in friendship whatever our differences and I realised that she had been deeply upset by a board of governor's decision relating to a pupil. It was a difficult moment which was going to test the school. Even during my tenure I would not have been able to go against a ruling from the governors and nor could Fafa, as we had jointly set up the governing body to oversee the school and an appeal procedure that had worked, all be it against her personal views.

I sympathised and appreciated her feelings but the ruling could not be overturned. The other problem was that individual personality differences had surfaced and had muddied the waters somewhat. I have completed much work with governing bodies in the UK since then and my first job was always to take the 'personality' out of the equation. It is wise and better to do so. In reality the problem should, perhaps, have never reached the stage that it did. I would never have allowed its progression. Diffusion at an early stage had been called for. It had not been applied.

However, there had been many times, in board meetings in particular, when I had disagreed with individual members and even the whole board but I had to abide by their decision. Founder or not, it was my job. Indeed my chairman and I had quite lively disagreements at times which have not affected our relationship or friendship. We were both passionately involved with the school. It was a source of great sorrow to me that Fafa decided to resign. She had worked so hard to make the school a success and contributed much to its foundation. After a long and proper interview process, Valerie Mainoo took over as principal and was able to build up a strong management team to support her with the realisation that the senior school staff needed much direct help and encouragement.

Her style of leadership is different from mine. I was a typical example of an independent school head that lived for his school and was very practical and deeply involved. I was passionate about all aspects of the Roman Ridge School but also recognised the practicalities facing us and firmly believed in directly helping my teachers with practical solutions especially in the classroom situation. My strengths lay not only in my leadership skills but as a good classroom practitioner. I was also, of course, a grey-haired revolutionary, which helped.

Valerie's style is light and quite intellectual yet she handles the school well and I am very impressed with her friendly and sympathetic treatment of the pupils and staff. She is a good communicator and gets on well with all stake holders in the school. It is a successful style and she also knows every pupil. We meet several times a year and we manage to discuss problems and management and she is well aware that I would prefer the management to be more hands-on in the classrooms with direct help and support for the staff who would welcome helpful and supportive intervention. I call it 'nudging' and it worked fantastically well in the junior school.

Our senior management team is designed to be top-heavy in our recognition that our staff require such help in many aspects of their teaching. The Roman Ridge School prides itself on the individual attention given to our pupils and we should operate the same policy with our staff. I once attended a senior Maths class in which the teacher was going through the exam paper with the whole class. Some pupils had scored very highly on the paper and I questioned why they should have to suffer the tedious process of going through the paper with the lower scoring pupils. They already knew and understood.

The teacher was wasting their time. The seminar method of staff training can also fall into this trap. You may be involving all staff in a session that many of them already know and are competent in that skill or subject matter. Individual 'nudging' is far more efficient and practical in most circumstances and competent knowledge and expertise can be passed on directly to those staff who are in need of it.

I flew out to Accra and the school to celebrate its tenth anniversary in June 2012. This was a wonderful and joyous occasion which I much enjoyed and it was wonderful to see all the pupils, staff and parents celebrating our first major milestone. It was combined with our annual prize-giving and graduation ceremony in which many of our founding pupils were graduating, which gave me much pleasure. The 'magic' of the Roman Ridge School was still intact and being carefully preserved. It was a wonderful occasion and I enjoyed it all the more because I did not have to organise it! I just sat back and enjoyed the delightful event.

I have always been aware of my unusual role at the school. Having seen and dealt with several examples of retiring heads meddling and interfering with the establishment from which they have retired, I was anxious not to be accused of same. As a founder, however, I do hold a role in the protection of our aims and ethos and I am still a member of the board of directors. The latter role is not onerous but the board now is actively seeking a way forward to develop and expand the school. I hold some, not inconsiderable knowledge of the educational world in general and am wary that there are many pitfalls in selecting the best way forward so keep an eye on any developments.

The board of directors and I am very keen that the school moves forward but not at the expense of our aims and ethos or the destruction of our special atmosphere. Valerie Mainoo handles my role and me very well indeed and though we may discuss problems and difficulties my response is generally based on normal good educational practice that I would offer my clients in the UK. I hope that my advice has been of some value.

My last visit was in June in 2016 and my main purpose was to visit the senior school speech and prize-giving day and graduation ceremony to officially launch the Ridge Readers Project that I and the pupils had started. The present literacy rate in Ghana is estimated to be 88% for men and 83% for women. Furthermore, a figure of 89% has been the number of children attending primary school education which means that a possible 11% receive no education at all.

Many Ghanaian children do not get the opportunity to read, speak or write good English and experience difficulties with comprehension. Many schools have a weak reading and oral culture. There is an obvious need for help in this area and our senior pupils have stepped up to the mark by accepting my challenge to provide assistance to local schools in Accra and also attempt to reach out to other groups such as the deprived, disabled, and disadvantaged children and adults as well as attempting to address the gender imbalance.

Our aims are simple. The pupils will take out suitable reading books to all of the above and spend valuable time with them helping with their reading and also discussing the contents. By this method we hope to improve reading, comprehension and oral skills all at the same time. The book becomes the property of the recipient, being clearly marked with the owner's name, with the hope that it will go back into families, thus widening its influence and effect. Many skills in Ghana are passed on among siblings and who knows, it might also reach illiterate adults as well. We had thought to start a registered charity for this project but this would entail much expense, so the school has agreed to keep the project in-house, and our promise to all donors is that:

ALL MONIES RAISED AND DONATED WILL BE USED SOLELY FOR THE PURCHASE OF BOOKS FOR THE PROJECT AND NOT FOR ANY OTHER EXPENSES AT ALL.

The project is now underway and further expansion is envisaged and planned. At the launch, the senior pupils were in Ridge Readers T-shirts and I performed the following rap (another of my many talents):

> *So everybody now, you've just gotta listen*
> *For these pupils here are gonna make Ridge Readers glisten.*
> *Ridge Readers now heading down the track*
> *And we're all determined to make it crack.*
> *We're gonna teach the reading. Provide the WOW!*
> *Teaching weak readers the what and the how.*
> *We're gonna talk the talk and strut our stuff*
> *It ain't gonna be easy—it's gonna to be tough.*
> *We're out to teach the love of reading*
> *And our Sixth Form is good at succeeding.*
> *We're gonna provide the books and that's the mission,*
> *All delivered with verve and passion.*
> *I-L-L-I-T-E-R-A-C-Y you'd best watch out*
> *The Roman Ridge Readers are out and about.*

The pupils are responsible for purchasing the books in Accra where they are much cheaper than in the UK and many of them have attended reading lessons to learn how to teach reading effectively. I have also been busy in the UK and have privately published an anthology of my poetry and short stories to raise funds for Ridge Readers. All proceeds from my art show also go towards the project. Long may it flourish and thrive.

While in Accra, I was roped in by Valerie Mainoo on the panel to select the Teacher of the Year. I thought she only required me for a morning and that would be it. How wrong I was. The process was akin to a triple marathon and went on for day after day leading me to remark that the Vatican spends less time to choose a pope than the Roman Ridge School to choose the Teacher of the Year Award. I am all for the award and all aspects of selection have to be seen to be fair but I did think the process was somewhat long-winded and I had to inform the panel that I was only in Accra for a couple of weeks and not for the rest of the summer. I began to think that I would never see England again.

I met up with members of the senior management team over the last half-term while they were In London visiting independent schools in the Croydon area. I do always make the effort to see staff on such visits but I always tease Gordon that I cannot travel south of the river. The West End is my stamping ground and extends to the British Museum and Library and that's it. I do make occasional exceptions to visit Lara and her new baby, Alexander Mustapha, who officially live in south Chelsea but I reckon it is really Battersea. I am always glad to return to the safety of central London.

Perhaps it is not wise for a former headmaster to also admit that I spend some time in Soho for this is still the heart of the British Film Industry and various companies and national organisations are based there as well as several studios. It also where I do my 'voice-over' work.

Chapter Thirty
You Can Choose Your Friends but You Cannot Choose Your Family

I have made mention of my extended family throughout this book and its members, and their activities and antics, have been largely included. Some autobiographies have been known to upset family members and friends and I do hope that this is not the case with mine. When I first started writing, I was assured by my solicitor that I could not be sued if I adhered strictly to the truth and that made several relatives even more worried and concerned. My favourite cousin (yes, Christine, it is now firmly in print and cast in stone and what is more, it is the truth) has a sign in her house which is a simple home truth:

Remember, as far as any visitors are concerned, we are a normal, happy family.

I do believe that we do conform to the term of a normal family though like all others we have had our problems and difficulties, which is probably true of most families anyway. My immediate family, namely the Foords and the Allens, have been largely described and most of their older members have passed on. Their children, my cousins, are mostly still active and I do make every attempt to stay in touch and I have contacted most of them for the research on this book.

I suppose that I have a much larger 'family' in terms of my ex-pupils, friends and acquaintances with whom I do try to maintain contact. There are many, however, that I have not seen for years and I intend to do something about that so please be warned if I have not been in touch for a long while. I may be on my way. Perhaps you should organise an immediate long holiday?

While abroad, I could obviously not devote as much time to family connections as I would have wished and I saw my cousins but rarely. Upon my return to the UK, however, and with the return to Bristol by my mother, I have been able to take a fuller part in family life. Now my present cousins are mostly in a cluster to the north of Bristol and then there are outposts in Derbyshire, Herefordshire and Cornwall, where my niece and her family also live close to my brother. I have visited the outposts but we mostly meet up in the Bristol area and for family functions, parties and special occasions.

There is a central cousin, Christine, around whom we all revolve. She is the organiser, social secretary and conduit for us all and it is she, and she alone, who knows what each cousin is doing at any given moment, their state of health, their state of mind, their movements and their activities. Quite amazingly, she also carries out this role for her own extended family of brothers and sister, children and grandchildren, aunts and uncles and her friends and even their pets including rabbits, cats, dogs and goldfish. She is devoted to the care of her disabled son, Neil, who has very limited mental capacity and a multiplicity of other problems but who is very much the centre of family life in her house at Frampton Cotterell, the door of which is also open to all, each day and every day.

Christine never stops and is an expert at multitasking which she has down to a fine art. Indeed, she could run training courses on the subject for she is ultra-qualified and extremely competent. Just watching her during the day makes me feel tired as she rushes here, there and everywhere. Compared to Christine, I am beginning to think that Jesus was an amateur in his miracle of 'feeding the five thousand' with a few loaves and fishes for she cooks for us all and provides delicious food day in and day out. The large dining room table is already too small for the family and guests and a larger one is contemplated, I understand.

Perhaps the mention of dining room tables is an appropriate juncture to introduce my cousin John, who hails from the Helliker family, which appears to be massive, judging by the many relations I have met at parties. Now, in truth, I cannot be sure whether it was their present table that was the subject of this tale but I do know that John was despatched, one morning, some time ago, to purchase a dining room table. He returned with…a red telephone box which now stands at the end of their garden. I would not be so unkind as to suggest that this was typical of my cousin John.

His lackadaisical approach to life is a source of great amusement to us all to the extent that I was thinking of bringing down a film crew for a week to follow him around his various activities, for John is also very forgetful and quite disorganised which leads to much fun and hilarity within the family. His list of DIY jobs is never ending, not because there are so many of them, but purely because he never completes any of them.

His kitchen has been a work in progress for many years, the tidying of his garage can be likened to the Herculean task of cleaning out the Augean stables, his electric front gates still await electrification, a dripping tap has dripped for well over two years, etc.

To top all this, my cousin John holds a licence to collect scrap and the front of the

house is often littered with discarded fridges and washing machines, wooden pallets, old furniture and rolls of wire which is often mixed with the wood he collects for the wood burners in the house. John also uses chainsaws. I use the term loosely because he ignores every safety guideline in the book and wears no protective or safety gear at all. I fully expect him to lose a hand or limb or be ripped by a broken chain, for his handling of them can be likened to the crazed characters in The Texas Chain-Saw Massacre (having seen chainsaw accidents when working in the timber industry, I do worry about him).

There is a side entrance to the house which John shuns. Everything comes through the front door, through the lounge and kitchen and out into the garden, be it wood, metal, chainsaws, tools, pallets, boxes, cement, timber or rubble and vice versa. John is lucky, however, as he has a senior management team and an all-female one at that. Let there be no gender inequality with my cousin John who has five sisters anyway.

The team tasked with keeping John on course with his somewhat haphazard working practices? Tina, his daughter, 'Neats' his auntie and Christine, his wife—and what a formidable team they are. Christine deals with the formulation of his list of 'things to do', 'Neats' helps him with the actual work and keeps him on track while Tina is held often in reserve as an ultimate deterrent.

If Tina is called in then John starts to shake. She is mercilessly and coldly efficient and has the power to get things done. John knows that he is in trouble and had better shape up quick, or else. Occasionally, John calls in outside consultants. That is to say that his Uncle Dave comes down from Cambridge to advise and help him out.

John's work pattern is somewhat erratic and would bemuse a time and motion study expert as he spends a lot of 'time' with no motion. I think that he deliberately carries everything through the house so that he can have a quick rest on the sofa between trips, where he often rests his eyes for indefinite periods. You can always find my cousin John. If you follow the trail of discarded gloves, hats, tools, car keys and shoes. You will generally locate him at the end of it.

His wallet lives almost permanently on the big table. It is not worth pinching, for it rarely, if ever, contains money and what would anyone do with a Co-op Credit Card? I ask you. I think it is only accepted by about two retail outlets in the whole of England. For all his foibles and erratic habits, John is remarkably consistent with his TV viewing and he likes wildlife programmes, the problem being that he seems to watch the same two programmes day in and day out.

One pride of lions in the Luangwa Game Park in Zambia is contemplating serving an injunction on him for a constant breach of their privacy and the gnus of the Serengeti are alleging harassment as well! They were also wondering why he was watching them again with his eyes closed? They think it impolite. Respect, John, respect please. Lions have feelings as well, you know. Above all, John is kind and will do anyone a favour. You understand that I could not recommend him for any DIY work you were contemplating but he does help many members of his family and friends in a variety of ways.

There is a rumour within the family that the pope is considering Christine for a sainthood having coped with John for so many years, let alone with everyone else in the family. Christine is not so keen. The first requirement for a sainthood is that you have to die and none of us want that. Where would we eat then?

I have always told Christine that if she stopped feeding us, the house would quickly empty and she might have a quieter life. She enjoys the hustle and bustle, however, and seems to enjoy her present lifestyle. I think that she is a modern saint for she is always prepared to help out anyone and has been incredibly kind and patient with my mother whom she visits often at Woodlands. We would all be lost without her. It is pleasant to be back in the family again and I hope the feeling is mutual as I really enjoy the banter, the children who are now really young adult kids (YAKS), all the cousins, John's antics and mishaps and Christine's cooking. I am there mainly at weekends but I do get time for myself during the week when I have an extensive work programme in Devon and Somerset.

My present pattern of life is quite enjoyable. When I gave up my flat in Devon, I lived in a caravan in Somerset for some time which was really so that I could explore the housing market in the area. I have now been at the site for some years. When my mother moved to Bristol, I bought a boat as a base there. It was half a narrowboat called Indaba and was moored in a marina between Bristol and Bath. I used her mainly at weekends when I was visiting my mother and was often on her during the holidays.

Narrowboats are notoriously difficult to drive and it was difficult to handle by myself so I eventually sold her and purchased a Fairline Sunfury cabin cruiser with twin 150hp inboard engines which is now moored in central Bristol. At weekends I sleep on her and return now to my Chausson Welcome 88 motorhome in Othery in Somerset during the week.

I like compact living and enjoy the lifestyle that I have and I am very mobile with the motorhome which can tow my little Skoda Citigo wherever I move. I do trips to my favourite haunts including, Brecon, Norfolk, Cornwall and Devon, calling in on friends on the way.

But what of my future plans? I have forced myself to think of what might be to come

in the next few years. My immediate target will be to get this book issued. As it is very much a personal memoir, the chances of publication are slim but I have enjoyed writing it. It has been fascinating reliving the main events and milestones of my life and recalling friends and colleagues. It has been most cathartic and therapeutic in many respects and I hope any criticisms it contains of people, institutions and methods will not cause offence.

My life has contained a varied pattern of good and bad, and problems and difficulties, which have all contributed to my old present beliefs, views and standards. I have always told my pupils that, 'We all learn by our mistakes'. and that is true of me. I have been ashamed of some of my actions and very proud of others. Mistakes, errors, wrong thinking and failings have all contributed to the good things in my life.

I am stronger and more tolerant because of them. I used to get quite angry and impatient in my youth and early teaching and now this has been replaced by a more placid and patient me with a steely determination almost to the point of controlled obstinacy. I do not get angry. There is no point and I have endless patience. All this has been learned over a long time; the same period that has continued to develop my sense of humour. I once told my mother, many years ago, that I was looking forward to growing old because I could then say and do whatever I liked.

'You do that already, dear,' she replied.

Mothers are very wise.

I do not feel old, though I do creak now and again, and my faculties have not started to desert me yet. Dealing with mostly students nowadays helps keep me young at heart and should that fail then my young cousins will fill the gap. I do tend to tease them somewhat and they respond in like so we have many lively conversations. I will always back them financially whenever possible if they participate in activities such as ballet, dance, sport and athletics and also treat them to special days involving adventure activities such as climbing, abseiling, canoeing and sailing. I will not back them for anything electronic!

As a senior citizen and a holder of a bus pass, I do reserve the privilege to gently rant and indulge in good old-fashioned raillery. I dislike the modern trend of menus in public houses offering gastro food which is always presented in a pretentious manner with descriptive prose that could well grace an engagement card and on peculiarly-shaped plates and dishes or slates and planks. I call it 'ghastly food' and avoid it like the plague as I do the curry and Chinese restaurants simply because their food is not genuine. Real Chinese and Indian food is nothing like that served in most establishments. That is an easy one to avoid anyway because I am not allowed to eat hot spiced food.

Rocket is another rant. Why does EVERYTHING have to be served with rocket nowadays? You cannot order or buy a sandwich without rocket. With all of the above, I challenge friends and relatives to name a good restaurant in north Bristol or south Gloucestershire. There are none.

I dislike the cult of personalities. I have no interest whatsoever in so-called 'stars' or celebrities to which TV and the press pay so much homage and devote extensive coverage about their personal lives. Who wants to know about their marriages, divorces and sex lives? And who are they to offer political advice? I never elected them as my representative. They are entitled to their views, of course, but not because of their fame.

As for their moaning and wailing about their abuse, drug addiction or drink problem due to their fame, I just do not want to know. I am sympathetic to the present problems of our society today and often deal with them in the course of my work. Pampered, overpaid celebrities, however, should sort out their problems in private not through the media. In family conversations involving the 'soaps' I often have to remind my relatives that it is not true. It is fiction though it is not helped by the press covering all 'soaps' as news.

I also deplore modern employment methods which are based on targets and appraisals and the ticking of boxes. I do not object to the aim but I do object to the method of its implementation as it is mostly improperly applied and the process becomes a cumbersome exercise with little real meaning or value. They are completed because they have to be, not through any genuine attempt at assessment. In my experience many of the assessors are highly qualified and trained but have no experience and/or common sense. I approve of sensitive, ongoing and continuous appraisals conducted by a sympathetic appraiser who knows the member of staff well and can also advise and help the employee being assessed in his work. It should not be a remote, meaningless chore.

In my politics, I lean to the right and I do follow political events keenly at home and abroad. I do wonder about two concerns. Most politics is confrontational—one party against another. Now, in offering possible solutions to problems in the educational world, my first task was to remove any confrontation from the equation. Confrontation is no help to any situation. In my naivety, perhaps, what would happen if it were removed from politics? Would that be at all possible? To have a country run using cooperation?

I do also wonder about the Welfare State. In many respects, it is a wonderful thing and I have received many of its benefits. Can we go on affording it with its ever-increasing demands and rising costs? I have noticed with interest young people's present views. They now expect the Welfare State to provide everything for them as a right. I question the wisdom of this view.

Common sense is fast disappearing from our world as are practical and human skills. As we race into an age of IT, social media, electronic devices and robots has anyone stood back to take a long hard look along the road to our final destination? I consider that my generation has let down the youth of today quite badly but the younger generation have no clear idea of what is in the future. It is rather late but I suggest that we are heading for an Orwellian Society fast. We had better watch out.

As an English teacher, I also much deplore the use of the word 'like' that seems to be interjected after every four or five words when the modern generation deigns to look up from their smartphones and talk, which, I suppose, is unusual anyway. How on earth did it insidiously creep into the nation's vocabulary? Having made its successful clandestine entry, it has now been successfully followed by 'you know', 'kind of', 'sort of', 'actually', 'to tell the truth' and 'honestly speaking'—the latter two presupposing that the person does not generally tell the truth or is not normally honest? Has etcetera now been replaced by 'and stuff like that', 'I mean', 'does that make sense?' 'OK?' 'Wot ever!'.

I also abhor the overuse of the demonstrative adjective 'there', much beloved by TV commentators and announcers. 'We now hand you over to our correspondent in London there.' Why? I am fed up with these constant repetitions that set me on edge every time I hear them, though I do have fun with my young cousins challenging them to speak for two minutes without saying the word 'like'. They have yet to succeed. 'You know', 'it's kind of' 'sort of' 'like' difficult 'actually'.

I have already hinted that I am heading for a change in my life again. I am actively in the process of clearing my decks at present and closing down my contracts and giving notice where necessary. I am leaving my London contacts open for the meantime simply because I can earn more in two or three days filming than I earn in a couple of months in the South West.

By the summer I will have no real obligations and can then decide on what to do. I do intend to visit friends and relations at home and abroad so please take this as a signal of my intent if we have not been in touch recently. I also wish to visit places in the UK that are on another BID list including Hadrian's Wall and Chester. I have been to Chester before but never got to see the walls and the Roman Army Museum. Sutton Hoo is another must and I would like to walk one or two of the ancient trackways of England.

I intend to continue with my fundraising for Ridge Readers and no doubt I will be out in Ghana at some time exploring one or two further projects that I have in mind. I also intend to spend a little more time with my mother in Bristol. Cousins be warned. This process has started and I will be free latest at the beginning of July. I hesitate to use the word retiring as I hope to continue to be very active. Who knows, perhaps one of those unusual opportunities might jump into my lap again, which I have been lucky to experience throughout my life?

Epilogue

It is with great sadness I must announce that my mother passed away peacefully during the final editing of this autobiography on Good Friday, 30th March, at the wonderful age of 98.

She will be much missed by me and the whole family who loved her dearly and who much enjoyed her company and sense of humour. She was a strong, patient, courageous and loving mother and she and I enjoyed a very strong relationship and a special bond throughout the whole of my varied life. I will be eternally grateful for her constant love and support in all my endeavours.

She will be particularly remembered for her support and involvement in fundraising for cancer research over a period of some thirty years. She was also a great supporter of the Roman Ridge School and had a great interest in the children and their academic success.

May she rest in peace.

Mum
1919–2018

Finis!